TECHNICAL LIBRARY
NAVAL SURFACE WEAPONS CENTER
DAHLGREN LABORATORY
DAHLGREN, VA 22448

High Speed Pulse Technology

VOLUME III

CAPACITOR DISCHARGE ENGINEERING

High Speed Pulse Technology

I CAPACITOR DISCHARGES, MAGNETOHYDRODYNAMICS, X-RAYS–ULTRASONICS
II OPTICAL PULSES, LASERS, MEASURING TECHNIQUES
III CAPACITOR DISCHARGE ENGINEERING

VOLUME III
CAPACITOR DISCHARGE ENGINEERING

By Frank B. A. Früngel

IMPULSPHYSIK GMBH.
HAMBURG, GERMANY

1976

ACADEMIC PRESS · New York San Francisco London
A Subsidiary of Harcourt Brace Jovanovich, Publishers

COPYRIGHT © 1976, BY ACADEMIC PRESS, INC.
ALL RIGHTS RESERVED.
NO PART OF THIS PUBLICATION MAY BE REPRODUCED OR
TRANSMITTED IN ANY FORM OR BY ANY MEANS, ELECTRONIC
OR MECHANICAL, INCLUDING PHOTOCOPY, RECORDING, OR ANY
INFORMATION STORAGE AND RETRIEVAL SYSTEM, WITHOUT
PERMISSION IN WRITING FROM THE PUBLISHER.

ACADEMIC PRESS, INC.
111 Fifth Avenue, New York, New York 10003

United Kingdom Edition published by
ACADEMIC PRESS, INC. (LONDON) LTD.
24/28 Oval Road, London NW1

Library of Congress Cataloging in Publication Data

Früngel, Frank B   A
   High speed pulse technology.

   Translation of Impulstechnik.
   Includes bibliographies and index.
   CONTENTS.–v. 1.   Capacitor discharges.   Magnetohydro-
dynamics.   X-rays.   Ultrasonics.–v. 2.   Optical pulses.
Lasers.   Measuring techniques.–v. 3.   Engineering of
capacitor discharges.
   (Electricity)   I.   Title.
   TK7835.F7413   621.3815'34   65-16665
   ISBN 0–12–269003–6 (v. 3)

PRINTED IN THE UNITED STATES OF AMERICA

Contents

FOREWORD	ix
PREFACE	x
CONTENTS OF PREVIOUS VOLUMES	xi

A. The capacitor, its characteristics and its applications in discharge circuits

1.	Capacitors in Discharge Circuits in General, Types of Capacitors	1
2.	Behavior of Insulators	4
3a.	Behavior of Gaseous and Liquid Dielectric Materials	12
3b.	High-Voltage Dielectric Fluids	17
4.	The Vacuum Capacitor	19
5.	Specifying a Spark Discharge or Flash Lamp Capacitor	19
6.	Laser Discharge Capacitors	25
7.	Pulse Capacitors for Soft- and Hard-Tube Circuits	29
8.	Measurements of Ultralow Inductance (nH) Capacitors	32
9.	High-Voltage, Low-Inductance Capacitors for Marx Circuits	45
10.	The Parallel Plate Capacitors for Superposed Frequencies	47
11.	Big Electrolytic Capacitor Banks	49
12.	Tantalum Capacitors	50

B. Switching means

1.	Controlled Gas Discharge Tubes	52
2.	Controlled Semiconductor Switches	61
3.	Push–Pull Switching Circuits for Rectangular Pulses	63
4.	Modern High Power Switching Gaps	69
5.	Liquid and Solid Gaps	74
6.	High-Voltage Vacuum Contactors	75
7.	The Lightning Protectors	80
8.	Laser-Controlled Spark Gaps	84
9.	Quenched Spark Gaps	88
10.	Crowbar Switches	93
11.	Other Unusual Switching Gaps	95

C. Line conductors

1.	The Influence of Impedance and Its Transformation by Connecting Lines	113
2.	The Coaxial Low-Impedance Water Line and the $CuSO_4$ Output Resistor	122

D. Conversion of capacitor energy into current impulses

1. Direct Discharge through a Conductor — 126
2. Transformed Discharges for Highest Currents at Low Voltage — 142
3. Subnanosecond Pulse Generators — 144
4. Inductive Energy Storage Systems Applied to Extend the Duration of Current Pulses from Capacitor Banks — 145

E. Conversion of capacitor energy into voltage impulses and its practical applications

1. Very Fast Voltage Pulse Generators for Scientific Applications — 154
2. Kerr Cell and Pockels Cell Pulsers and Related Techniques — 165
3. Pulsers for Laser Excitation and Flash Photolysis — 180
4. High Voltage Pulse Transformers and their Application to Spark Tracing (Aerodynamics) and to X Ray, Electron, and Light Pulses — 200
5. Megavolt Pulse Techniques: Lightning Flash and EMP Simulation — 215

F. Conversion of capacitor energy into x-ray flashes and beams of electrons, ions, and neutrons

1. New Methods for Generating X-Ray Flashes (Debye–Scherrer and Laue Patterns, Plasmas and Lasers) — 229
2. New Designs and Applications of X-Ray Flash Equipment for Motion Analysis — 241
3. Hard X-Ray (Gamma-Ray) Flashes and Electron Beam Pulses of High Energy — 279
4. High-Energy Ion Generation — 288
5. Neutron Flashes for Material Testing — 292

G. Conversion of capacitor energy into heat

1. Impulse Welding, Direct Capacitor Discharge — 298
2. Welding by Transformed Capacitor Energy — 303
3. Fast Metallic Phase Transformations by Current Pulses — 312
4. Pulse Hardening of Carbon Steel — 313
5. Exploding Wires and Their Applications; Exploding Wire Shutters — 320
6. High Temperature Plasma Generation by High Energy Capacitor Discharge — 328
7. Plasma Heating by Laser Beam Energy — 337

H. Conversion of capacitor energy into magnetic fields

1. Pulsed Magnetic Fields, Concentration — 349
2. Electromagnetic Plasma Propulsion — 355

3. Charging of Permanent Magnets	359
4. Faraday (Magnetooptical) Effect and Its Use	360
5. Capping and Uncapping Shutters	362
6. Magnetically Driven Hypervelocity Macroparticle Accelerators	365
7. Magnetically Driven Gas Accelerators	370
8. Megagauss Fields	371

I. Conversion of capacitive stored energy into acoustic pulses

1. Acoustic Pulses	380
2. Acoustic Pulses and Electroceramics	394
3. Advanced Sonar Techniques	401
4. Underwater Shock Waves from Sparks	409
5. Shock Waves in High Pressure Physics and Metal Forming	420

BIBLIOGRAPHY	438
INDEX	483

Foreword

I know that I will be joined by many research workers in welcoming Dr. Frank Früngel's third volume on high speed pulse technology. His first two volumes, which appeared ten years ago, were a monumental work of literary review containing some 1006 references. Früngel not only went into great detail on his own voluminous field of pulse applications, but gives his interpretation and review of the mountainous information of others. Now, the third volume catches up with developments in high speed pulse technology that have occurred since 1965.

His bibliography has now gone from 1006 to a grand total of 2006 references! This alone is an extremely valuable item to aid those who wish to work in pulse technology. Many of the articles are in obscure journals, do not pass into the general stream of information, and are almost impossible to find.

The chapters in Volume III follow almost exactly those of Volumes I and II. The reader will note that Chapter A, on the capacitor, is greatly expanded over the previous treatment. This is due to the tremendous importance of the capacitor as an energy storage element for almost all pulse systems. Früngel has done a very commendable job of considering all of the factors that influence the type of capacitor that is considered for a specific task in the pulse domain.

The remainder of the book is an update of pulse technology as viewed by Dr. Früngel. I am impressed by the range and complexity of this field in the busy ten years that have just passed.

Cambridge, Massachusetts HAROLD E. EDGERTON
September, 1975 Massachusetts Institute of Technology

Preface

Volume III of "High Speed Pulse Technology" was written in response to numerous requests for an updated edition of Volumes I and II that would take into account the vast amount of research done from 1964 onward. As a completely new edition would have been extremely voluminous, Volume III has been produced as an updating supplement to Volume I. Even so, it was necessary to present most of the research in the form of extracts or summaries of papers in order to restrict the new volume to a reasonably handy size. Only papers likely to be difficult of access to most readers have been reported more fully.

The new volume is directed mainly at scientists, especially physicists, and electrical and other engineers working on the production and practical application of capacitor discharges for the generation and utilization of high speed pulses of energy in different forms, such as electric current and voltage, x-rays, gamma rays, heat, beams of electrons, neutrons and ions, magnetic fields, sound and shock waves in gases and liquids. Applications dealt with include impulse hardening of steel, ultrapulse welding of precision parts, x-ray flash technology, ultrafast image converters, exploding wire shutters and light sources, electromagnetic shutters, flash photolysis, spark tracing in aerodynamic and automotive research, high energy electromagnetic pulse generation, plasma physics, magnet charging, magnetically driven gas and particle accelerators, acoustic echo techniques for remote atmospheric sensing, sonar, and shock waves in high pressure physics and metal forming.

It is hoped that the book will also prove to be a useful reference work for the teaching staff and students of universities and technical institutes and that research and development scientists and engineers will gain inspiration from this up-to-date summary of work in fields allied to their own specialities.

Chapter headings and subheadings are as closely aligned with those of Volume I as is practicable.

I wish to thank my colleagues Geoffrey J. W. Oddie and H. G. Patzke for their valuable help in preparing this book.

Contents of previous volumes

VOLUME I

A. The capacitor, its characteristics and its applications in a discharge circuit

1. Capacitors in Discharge Circuits in General and the Equivalent Circuit Diagram
2. Characteristics of Dielectric Material
3. Tables of Properties of Dielectric Materials Used for Capacitors
4. Specifications and Standards of Capacitors
5. Faults Which May Occur in Fixed Capacitors
6. Basic Combinations of *RCL* Elements in the Circuitry
7. Calculations and Layout of Capacitors for Discharge Operation
8. The Capacitor with Large Energy Content for Powerful Single Discharges
9. The Impulse Capacitor in Cyclic Charging Operation
10. Periodic Discharge Operation
11. Capacitor Battery for Continuously High Output in Discharge Operation (Impulse Welding Operation)
12. Capacitors Having Extremely High Insulation Values
13. Capacitor Lines with Constant Impedance per Unit of Length

B. Switching means

1. Thyratrons
2. The Ignitron
3. The Multivibrator
4. Symmetrical Switch Tubes with Mercury Filling
5. Fixed Spark Gaps
6. Moving Spark Gaps
7. The Lightning Protector
8. The Actuation of Magnetic Contactors at Predetermined Phase Angles
9. Quenching Spark Gaps

C. Line conductors

1. The Influence of Impedance
2. The Overhead Line for High and Highest Voltages with Large Conductor Radius
3. The Coaxial Cable
4. Energy Considerations
5. Sandwich Lines
6. Compensation Conductor Forms

D. Conversion of capacitor energy into current impulses

1. Direct Discharge through a Conductor
2. The Transformed Discharge for Highest Current Peaks

E. Conversion of capacitor energy into voltage impulses

1. The Cascade Circuit
2. The Nondistorting Pulse Transformer
3. High Peak Power Transformers with Differentiating Characteristic
4. Impulse Setups for Cyclic Operation and Very Short Impulse Duration
5. Pulse-Forming Network
6. Cable Discharge Lines and Pulse Generators
7. Spark Gap Triggering Device by Marx
8. Ignition Transformer for Internal Combustion Engines
9. Voltage Impulses in Extensive Networks
10. Automatic Overhead Line Protection by Means of Periodic Capacitor Discharges
11. Lightning Flashes in Nature as Capacitor Discharges
12. Steepness of Voltage Shock Pulses
13. The Operation of Kerr Cells by High-Voltage Impulses
14. Biological Applications of Voltage Impulses in Electrical Fishing

F. Conversion of capacitor energy into x-ray flashes and neutrons

1. Physics, Design, and Circuitry in X-Ray Flashing
2. X-Ray Irradiation of Subjects in Mechanical Motion
3. Capacitor Discharges in Feeding Ion and Neutron Sources

G. Conversion of capacitively stored energy into heat

1. Impulse Welding, Direct Capacitor Discharge
2. The Transformed Capacitor Discharge in Welding Engineering
3. Electrodynamic "Cooling" by Particle Ejection in Capacitor Impulse Welding
4. Microinduction Hardening—Application of Instant Heating by a Capacitive Energy Impulse
5. High-Frequency Heating by Rapid Impulse Sequences—the Blast Spark High-Frequency Generator by Marx
6. Annealing of Wires by Means of Capacitor Discharge Pulses
7. Exploding Wire Discharges
8. High-Temperature Plasma Generation by High-Energy Capacitor Discharges

H. Conversion of capacitively stored energy into magnetic fields

1. The Creation of Very Intensive Magnetic Fields for Short Duration
2. Application of Strong Magnetic Shock Fields in Nuclear Physics
3. Capacitor Discharges in Magnetic Plasma Generation and Acceleration
4. Capacitor Discharge Magnetizers
5. The Magneto-Optical Shutter—Application of the Faraday Effect
6. Metal Forming with Pulsed Magnetic Fields

I. Conversion of capacitively stored energy into acoustic impulses

1. Conversion by Electroacoustic Converters and Applications
2. Air Impulse Sound
3. Shock Sound by Underwater Capacitor Discharges
4. Applications of High Intensity Shock Group
5. The Photography of Discharge Sound Shocks and Its Technical Application

J. Material working by high-frequency capacitor discharges (spark erosion)

1. Facts and Purposes of the Mechanism of Spark Erosion
2. Electro-Erosion Machining of Metals
3. Shaping Metals by Electrical Explosion Shock Wave

BIBLIOGRAPHY
AUTHOR INDEX
MANUFACTURERS INDEX
SUBJECT INDEX

VOLUME II

K. Light flash production from a capacitive energy storage

1. Basic Considerations and Principles of Spark Light Photography
2. The Guided Spark (Sliding Spark)
3. Capillary Spark
4. The Air Spark under Atmospheric Conditions
5. The Free Spark in High Pressure Atmosphere
6. Physics, Design, and Application of Spark Chambers
7. The Controlled High Frequency Capacitor Discharge for Light Flash Generation
8. The Production of High-Voltage, High-Frequency Sparks and Their Application in the Field of Aerodynamics
9. The Indirect Light Flash Produced by Means of Capacitor Discharge with Laser
10. Industrial Equipment for Laser Flashing
11. Consideration of Laser and Spark Light Illumination and Signaling

L. Signal transmission and ranging systems by capacitor discharges and lasers

1. Claims for Light Transmitting System
2. Time and Spectral Characteristics of Light Pulses for Signaling Transmission
3. Calculation and Layout of Transmission Systems with Spark Light
4. Experimental Documents and Results
5. Application of Light Impulse Signal Transmission
6. Optical Ranging Systems Using Laser Transmitters

M. Impulse measuring techniques

1. High Voltage Power Supplies
2. Impulse Oscilloscopy
3. Impulse Measurements by Means of Spark Gaps
4. The Measuring of Impulse Currents. The High-Current Shunt
5. Voltage Measuring with Voltage Dividers
6. The Measuring of Impulse Magnetic Fields by Means of Hall Probes
7. Measuring Technique for Light Impulses
8. Optical Measurements of High Speed Thermal Processes
9. Measuring and Counting of X-Ray Flashes
10. The Measuring of Sound Impulses and Shock Waves
11. Conversion Factors of Various Energy Units

BIBLIOGRAPHY
AUTHOR INDEX
MANUFACTURERS INDEX
SUBJECT INDEX

A. The capacitor, its characteristics and its applications in discharge circuits

1. Capacitors in Discharge Circuits in General, Types of Capacitors

The electrical capacitor is still the most important storage reservoir for electrical energy which will subsequently be discharged at a high rate into the consumer circuit to produce a quick flow of energy. The weight and cost of the power packs are mainly due to the capacitors [1529 f].*

Capacitors consist of metallic plates, or foils, with an insulating material called a dielectric between them. Paper saturated with oil is commonly used as a dielectric material, while other capacitors use a thin layer of plastic. When a voltage is applied to the capacitor plates, electrical charges flow to the plates, and the dielectric is subjected to an electrostatic field. Electrical energy is required to charge the capacitor to a given voltage according to the important relationship

$$\text{energy stored in a capacitor} = CE^2/2 \quad \text{W-sec}$$

where E is the voltage across terminals in volts, and C the capacitance in farads. The capacitance C is a number indicating the ability of a capacitor to store electrical charge per volt of applied voltage. A capacitor with a large capacitance will store more energy than one with a smaller capacitance at the same voltage. As will be discussed later, the light output from a flash lamp is a function of the energy from the capacitor and, therefore, can be proportional to the capacitance. By obtaining a suitable capacitor at a given voltage, the amount of light from a flash lamp can be predetermined when the efficiency is known.

Edgerton [1529] presents some useful equations about the dimensioning of capacitors. The factors that influence the design of a capacitor, which were presented in Volume I, are now briefly reviewed.

---

* From "Electronic Flash Strobe," by H. E. Edgerton. Copyright © 1970 (McGraw-Hill Book Co.). Used by permission of McGraw-Hill Book Company.

Capacitance

The capacitance of any capacitor can be calculated from the equation

$$C = 0.0884 \times 10^{-6}(KA/d) \quad \mu F$$

where A is the area of the dielectric in square centimeters, d the thickness of the dielectric in centimeters, and K the dielectric constant. From this relationship it is observed that a large capacitance requires a large area of a thin dielectric which has a high dielectric constant.

The thickness of the dielectric is one of the most important factors to the capacitor manufacturer. If the material is thin, a large capacitance in a given size container is obtained. However, the maximum allowable voltage before electrical breakdown occurs is reduced. An engineering balance in design must be reached, in which the thickness is as small as possible without experiencing an objectionable number of short-circuited capacitors due to electrical breakdown.

a. *Facts for Choice of the Most Suitable Capacitor*

For each capacitor application, the factors that influence the specifications are

(1) voltage stress of the dielectric,
(2) capacitance,
(3) type of duty cycle,
(4) life required,
(5) internal series inductance.

The factors to be taken into account in selecting types for particular purposes are: voltage, capacitance, inductance, repetition rate, and lifetime under charged dc conditions or reckoned in number of discharges. A capacitor charged to, e.g., 1000 V may be discharged in a nonaperiodic manner. Then the voltage will pass zero and reach a controversial polarity of, e.g., 500 V. The percentage of this voltage versus the charging voltage is called reversal percentage, which, in this case, is 50%. Here, the voltage reversal percentage is very important. The dielectric molecular dipole stress is proportional to the sum of voltage and reversal voltage. A capacitor with 3 kV charging voltage and 2 kV voltage reversal must be dimensioned with respect to life, like a 5-kV capacitor without reversal and the price ratio in this very case would be $3^2:5^2 = 9:25$.

Capacitors for electronic flash lamp operation are usually highly stressed since the energy per pound is proportional to the square of the voltage. The rating for life may be a function of the time that the capacitor is held at full voltage awaiting a flash. However, if the flash lamp is a low-resistance lamp, then the discharge circuit may be oscillatory and the dielectric will be under greater stress than if a nonoscillatory discharge is used. Furthermore, the large peak current may cause deformation and even burning at the tabs inside the capacitor at the foil connections.

Capacitors are made of large sheets of very thin aluminum foil wound with thin layers of paper or other dielectrics, such as plastic films, and then pressed into a compact form. One important type of capacitor uses very thin metal that is sprayed or evaporated onto the dielectric. Some of these capacitors are self-healing since an electrical breakdown causes the metal to evaporate away from the defect. These are called MP, meaning metal–paper type. The MP capacitor can be used in flash equipment of 100 μsec flash duration and longer.

Capacitor sections are inserted into metal cans which are exhausted of air and filled with oil or other material before sealing with solder. The dielectric constant of paper, oil, glass, and other materials is from 2 to 10 times greater than that of air. The dielectric constant is a function of temperature in most materials. For example, the capacitance may be greatly reduced when the liquid dielectric freezes.

b. *Comparison of Capacitors*

Table A1-1 gives the approximate factors for several types of capacitors. The whole story for selection of a specific capacitor for any duty is not self-evident from the table. One factor not given is the minimum practical voltage of each type of capacitor. Paper capacitors, for example, cannot be made with a very thin dielectric because of the quality control problem of manufacturing the paper without defects. Several layers of paper are used. Electrolytic capacitors are widely used because of their operation at the relatively low voltage of 450–500 V. The MP type of capacitor is finding many uses because of its ability to store energy and to self-heal some of the electrical breakdown.

The internal inductance of a capacitor is a very important factor if short energy pulses are required. The inductance can be reduced by the use of multiple connecting tabs or by complete edge connections of the metal foils.

TABLE A1-1

APPROXIMATE COMPARISON OF THE SPECIFIC COST, WEIGHT, AND DENSITY OF THREE TYPES OF CAPACITORS

Type of capacitor	J/lb	J/in.^{3}	Temperature range (°C)	$/J
Oil paper	25	1.0	−55 to +85	0.10
Mylar MP	60	2.5	−55 to +60	0.30
Electrolytic	100	5.0	−10 to +65	0.10

2. Behavior of Insulators

Each capacitor needs insulators. Their design has to meet the particular need of the discharge circuit: low inductance; high discharge current; the outside surface well-dimensioned to the environmental conditions such as temperature, humidity, and air pressure; the inside surface adapted to the capacitor design and material, such as vacuum oil, electrolytic liquids, or solid-state insulators.

Vlastós [1622] studied the industrial and saline pollution of insulators. The complexity of the problem, however, due to the simultaneous influence of many controlled and uncontrolled parameters, suggests that further basic research should be done for a better understanding of insulator pollution and particularly of the breakdown of clean and polluted insulators (Table A2-1).

To study the mechanisms which lead to breakdown under different insulator–surface conditions and under different voltage–time relations, Vlastós performed an experiment in which string insulators and simplified insulators in the form of disks or cylinders are used. He used high-speed photography as a diagnostic tool by applying image intensifiers and ir detectors in studying predischarge phenomena on insulator surfaces (see [1622], Chapter K13).

Widmann [1623] designed a creepage discharge system with insulating materials between electrodes which are considerably different with regard to their dielectric strength. After the initial voltage is exceeded, creeping discharges occur along the surface of the insulating material of the higher breakdown field strength, without the total flashover necessarily occurring at the same time. The report contains measuring results with creepage discharge arrangements of cable insulating paper and pressboard in transformer oil tested with impulse voltage rise and decay times of 1.2/50 μsec.

TABLE A2-1

ELECTRICAL CHARACTERISTICS OF STRING PORCELAIN AND GLASS INSULATOR UNITS[a]

	Porcelain[b] (in kV)	Glass[b] (in kV)
50% impulse flashover voltage, positive	120	160
50% impulse flashover voltage, negative	125	150
Dry, 50 Hz flashover voltage	80	107
Wet, 50 Hz flashover voltage	48	57
Dry, one-minute, 50 Hz withstand voltage	70	100
Wet, one-minute, 50 Hz withstand voltage	40	50

[a] Vlastós [1622].
[b] Creepage distance for porcelain: 280 mm; for glass: 510 mm.

a. Flashover Voltages

With bare electrodes, the flashover voltage of the creepage discharge arrangement increases almost linearly with the length of the creepage distance but nonlinearly with the thickness of the creepage distance. While the creeping spark inception voltage is mainly dependent upon the electrode form, especially the electrode radius, the creep-flashover voltage is practically independent of the electrode form.

Mosch et al. [1610] investigated channel inception time and breakdown time as criteria for long-time behavior of solid insulations. When investigating the long-time behavior of solid insulations, it is useful to measure not only the breakdown time but the treeing (channel) inception time. Thus, the influence of several parameters, such as size of the defect (radius of the needle-point being used to introduce the treeing process), distance of electrodes, impurities, formation of the partial discharges (treeing), etc., can be determined more precisely. Investigations of Plexiglass and epoxy resin insulations have shown that with extremely inhomogeneous electrode arrangements, the treeing inception time is essentially less than the breakdown time, i.e., most of the time is required for forming the partial discharge channels.

b. Treeing Breakdown in Plastic Insulators

Nawata et al. [1611] measured the voltage and temperature dependence of treeing breakdown in plastic insulators. Recently, many kinds of synthetic polymers have been used in electrical insulation systems, because of their remarkable physical and chemical properties as insulators. On the other hand, when used in rather intense electric fields, these polymers exhibit various problems in electrical insulation properties, such as the treeing breakdown (dielectric breakdown by channel propagation) which is regarded as a precursor of the ultimate complete breakdown of the insulation system (nine other references are given in this paper). The treeing breakdown in polyethylene and polymethyl methacrylate has been investigated for the electrode arrangement of needle-plane. The effects of temperature and applied voltage on the induction period to the inception of treeing and on its growth rate after the tree inception were studied by Nawata et al.

c. Mica Bonding

Similar work has been done by Ryder et al. [1613] on partial discharge experiments on small samples of epoxy-resin-bonded mica insulation. Surface discharge experiments have been performed in nitrogen and helium at NTP, and internal discharge experiments in dry air at NTP and hydrogen at 65°C and 310 kN m^{-2}. Lifetimes in excess of 25 yr were recorded, at which point experiments were terminated. All mica-based materials delaminated to some extent, forming natural cavities. Within these cavities white dentritic deposits were formed which were determined to be the result of discharge action on mica. In general, the materials made from small mica flakelets appeared to withstand the discharge attack better than large flake materials, although all materials had lifetimes of at least 25 yr.

The influence of mechanical stress on the growth of predischarge channels in epoxy resin has been studied by Schirr [1614]. The allowable mechanical stress in view of the electrical behavior for the design of epoxy-resin-insulated high-voltage equipment is very much smaller than the material strength discovered in laboratory measurements with test samples. For long-time behavior, it was felt that the technological state of the material had a significant effect on the electrical strength. Special thought was given to the effect of mechanical stress in the insulating material. Stress may be caused by shrinking during the curing and cooling process of the insulation.

The experiments show that mechanical stress makes the material anisotropic with respect to its electrical strength. For the test samples used in this investigation, superposed pressure increases the breakdown strength compared to the case without mechanical stress, while internal tension decreases the breakdown strength.

d. *Erosion Resistance, Weatherability*

Penneck and Swinmurn [1618] deal with polymeric materials for use in polluted high-voltage environments. At the present time there is a growing interest in polymeric electrical insulation materials for such high voltage uses as cable terminations, insulated cross arms, and suspension insulators, but most polymers are not suitable for use at high voltage in polluted atmospheres, due to their susceptibility to track. This paper describes the development of nontracking erosion-resistant materials, together with details of natural and accelerated weathering studies. Inclined plane tracking voltages for a wide variety of filled and unfilled polymers have been measured and are presented here, together with correlations with chemical structure and degradation modes.

However, most organic polymers are not suitable for use in high-voltage conditions in polluted atmospheres. In such cases moisture or fog, together with salt dust particles, ionic pollution, etc., cause small leakage currents to flow across the surface of an insulator under electrical stress. These small currents, which may last for several seconds, cause a rise in temperature with consequent moisture evaporation. Ultimately a dry band is formed. Sparks or surface discharges across the dry bands reach very high temperatures and cause degradation of the polymer, often with the formation of a conducting carbonaceous path. Once commenced, the carbonaceous deposits usually extend quickly in dendritic fashion until the insulation fails ultimately by

2. BEHAVIOR OF INSULATORS

progressive creepage tracking. An alternative failure mode is by erosion, where small parts of the insulator surface are removed by the discharges, with no conducting deposit being formed and virgin material being exposed. Erosion is a comparatively random process and failure times are considerably greater than for tracking.

In general, it has been shown in model experiments that the greatest damage occurs if the discharge current is less than 20 mA, although in the case of a bisphenol epoxide resin, the maximum rate was observed at 1.5–2.0 mA. The small value of the discharge current causes very high heat flux in very small areas and thus breakdown occurs.

There are a number of characteristics a polymeric insulating material must possess in order to be a candidate for use:

(a) nontracking in service,
(b) low erosion rate,
(c) toughness (i.e., nonbrittle),
(d) weatherability—retention of electrical and mechanical properties after weathering, and
(e) wide range of continuous operating temperatures, e.g. from -40 to $+120°C$, to be acceptable in most climatic conditions.

The term weatherability refers to the ability of a material to provide long-term service life under complex and variable conditions which tend to degrade that material. Such variables are uv radiation, water, temperature, oxygen, and miscellaneous atmospheric pollutants such as sulfur dioxide, soot, dust, and ionic impurities. A complicating factor is that almost all these variables except oxygen are widely and erratically dependent on geographical location and season.

e. *Influence of Mechanical Stress*

Kindij [1608] also describes the influence of mechanical stresses on electrical breakdown and finds it greater than that of any other known cause for operation near the elastic limit. Treeing and breakdown are always perpendicular to the direction of the mechanical stresses.

A paper by Németh [1617] on long-time behavior shows the polarization of long-time constants in insulating materials. He determines the time constant distribution by measuring the absorption current. Polarization in dielectrics can be considered as the sum of a number of elementary polarizations of various relaxation time constants and intensities. More complex are the field strength effects in two-dielectric arrangements (Weiss [1565]). Based on Maxwell's image charge method, Weiss has developed a reliable method for the calculation of electric fields for nearly any given arrangement of two dielectric materials. Since a dielectric boundary has the same effect as a surface charge, not only the real surface charges of the electrode but also the fictitious surface charges of a dielectric boundary can be computed by this new method. Both kinds of charges are being replaced in computation by a finite number of discrete charges.

For the case of two electrodes embedded in dielectrics, the effect may be called the "embedding effect." If the angle α between the electrode and the dielectric boundary is less than 90°, the field lines starting from the electrode in medium I are refracted at the dielectric boundary as if the electrode has a sharp angular bend at the corner point P. For this reason the field strength at P is infinitely high. If α is greater than 90°, the field lines starting from the electrode in the medium are refracted at the dielectric boundary as if the electrode has another kind of angular bend at P. This case must be compared with the inside of a corner and for this reason the field strength is zero at P. If α is exactly 90°, the field strength at P has a finite value which is different from zero. The embedding effect demonstrates that in the construction of high-voltage insulators, an angle between the surface and the conductor of less than 90° is to be avoided.

f. *Ultraviolet in the Predischarge*

The very early phase of a surface discharge emits a very small quantity of ultraviolet radiation. This can be easily detected by applying a method given by Früngel and Ebeling [1568]. The number of uv photons, which increases in proportion to the ionization ratio in a certain range, is measured with a "daylight-blind" gas-filled photocell and amplified into a direct current readout. The uv part of the spectrum below 3000 Å is absorbed by the O_3 layer in the upper atmosphere. Therefore cadmium is chosen as the photocathode material because of its long-wavelength cutoff at 4.1 eV or 3000 Å. Near ground level the amount of atmospheric O_3 is so low that the registration sensitivity is sufficient, with a very simple optical and electronic arrangement, to measure corona currents of 1 μA at distances far greater than 10 m. For experimental work, use was made of a meteorological device, the "Fumosens," which measures fog density by registration of the scatter of an additional uv light source.

g. *Very Low Temperatures*

An investigation into internal discharges in artificial air-filled cavities in impregnated paper at temperatures down to that of liquid nitrogen (space applications) is described by Hossam-Eldin *et al.* [1607] (see Fig. A2g-1). It has been shown that when the temperature is lowered, the repetition rate and the magnitude of the discharges under direct-voltage conditions are greatly reduced and the electric strength of the insulation is correspondingly increased.

Improvements are also obtained in the electric strengths with alternating and surge voltages. A high-voltage, direct-current, impregnated-paper-insulated cable cooled with liquid nitrogen is considered and calculations indicate that such a cable operating at ± 500 kV with a rating of 5000 MW should be technically feasible.

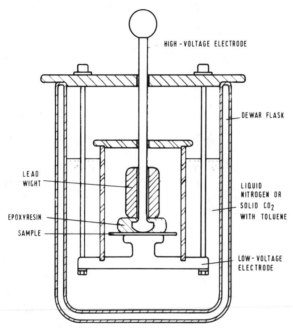

FIG. A2g-1. Test cell of Hossam-Eldin and Pearmain [1607].

In an experiment made at temperatures approaching that of solid carbon dioxide, the high-voltage electrode was surrounded by moulded epoxy resin to prevent edge discharge. When using liquid nitrogen the epoxy resin was omitted and the liquid nitrogen itself acted as an immersion medium. A 100-channel pulse-height analyzer was employed to record the discharges. The paper, which had been made wholly with deionized water, and the impregnating compound were similar to those previously used and the test samples were prepared in the same manner, each sample consisting of three sheets of paper with the central one punched with a cylindrical hole of 2.0 mm diam.

The impulse breakdown voltage and electric strength were measured using $1/50$ μsec impulses having a negative polarity. The first impulse had a voltage of approximately one half of the expected breakdown voltage and the voltage was increased in steps of $\sim 3\%$, with one impulse at each step. Mainly for the construction of high-voltage, very-high-energy cable capacitors, the results may be interesting: Some of the most important characteristics of a high-voltage, direct-current, impregnated-paper-insulated cable cooled with liquid nitrogen have been considered and calculations indicate that such a cable should be technically feasible. Many aspects of the design, installation and operation, the refrigeration plant, and the dimensional changes of the conductors, metallic sheaths, steel pipes, and insulation during cooling require further detailed attention and the economics of the cable system would have to be carefully assessed. However, the advantages to be gained by cooling the insulation as well as the conductors should provide some additional incentive towards the further study and development of a cable capable of transmitting power in the amounts that may well be required in the near future.

h. *Internal Discharges in Paper*

Internal discharges in paper insulation under direct-voltage conditions have been measured by Badran *et al.* [1606]. An investigation into the repetition rate and magnitude of internal discharges in artificial air-filled cavities in impregnated paper is described with particular reference to the effects of the time of voltage application, electric stress, temperature, cavity dimensions, and type of paper. It has been shown that under high electric stress, the discharges can lead to the breakdown of surrounding insulation. Measurements of the volume resistivity of impregnated paper are reported and their relevance to discharge phenomena discussed. The behavior of synthetic-resin-bonded paper bushings, oil-impregnated paper capacitors, and solid-type cable have been studied at voltages up to 150 kV dc and their discharge characteristics are compared. It is concluded that a discharge test provides a valuable indication of the quality of electrical insulation for use under direct-voltage conditions.

The number of discharges greater than 10 pC observed during the first 100 sec after the application of various voltages to the original cable and to the cable with the reduced thickness of insulation are shown. The cable with the reduced thickness of insulation had a lower discharge inception voltage, but at high voltages, in spite of the larger stresses to which it was subjected, the cable with the reduced thickness of insulation showed fewer discharges, presumably because of the smaller volume of insulation tested.

Phenomena with impregnated paper are very similar to those reported for polythene. The principal discharge characteristics of synthetic-resin-bonded paper bushings, oil-impregnated paper capacitors, and solid-type cable have been determined. Discharges in the bushings at ambient temperature were extinguished at voltages up to a value of 1.8–2.5 times the peak-rated alternating voltage, within times up to several hundred hours. During heat cycles, voltages of about 1.8 times the peak-rated alternating voltage were needed to produce discharges having a magnitude in excess of 100 pC. In the cable, discharge extinction occurred under constant voltages sufficiently high to produce a maximum stress of 8 times the peak working maximum alternating stress. These findings apply only to the particular types of bushing and cable tested, and other types may well demonstrate different characteristics.

An alternating voltage test does not give a reliable guide to the future behavior of insulation under direct voltages, and it is suggested that the techniques which have been described by Badran *et al.* form a means for assessing the quality of the insulation of equipment designed for use under high direct voltages.

i. *Polyethylene at Cryogenic Temperature*

Dielectric breakdown of polyethylene films at cryogenic temperature is also discussed by Ieda *et al.* [1609]. In order to know the breakdown mechanism, they investigated the dependences of electric strength upon applied voltage waveform, specimen thickness, and temperature of polyethylene films in the cryogenic temperature region. In addition to blank polyethylene, pyrene-doped polyethylene is also used to study the interaction of the high-energy electrons accelerated by the applied field with the π electrons in the pyrene molecules which have rich excited-energy levels.

j. *Voltage Stabilizers*

Recently, additives called "voltage stabilizers" have been often mixed into insulators to improve the electrical properties. However, the mechanism of this effect is not yet clear. Low-density polyethylene doped with surfactant (AS-1), which is one of the additives, has been investigated to examine this effect fundamentally.

The electric strength, which decreases with increase in film thickness, is an increasing function of temperature. The introduction of pyrene as an impurity into polyethylene increases the electric strength. The temperature dependence excludes the thermal breakdown process from the predominant mechanism. The variation of electric strength with film thickness and the impurity effect lead to an avalanche process as an operative breakdown mechanism. Narayana Rao [1612] discusses breakdown mechanism in plastic insulation under the influence of corona discharges, and states that in cases of corona discharges in gases like N_2, O_2, and SF_6, as well as for multilayer arrangement of such foils, it is either

(a) a reversible superficial mechanism occasioned by the developing surface or oxide layers;

(b) an irreversible volume mechanism arising out of the chemical effects aggravated by ambient humidity; or

(c) a reversible volume mechanism which is controlled by the combined effects of diffusion and reactivation processes inside the material.

k. *Lichtenberg Figures in Transformer Oil*

A study of Lichtenberg figures in transformer oil with impulse potentials was made by Balakrishna and Govinda Raju [1615] at the Department of High Voltage Engineering of the Indian Institute of Science, Bangalore, India. Using the Lichtenberg figure technique, the characteristics of streamers in a point-plane gap in transformer oil under impulse potentials of both positive and negative polarities have been studied. Data are presented on the number of streamer branches and their spatial distribution about the axis of the gap, and a potential fall at the tip of the streamer as it crosses the gap. The velocities of streamers, which are measured by applying chopped impulses of various durations, are 0.8–2×10^5 cm/sec.

Rumeli [1620] reports on ways of preventing discharge growth over polluted insulating surfaces. Contaminants on the surface of high-voltage insulators often cause flashovers which may lead to interruptions in energy transmission and result in economic losses. In this paper, the use of insulators having conducting rings placed at appropriate places on their surfaces is suggested as an effective method to combat flashover. The insulator with conducting rings on its surface is called an "R-insulator." The superior flashover performance of R-insulators is demonstrated by comparative experiments on models.

3a. BEHAVIOR OF GASEOUS AND LIQUID DIELECTRIC MATERIALS

A German paper by Beyer and Bitsch of the Schering Institute for High Voltage Engineering [1616] deals with the influence of different gases and moisture on the electrical strength of insulating oils in an inhomogeneous field. The partial discharge power at a certain point depends on the kind of gas and the degree of saturation. According to Henry's law, the Ostwald absorption coefficient of various gases and moisture in various oils was measured as a function of the temperature.

The more easily a gas can produce charge carriers, the more unstable the complete system becomes. Therefore, oils saturated with electropositive gases with higher ionization energies (to produce positive ions and free electrons) are more stable than oils saturated with electronegative gases with lower attachment energies (to attach a free electron and to produce a negative ion). Also, the nonuniformly stressed oil is more stable when partly saturated than when completely saturated. Moisture in this instance can also be regarded as a gas. The results show clearly that for the electrical strength of a nonuniformly stressed oil, the events in the gaseous phase are decisive. In the case of moisture in solution, such a gaseous phase may be assumed. Table A3a-1 shows events which may occur with the more usual gaseous components in oil.

3a. BEHAVIOR OF GASEOUS AND LIQUID DIELECTRIC MATERIALS

TABLE A3a-1

PAIRS OF CHARGE CARRIERS FOR DIFFERENT GASES

Gas	Event	Ionization energy (eV)
SF_6	$SF_6 + e \rightarrow SF_6^-$	~2
O_2	$O_2 + e \rightarrow O + O^-$	~3
C_2H_2	$C_2H_2 \rightarrow C_2H_2^+ + e$	11.3
CO_2	$CO_2 \rightarrow CO_2^+ + e$	13.7
Ar	$Ar \rightarrow Ar^+ + e$	15.7
N_2	$N_2 \rightarrow N_2^+ + e$	16.8
H_2	$H_2 \rightarrow H + H$	4.5
	$H \rightarrow H^+ + e$	13.5
H_2O	$H_2O \rightarrow H_2O^+ + e$	12.5

a. SF_6 Breakdown

An insulating gas of increasing importance is sulfur hexafluoride, SF_6. Often it is simply used instead of oil, or mixed with nitrogen under high pressure. The process of SF_6 breakdown is entirely different from that of simple gases like N_2 or air.

The condition for SF_6 breakdown in slightly nonuniform fields is discussed by Mosch and Hauschild [1602]. In slightly nonuniform fields, stable discharges are not present; only in extremely nonuniform fields do stable partial discharges exist before breakdown. Therefore, stable partial discharges mark the transition from slightly to extremely nonuniform fields (partial discharge transition). The maximum electrical field strength necessary for breakdown depends on the electrode radius and the distance between the electrodes, and can be calculated from the dielectric strength by using characteristics found experimentally, The dielectric strength is given for ac (50 Hz), dc, and impulse voltages (1.2/50 μsec rise/decay times), as well as switching surges (250/2.500 μsec rise/decay times). In this way it is possible to calculate the breakdown voltage for insulation arrangements having electrodes where the maximum electrical strength occurs on a point or a short line.

Oppermann [1603] studied the well-known law of Paschen and its validity for SF_6. It is very important to keep the electrodes well polished and clean. Generally speaking, at 3 bars SF_6 insulates similarly to oil. The limits of Paschen's law are an essential function of the gas density, the gap spacing, the electrode material, and also the voltage mode. The values of these dimensions at the limit of Paschen's law and their interdependences were determined. The result shows that the electrical breakdown field strength at the limit of Paschen's law for each mode of voltage is constant and does not depend on the other test parameters.

The SF_6 electrical breakdown in a uniform field was measured by Hasse [1601] for gap spacings of 1–2 cm. The experiments were performed with near-rectangular high-voltage pulses of variable magnitude and duration at pressures from 250 to 760 Torr. Using variable magnitudes and durations of the voltage pulses, it was possible to obtain the discharge formation at different stages of development and to record it simultaneously by oscilloscopic and photographic means. Statistical scatter of the discharge permits the observation of the different stages of discharge development in the extremely short time from beginning to total breakdown.

The experimental results are: The discharge development in SF_6 always starts at the cathode, which is contrary to the discharge in air; partial sparks in SF_6 are observed only ahead of the tip of the streamer channel developing from the cathode, while in air the whole predischarge trace is filled with separate parts of sparks which merge quickly and thus cause a voltage collapse without step; and the voltage collapse in SF_6 is marked by a sharp step.

SF_6 is one member of the family of electronegative gases which trap free electrons. A summary of this particular subject has been given by Raether [1708]. In a paper by Teich and Sangi [1605], the range of available data on the electronegative gases SF_6, CO_2, CCl_2F_2 is extended, especially for high values of E/p and low pressures. Drift velocities of electrons and positive ions as well as ionization and excitation coefficients can be described by simple equations with empirical constants which are tabulated. Electron detachment is directly observed in SF_6. Excitation data are compatible with feedback coefficients up to medium values of E/p; at high E/p, positive ion feedback is found. Excitation of SF_6 is independent of E/p, but emission is pressure-dependent due to quenching. Solid spacers designed not to distort the field cause very little reduction of breakdown voltage. Emission of uv radiation is utilized in a corona discharge detector.

Steady-state measurements with radioactive primary current sources or uv irradiation of the cathode were used as well as "dynamic" methods of avalanche current observations, with either single or multiple primary electrons; the latter were provided by high-intensity uv flash lamps of short pulse duration (10–15 nsec). The gases investigated were SF_6, CO_2, and CCl_2F_2 (Arcton 12). Fairly extensive data of some parameters in CO_2 (mostly from steady-state measurements) are already available and could be used to check the validity of avalanche methods employed here.

The influence of conducting particles on the ac breakdown of compressed SF_6 has been investigated by Cookson and Farish [1598] for two coaxial systems using voltages up to 450 kV rms at pressures up to 1.8 MNm^{-2}. For filamentary particles of 0.1 or 0.4 mm diameter wire, there was a maximum breakdown voltage at 0.5 MNm^{-2} which became more pronounced with

3a. BEHAVIOR OF GASEOUS AND LIQUID DIELECTRIC MATERIALS

increasing particle length and decreasing particle diameter. With spherical particles there was a critical sphere size which gave the lowest breakdown voltage at a given pressure. Tests with wires and spheres fixed to the inner conductor, to simulate breakdown initiated by field enhancement at particles migrating to the inner conductor, gave ac and impulse results which differed from the free-patricle ac breakdown voltages. Calculations of the motion of spherical particles and of the energy in microdischarges at the electrodes have been correlated with the breakdown measurements. (There are 11 other references in this paper; see especially [1709, 1711].)

b. *High Speed Observations*

New methods for observing rapid events, such as the use of gated-image intensifiers and sensitive photomultipliers, led to a better understanding of breakdown phenomena in simple air; Banford and Tedford [1597] also observed the growth of electron avalanches in filtered nitrogen in uniform fields at values of *pd* (pressure × distance) up to ~14,000 Torr cm, corresponding to sparkover voltages of ~500 kV, dc. The data obtained have allowed conclusions to be made regarding the mechanisms leading to sparkover. The current pulses of electron avalanches are observed by means of an oscillograph connected across a resistor in series with the stressed discharge gap (electrical method). The shape of the observed pulse depends on the value of the series resistor, which, in the present experiments, was adjusted to give the so-called "balanced" pulse condition. Photomultipliers may also be used in observation of individual avalanche pulses (optical method), and it has been shown that the rise of the electron current recorded by this means is identical with that recorded by the current measurement technique. The photomultiplier method has inherent advantages of sensitivity at reduced noise interference, and was used widely in the present investigation. There is no evidence to suggest a transition of breakdown mechanism in air to one which is independent of cathode secondary emission, as had previously been suggested.

c. *Positive Discharges in Atmospheric Air*

A paper by Gallimberti and Stassinopoulos [1589] refers to the development of positive discharges in atmospheric air. The positive rod–plane gaps 30 cm in length operated under impulse voltages with long tails and with fronts of 2.3–20 μsec. The experimental results showed that such gaps under impulses of sufficient amplitude breakdown either within a few microseconds from the start of the impulse, or with time lags of the order of 100 μsec. The parameters govern the various mechanisms of the discharge and suggest a qualitative model of the phenomenon.

d. Constructing Pressured Air Capacitors

In constructing pressured air capacitors, it is also important to consider the influence of air conductivity on corona and breakdown voltage. Flieux et al. [1599] emphasize the importance of this often neglected parameter. The breakdown voltage of a point–plane gap is studied when a positive impulse voltage is applied to the point.

The radius of curvature at the tip of the point is 0.1 cm, the gap length is 12 cm, and the duration of the initial front of the impulse voltage is 130 μsec. It is shown that the dispersion in the breakdown voltage is mainly due to the erratic value of the voltage at which the first corona pulse occurs. The mean

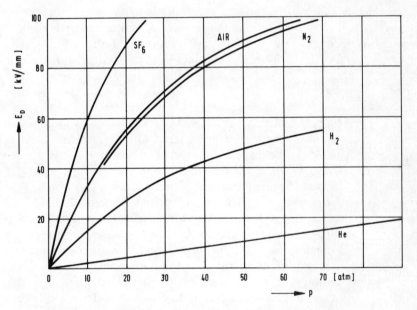

FIG. A3a-1. Breakdown voltage E_D of some gases in relation to the pressure.

value of this inception voltage can be varied by modifying the density of negative ions and this modification is achieved by means of an auxiliary corona created by a thin wire in the area where testing took place. When the impulse voltage has a constant crest value, the mean inception voltage decreases and the breakdown probability increases if the negative ion density is increased. For example, for a crest value of 72 kV, the breakdown probability can be increased from 0 to $\sim 100\%$ only by varying air conductivity.

Gaenger [1600], one of the pioneers in measurements of the insulating behavior of pressurized air studied the electric strength of air with high alternating and impulse voltages and pressures up to 100 kg cm^{-2}. Measurements of the breakdown strength of highly compressed air in a quasi-homogeneous field with alternating and impulse voltage up to 500 and 900 kV are reported. The highest measured strengths were of the order of 120 kV mm^{-1}. Since the gas and the test cell had not been cleaned with extreme care, the values obtained were definitely lower than would be expected with an absolutely pure gas. Figure A3a-1 shows some practical results.

3b. High-Voltage Dielectric Fluids

For impregnating and filling cables and capacitors, the usual transformer oil is being replaced more and more by special dielectric liquids. The properties of four kinds of liquids are given below [1904]. The popular dielectric fluids are for use in high-voltage power supplies, cables, and transformers, as well as in capacitors. More expensive liquids are appropriate for use in power supplies, cables, and capacitors where ultrahigh stresses are involved. Such fillings must be used in a sealed environment since they are quite hygroscopic and cannot be exposed to atmospheric conditions without degradation. Liquids of another type are nonflammable and are primarily high-voltage ac dielectric materials for high-voltage transformers and ac power factor correction capacitors.

Still another dielectric fluid is specially designed for operation in high-temperature applications and is capable of operating in environments up to 225°C. It is used primarily as an electrical insulating fluid in electrical circuits, transformers, cables, power supplies, and capacitors. High Energy, Inc. manufactures all these kinds of dielectric fluids. Type Q 2101 is a high-quality, medium viscosity oil with low dielectric loss, freedom from gas evolution, and absence of wax. The fact that it has a low hygroscopic factor and is easily handled without contaminating, makes it an excellent general-purpose dielectric. Type Q 2102 is a specially processed, high-grade natural oil that is light in color, has a low fatty acid content, and is chemically composed of 90% triglyceride of ricinoleic acid. It is insoluble in water but is soluble in alcohols, esters, ethers, and ketones.

The principal characteristics of type Q 2103 fluid are its low loss factor in 60-Hz applications, its high dielectric constant, and its nonflammability. It meets all NEMA and military specifications for fire-resistant fluids. Its basic composition is polychlorinated biphenyl. Type Q 2107 is a clear, stable fluid with a very flat viscosity/temperature curve. It provides excellent dielectric properties throughout an extensive range of temperatures and frequencies and resists degradation under extremes of operating conditions. (See Table A3b-1.)

4. The Vacuum Capacitor*

A very high vacuum is a hard competitor of high-pressure gases because it avoids by its nature any kind of avalanche up to field strengths at which the electrode surface begins to emit electrons. High-vacuum capacitors are also available in variable form, e.g., for a capacitance range of 10–500 pF up to 10 kV [1181]. The permissible average current can amount to 40 A. Due to their freedom from dielectric losses, they can be used in ultrarapid event discharge circuits. The highest commercially available capacity is 5000 pF at 5 kV peak rf working voltage and the typical range of variation is usually 1:50 or 1:100. Types are also available for up to 30 kV peak rf voltage. Variable types have capacitances of 6–600 pF, fixed capacitors up to 500 pF.

5. Specifying a Spark Discharge or Flash Lamp Capacitor

Spark discharge circuits and flash lamp applications require capacitors with voltages anywhere from 1000 to 35,000, and energy levels from 1 to 5000 J/capacitor. Lifetime requirements vary from a few flashes to more than 10^{11}. Early in his design consideration, the engineer should come to some general decisions on capacitance value, lifetime requirement, and size of each capacitor. When specifying, he should also give the desired reliability and a description of the environment and of all relevant operating conditions.

Hayworth [1901] presented some rules of thumb about lifetime. A capacitor's lifetime is a function of many dependent and independent variables whose details are best left to the capacitor engineer. The user should remember, however, that lifetime is dependent upon some exponential power of the voltage, and upon frequency and reversal (Q). All of these values should be given by the capacitor manufacturer. Typically, the lifetime of a paper-dielectric capacitor varies as the voltage to the 5th power; a polyester-film dielectric, to the 7.5th power; and according to preliminary evidence, polypropylene dielectric follows a 9th-power law for discharge applications.

The strongest determinant of capacitor lifetime, after the charging voltage, is the percentage voltage reversal, the ratio of the first negative voltage peak to the dc charging voltage, shown in Fig. A5-1. If the capacitor is expected to operate under varying conditions, the specifier should spell these out carefully. As a guideline, life is proportional to $Q^{-2.2}$, where Q is the value for the circuit. The percent reversal is related to Q by the formula % rev = $\exp(-\pi/2Q)$, or for a waveform photo, Q is approximately πN, where N is the number of cycles required for the wave envelope to decay to $1/e$ (approximately 1/3) of its initial value. Frequency is another determinant of

---

* See also Chapter B6.

5. SPECIFYING A SPARK DISCHARGE OR FLASH LAMP CAPACITOR

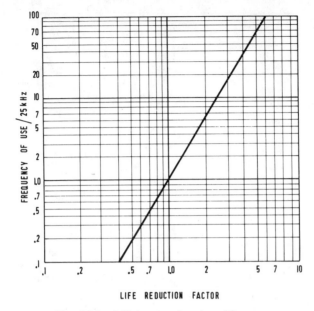

FIG. A5-2. Lifetime as a function of frequency.

Mica and synthetic mica papers have excellent electrical properties, but usually suffer from low energy density and the highest cost. Since the micas are usually encased in resin, they have superb environmental properties and operate at over 200 C.

a. *Off-the-Shelf Cases Cost Less*

Any capacitance value can usually be made to fit any case, but the capacitor may have a lifetime of only one pulse. For best reliability, allow the largest size possible and demand to know the expected lifetime of the model under the specified conditions. Choosing the case to fit the available space is an expensive approach. A great variety of off-the-shelf cases are available, all much less expensive than a custom container. A case can be as simple as a wrap of tape with epoxy-filled ends, an ideal choice if cost is the only consideration. For maximum power it may have radiating fins or water-cooling coils, or depend on an external heat sink. The benefits of various heat-transfer techniques are compared in Fig. A5a-1.

A coaxial capacitor has a hole through the center, into which the flash lamps are placed. With good external heat sinks, a properly designed coaxial capacitor can operate up to several thousand pulses per second.

In each voltage range, bushings are available in a variety of styles and prices. The low-inductance types over 10 kV are probably too expensive

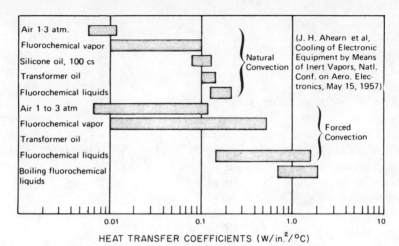

FIG. A5a-1. Heat-transfer techniques compared [1901].

unless strip-line connections are to be used. Bushings must be chosen with regard to operating voltage and environment (altitude, rain, dust, etc.) so this information must be passed on to the supplier.

Most containers are made of lead-coated steel (terneplate), or of glass, ceramic, or copper. The newest container for applications over 25 kV is a welded case of acetal plastic which provides built-in insulation and resistance against corrosion, but at a premium cost.

The user should list any limitation on size or weight, including outline drawings if the dimensions are not standard; list dimensional tolerances, bearing in mind that closer tolerances cost more; and include information on special paints or color-coding required.

b. *Predicting Losses*

A common cause of capacitor failure is overheating, due either to an unexpectedly high ambient temperature or to internal losses that cause thermal runaway. A cure can nearly always be found. If additional heat sinks or more external cooling cannot be applied, it is necessary to order a capacitor with higher ambient temperature rating or one with lower losses. If the package must remain the same size, the designer must choose a better dielectric at a sizable cost increment.

Unexpected losses can alter the circuit performance even if they are insufficient to damage the capacitor. It is useful, therefore, to be able to calculate these losses from the manufacturer's data. The power loss in any capacitor, whether ac or dc, is given by $P_L = I_{rms}^2 R$, where R is the equivalent series resistance. This resistance has only a mathematical reality, and results in a loss which is, in effect, that of series resistance (connections, leads, foils, and

5. SPECIFYING A SPARK DISCHARGE OR FLASH LAMP CAPACITOR

terminals), shunt resistance (corona, leakage), dielectric losses, and other losses.

For most applications, it is sufficient to assume that the dielectric losses predominate, and that these losses are independent of all variables except frequency and temperature. A good estimate of losses, therefore, can be based on the manufacturer's data on dissipation factor, which are almost always taken under simple, low-voltage, sinusoidal conditions. The equivalent series resistance is then given by $R = DF/2fC$, where the DF is taken at the temperature and frequency of interest. For nonsinusoidal waveforms, using the characteristic or predominant frequency of the signal gives answers accurate enough for our purposes.

The value of the root-mean-square current can be difficult to determine for a complex waveshape. Direct measurement is always best, but some calculated values are shown in Fig. A5b-1 for a few commonly encountered situations. Using the rms current from Fig. A5b-1 for the repetitive, damped discharge gives the interesting answer for power loss:

$$P_L = DF \times \tfrac{1}{2}CV_0^2 \times Q/T$$

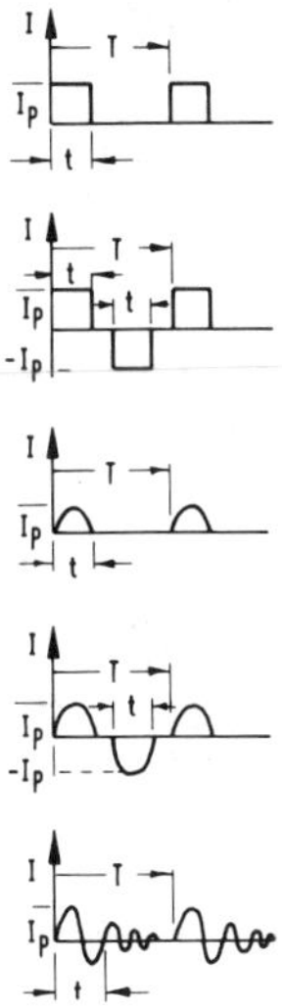

$$I_{RMS} = I_P \left[\frac{t}{T}\right]^{1/2}$$

$$I_{RMS} = I_P \left[\frac{2t}{T}\right]^{1/2}$$

$$I_{RMS} = I_P \left[\frac{t}{2T}\right]^{1/2}$$

$$I_{RMS} = I_P \left[\frac{t}{T}\right]^{1/2}$$

$$I_{RMS} = V_0 W_0 C \left[\frac{tQ^3}{\pi T(1+4Q^2)}\right]^{1/2}$$

$$\cong CV_0 \left[\frac{W_0 Q}{2T}\right]^{1/2} \quad Q > 2$$

$$Q = \frac{2\pi E}{\Delta E}\bigg|_{cycle} = \pi N_{1/2} = \frac{2\pi}{1 - \left(\frac{V}{V_0}\right)^2}\bigg|_{cycle}$$

$$= \frac{2\pi}{1 - e^{-\frac{2\pi R}{W_0 L}}} \cong \frac{W_0 L}{R}$$

FIG. A5b-1. RMS currents for complex waveshapes.

which also shows that the energy loss per discharge is just the DF times the initial peak stored energy times the circuit Q.

It can be shown in a similar manner that the energy dissipated in charging a capacitor is as given in the following cases:

Resonant charging $\quad\quad\quad\quad E_L = DF \times \tfrac{1}{2}CV_0^2 \times 8/3$
Constant current charging $\quad\quad E_L = DF \times \tfrac{1}{2}CV_0^2 \times 2/\pi$
Resistive (exponential) charging $\quad E_L = DF \times \tfrac{1}{2}CV_0^2 \times 1/\pi$

In these equations the capacitor energy is calculated at peak voltage. Note that this is not the energy lost in the charging circuit, only that portion lost in the capacitor. Using the total power loss and the radiating area of the case, and with temperature-rise data from Fig. A5b-2, the designer can calculate the power loss per square inch.

As a final example, let us consider the case of a 30-μF capacitor operating at 5 kV into a flashlamp at 5 pulses/sec with a Q of 2 (45% reversal). This capacitor is $4 \times 5 \times 7$ in. and thus has a radiating area of about 150 in.^{2} Its

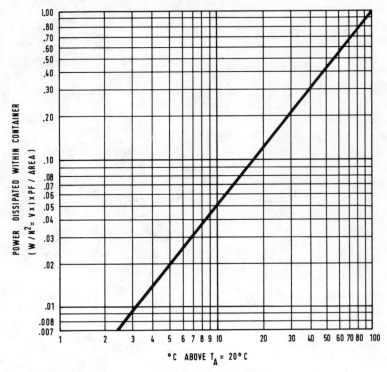

Fig. A5b-2. Relation of temperature rise to power dissipation within the container of a capacitor (military standard 198-A).

DF is 0.006 and its upper temperature limit is 65°C. Therefore the power loss is

$$P_L = 0.006 \times 375 \times (2/0.2) = 22.5 \text{ W}; \quad P/A = 0.15 \text{ W/in.}^2$$

$$\Delta T \cong 25°C \quad \text{or} \quad T_{\text{case}} = 45°C$$

It would be unreasonable to expect to increase either the voltage or the firing rate significantly without using a capacitor of lower loss or higher temperature rating. This information shows the necessity of specifying correctly the repetition rate, the duty cycle, and the dissipation factor. Typical values of DF are: polyester film and paper, 0.005; paper alone, 0.005; paper and polypropylene, 0.0007; polypropylene alone, 0.0003. Measuring of low inductances is discussed further in Chapter A8.

6. Laser Discharge Capacitors

Laser discharge storage capacitors are designed for overdamped, non-ringing applications such as: laser applications, photoflash, spot welding, xenon lamp discharge, and other discharge applications where the primary load is of a resistant nature. This is the softest and nearly ideal case of a discharging capacitor. Two typical programs for the manufacture of such capacitors may be presented. High Energy, Inc. [1904] has designed and manufactured a type of capacitor for critically damped discharge circuits. It is constructed of multilayers of high-density Kraft paper impregnated with castor oil. The capacitor sections are of multitab construction. The entire current path of the capacitor has been designed not only to withstand the high amperage but also to withstand the mechanical shock of discharge. The capacitor has been geometrically designed for magnetic field cancellation which affords an extremely low inductance capacitor (see Table A6-1, Fig. A6-1).

The cases are rectangular cans fashioned of heavy gauge steel which is heliarc welded to ensure maximum strength and durability, and they are finished in a high-quality gray enamel paint. The cans have been designed, in case of an internal faulty discharge, to withstand approximately ten times the energy content of the individual capacitor during a fault condition without erupting. The terminals are porcelain-to-metal sealed bushings capable of continuous function at temperatures up to 200°C. This is possible by utilizing advanced techniques employing a bonding alloy having a melting temperature over 300°C. This insures a high degree of mechanical strength to withstand the heavy shock involved in the discharge currents.

The capacitor sections have either tab or extended foil construction,

TABLE A6-1

TECHNICAL DATA FOR LASER CAPACITORS[a]

Part No.	Energy (J)	Capacitance (μF)	Voltage (kV)	Size				
				A	B	C	D	E
CJL101	85	27	2.5	$3\frac{3}{4}$	$4\frac{9}{16}$	6	$1\frac{1}{2}$	2
CJL102	100	8	5	$3\frac{3}{4}$	$4\frac{9}{16}$	6	$1\frac{1}{2}$	2
CJL103	175	56	2.5	$3\frac{3}{4}$	$4\frac{9}{16}$	$9\frac{5}{8}$	$1\frac{1}{2}$	2
CJL104	200	16	5	$3\frac{3}{4}$	$4\frac{9}{16}$	$9\frac{5}{8}$	$1\frac{1}{2}$	2
CJL105	300	96	2.5	4	8	$9\frac{1}{2}$	$2\frac{1}{4}$	$4\frac{1}{2}$
CJL106	350	28	5	4	8	$9\frac{1}{2}$	$2\frac{3}{4}$	$4\frac{1}{2}$
CJL107	450	144	2.5	4	8	$12\frac{7}{8}$	$2\frac{3}{4}$	$4\frac{1}{2}$
CJL108	500	40	5	4	8	$12\frac{7}{8}$	$2\frac{3}{4}$	$4\frac{1}{2}$
CJL109	500	10	10	4	8	$12\frac{7}{8}$	$2\frac{3}{4}$	$4\frac{1}{2}$
CJL110	850	272	2.5	$4\frac{1}{8}$	$13\frac{1}{2}$	$12\frac{7}{8}$	$2\frac{3}{4}$	$6\frac{3}{4}$
CJL111	900	72	5	$4\frac{1}{8}$	$13\frac{1}{2}$	$12\frac{7}{8}$	$2\frac{3}{4}$	$6\frac{3}{4}$
CJL112	900	12	10	$4\frac{1}{8}$	$13\frac{1}{2}$	$12\frac{7}{8}$	$2\frac{3}{4}$	$6\frac{3}{4}$
CJL113	1500	120	5	$5\frac{1}{8}$	$13\frac{1}{2}$	$15\frac{1}{2}$	$2\frac{3}{4}$	$6\frac{3}{4}$
CJL114	1500	30	10	$5\frac{1}{8}$	$13\frac{1}{2}$	$15\frac{1}{2}$	$2\frac{3}{4}$	$6\frac{3}{4}$
CJL115	3000	240	5	$7\frac{1}{4}$	14	22	$2\frac{3}{4}$	8
CJL116	3000	60	10	$7\frac{1}{4}$	14	22	$2\frac{3}{4}$	8

[a] High Energy, Inc. [1904].

depending upon the peak discharge currents and inductance desired of the individual capacitor. The capacitor is designed and manufactured to withstand the highest voltage strength with maximum safety. The specially treated capacitor grade castor oil is capable of increasing the life expectancy approxi-

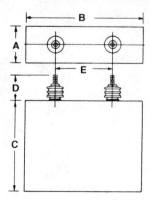

FIG. A6-1. Sketch showing dimensions given in Table A6-1.

mately 50 times over conventional dielectrics. The life expectancy of a capacitor system is a function of the peak charging voltage applied to the capacitor (see Fig. A6-2).

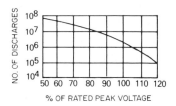

FIG. A6-2. Effect of percent of rated peak voltage of laser capacitors on number of discharges obtainable. Temperature: 25°C; voltage reversal: 0.

Laser capacitors for high-voltage energy storage purposes are also made by CSI (Capacitor Specialists, Inc.) [1903]. They claim long dc life, which permits holding capacitors charged. The capacitors permit high peak currents (the standard is 35,000 A) and have low inductance (10–40 nH). Data for some capacitors are shown in Table A6-2; the temperature range is −5 to 55°C. (See Figs. A6-3, A6-4a, b, c.)

TABLE A6-2

CAPACITOR RATINGS

Capacitance (μF)	Voltage (kV)
100	2
100	3.5
70	7.5
10	15
2.0	25

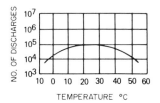

FIG. A6-3. Effect of temperature on number of discharges obtainable. The maximum temperature range of types CJE, CJL, and CJS capacitors is 10–40°C.

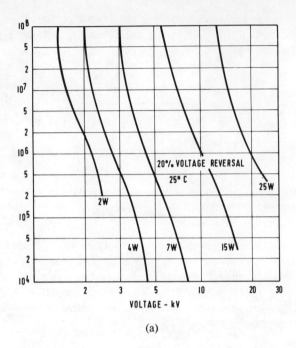

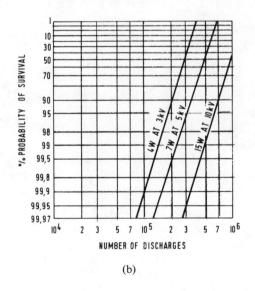

Fig. A6-4a, b. Life, voltage and number of discharges: (a) discharges vs voltage at 20% voltage reversal; (b) probability of survival vs number of discharges.

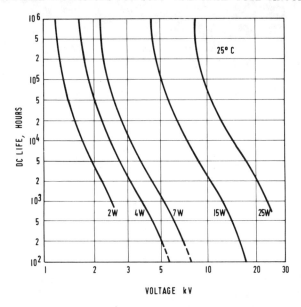

Fig. A6-4c. Life, voltage and number of discharges: (c) dc-life in hours vs voltage at 25°C.

7. Pulse Capacitors for Soft- and Hard-Tube Circuits

a. *High Repetition Rate Pulse Capacitors*

Pulse capacitors are intended for high repetition rate pulse-forming networks. These capacitors are designed to be fully charged and discharged with up to 20% voltage reversal, the most common requirement for soft-tube modulator PFN-type capacitors. They may also be used in energy discharge circuits where extremely high repetition rate pulsing is required.

For this kind of operation the capacitors have been designed to operate in an ambient temperature of up to 40°C and, under full rating, to have a maximum of 25°C temperature rise. They may be overrated up to 25% in terms of repetition rate or given voltage. Forced air cooling is required. For special applications, higher velocity air may be used for increased ratings. Fins may also be used to increase cooling efficiency and, if necessary, water-cooled units may be supplied to meet pulse requirements.

b. *Low Repetition Rate Capacitors*

For low repetition rates, such capacitors are constructed of paper and mineral oil with extended foil construction; for high repetition rates, High Energy, Inc. uses a plastic film and proprietary fluid affording an extremely

TABLE A7a-1

SOME TYPICAL SOFT-TUBE PULSE CAPACITORS[a]

Part No.	Capacitance (μF)	Voltage (kV)	pps	Part No.	Capacitance (μF)	Voltage (kV)	pps
PL101	0.001	10	250	PL165	0.01	20	250
PL104	0.001	10	2000	PL168	0.01	20	2000
PL105	0.001	20	250	PL169	0.01	40	250
PL108	0.001	20	2000	PL172	0.01	40	2000
PL109	0.001	40	250	PL209	0.05	10	250
PL112	0.001	40	2000	PL212	0.05	10	2000
PL161	0.01	10	250	PL217	0.05	40	250
PL164	0.01	10	2000	PL220	0.05	40	2000

[a] High Energy, Inc. [1904].

low dissipation factor (Table A7a-1). Regardless of repetition rate, all capacitors are mounted in all-heliarc-welded containers and are conservatively designed from both a voltage and heat-dissipation standpoint. Containers are mostly rectangular cans, they are fabricated from heavy-gauge steel to assure maximum strength and durability. Base seams are welded to ensure against leaks. A high-quality finish protects against high humidity and/or salt conditions.

c. Hard-Tube Modular DC Capacitors

Hard-tube modular dc capacitors are specifically designed for large capacitor banks. They are built to withstand occasional "crowbarring" but not to exceed one "crowbar" per minute over a one-hour period. The low-inductance characteristics of the type CE capacitor ensures a fast risetime; extremely heavy current paths will handle the normal pulse currents as well as the "crowbar" currents encountered in hard-tube modular circuitry.

This capacitor therefore has applications for impulse generators, high-voltage power supplies, voltage doubling circuits, communication receivers and transmitters, x-ray equipment, and energy storage along with other specific electronic applications [1904]. The temperature range is $-55°$ to $+85°C$. Operation up to $115°C$ is possible if done at 75% of the rated voltage (Table A7c-1). The life is 44,000 h with 96.8% survival at 90% confidence level. The test voltage is usually two times rating plus 1000 V for one minute and the insulation resistance is 2×10^5 M$\Omega \times \mu$F at 25°C. The peak ripple voltage added to the dc voltage should not exceed the rated voltage. Such capacitors use multilayers of quality Kraft capacitor tissue and pure, high conductivity, aluminum foil. The multitab construction design enables the capacitor to withstand the highest currents encountered in the applications for which it is designed (Table A7c-2).

TABLE A7c-1

PULSE REPETITION RATE AND PULSE DROOP OF HARD-TUBE CAPACITORS[a]

Pulse repetition rate/second	Pulse droop not to exceed % of rated voltage
60	20
120	15
400	12
1000	7
2500	4
5000	3
10,000	2
15,000	1.5

[a] High Energy, Inc. [1904].

TABLE A7c-2

SMALLEST AND GREATEST HARD-TUBE CAPACITORS FOR 10–60 KV[a]

Part No.	Voltage (kV, dc)	Capacitance (μF)	Part No.	Voltage (kV, dc)	Capacitance (μF)
CE101	10	0.1	CE147	25	2.0
CE110	10	10.0	CE148	30	0.1
CE111	125	0.1	CE152	30	1.0
CE122	15	0.1	CE158	40	0.1
CE132	15	5.0	CE161	40	0.75
CE133	20	0.1	CE165	60	0.05
CE140	20	3.0	CE168	60	0.5
CE141	25	0.1			

[a] High Energy, Inc. [1904].

d. *Plastic Case Capacitors*

Plastic case capacitors are designed for such high energy, fast discharge, high-reversal applications as nuclear research, high-temperature investigations, plasma generators, arc discharge, ballistic accelerators, exploding wires,

TABLE A7c-3

TEN TYPICAL CAPACITOR BANK COMPONENTS IN RECTANGULAR PLASTIC CASES

Part No.	Energy (J)	Capacitance (μF)	Voltage (kV)
CPE101	1000	5.0	20
CPE105[a]	5000	25.0	20
CPE201	1000	1.25	40
CPE205[a]	5000	6.25	40
CPE301	1000	0.8	50
CPE305[a]	5000	4.0	50
CPE401	1000	0.2	100
CPE405[a]	5000	1.0	100
CPE501	1000	0.12	125
CPE505[a]	5000	0.64	125

[a] Life expectancy: 40,000 shots.

ion propulsion, hypersonic wind tunnels, metal forming, and other phases of magnetohydrodynamics (Table A7c-3).

This type of capacitor is constructed for repeated oscillatory discharging. It is constructed of multilayers of high-density Kraft paper impregnated with castor oil. The capacitor sections are of extended foil-swaged construction. The entire current path of the capacitor has been designed to withstand not only the high amperage but also the mechanical shock of discharge. This capacitor has been developed geometrically for magnetic field cancellation and has an extremely low inductance.

The rectangular cases are fabricated of a thermoplastic material and are welded to ensure maximum strength and durability. They are designed to withstand approximately ten times the energy content of the individual capacitor during a fault condition without erupting. The terminals are either copper or aluminum bus capable of conducting the high currents encountered during discharge. Busbars are $\frac{1}{4} \times \frac{3}{4}$ in. and approximately 4 in. long and are O-ring sealed to ensure leak-proof performance.

8. Measurements of Ultralow Inductance (nH) Capacitors

As shown in Chapter A5 and 7, a wide variety of techniques can be applied when internal inductances greater than 50 nH are to be measured. In most of these, the self-resonant frequency of the capacitor is measured by means of a calibrated rf generator or the time-base of an oscilloscope. Hayworth [1940] presented a paper on differentiating a nanohenry from a microfarad.

8. MEASUREMENTS OF ULTRALOW INDUCTANCE (NH) CAPACITORS

Figure A8-1 is a block diagram of a test arrangement that can be used for "standing-wave method" measurements. Generally, the input impedance of the test capacitor will be considerably less than the source impedance of the rf oscillator. To improve the impedance match, an autotransformer can be employed. It should have a total inductive reactance approximately equal to the oscillator output impedance, at the lowest frequency expected in the measurement. The coil usually can be made up, when needed, from No. 12 or No. 14 AWG insulated solid-copper wire. The load tap, to couple the test signal to the test capacitor, should include $\sim 10\%$ of the coil's turns.

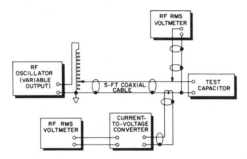

FIG. A8-1. Block diagram of an arrangement for measuring the series-resonant properties of a capacitor. The current probe, the current-to-voltage converter, and its associated rf voltmeter, monitor the rf-current drive to the capacitor, so that the oscillator's output can be adjusted to maintain constant current as its frequency is varied. The oscillator's high output impedance is coupled to the low impedance of the test capacitor by means of an autotransformer. The upper rf voltmeter, which is used to observe the voltage dip associated with self-resonance of the capacitor, can be replaced by a suitable oscilloscope, if available.

The drive signal should be transmitted to the capacitor through about 5 ft of RG-8 coaxial cable. Care must be taken to locate all leads and circuit components for minimum cross coupling. The movable tap on the test-capacitor input lead provides a means for establishing whether or not a frequency that exhibits minimum-voltage response is the true resonant frequency. While maintaining a constant input-current level, the oscillator frequency is varied to determine the lowest frequency that exhibits a minimum-voltage indication across the test capacitor. Then a second measurement is performed with the voltmeter tap relocated so as to include less inductance (i.e., moved closer to the capacitor). If this next resonant frequency is higher, then the true resonant frequency is being approached. The procedure is repeated until the highest possible resonant frequency (in the fundamental mode) is obtained. This frequency, and the known value of capacitance, can be used to compute L_s, the effective series inductance of the capacitor.

If available, a high-gain oscilloscope with suitable bandpass characteristics can be employed for the rf-voltage indicator, and may help to ensure that only fundamental-mode frequencies are recorded. In general, this standing-wave method is useful for inductance values as low as 50 nH. Transient-response observations with an oscilloscope can also be employed to measure relatively high-inductance values. In one such method, an external short circuit is applied to the capacitor. A Rogowsky belt [1941] (sensitive current transformer) is employed to couple currents flowing in this short circuit to an oscilloscope so that the self-resonant frequency of the capacitor can be observed. To shock the test capacitor into a damped oscillation, a lower-value capacitor and a large external coil are selected that will resonate at a higher frequency than the test capacitor. In performing this measurement, the second capacitor is charged and then shorted through the coil when it is adjacent to the test capacitor. The position of the coil with respect to the test capacitor is critical. If it is too close, the resonant frequency of the test capacitor may be shifted from its true value. As a result, it is difficult to obtain reliable results with this technique.

A more common transient-response technique consists of short circuiting the test capacitor after it has been charged, to excite a damped self-oscillation. Consistent results are also difficult to attain with this technique. The effective resistance and inductance of the spark path are not predictable, and, in addition, the inductive contribution of the external short circuit can be significant and difficult to evaluate. Very low inductance (less than 40 nH) can be achieved in capacitors that are specifically designed for this purpose. Measurement of these small inductances is complicated by several factors. External devices, that must be connected to facilitate such measurements, may add external inductance comparable to or greater than the capacitor's self-inductance. Resonant-frequency test results are difficult to interpret. At the higher test frequencies that are commensurate with such low inductances, capacitor windings exhibit the properties of a distributed constant delay line. As a result, wave-reflection phenomena distort the wave patterns that are observable with an oscilloscope.

It becomes necessary, then, to distinguish between effects attributable to the capacitor's series inductance, L_s, and the transmission-line properties of the capacitor's structure. L_s represents the electrical properties of the terminals, lead wires, interwinding spaces, and other field-filled voids associated with the capacitor windings. The winding, on the other hand, exhibits the properties of pulse time, impedance, and capacitance, which can be related to the energy-storage properties of a transmission line. With the aid of an external variable inductance, a transient-response technique has been used to determine series-inductance values as small as 400 pH, in spite of these complex characteristics of the capacitor winding.

a. Variable-Inductance Technique

While it may not be possible to isolate the internal inductance L_s, in order to measure it directly, it is possible to infer its value, from measurements performed with the device shown in Fig. A8a-1. A coaxial shorting element is constructed, whose inductance can be varied in well-defined steps, so that the difference in inductance between any two positions can be accurately calculated. When connected directly to the capacitor's coaxial terminals, this calculable variable inductor is effectively in series with the internal inductance. Also, it exhibits a simple geometry for the computation of inductance differences corresponding to the various positions of the conductive disk, that connects the center conductor to the outer tube.

For each measurement, inductance differences so computed may require a "skin-effect" correction. This will be determined by the dominant frequencies present in the transient wave; as higher frequencies are involved, the effective dimensions of the shorting device will increase. With appropriate high-frequency equipment, it is possible to measure, and thus verify, the computed inductance differences.

While it is not possible to produce a zero-inductance value with the device, the minimum value should be as small as possible. It is convenient to be able to increase the value of the external inductor to about five times that of L_s. The adjustment range required usually can be approximated from an estimate based upon the capacitor's geometry. A short air gap is introduced in the center conductor, to start a transient current. To perform each step of the measurement, the voltage across the gap is increased until dielectric breakdown occurs in the gap. Coupled into a magnetic-field probe, the resulting transient current is observed, and photographed, as it appears on an oscilloscope. The procedure is repeated for at least five discrete values of the variable inductor. Greater

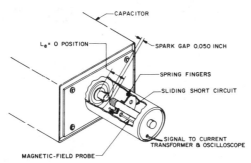

FIG. A8a-1. Mechanical diagram of a calculable variable inductor, that can be employed to measure internal inductance as small as 400 pH. This external inductor is adjusted by changing the position of the sliding disk in the coaxial stub. A "B-probe" or magnetic pickup is employed to monitor the current decay in the composite circuit, after a transient current is initiated by a spark discharge across the center-conductor gap.

precision can be achieved by increasing the number of inductance steps and observations to 10 or 15.

To obtain clear records of the transient currents, the signal-to-noise ratio should be increased as much as possible by adjusting the number of turns of the magnetic-field probe. Also, the use of a current transformer (Rogowsky belt) can be helpful. However, due consideration must be given to frequency response, so as to preserve the waveform coupled into the probe. The output signal from the probe is proportional to dB/dt, the time rate of change of flux density, which is proportional to dI/dt, the time derivative of the transient current.

A typical transient waveform is shown in Fig. A8a-2. The portion of this curve that follows the maximum peak is a voltage-decay curve for the particular inductance value of the circuit at the time, combined with the transient characteristics of the capacitor winding. The capacitor-winding characteristics are constant, for all of the measured transients. By extending the decay curve for each observed transient back to the zero-time line (initiation of the spark), a series of "h" values, each corresponding to a different inductance value, are obtained where

$$h = (dI/dt)t = 0$$

If the reciprocal values, h^{-1}, are now plotted against the inductance values, as shown in Fig. A8a-3, a straight-line plot will be obtained if the measurements have been made carefully. This line can be extended back through the ordinate line—which corresponds to the minimum inductance setting of the calculate

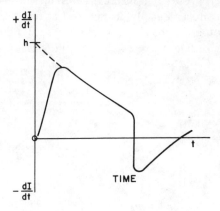

FIG. A8a-2. Time plot of a transient response of a typical low-inductance foilwound capacitor. The magnetic probe shown in Fig. A3a-2 produces an output that is proportional to dI/dt. Extension of the current-decay portion of the curve yields a hypothetical value, h, for this variable at $t = 0$. A series of these values are obtained, for known changes in the value of the variable inductance of Fig. A8a-1. The "h" values are then employed to construct a solution plot, such as that shown in Fig. A8a-3.

8. MEASUREMENTS OF ULTRALOW INDUCTANCE (NH) CAPACITORS

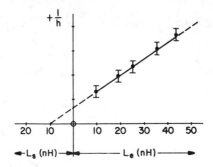

FIG. A8a-3. Plot of the reciprocal values of "h," obtained from a series of transient records such as the one shown in Fig. A8a-2. Inductance values (L_e) of the external inductor of Fig. A8a-1 are the units of the right-hand abscissa. The origin corresponds to the minimum-inductance position of the variable inductor, and L_s values, plotted to the left of the origin, represent inductance contributed by the capacitor and its terminals.

inductor—to an intercept with the abscissa ($h^{-1} = 0$). There will be the value of the series inductance, L_s, behind the shorting device's minimum inductance position.

The curve of Fig. A8a-3 shows a resultant value of L_s of ~ 10 nH, which is typical for an extended-foil capacitor. It should be noted that the dimensions of h, dI/dt, and h^{-1} are arbitrary and need not be known. It is not, therefore, necessary to calibrate the probe system.

As a practical application, a capacitor with pressurized air insulation is described by Freitag and Schiweck [1581]. With the center electrode and the high-voltage electrode short-circuited, a compressed-gas capacitor up to ~ 800 kV has been realized which is free from loss or proximity effects. In this capacitor, the measuring capacitance can be increased to ~ 600 pF by connecting the intermediate electrode with the high-voltage electrode and by

TABLE A8a-1

DERATING DATA

Maximum voltage	5 kV
Capacitance	32 μF
Energy	400 J
Weight	3.5 lb
Maximum current	40,000 A
Inductance	10×10^{-9} H

TABLE A8a-2

DC LIFE DATA (90% RELIABILITY)

Voltage (kV)	Life (h)
5.0	2
4.5	6
4.0	20
3.5	70
3.0	200
2.5	1000
2.0	4000
1.5	20,000

connecting the insulated center portion of the low-voltage electrode with the remaining parts of the low-voltage electrode.

The Maxwell Company (San Diego, California) [1529 f] gives operating data for one of their Mylar plastic film capacitors (No. 5C32MN) (Tables A8a-1 and 2).

b. *Banks for High-Temperature Plasma Machines*

Especially for the generation of high-temperature plasmas, an extremely high flow of energy is needed to achieve the highest peak power values. Capacitor banks of very low inductance are therefore required but, to reduce cost, the lifetime may be limited to the number of shots that can be reasonably expected (see Table A8b-1).

Advanced Kinetics [1706] is engaged mainly in making large capacitor banks for application to pulsed magnetic fields, plasma physics, laser research, arc tunners, shock tubes, underwater sparks, and metal-forming. Systems are available with an energy storage from 3000 to 100,000 J in 3000-J increments (larger systems on request), storage at 5, 10, 15, or 20 kV, selection of a maximum repetition rate (model A: 1 pulse/5 min; model B: 1 pulse/min; and model C: 6 pulses/min). Output via high-voltage, low-inductance coaxial cables, safety interlock circuitry, short circuit output currents up to 100 kA/3 kJ unit. These high-voltage capacitor energy storage systems deliver a current of fast rise time into inductive and resistive loads. The internal inductance of each 3000-J unit is about 70 nH at the output cables, and less in larger energy units.

TABLE A8b-1

HIGH ENERGY ELEMENTS OF LOW-INDUCTANCE CAPACITOR BANKS[a]

Part No.	Energy (J)	Capacitance (μF)	Voltage (kV)	Inductance (μH)
CJE101	3000	15	20	0.055
CJE102	3000	60	10	0.055
CJE103	3000	240	5	0.055
CJE104	2000	10	20	0.05
CJE105	2000	40	10	0.05
CJE106	2000	160	5	0.05
CJE107	1500	7.5	20	0.045
CJE108	1500	30	10	0.045
CJE109	1500	120	5	0.045
CJE110	1000	5	20	0.035
CJE111	1000	20	10	0.035
CJE112	1000	80	5	0.035

[a] High Energy, Inc.

8. MEASUREMENTS OF ULTRALOW INDUCTANCE (NH) CAPACITORS

c. *Rise Time of the Current*

The current rise time is 3 μsec for each 3000-J unit with shorted output cables. The rise time of a complete system in operation is governed by the sum of load and internal inductance. The storage unit has to combine the following items:

(a) Low-inductance storage capacitors of long lifetime;
(b) System completely coaxial;
(c) Switch tube housed in dry dielectric high-voltage-insulated coaxial housing with 70 nH inductance/3 kJ unit (including capacitor, housing, and cables to exit port);
(d) Switch tube cooling providing either convection cooling in well-ventilated cabinet, controlled forced air ducted into coaxial housing, or controlled and metered liquid coolant with thermal contact as switch tube;
(e) High-current output available from high-voltage, low-inductance cables; length beyond cabinet porthole to be specified by customer;
(f) Emergency discharge switch interlocked with all doors and panels and deenergized on power failure.

(See Chapters G and H, this volume and Volume I.)

d. *Very Large Plasma Generation c-Batteries*

In 1964 Jahoda *et al.* [1219] published a paper on a plasma experiment with a 570 kJ capacitor bank to operate theta-pinch discharges (Fig. A8d-1).

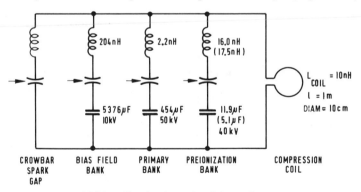

FIG. A8d-1. Circuit schematic of the Scylla IV system.

The purpose and essential features of the capacitor banks are as follows:

(1) A 10-kV, 280-kJ bank provided a bias magnetic field B_0 to uniformly permeate the preionized plasma. The magnitude of B_0 was variable between 0 and 12.6 kG with a rise time of 55 μsec. Further studies of the effect of trapped field on the plasma heating process could be carried out with either

negative or positive B_0. The B_0 bank was connected to the collector plates with ignitron switches and 30 m lengths of RG 17/U cable to provide appropriate isolation from the high-voltage system.

(2) A 40-kV, 10-kJ preionization (PI) bank with low inductance and short rise time was used to ionize the deuterium gas. This bank, consisting of 14 capacitor spark gap units, produced a 330 kHz oscillating magnetic field with peak magnitudes up to 10.6 kG and maximum induced E_0 voltage of 0.4 kV/cm just inside the discharge tube wall. Although this bank operated at 40 kV, 120-kV capacitors were used in order to withstand the transient voltage doubling from the primary capacitor bank. The spark gap switches were of the four-electrode design with an oil-insulated return and were triggered simultaneously by a single 75-kV spark gap capacitor unit.

(3) A 50-kV, 570-kJ primary bank with very low inductance (2.2 nH) provided the fast rising magnetic field, which had been characteristic of the previous Scylla devices, for shock heating and compression. Previous Scyllas had utilized 85-kV capacitor banks to produce approximately 40 kV across the compression coil terminals. However, the development of a low-inductance 50-kV, 2-μF capacitor, combined with an improved four-electrode spark gap enabled the new 50-kV bank also to develop 40 kV across the coil terminals. The primary bank consisted of 216 2.1-μF, 50-kV capacitors with individual spark gap switches. These switches were the four-electrode type with oil insulation between the electrodes and the current return. Six special low-inductance coaxial cables connected each spark gap capacitor unit to the collector plate system. The 216 spark gaps were triggered simultaneously by a master 75-kV, 0.3-μF trigger capacitor. Bank energy isolation was provided by connecting each group of six capacitors to the dc charging bus through a 1500-Ω, 1-kW resistor in series with a 7.5-A fuse. The primary bank transfered 82% of its voltage and energy to the 1-m coil. This produced a peak magnetic field in the coil of 93 kG with a rise time of 3.7 μsec. The maximum azimuthal electric field just inside the discharge tube wall was 1.1 kV/cm.

A fourth capacitor bank was being added to the Scylla IV system to extend the magnetic compression field of the primary bank both in magnitude and time. This "power crowbar" bank had 3 MJ of energy storage at 20 kV. This bank was expected to increase the magnetic field in the 1-m coil to more than 200 kG with a half-period of ~ 50 μsec. Today batteries with more than 2 MJ are being used for plasma experiments.

e. *Small Nanosecond-Discharge Capacitors*

Low-energy ultrafast capacitors are often used for producing nanosecond light flashes. The spark gap is incorporated in the capacitor itself. The candle-power (CP) peak light is approximately proportional to the peak of current voltage [1529]. (See Table A8e-1.)

8. MEASUREMENTS OF ULTRALOW INDUCTANCE (NH) CAPACITORS

H. Fischer has developed a series of very low inductance spark sources.* He quotes the following performance:

Type PL-103   C = 3.4 nF,   V = 4.2 kV;   duration 25 nsec (1/2 light).
Type Sm-6   C = 0.27 nF, V = 3.5 kV;   duration 7 nsec (1/2 light).

A paper by Yguerabide discusses the generation and detection of subnanosecond light pulses which are used to study the luminescence of different materials.

TABLE A8e-1

EXPERIMENTAL CAPACITORS FOR SHORT FLASH PRODUCTION

Capacitance (pF)	d (in.)	D (in.)	E (in.)	Intensity (CP)	Duration (nsec)	Output (Hz)
70	0.007	0.625	1.00	2600	10	30×10^{-6}
167	0.007	0.625	1.25	11,500	15	200×10^{-6}
340	0.007	0.625	1.75	72,500	25	2000×10^{-6}
580	0.007	0.500	2.325	82,500	45	4000×10^{-6}

The conditions of gas pressure, capacitance, and gap width necessary to achieve fast light pulses have been established for several gases and mixtures of gases with oxygen. The experimental results indicate that high pressure (10 atm or larger) and minimum stray capacitance are required for lamps filled with pure gas. A lamp filled with hydrogen, and satisfying these requirements, yields light pulses with 0.35 nsec risetime and 0.5 nsec half-width. The time characteristics of the fast pulses are not affected by gap width of 0.1–0.6 mm. Lamps containing a mixture of a gas and oxygen produce similar light pulses at total pressure much less than 10 atm.

Other sources of 1 nsec duration are described by Malmberg (hydrogen discharge lamp) [1760], Kerns, Kirsten, and Cox (mercury) [1761], and D'Alessio et al. (mercury) [1762]. The light output of these extremely short flashes is very small and apparently cannot be used in conventional photography, even for the back-lighted silhouette type. Perhaps with an image converter there may be a useful combination.

The Fischer Nanolite 1058 is one of the more frequently used nanosecond capacitors operating with a built-in spark gap (see Table A8e-2).

---

* Commercially these nanosecond lamps are called "Nanolite," manufactured by Impulsphysik, Hamburg 56, W. Germany.

TABLE A8e-2

CHARACTERISTICS OF FISCHER NANOLITE[a]

Characteristics	Basic units 8N18-	8N11-	8N8-	(4N9-)
Flash duration (halfwidth)	18 nsec	11 nsec	8 nsec	9 nsec
Rise time (10–90%)	4 nsec	3 nsec	2 nsec	2 nsec
Capacity ($\pm 10\%$)	3 nF	1.5 nF	850 pF	500 pF
Energy per flash	25 mJ	14 mJ	9 mJ	5 mJ
Repetition rate (with 12/13)	10.000 fps	15.000 fps	20.000 fps	25.000 fps

[a] Impulsphysik GmbH, Hamburg [1058].

f. *Other Rapid Discharge Capacitors*

Tobe Deutschmann Laboratories [1007] developed five types of rapid discharge capacitors. Capacitors are made in the form of backs of foils, coaxial disks, parallel plates, coaxial tanks, and for extrahigh voltages, several such units in series.

Foil capacitors are constructed to withstand the high peak currents and severe impulsive forces produced during rapid discharge. Extended foil construction and a recently developed highly conductive solder assure mechanically sound low-resistance connections capable of conducting high peak currents. Series–parallel connections of internal capacitor sections provide uniform stress distribution and corona-free operation. Tobe Deutschmann alumina-filled epoxy bushings fulfill the dual requirement of strength and superior electrical properties. Foil capacitors have a realistic usable self-inductance of 5 nH. A 10-in.-wide strap shorting the capacitor terminals across the barrier insulator has an inductance of 2 nH. The total circuit inductance of the shorted capacitor is 7 nH, which may be easily measured using conventional methods. Tobol C-impregnated, high-density Kraft paper assures 10,000 discharge life expectancy at rated voltage and 80% voltage reversal.

Low-inductance-capacitor discharge switches designed to mate directly with this model capacitor and having comparable inductance- and current-carrying capabilities are available. Final electrical tests of all Tobe Deutschmann capacitors include a high-potential test at 125% rated voltage for 1 min and 10 short-circuit discharges at rated voltage in addition to the normal static bridge measurements.

g. *Coaxial Disk and Parallel Plate Series Capacitors*

Coaxial disk and parallel plate series capacitors are constructed by laminating a number of low-impedance annular striplines, connected in

TABLE A8g-1

RAPID DISCHARGE ENERGY-STORAGE CAPACITORS[a]

Foil capacitor type	Capacitance (μF)	Voltage (kV)	Inductance (nH)	Energy (J)
ESC 248A	15.0	20	5	3000
ESC 247D	0.5	20	1.0	100
ESC 267-1	10.0	10	1.0	500
ESC 276	1.0	100	60	5000
ESC 249B	1.0	120	90	7200

[a] Some of the types of Tobe Deutschmann [1007].

parallel. Using a high-quality silicone-impregnated mica sheeting as the dielectric and a capacitor plate whose thickness is greater than twice the skin depth of penetration at self-frequency results in $Q > 250$ at 5 MHz. These capacitors can be operated at repetition rates up to many thousands of pulses per second with an appropriate derating of peak charge voltage from the low-duty cycle ratings listed in Table A8g-1.

The transit time of the individual lines is 3.4 nsec and virtually all inductance is developed at the terminal where the individual plates are interconnected. Parallel plate series capacitors have a self-inductance lower than any other commercially available energy storage capacitor. The self-inductance of this series capacitor is ~ 0.5 nH. The inductance of a wide strap shorting the capacitor across the barrier insulator is typically 1 nH. Self-inductance can be further lowered by modifying the terminal and insulating barrier, if required, at a nominal increase in manufacturing costs.

h. *Extrahigh-Voltage Series Capacitors*

Extrahigh-voltage series capacitors are constructed to withstand the high peak currents and severe impulsive forces produced during rapid discharge. Alumina-filled epoxy terminal bushings are used exclusively to provide mechanical properties superior to conventional wet process porcelain. With a coaxial return sheath around the terminal insulator, there is adequate creepage protection for operation under oil or pressurized gas.

Extrahigh-voltage capacitors have a realistic and usable self-inductance of 30 nH. The specified inductance includes the inductance of a copper sheath in close proximity to the terminal insulator which shorts the capacitor for the purpose of measurement. The high-voltage terminal plate and the mounting hardware can be modified to facilitate interconnections to allied switches and transmission lines. As many as 20 series capacitor sections are used in construction to assure corona-free operation.

i. *Coaxial Tank Series Capacitors*

Coaxial tank series capacitors are constructed to withstand the high peak currents and severe impulsive forces produced during rapid discharge. Extended foil construction and a recently developed highly conductive solder assure mechanically sound low-resistance connections capable of conducting high peak currents. Series connections of internal capacitor sections provide uniform stress distribution and corona-free operation. Electrical connections to the case side of the capacitors are normally made in a ring of studs arranged around the outer edge of the capacitor at 18 . Threaded holes can be supplied in place of studs if flush mounting is required. The number of studs connected to the high side of the capacitor varies with electrical requirements. Normally, four studs are used on large diameter capacitors and single studs on smaller units. Recessed wells in the top insulator can be provided for insertion of switches on special order.

Table A8g-1 shows examples of each of the mentioned types, with their characteristics.

j. *EMP (Electromagnetic Pulse) Devices*

Ultrafast capacitor discharge circuits are needed also for the generation of the so-called EMP (see this volume, Chapter E 15, 16). In 1970, the Ion Physics Corp. [1701] finished several large EMP simulator systems. The general design considerations for ultrarapid discharge capacitors are:

1. The geometry of the store should be compatible with the output pulse requirement.
2. It should occupy the smallest volume, i.e., have the highest energy density, consistent with the required reliability. This not only makes general economic sense, but also ensures that its energy discharge characteristics are most acceptable to modern specifications.
3. The energy charging efficiency should be as high as possible, unless this implies unacceptable complexity.

The energy stored per unit volume in the electric field E between two electrode structures in a dielectric medium of permittivity ε is given in MKS units by

$$U_e = (\varepsilon E^2 / 2)$$

For parallel electrodes having an effective area A, and spacing d, integration of the energy distribution over the volume between the electrode gives the total stored energy.

$$W_e = (1/2)(\varepsilon A/d) V^2$$

where V is the total voltage between the electrodes. The total stored energy can then be expressed as

$$W = (1/2)CV^2$$

where C is the capacitance between the electrodes.

Energy is therefore most commonly stored as electric charge in a capacitor, which may be "lumped," in the form of impregnated assemblies of dielectric/metal foil windings, or "distributed" in the transmission line form. These capacitors may be obtained in conveniently sized units, and an energy storage bank is most often formed by a series/parallel arrangement of such units. The energy densities which have been achieved with this method are typically 2–3 J/in.^{3}, i.e., ~ 150 kJ/m^3.

9. High-Voltage, Low-Inductance Capacitors for Marx Circuits

Special ultralow-inductance capacitors have been designed primarily for Marx generator circuits in which extremely low-inductance, fast discharge, nonreversing capacitors are required. (High Energy, Inc. [1904].)

All ultralow-inductance capacitors fall into two general categories: a barrier version and a slotted version. Although the basic dielectric configuration is the same in both design, the barrier version, which is less expensive, is characterized by an inductance of less than 20 nH. On the other hand, in the slotted design, which is more difficult to build and therefore somewhat more expensive, a capacitance of less than 9 nH can be achieved up to 100 kV.

The barrier version is designed to withstand its rated voltage without additional insulation. This unit is intended for Marx circuits where a single capacitor is required for each stage of the Marx. The slotted version requires the insertion of Mylar or other high-voltage insulation into the slot between the high-voltage rail outputs. This allows any number of capacitor units to be parallel to a common rail switch.

The capacitor sections are constructed of a unique plastic film, paper and impregnating oil combination together with a new winding technique developed by High Energy, Inc. to provide an extremely low-inductance design. Ultralow-inductance capacitors with an energy density up to 200 J/kg, are manufactured by High Energy, Inc. [1904] (see Table A9-1).

The ultralow inductance makes these capacitors ideal for service in Marx generators, line modulators, high-voltage plasma research, and particle accelerators. The Magnum capacitor is especially designed for ultralow inductance at ultrahigh voltages. Special proprietary techniques are used in the internal bus–foil connections, and section wrapping to ensure field cancellation (Fig. A9-1) (Table A9-2).

TABLE A9-1

TYPICAL SPECIFICATIONS

Part No.	Voltage (kV)	Capacitance (μF)	Energy (J)
(a) Type UL parallel			
UL101	75	0.1	280
UL103	75	0.5	1400
UL105	100	0.1	500
UL109	125	0.1	780
(b) Type UL slotted			
UL201	75	0.1	280
UL204	100	0.1	500
UL206	125	0.1	780

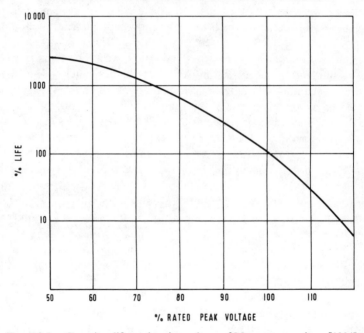

FIG. A9-1. Capacitor life vs charging voltage of Magnum capacitors [1904].

The capacitor sections, which are either paper or paper and plastic depending on application, are housed in a rigid plastic case. These capacitors use a separate rail assembly with an outer and inner bus, or collector, plate. This unique rail assembly, together with an ingenious "tabbing" arrangement, is

10. PARALLEL PLATE CAPACITORS FOR SUPERPOSED FREQUENCIES

TABLE A9-2

SOME TYPICAL MAGNUM CAPACITORS

Part No.	Capacitance (μF)	Voltage (kV)	Energy (J)
ML102	0.10	50	125
ML104	0.10	75	280
ML105	0.01	100	50
ML106	0.05	100	250
ML107	0.01	125	78

parallel to each section, and thus cuts the inductance of the capacitor in half in comparison with conventional parallel plate transmission methods, and the external bus portions are O-ring sealed to ensure leak-proof performance.

The capacitors are claimed to receive a considerably longer high-vacuum cycle than do conventional capacitors. Specially processed oil which has been filtered for particulate matter, ionically filtered and hard-vacuum degassed to remove any dissolved gases, is introduced into the capacitor after a 24-h high-vacuum drying period. In this way, the possibility of corona due to air bubbles or entrapped particles is eliminated (Table A9-3).

TABLE A9-3

HIGH-VOLTAGE CAPACITORS[a]

Model	μF $\pm$ 10%	Voltage (kV)	Av. Life	Energy (J)
100W160	0.15	100	10^5	750
100W330	0.015	100	10^6	90
70W331	0.2	70	10^6	500
60W257	0.2	60	10^6	360
50W334	0.3	50	10^6	375
35W337	0.6	35	10^6	360
35W328	0.025	35	10^8	15

[a] Capacitor Specialists, Inc. [1903]. All have less than 15 nH inductance and low-inductance output.

10. THE PARALLEL PLATE CAPACITORS FOR SUPERPOSED FREQUENCIES

Capacitor discharges are often oscillatory. The heat generated as a result of dielectric loss necessitates very effective water cooling. Like the first capacitors of 100 yr ago, which replaced the Leyden jar by glass plate backs, the parallel plate capacitor has now been specifically designed to provide long, trouble-free service under high frequency, high KVAR loading. This is accomplished through a unique design innovation which guarantees better cooling, and thus, lowers operating temperature. It ensures an absolutely rigid construction, thus eliminating cantilever problems which lead to vibration and high-frequency parasthetics. In addition, this rigid construction ensures a uniform voltage gradient throughout the capacitor, and eliminates high-stress areas encountered in conventional designs. The capacitor uses a high-frequency water-cooled terminal which is internally connected to a massive heat exchanger which cools the entire capacitor by both conduction and convection. (Table A10-1.)

Standard water-cooled terminals are constructed with a special ceramic affording a low loss at high frequencies. The internal assembly consists of two sets of plates, a hot plate group which is assembled to the water-cooled terminal, and the cold plate group which is assembled to the case. The unique structure orientation affords superior cooling and a uniform voltage gradient not found in other parallel plate capacitors.

The capacitor is vacuum-impregnated with an ionically purified dielectric fluid insuring a high-corona starting voltage and a long trouble-free life. It has a $Q > 3000$ in the frequency range of 50 kc to 2 Mc. The parallel plate capacitor can be operated without water so long as the case temperature does not exceed 65°C. It is recommended that at least $\frac{1}{2}$ gal/min be provided when operating the capacitor at full rating. Higher cooling rate needs water pressure as follows:

Gal/min	psi
0.5	0.4
1.0	1.2
1.5	2.4
2.0	3.8

Another low-inductance capacitor is the so-called type FC (flat capacitor) (High Energy, Inc. [1904]). This is highly specialized for use in Marx

TABLE A10-1

SOME PARALLEL PLATE CAPACITORS (RATINGS AT 540 kHz SINUSOIDAL)

Part No.	Maximum (kV rms)	Capacitance (μF)	Maximum current	KVAR
PP101	5.4	0.014	256	1382
PP103	8.5	0.010	289	2456
PP106	10.3	0.0070	245	2523
PP112	15.0	0.0023	117	1755

11. BIG ELECTROLYTIC CAPACITOR BANKS

generators, line modulators, high-voltage plasma research, particle accelerators, and impulse generators where versatility in design is of critical importance. It might be referred to as a "building block" capacitor since two or more units can be combined to perform the function of a single unit. It is an extremely rugged capacitor and can withstand severe mechanical shock and vibration environments. It has abnormally low moisture absorption characteristics and is compatible with a sulfur–hexafluoride atmosphere. The flat capacitor is housed in an extruded thermoplastic case with welded thermoplastic ends. This welded construction, together with O-ring sealed busbars, ensures a virtually leak-proof system. External connections are by rail-type busbars for high-current, low-inductance applications and are located at both ends. (See Table A10-2.)

TABLE A10-2

SOME TYPICAL FLAT CAPACITORS[a]

Part No.	Voltage (kV)	Capacitance (μF)	Energy (J)	Length (L)
FC101	37.5	0.1	70	$4\frac{1}{4}$
PF102	37.5	0.2	140	$8\frac{1}{2}$
FC103	37.5	0.5	350	22
FC201	50	0.05	62.5	4
FC202	50	0.1	125	$7\frac{1}{2}$
FC203	50	0.2	250	15
FC301	62.5	0.05	97.5	6
FC302	62.5	0.1	195	12
FC303	62.5	0.2	390	24

[a] 100% of the rated voltage applies to an entire life of 100 shots, 120% to 10 shots, 70% to 1000, 50% to about 3000 shots. High Energy, Inc. [1964].

11. BIG ELECTROLYTIC CAPACITOR BANKS

Low-Loss Electrolytic Capacitor Banks

The electrolytic capacitor is the most inexpensive type for very low repetition rate, high energy, low voltage capacitor banks. A special electrolytic capacitor storage bank (type HPG-E) is available from Adkin [1705]. Its applications are energizing high field magnet coils, transient discharging into resistive loads, generating underwater sparks, powering flash lamps, and storing energy from 5000 to 50,000 J at 450 and 1350 V. (Larger systems are available in multiple modules of 25,000 J.) The repetition rate for the following models:

Model A 1 pulse/6 min max
Model B 1 pulse/3 min max
Model C 1 pulse/min max.

The storage system HPG-E provides a low-voltage capacitor bank suitable for application where a pulse of high current is desired for times in the order of 100 μsec–10 msec. The output currents and discharge times obtainable depend on load inductance and resistance. Electrolytic banks must be operated in the critically damped or overdamped mode to avoid deleterious effects on the dielectric due to voltage reversal. Therefore, the condition $R^2 > 4l/C$ must be observed for this purpose, and an adjustable damping element can be introduced into the discharge circuit to achieve the desired mode if the load parameters are not appropriate. Basic tables are provided with each bank to allow the operator to determine in a simple manner the parameter for the critically damped configuration. Protecting high-current diodes are provided to bypass any backswing that may occur. These devices are necessary if capacitor lifetime is to be maintained.

The storage capacitors are rigidly mounted on low-inductance rails which are removable for inspection. Conversion from the parallel mode (450 V) to the series mode (1.350 V) is accomplished by a simple change in the transmission line system. Energy is switched from the bank by means of water-cooled (model C), high-current ignitrons; 25,000 J banks have four ignitrons; 15,000 J banks have two; and 5000 J banks have one. All ignitrons are protected by temperature-actuated interlocks.

The storage bank is charged and triggered by solid-state circuit. All systems are interlocked to insure personnel safety and reliable operation. Output current is available via 4, 8, or 12 (depending on storage capacity) low-inductance coaxial cables which extend 5 ft from the rear of the cabinet. Magnetically swaged cable connectors are fitted on the output cables to accommodate coupling to any Adkin coil or energy header. Constant load resistance can be assured by the addition of a load, cooling system and a load temperature monitoring system. Both of these systems are installed inside the bank cabinet.

12. Tantalum Capacitors

For extremely hard environmental conditions (military or space), tantulum capacitors are used. Their remarkable internal resistance makes them unsuitable for very high peak power, but for generating microsecond pulses at rather low voltage, especially in space applications, they should be considered as useful components.

Fansteel "VP" tantalum capacitors [1008] are noted for their electrical stability over a wide range of operating temperatures, indefinitely long shelf life and high-dielectric resistance. The porous tantalum anode is permanently sealed in an uninsulated fine silver case which serves both as cathode and container for the electrolyte. The tantalum oxide film is the most stable dielectric (chemically and dielectrically) known. This property combined with the large surface area of the porous tantalum anode results in extremely high capacity. Electrical leakage is so low that it can be considered negligible for most applications.

In addition to high performance and reliability, the "VP"-type also features exceptional resistance to the effects of vibration and physical shock. This characteristic—along with high-capacity ratings in the smallest physical sizes obtainable—makes the "VP" capacitor ideal for use in mobile and airborne electronic equipment. Its ability to withstand the effects of acceleration and impact shock particularly qualify it for use in missile and rocket components, even under reduced pressure at high altitudes.

B. Switching means

1. Controlled Gas Discharge Tubes

a. *Thyratrons*

To update the survey of available thyratrons suitable for heavier duty cycles, it should be mentioned that the English Electric Valve Co. Ltd. [1010] manufactures hydrogen thyratrons, type CX 1154, having 40 MW peak power output at 2500 A and 40 kV peak forward voltage (see data in Table B1a-1). They are intended mainly to feed radar pulse modulators.

Table B1a-1

Data for Hydrogen Thyratron CX1154[a]

Pulse repetition rate	500 pps	50,000 pps
Peak network voltage	30,000 V	10,000 V
Pulse width	2.0 μsec	0.4 μsec
Network impedance	7.5 Ω	50 Ω
Peak anode current	2000 A	100 A
Average current	2.0 A	2.0 A
Peak output power	30.0 MW	0.5 MW
Average power output	30.0 kW	10.0 kW
Trigger voltage	275 V	275 V
Trigger impedance	300 Ω	300 Ω
Negative bias	-25 V	-50 V
Jitter	0.005 μsec	0.005 μsec
Reservoir range	$\pm 10\%$	$\pm 10\%$
Heater voltage	3.0 V	6.3 V

[a] English Electric Valve Co. Ltd.

For extreme environmental conditions, EG&G produces ceramic thyratrons [1009, 1010]. Typical ratings are shown in Table B1a-2. For heavy-duty cyclic operations, the circuit of Fig. B1a-1 is normally used. Radar pulse drivers generally use a network consisting of inductors and capacitors in place of the simple capacitor used in most strobe circuits since a square waveform output is desired for the radar pulse oscillator. Numerous experimenters have attempted to use similar pulse networks to obtain a square pulse of light from an electronic flash lamp.

1. CONTROLLED GAS DISCHARGE TUBES

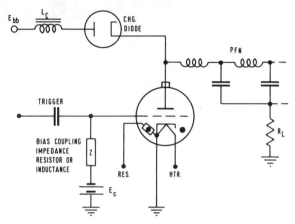

FIG. B1a-1. Modulator schematic.

Edgerton [1529 o] recommended a circuit to feed high speed photographic flash tubes from a hydrogen thyratron modulator. At the moment, the highest anode peak forward voltages (of up to 70 kV) are handled by the deuterium-filled thyratrons made by General Electric Co. Ltd. [1633]. This range of metal-bodied deuterium-filled thyratrons is designed for use in linear accelerator and high power radar applications.

TABLE B1a-2

DATA FOR SOME EG & G CERAMIC HYDROGEN THYRATRONS FOR EXTREME ENVIRONMENTAL CONDITIONS

	E2830	E3193	E3194	E3213
Peak anode voltage (kV)	20	35	35	70
Peak anode current (A)	500	2000	5000	2500
Average current (A)	0.6	5.0	15.0	1.5
Rate of rise of current (A/μsec)	2.000	7.500	10.000	7.500
Jitter (nsec)	2	5	5	5

All four tubes have the following important characteristics:

(1) The reservoir system is controlled by a barretter and thermistor (incorporated within the tube cap or envelope) to give constant gas pressure over a wide range of ambient temperature and supply voltage. Consequently, the reservoir voltage may be varied $\pm 7\frac{1}{2}\%$, or the ambient temperature from $\sim -40°$ to $+100°C$ without a significant change in the tube pressure. The characteristics, therefore, do not alter.

(2) All tubes are filled with deuterium which has a higher Paschen sparking potential than hydrogen. The E3193 and E3194 are single gap tubes which operate at 35 kV, compared with similar hydrogen-filled tubes which cannot exceed 25–30 kV. The E3213 is a double gap tube which will operate at 70 kV compared with the 50 kV of its hydrogen counterpart.

(3) The cooled grid gives good delay time drift characteristics. The amount of drift will vary depending upon the rating and upon the method of measurement. Typically, the delay time is 200 nsec and the drift is ± 20 nsec over a wide range of operating conditions.

(4) The cooled grid permits a high Pb factor. The Pb factor permissible will vary widely depending upon the other operating conditions but the E3213, for example, will operate at $Pb = 160 \times 10^9$.

The demand for higher power thyratrons for linear accelerator and high power radar applications has led to a completely new approach to design techniques. The M–O Valve Co. Ltd., an established leader in this field of technology, produced a 40-kV 10,000-A tube in 1961, and is now introducing a range of compact single and double gap tubes with operating voltages of 20, 35, and 70 kV.

The M–O V design is centered around the metal body philosophy. This facilitates the cooling of the grid and enables the gas density of the tube to be kept constant thus contributing to compactness and well-controlled triggering characteristics at high powers. The delay time drift, for instance, is about one-tenth of that in a glass tube and about one-third of that in a conventional ceramic tube. The ability to cool the grid effectively eliminates grid emission,

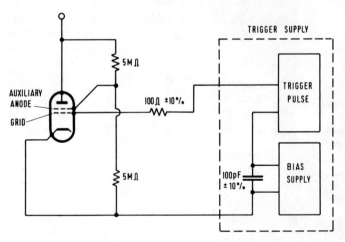

FIG. B1a-2. Grid and auxiliary anode circuits of high-voltage, high-current M1O Valve Co. Ltd. thyratrons.

and enables the tube to be run at high power density which in turn results in very high power tubes of extremely compact construction. The fact that the anode and the cathode support are constructed as part of the envelope also contributes to the high power characteristics of the tube. The cathode is a specially designed type which gives a life in the order of 5000–20,000 h.

Notes

(1) It is recommended that the reservoir be supplied by a separate transformer from that supplying the cathode heater.

(2) The auxiliary anode shall be maintained at half the anode voltage ($\pm 10\%$) by means of a resistance potentiometer with a total resistance of 10 MΩ.

(3) For direct switching the forward anode voltage must not exceed 40 kV; it should be ascertained that this voltage is not exceeded due to transients.

(4) The peak inverse voltage shall not exceed 20 kV for the first 25 μsec following the cessation of anode current.

Fig. B1a-3. Some multistage high-voltage hydrogen and deuterium ceramic thyratrons.

(5) This applies for pulse lengths (t_p) not exceeding 5 μsec. For greater pulse lengths peak current must not exceed $2500(5/t_p)^{1/2}$ (t_p in μsec).

(6) For operation the valve shall be immersed in hydrocarbon transformer (HT) oil. The container shall be sufficiently large to permit free convection.

(7) The temperature specified is the temperature of the oil measured 2 in. away from the valve level with the grid/auxiliary-anode insulator, measured at any fixed rating between the fifth minute after application of HT and any subsequent time.

(8) Details of the circuit to be used for connecting the trigger supply to the grid are given in Fig. B1a-2. The trigger pulse voltage specified is to be measured with the grid disconnected.

(9) Transients up to 25% of the anode voltage may appear at the grid during the rise time of the anode current pulse. The grid circuit insulation should be designed to withstand this transient voltage, which is normally less than 100 nsec in duration.

(10) The effective impedance of the required trigger supply is indicated for each tube, both during the triggering period, when current is flowing from the trigger supply to the valve, and during the recovery period when current is flowing in the reverse direction.

b. *Krytrons*

This is a cold cathode switch tube of extremely small dimensions. The main features of Krytrons are capability in radiation environments, reliable firing with no warm-up, high peak current, high-voltage hold-off, short delay time and low jitter, operation over wide temperature range, compactness and light weight. Its applications are: exploding bridge wire systems for missile stage separation, motor ignition, arming and fusing nanosecond pulse generators, radar beacon modulators, trigger transformer primary switch for triggering xenon flashtubes, triggered spark gaps, ignitrons and spark chambers, and gallium arsenide cell switchs. (See Fib. 1b-1.)

The Krytron system has been developed by EG&G. Another typical circuit for nanosecond pulse generation by two Krytrons is given in Fig. B1b-2. V_1 and V_2 must be spaced at least 1.5 in. apart in order to prevent the firing of V_1 from bypassing the delay network and firing V_2 at time zero. In applications where space is a problem, the tubes may be spaced closer together by placing a metal shield between the two tubes.

The output pulse width is determined by the length of time it requires to build sufficient voltage on the grid of V_2. The pulse width, therefore, is controlled by both the setting of R_6 and amplitude of the output pulse. In the event supply voltage is lower than 5000 V, it will be necessary to modify the RC time constant of the delay network in order to maintain the specified output pulse width range. For more details see [1009] and Table B1b-1.

1. CONTROLLED GAS DISCHARGE TUBES

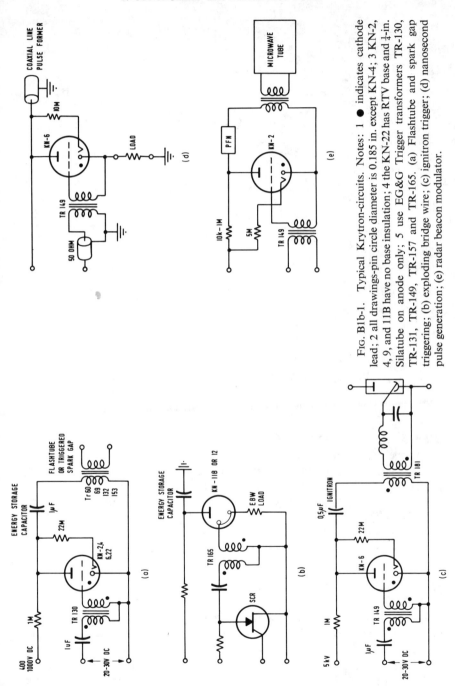

FIG. B1b-1. Typical Krytron-circuits. Notes: 1 ● indicates cathode lead; 2 all drawings-pin circle diameter is 0.185 in. except KN-4; 3 KN-2, 4, 9, and 11B have no base insulation; 4 the KN-22 has RTV base and ¼-in. Silatube on anode only; 5 use EG&G Trigger transformers TR-130, TR-131, TR-149, TR-157 and TR-165. (a) Flashtube and spark gap triggering; (b) exploding bridge wire; (c) ignitron trigger; (d) nanosecond pulse generation; (e) radar beacon modulator.

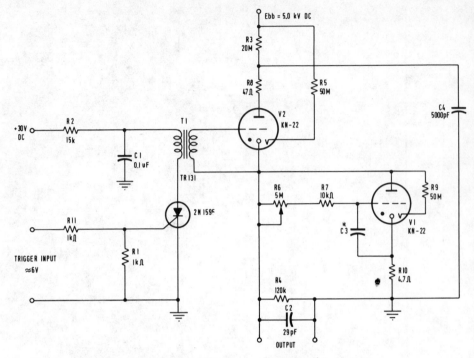

Fig. B1b-2. Schematic diagram, Krytron pulse generator.

Often in such tubes a low-intensity beta-radiator like ^{63}Ni is provided to have sufficient electrons continuously available. The quantity of this isotope (electron energy ~ 60 kV) and its susceptibility damage make the lifetime of these tubes uncertain. They should be used mainly where other devices cannot be used because of a lack of heating power or because the environment is unfavorable, e.g., inside projectiles or in explosive gases. The nickel isotope is available in the form of thin foil. Normally, an amount of 10 μC is sufficient, and in most countries this is available without the otherwise needed formalities for acquiring radioactive isotopes. The nickel foil is pulse-welded to a piece of nickel wire which is in turn welded to one of the electrode supports of the gap or tube.

c. *Ignitrons*

The development of ignitrons from 1965 (since the publication of "High Speed Pulse Technology," Volume I) up to the present has been remarkably slow. Peak current is limited to $\sim 10^5$ A, since the mercury vapor ions at higher specific current densities tend to damage the anode by pinching or point burning. Table B1c-1 shows some types produced by the English Electric

TABLE B1b-1

TECHNICAL DATA FOR EG&G KRYTRON TUBES

Tube type	Anode operating range		Peak current Max[b] (A)	Pulse duration Typical[c] (μsec)	Grid trigger		Firing characteristics		Keep-alive current Typical (μA)	Life data			
					Trigger amplitude Min[d] (V)	No-fire test Min[e] (V)	Delay Max[f] (μsec)	Jitter Typical (μsec)		Operations (typical number)	Anode voltage (V)	Peak current (A)	Duration (μsec)
	Min[a] (V)	Max (V)											
KN 1	500	3000	300	5	300	60	0.4	0.04	50	2,000,000	2000	40	1
KN 2	300	4000	500	5	200	75	0.2	0.02	50	7,000,000	2000	40	1
KN 4	400	5000	2500	10	250	90	0.3	0.03	50	25,000	1200	270	6
KN 5	700	3000	2000	10	300	75	0.4	0.04	50	50,000	3000	1000	1
KN 6	700	5000	3000	10	250	90	0.25	0.03	50	100,000	5000	1000	1
KN 6B	700	10,000	3000	10	250	90	0.30	0.03	50	150,000	5000	1000	1
KN 10[g]	800	3500	1000	2	500	90	0.6	—	0	500	1500	1000	1
KN 12[h]	1000	5000	1000	2	1000	90	0.4	—	0	500	1500	1000	1

[a] Minimum operating voltage can be extended downward to a potential approaching zero by increasing the trigger voltage (see operating notes).
[b] Peak current ratings are tested values. Operation at higher values must be accompanied by reduction in pulse width (see operating notes).
[c] Pulse duration listed is typical. Operation at longer pulse widths must be accompanied by reduction in peak amplitude.
[d] Minimum trigger is measured for an anode operating voltage of 1 kV.
[e] The no-fire test indicates the trigger noise rejection capability.
[f] Delay time is measured at an anode potential of 3000 V with a 500 V peak trigger and 50 μA keep-alive current; excepting KN 10 and KN 12 that are three-element tubes. The KN 12 delay time is measured with a 1000-V trigger potential.
[g] Low pressure krytron.
[h] Vacuum krytron.

TABLE B1c-1

IGNITRONS—CAPACITOR DISCHARGE, PULSE DUTY[a]

Type	American equivalent	International letter size	Maximum ratings					
			Peak forward anode voltage (kV)	Peak reverse anode voltage (kV)	Peak anode current (A)	Mean anode current (A)	Ampere-sec/pulse (A·sec)	Duration of current (msec)	Pulse frequency (pps)
BK178	—	D	25	25	100,000	40	200	150	0.2
BK194	—	E	25	25	100,000	80	400	150	0.2
BK416/7703	7703	A	20	20	100,000	[b]	[b]	[b]	0.03
BK428	—	A	20	20	100,000	0.75	20	100	0.03

[a] English Electric Valve Co. Ltd.
[b] Rated for 20 μsec oscillatory current discharges.

2. CONTROLLED SEMICONDUCTOR SWITCHES

Valve Co. Ltd. [1010] which are specially developed for high current capacitor discharge pulse duty.

For most applications at low repetition rates, water cooling of ignitrons can be done with oil instead of the conductive water. Normal transformer oil is suitable, if a small oil-resistive pump provides the circulation. The recooling of the oil, if necessary, can be done by a heat exchanger and an air fan or for heavy-duty operation, by a flow of cooling water.

2. Controlled Semiconductor Switches

a. *Thyristors*

A main competitor of the thyratron is the thyristor, a semiconductor device. It needs a small starting energy, like the thyratron, and in its circuitry a quenching mode must be provided to end the conductive state after decay of the discharge current. A thyristor might be expected to have an unlimited life since there is no ion bombardment to limit its normal functioning. On the other hand, other factors in practice combine to set a definite limit to its lifetime. During 1971 Schwickardi [1156] made an examination of 72 power line thyristors, and his report has the following summary: A number of 72 power line thyristors of nine well-known producers in Europe, the United States, and Japan were tested within five years with special test equipments under practical circuit conditions.

After the initial mechanical, thermal, and electrical measurements, periodic tests were made to find out the degradation and the catastrophic failures. In order to compare test results with a standard reference, it was necessary to develop technological and physical analysis methods on a reference thyristor unit of energy brand. Therefore, the defective thyristors were opened and the alloy-diffused SCR pellets were examined technologically and microscopically after grinding and etching them.

Most of the reverse or blocking voltage was caused by faults on the surface as well as in the bulk of the silicon semiconductor. The correlation between the electrical phenomena and the technological faults are described. Other faults were noticed in the alloying and solder contacts. The third type of important failure showed itself in the triggering behavior of the units. There are a number of basic and complicated mechanisms leading to semiconductor device failure. The design of the devices, the care with which they are manufactured and the operating conditions have a great deal of influence on the characteristics and the failure behavior in long-term life tests. These final results are discussed and the phenomena are compiled in a reliability hypothesis on thyristors.

The peak inverse voltage of normal thyristors is something like 300 V, and often 1500 V can be obtained in selected devices. The 2500-A peak current may be at the moment a definite barrier. More important than the peak are the permissible rate of rise of the current and the time needed after quenching to make the thyristor ready for the next operation. This time is shorter for low

TABLE B2a-1

PROPERTIES OF SOME TYPICAL SILICON THYRISTORS[a]

	Types		
	BRX 49[b]	T3 N5 C00[c]	BT 119[d]
Periodic peak blocking voltage ($\pm$), V	400 (V_{DRM}, V_{RRM})	500 (V_{DRM}, V_{RRM})	750 (V_{DRM})
Mean forward current, A	3.6 I_{TRM} @ $T_i = 50°C$	5 I_{TAV} @ $T_C = 85°C$	—
Repetitive peak current, A	—	—	12 I_{TRM}
Surge forward current half-wave, A, @ 50 Hz	6 I_{TSM} @ $T_i = 125°C$	55 I_{TSM} @ $T_i = 125°C$	85 I_{TSM} @ $T_i = 25°C$
Mean forward current half-wave, res. load, mA, @ $T_{amb} = 25°C$	400 I_{TAV}	—	—
Crit. rate of voltage rise, V/μsec, @ $T_i \parallel 125°C$[e]	—	200 $S_{V_{crit}}$	—
Breakdown voltage at gate current zero	—	—	$V_{(BO)_0} > 800$
Forward voltage, V	$V_T < 1.7$ @ $I_T = 1$ A (pulsed)	—	$V_T = 2.3\ (<3)$, @ $I_T = 30$ A
Leakage current ($\pm$) @ V_{DRM} resp. V_{RRM}, nA	$I_D, I_R < 100$	—	—
Nom. load current half-wave res. load @ $T_{amb} = 45°C$	2.5 I_{TAV}	—	—
Off-state forward current @ V_{DRM}, μA	—	—	15 I_D
Holding current, mA	$I_H < 5$ @ $R_{GC} = 1$ kΩ	$I_H < 25$	$I_H = 30\ (<100)$
Turn-off time, μsec	—	$t_p = 17$	$t_p < 2.4$ @ $T_C = 70°C$
Trigger current	$I_{GT} < 200\ \mu$A @ $V_D = 7$ V	$I_{GT} < 20$ mA @ $V_D > 6$ V	$I_{GT} = 15\ (<40)$ mA @ $V_D = 6$ V

[a] See Ref. [1704].
[b] Small silicon planar thyristor.
[c] Cathode-triggered silicon thyristor in metal case.
[d] Fast silicon thyristor (for horizontal deflection circuits in TV receivers).
[e] When voltage is rising from 0 to 67% of V_{DRM} resp. V_{RRM}.

power types and can be as short as 5 μsec. The properties of some typical thyristors are given in Table B2a-1.

The limited current rise and the unavoidable jitter time of thyristors make them mainly suitable for capacitor discharge applications of "softer" kinds, e.g., feeding pulse transformers used for pulse welding (see this volume, Chapter G2); generation of other pulses to control spark gaps; or generation of strong magnetic fields using coils which limit by their inductance the peak current and rate of rise.

b. *Low-Voltage, High-Current Power Transistors*

The single chip lanar construction of large transistors of the NPN type enables the silicon transistor to operate as a switch or modulator of extremely fast response, for voltages of ~ 100 V. Because of its critical characteristics, only circuits recommended by the manufacturer should be used.

TABLE B2b-1

MAXIMUM RATINGS OF 250 A POWER TRANSISTORS[a]

	SDT 5810	SDT 5811	SDT 5812	SDT 5813
Collector-base voltage (V)	60	80	100	120
Collector-emitter voltage (V)	40	60	80	100
Emitter-base voltage (V)	10	10	10	10
Continuous collector current (A)	200	200	200	200
Peak collector current (A)	250	250	250	250
Continuous base current (A)	20	20	20	20
Peak base current (A)	25	25	25	25
Storage temperature (°C)		$-65-+200$		
Operating junction temperature (°C)		$-65-+200$		
Thermal resistance (Θ_{jc})		0.5 °C/W		
Power dissipation (100°C Case), (W)	200	200	200	200

[a] Solitron Dev., Inc. [1905].

Table B2b-1 shows four examples of 250 A peak current power transistors.

3. Push–Pull Switching Circuits for Rectangular Pulses

In the field of telecommunication, television, and laser technique, solid-state driving of electrooptic light deflectors at high frequencies is of increasing importance [1174].

The high voltages required for electrooptic crystals tended to impose severe limitations on the operating speed of digital light beam deflectors. The use of well known and other published techniques for switching these high voltages made waveforms far from ideal. This problem would have been alleviated if suitable new materials had been developed with low half-wave voltages, and if these had become available for use in deflectors. As no suitable new materials have appeared, acoustooptic deflectors have, on the whole, been used in electrooptic systems applications requiring high speed light deflectors.

Of the available electrooptic materials, KD^xP, which can be obtained in quantity with sufficient optical quality, remains the most practical for digital light deflectors. In this application, KD^xP crystals are used most often in the z-cut longitudinal mode as polarization switches. These alternate with suitable birefringent elements that produce the digital deflection. Although several laboratories have developed techniques for switching the necessary voltages, few reports of these techniques have been published.

The voltage waveforms that are required across the electrooptic crystals are ideally rectangular; approximations to this are of course acceptable. The edges must be fast; switching times compatible with those of integrated logic circuitry would be desirable. Similarly, switching rates in the megahertz region would be useful. Peak-to-peak voltages must be equal to the half-wave voltages $V(\lambda/2)$ of the electrooptic crystals, and voltage definition of better than 5% is necessary for satisfactory extinction in deflector stages. In fact, a number of switching systems can be devised, including some requiring more complex waveforms. These and the method of generating them are described in this paper.

The effective half-wave voltage of KD^xP in a digital electrooptic deflector depends on the wavelength of the light used, the crystal alignment, the temperature, the degree of deuteration of the crystals, and the type of electrodes used. In the region of 20°C, with light at 632.8 nm, using conducting stannic oxide electrodes on transparent substrates glued to the crystal, the half-wave voltage of KD^xP is between 3.8 and 4.2 kV. The circuits described in this paper will effectively switch these voltages with transition times of 200 nsec at rates up to 1.2 million times per second. They are all solid state, using transistors as the active devices. They are powered by a supply that uses solid-state voltage stabilization and allows individual adjustments to be made to the voltages switched across each stage in a multistage digital light deflector.

The high-voltage switching circuits use 450-V transistors. Although transistors are available with breakdown voltages as high as 1.5 kV, these are slower and not necessarily suitable in this application. The transistors used are stacked to multiply their breakdown voltage. Circuits have been developed in the past to do this, but these have required slowing down circuits to ensure simultaneous switching of all the transistors. The circuits described in this paper exploit several useful transistor characteristics, some of which are normally a drawback, and do not require slowing down circuits. These circuits also switch in a time that is a function of the transistors, rise time, not fall time, which is usually slower. The performance of these circuits makes them ideal for driving high speed light deflectors.

Twelve of these circuits have been used to drive ten stages of a digital light deflector and two modulators in a development model of a random access holographic computer memory. In this system, the deflector will change state in periods as short as 3 μsec. One of the modulators, which also uses z-cut KD^xP, operates continuously at 312.5 kHz, equivalent to a deflector stage operating at 625 kHz. Reliable operation of this system has now been obtained for over 500 h. The switch lifetime is expected to be similar to that of all normal low-voltage solid-state systems.

Rectangular pulses are often needed to operate Kerr or Pockels cells as optical shutters (see Volume I, Chapter E13).

a. *High-Voltage Switching Circuits*

The two features that make driving electrooptic crystals difficult are the high voltages necessary and the power dissipation implied by having to charge a capacitor to a high voltage and discharge it, at high frequency. Pulse transformer circuits minimize the high-voltage circuitry necessary, in principle using only low-voltage switching devices. But pulse transformer circuits suffer from ringing between the inductance of the transformer and the capacitance of the load. This phenomenon can be reduced in severity by the introduction of suitably timed high-voltage switches (see Fig. B3-1a). These, however, defeat the object of the exercise to avoid high-voltage switching.

It has been suggested that the second feature about the driving of electrooptic crystals, that of high power dissipation, can be overcome by the use of biased

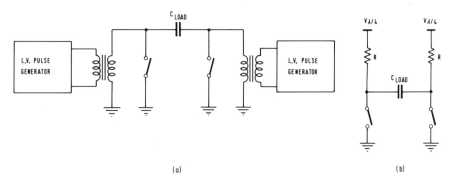

FIG. B3-1. (a) Pulse transformer drive circuit. (b) Resistor/switch bridge configuration.

B. SWITCHING MEANS

resonant circuits. These circuits use the current, circulating in a high Q resonant circuit, to transfer charge between the crystal and a tank capacitor. Component losses make these circuits impractical, and no success has been reported with them.

Other circuits have to switch effectively the whole half-wave voltage of the electrooptic crystals, and stored charge is lost, resulting in high power

TYPE	CIRCUIT CONFIGURATION	SWITCHING MODE AND PERIOD DEFINITION	BREAK-DOWN VOLTAGE OF SWITCHES	QUARTER WAVE OPTICAL BIAS?	POWER DRAWN FROM SUPPLY DUE TO SWITCHING OF LOAD CAPACITANCE (WATTS PER CYCLE)	POWER DRAWN FROM SUPPLY DUE TO SWITCHING OF STRAY CAPACITANCE (WATTS PER CYCLE)
1	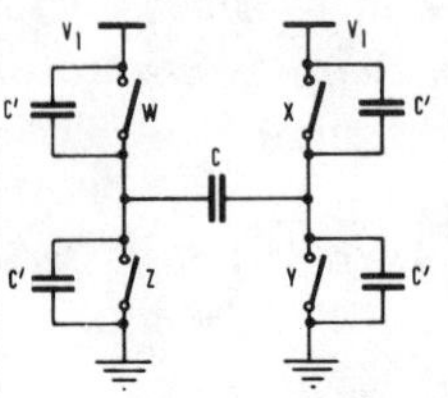	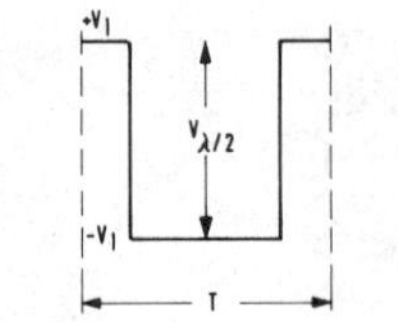	$V_1 = V_{\lambda/4}$	YES	$C(V_{\lambda/2})^2$	$C'(V_{\lambda/2})^2$
2	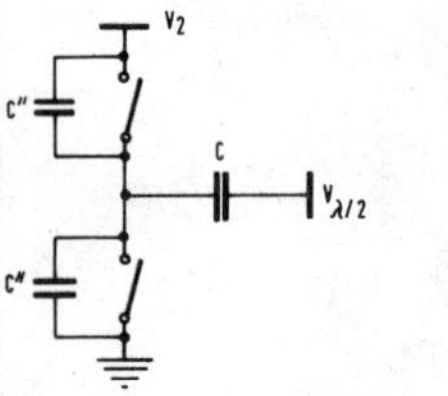	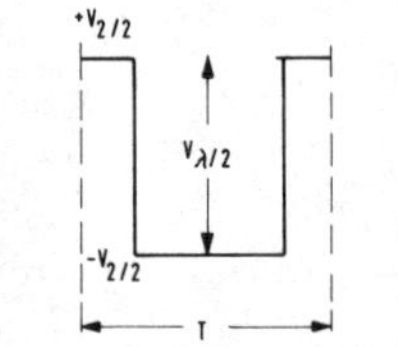	$V_2 = V_{\lambda/2}$	YES	$C(V_{\lambda/2})^2$	$2C''(V_{\lambda/2})^2$
3	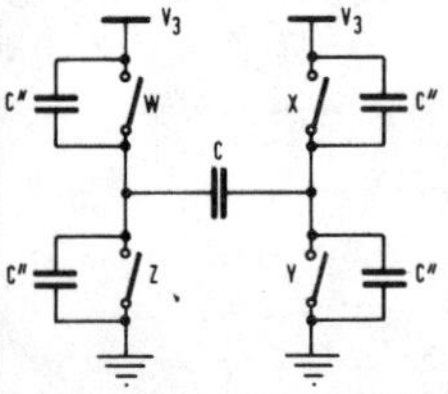	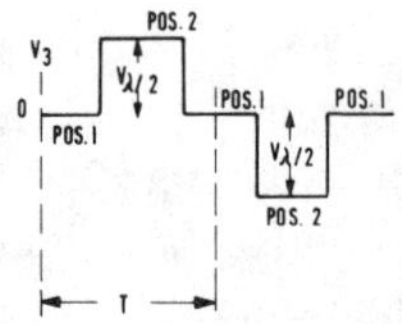	$V_3 = V_{\lambda/2}$	NO	$C(V_{\lambda/2})^2$	$2C''(V_{\lambda/2})^2$
4	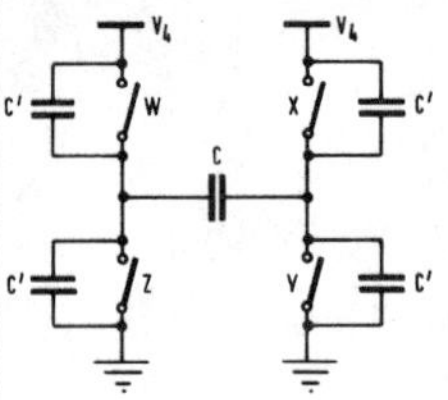	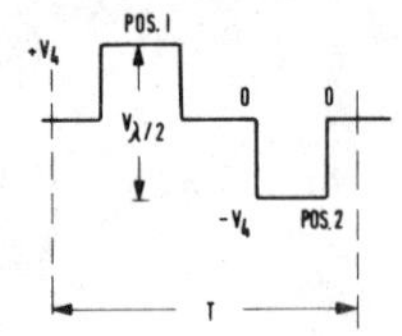	$V_4 = V_{\lambda/4}$	YES	$\frac{1}{2}C(V_{\lambda/2})^2$	$C'(V_{\lambda/2})^2$

FIG. B3-2. Some push–pull switch configurations to drive electrooptic crystals with half-wave voltage $V_{\lambda/2}$. A: all these configurations switch $V_{\lambda/2}$ across the crystal period is defined as including switching from position X to Y and back again; C is the capacitance of the load crystal of half-wave voltage $V_{\lambda/2}$; C' is the stray capacitance of a $V_{\lambda/4}$ switch; C'' is the stray capacitance of a $V_{\lambda/2}$ switch.

dissipation. Resistor switch combination can be used, possibly in a bridge configuration as in Fig. B3-1b. Rise and fall times depend on the values of the charging resistors. In a bridge configuration, one of the resistors is always connected across the supply. This gives rise to an astronomically high power dissipation in a fast system, when the resistor values have to be small. This problem is eliminated by the use of push–pull switch configurations. It has been difficult to make up these circuits using valves, but in the case of solid-state switches, an isolated drive is no problem. This has therefore been the solution adopted.

b. *A "2-kV" Push–Pull Transistor Switching Circuit*

Two of these circuits are used in antiphase in configuration 1 of Fig. B3-2 to drive electrooptic crystal stages in a deflector. Each circuit will, in fact, switch voltages greater than 2 kV, and two circuits are suitable for driving crystals with half-wave voltages as high as 4.4 kV. Each 2-kV push–pull circuit contains two chains of six 450-V transistors, type 2N3439. An individual transistor switch cell is shown in Fig. B3-3. These are cascaded as shown in Fig. B3-4.

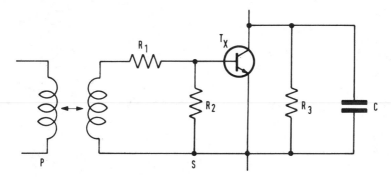

FIG. B3-3. An individual high-voltage transistor switch cell.

Features of the circuit are:

(1) Isolated transformer drive to each transistor, using linearly insulated ferrite toroids, 12.7 mm diam with two secondary windings per core. Primary to individual secondary winding capacitance is typically 3 pF.

(2) Resistors, R_1, help to control the charge injected into the transistor bases due to a current pulse in the transformer primary winding.

(3) Resistors, R_2, maintain a low resistance between the base and emitter at all times, inhibiting avalanche breakdown of the transistors.

(4) Resistors, R_3, define the effective OFF resistance of the transistors, eliminating the need for transistor selecting.

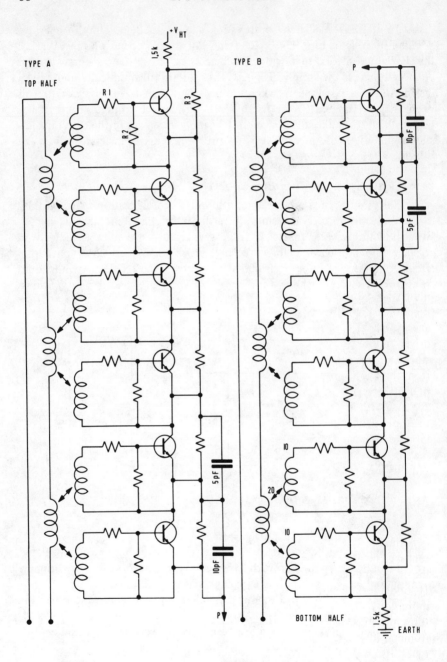

Fig. B3-4. A push–pull "2-kV" switch. Transistor are 3N 3439; $R_1 = 50\ \Omega$; $R_2 = 75\ \Omega$; $R_3 = 100\ \text{k}\Omega$; core transformers are linear ferrite toroids.

(5) Capacitors C^x, i.e., the 5 and 10 pF capacitors, compensate for distributed stray capacitance effects that tend to make transistors nearest the load junction switch the highest voltages. The capacitors effectively match the impedance of these stages.

(6) The particular transistors used have low junction capacitance values compared with other high-voltage transistors.

(7) Charge storage in the transistors maintains a switch chain in the ON state for a time after the charge pulse has occured. Continuous maintenance of the switch state is readily achieved by repeatedly pumping current pulses into the primaries of the associated pulse transformers.

(8) A useful property of push–pull circuits is that the turning ON of one switch stack actively pulls OFF the adjacent stack, which is maintained ON by charge storage. Thus, by suitable timing, charge storage is used such that transition speed is a function only of the rise time of one switch stack driving a capacitive load rather than a function of fall time or switch decay time. This rise time is very much faster than the decay time.

(9) Resistors may be included in series with the switch stacks without noticeably affecting the switch transition times. These resistors perform two functions. First, they dissipate some of the power due to the dumping of the energy stored in the load and circuit stray capacitances. Second, if logic or driver failure does occur, giving simultaneous switching of adjacent switch stacks, or if breakdown of the transformer insulation occurs, the current flow from the EHT supply is restricted. These resistors may typically have a value of 1.5 kΩ restricting the current through a push–pull 2-kV switch to below 700 mA without significantly increasing the rise time for load capacitance up to 20 pF. The value of these resistors must be adjusted for fast switching with large load capacitances. On the other hand, a large proportion of the power may be dissipated in these resistors, if desired, rather than in the transistors, but this means a sacrifice of fast rise time.

4. Modern High Power Switching Gaps

Although the triggered spark gap principle has been known for nearly a century, new types of circuit and construction are still being devised. By means of modern manufacturing methods like ceramic–metal-sealings and H_2 or deuterium as filling gases, two- and three-electrode spark gaps become a more conventional component for a wide range of environmental temperatures and high average power. Table B4-1 shows data in class A for ceramic-to-metal joint triggered spark gaps, in class B for lower duties glass-to-metal types. It can be seen that the typical range of voltage is in the area between 0.8 and 40 kV at energies between 50 and 10,000 J. The peak current is limited by

Table B4-1
Triggered Spark Gaps

Type No.	Applied voltage range kV dc E	Main static bkdn. kV dc E_t	Minimum trigger voltage[a] kV pulse $e_{trig\,min.}$	Energy (J)	Peak current[c] (kA)	Delay time[d] (μsec)
\multicolumn{7}{c}{Class A: Ceramic-to-metal}
TG-7	1.3–4.0	5.0	3.0–1.0	200	15	0.1
TG-88	1.3–4.0	5.0	3.0–1.9	200	15	0.1
TG-113[e]	1.3–4.0	5.0	2.5–1.6	100	10	0.1
TG-114	2.5–8.0	10.0	4.6–2.0	200	15	0.1
TG-121	0.8–2.0	2.5	3.1–1.8	150	15	0.3
TG-122	1.5–4.0	5.0	3.2–1.9	150	15	0.1
TG-123	2.3–6.0	7.5	3.8–2.0	150	15	0.1
TG-124	3.0–8.0	10.0	4.6–2.1	150	15	0.1
TG-125	4.5–12.0	15.0	6.3–3.5	300	20	0.1
TG-126	6.0–16.0	20.0	8.4–4.5	300	20	0.1
TG-127	7.5–20.0	25.0	9.7–5.2	400	20	0.1
TG-128	(10.0–24.0)[f]	30.0	(11.6–6.5)	2000	30	(0.1)
TG-129	(12.0–32.0)	40.0	(14.9–8.0)	2000	30	(0.1)
TG-130	(15.0–40.0)	50.0	(18.7–10.0)	2000	30	(0.1)
TG-143	0.8–2.0	2.5	3.1–1.8	50	10	0.2
TG-144	1.5–4.0	5.0	3.2–1.9	50	10	0.1
TG-145	2.3–6.0	7.5	3.8–2.0	50	10	0.1
TG-146	3.0–8.0	10.0	4.6–2.1	50	10	0.1
TG-151	2.7–6.0	6.8	3.7–2.0	200	15	0.1
TG-182	1.0–18.0	0–25	—	10,000	60	Var.
\multicolumn{7}{c}{Class B: Glass-to-metal}
TG-177	0.6–1.9	2.5	3.1–1.8	80	10	0.3
TG-178	1.5–3.5	5.0	3.2–1.9	80	10	0.2
TG-179	3.0–8.0	10.0	4.6–2.1	100	10	0.1
TG-180	4.5–12.0	15.0	6.3–3.5	120	10	0.1
TG-181	5.3–14.0	17.5	7.8–4.3	140	10	0.1

[a] Dual trigger gap.

[b] Minimum trigger values are given for mode A operation for the indicated applied voltage levels (typically 30 and 80% of E_Z). The trigger pulse has risetime and pulse width values as follows: $t = 0.5\ \mu$sec (10–90%) and $t = 3.0\ \mu$sec (50%). For high reliability circuits and for applications which require reduced delay times e_{trig} should be at least 150% of $e_{trig\,min}$.

[c] The energy ratings are for several hundred firings with a load of 0.3 Ω in series with the gap.

[d] The peak current is given as a conservative maximum current pulse which is approximately triangular or half sinusoid of 50 μsec half-width.

[e] The delay time values indicated are for mode A operation with applied voltage values of 80% of E and with e_{trig} equal to 150% of $e_{trig\,min}$.

[f] Values in parenthesis are tentative.

the design to 10–60 kA, the internal thermal loss is defined by the arc ionization voltage of hydrogen and the average current.

Newi [1596] reports about the variation of breakdown voltage of sphere gaps in air at negative impulse voltages. A good description of the breakdown behavior of a gap is its volt–time curve for linearly rising voltages. Gaps in atmospheric air with nearly uniform field have a constant time to breakdown provided the gap contains a sufficient number of free-charged particles for the initiation of the breakdown process. For this reason open-air sphere gaps with $s/D \leqslant 1$ (s = spacing, D = diameter) are used for high voltages as measuring device, for generating chopped test impulses or for protection from overvoltages. A different behavior was found with negative impulses generated with an impulse-energized ac transformer. E_t^* of the gap varied between 161 and nearly 300 kV with $S = 0.25$ kV/μsec. With ultraviolet light irradiation (uv-lamp), E is constant.

Other switching gap configurations are the wire–plane, disk–plane and rod–plane gap. Cookson and Farish [1588] reported that the corona onset and breakdown voltage characteristics of wire–plane, edge–plane and rod–plane gaps in atmospheric air have been measured for spacings up to 4 m with voltages up to 1 MV rms and 2 MV impulse. Dry alternating breakdown voltages for a fine-wire hoop–plane gap and for a disk–plane gap were substantially higher at moderate spacings than for the rod–plane system, due to the shielding effect of the ultracorona space charge; the average rms breakdown stress was 1.5 MV/m compared with 0.33 MV/m. Above a certain spacing, depending on wire size and hoop or disk diameter, the average breakdown stress decreased and breakdown voltages at the largest gaps approached the rod–plane values. The wet ac breakdown voltages for all hoop and disk electrodes corresponded to the rod–plane data. The curves of breakdown voltage versus positive impulse voltage rise time were identical for hoop, disk, and rod electrodes indicating that, even with rise times of ~ 600 μsec, there was insufficient time for formation of ultracorona space charge.

Finsterwalder et al. [1571] delivered a report about rotating symmetrical spark gaps for triggered chopping of high-impulse voltages. With different shape and distance of the electrodes of chopping and trigger gap, the triggered chopping of voltages could be achieved in the range 10–60% below the 50% breakdown voltage of the arrangement without triggering.

Bacchi and Pauwels [1011] report about a study of a very low jitter spark gap: Application to the realization of a high speed camera using a biplanar shutter tube with exposure time of 0.3 nsec. The design and construction of a new type of four-electrode spark gap intended to be fitted on pulse generators

---

* E_t is the static breakdown voltage.

with low jitter are described. Starting from the studies on a demountable spark gap, there is a sealed-off coaxial spark gap with 50 Ω characteristic impedance; for 15 kV pulses, the rise time is 9.5 nsec, the delay time after triggering is 7 nsec, and the jitter, 9.3 nsec. In order to realize a high speed camera using a shutter tube with 9.3 nsec exposure time, it is necessary to reduce the rise time of the pulse, while keeping the jitter low. The insertion of the four-electrode spark gap directly into the transmission line, the reduction of the interelectrode gaps and the miniaturization of the device led to rise times in the range of 0.15 nsec.

In combination with this principle, the spark chamber takes the form of a veritable end piece of a coaxial line, as shown in Fig. B4-1. Summarily, the spark chamber is composed of the following elements: two coaxial connectors, constituted by glass–metal passages which function in addition as supports for the electrodes of the primary spark gap; and two passages for positioning

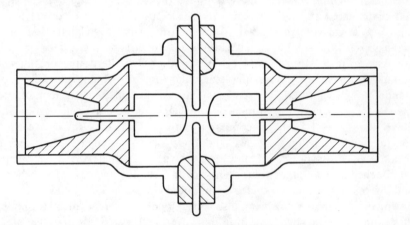

Fig. B4-1. Cross section of sealed spark chamber.

the rods of the auxiliary spark gap. These four passages establish the electrical insulation, as well as the means of maintenance of pressure constancy within the sealed spark chamber. The distance between the principal electrodes is divided into two space sectors, of $\frac{3}{10}$ mm each, by the electrodes of the auxiliary spark gap, the distance of the principal spark gap being fixed at $\frac{5}{100}$ mm. For a feed voltage of 30 kV the spark chamber is pressurized at approximately 20 bars of nitrogen.

The work described by the author demonstrates that it has become possible to improve remarkably over the performances of spark devices with three electrodes, utilized up to the present by the authors. The increasing number of difficulties shows that the results obtained are close to the limits, permitted

4. MODERN HIGH POWER SWITCHING GAPS

by the technique of ultrafast spark gaps in use at present. One may still try to reduce slightly the length of the formation line and obtain a quasi-triangular pulse, where the width at half-height will be 150 psec.

Vollrath [1703, pp. 78 ff, 82 ff] reports about the growth of the avalanche between anode and cathode up to the final phase of the fully conductive channel, and it can be seen that there are a large number of superposed factors such as speed of the electrons, the ions, and the shock front. During the nanosecond phase at the beginning of the conductivity of a nanosecond switching spark gap, all these parameters influence the operation and may, through the applicable physical laws, set a limit on either the rise speed of the current or the loss of energy in the plasma. Reference [1703, pp. 101 ff] reports especially on the recovery time to restore the original breakdown strength.

Rodewald [1577] reports about a new triggered multiple gap system for all

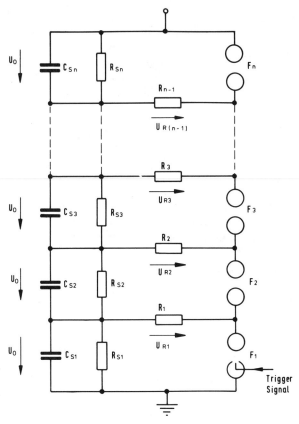

FIG. B4-2. Multiple spark gap with Ohmic capacitive voltage control (C_{sn}; R_{sn}) and generation of the breakdown voltage pulses on the resistances R_n.

kinds of voltages. A multiple gap system is described which consists of a chain of series-connected gaps. The voltage distribution along the gap chain is controlled by capacitors. Therefore, the grading is effective for fast-rising impulses as well as for ac (50 Hz) or switching impulses; with high Ohmic parallel resistors to the grading capacitors, the system can also be used for dc.

The chain of the gaps and the chain of the grading elements are connected together by one resistor in each link. After the triggering of the first gap voltage drops occur across the chain-connecting resistors. The essential point in the behavior of the arrangement is that the voltage drops along the chain-connecting resistors occur as overvoltages on the individual gaps. Because the overvoltages are high ($>100\%$), the system has a wide trigger range. It is possible to make parallel sparks in parallel gap chains in order to reduce the inductance of the switch.

5. Liquid and Solid Gaps

For replaceable single shot operations at extremely high peak power up to 10 TW, spark gaps have been designed which have a solid-state insulator instead of vacuum or gas. Dokopoulos and Lochter [1628] report on the use of such solid dielectric switches in plasma physics. For plasma physics experiments and in particular for magnetic compression systems, nanosecond fast switches with low internal inductance are required. They are mostly used for switching capacitors, charged up to several 100 kV, on a coil. In addition "crowbar" switches are required, i.e., switches which are able to short circuit the coil at current maximum in order to maintain a long-duration current pulse in the coil. Switches with solid dielectrics have a lot of advantages compared with spark gaps, in particular shorter switching time, lower inductance, and higher operation voltage (up to 500 kV).

Reference [1628] describes the development of nanosecond fast solid dielectric switches for several 100 kA, triggered by field distortion, i.e., by changing the shape of the electrical field. Contact is made by a great number of arcs. The operation voltage is up to 500 kV for capacitor discharge switching and up to 250 kV for crowbar switching. The report describes further a solid dielectric mechanical switch, driven by an exploding aluminum foil. Due to the metal-to-metal contact, the device has a contact resistance of some 10 $\mu\Omega$ and is able to carry currents of several megampere during a time of some milliseconds.

The same authors together with Steudle [1629] presented another unusual paper about the application of water for insulating high-voltage pulse systems. Comparing insulators for pulse voltages, one can find that water insulators have some advantages, such as high breakdown strength, high dielectric constant, very low costs and easy maintenance. One of the important properties is that water provides a voltage grading in inhomogeneous field, due to the increase of its conductivity with increasing electrical field. Therefore, for pulse breakdown tests of solid dielectrics, water is used to avoid flashover, i.e., to cause the breakdown between the test electrodes.

Water-insulated delay lines or capacitors are used for production of megavolt pulses on specific arrangements, e.g., diodes generating high-energy electron beams or x rays. In these systems water is the dielectric for storage of electrostatic energy, which will be transmitted on the specific apparatus by means of switches, e.g., spark gaps. The maximum energy density $\varepsilon E^2/2$ which can be stored is much higher than in other liquids (e.g., oil), because of the higher dielectric constant ε and the high breakdown strength E. A disadvantage of water capacitors or delay lines is that they must be pulse charged (e.g., by means of a Marx generator or any other capacitor arrangements), because the water conductivity causes a discharge. To get a high efficiency the charging time has to be much shorter than the discharging time. Therefore the inductance of the charging circuit must be sufficiently small in order to achieve a high efficiency, i.e., a high charging voltage.

6. HIGH-VOLTAGE VACUUM CONTACTORS

In February 1965, Toshiba marketed a vacuum switch applicable for motors. Subsequent research and development perfected mass-produced, high-voltage vacuum contactors available for circuits of 3.3 kV which use a ceramic tube suitable for frequent switchings at low surge values. This was followed in September 1969 by completion of a contactor available for 6.6-kV main-circuit voltage and one available for circuits of 3.3 kV with main-circuit current rating of up to 450 A. Addition of models meeting the specifications of all countries has completed this series of products, of which 7000 units have already been delivered, to the complete satisfaction of all users.

Ohwada and Ichihara [1917] checked the switching surge characteristics of these vacuum contactors under various service load conditions. It has been confirmed that the surges are low enough to cause no trouble in actual service. The reliability of these contactors has been verified by test and the delivered products have been totally free from troubles. These contactors are highly reliable and available for a wide range of uses.

There are two types of main assemblies: for voltages of 3.3 kV, and for those of 6.6 kV. The former weighs 20 kg and the latter 36 kg. Compared with an air contactor of the same rating, both the weight and the volume are $\frac{1}{3}$ to $\frac{1}{5}$. Being housed in a vacuum tube, the main circuit switch is free from influence of dust, salt air, harmful gases; and needs hardly any maintenance, inspection, and repair until its specified life expires. Though contained in a vacuum vassel, it exhibits excellent breaking capacity. There is no arc hazard. Even when low currents are switched, a low chopping level is maintained. Thus, the contactor is dependable in whatever service it is applied. Contactors available for circuits of up to 3.3 kV can be used for up to 450 A, while those available for circuits of up to 6.6 kV can be used for up to 300 A. In addition to the normally energized type, there is available an instantaneously energized type with a latch mechanism. The applicable range of the contactors seems to be extremely wide. Table B6-1 shows the specifications of these vacuum contactors.

TABLE B6-1

SPECIFICATIONS OF TOSHIBA HIGH-VOLTAGE VACUUM CONTACTORS

Item	CV432H-GAT CV432H-GATLI	CV432H-HAT CV432H-HATLI	CV461 J-GAT CV461 J-GATLI
Rated voltage (V)	2300–3300	2300–3300	4600–6600
Rated current (A)	300	450	300
Interrupting current (A)	4400 (symmetrical three phase)		
Short-time current	4400 A-2 sec	8800 A-0.5 sec	
Intensity to short-circuit current (peak value)	40 kA–0.5 cps		65 kA–0.5 cps
Frequency of operation	Continuous excitation type 1200 times/h Instantaneous excitation type 300 times/h		
Mechanical life	2.5 million (latch type 0.25 million)		
Electrical life	0.25 million		
Auxiliary contact	2a 3b or 2a 2b (latch type)		
Weight (kg)	20		36
Standard operating voltage	ac 100, 110, 115, 200, 220, 230, 240 V dc 80, 90, 100, 110, 115, 160, 180, 200, 220, 230, 240 V		
Closing current	ac, dc 110 V–4 A		ac, dc 110 V–9.2 A
Holding current	ac, dc 110 V–0.63 A		ac, dc 110 V–1.2 A
Tripping current	dc 110 V–3.2 A		dc 110 V–3.2 A
Impulse withstand (kV):			
Between live parts and the earth	45		60
Between main conductors (contactor closed)	45		60
Between poles	30		45

Structural consideration for combination with a power fuse or with a draw-out mechanism is fully made in designing. Accordingly, panel housing is easy and multilayer panel construction is possible. Vacuum interrupters are produced on an automated mass-production basis. Particularly, use of a continuous forming furnace assures a production capacity which will fully meet a substantial increase in future demand. The other parts and the assembling procedure are designed with full consideration of mass productivity.

6. HIGH-VOLTAGE VACUUM CONTACTORS

Being mass-produced on a conveyor line, standard products are low in cost.

In contrast to the gas-filled spark gaps, a family of triggered vacuum gaps has been reported by Lafferty [1222]. Characteristics of a sealed vacuum gap are described and the difficulties encountered in applying this gap as an overvoltage protection device are discussed. It is shown how these difficulties can be ameliorated by the use of gas-free electrode materials and by triggering the gap when breakdown is required. Several methods of triggering are discussed and some practical triggering devices are described that inject minute quantities of ionized hydrogen into the gap. The hydrogen is eventually recovered by the use of a titanium hydride getter. It is shown that breakdown of the gap can be accomplished in $<\frac{1}{10}$ μsec by first producing a glow discharge that is rapidly transformed into a metallic vapor arc. Properties of the metallic vapor arc are described which have an effect on the characteristics of the vacuum gap. A number of practical sealed-off triggered vacuum gaps are illustrated. These are used to carry microsecond capacitor discharge currents and 60-cycle power line currents for $\frac{1}{2}$ cycle. The operating voltage range is from a few hundred volts to 100 kV. The advantages of vacuum gaps over gas-filled gaps are given and a number of overvoltage protection and switching applications are listed.

To operate the trigger, a positive voltage pulse is applied to the trigger lead. The ceramic gap breaks down and an arc is established between the titanium hydride electrodes, thus releasing hydrogen and titanium vapor which are ionized and sustain the discharge. Expansion and magnetic forces produced by the discharge current loop drive the plasma out of the conical recess into the main gap. As the plasma spreads out in the main gap, a high-current glow discharge is first established between the main electrodes. The glow is transformed into an arc as cathode spots are established on the negative main electrode with release of metallic vapor. Measurements indicate that, with peak current pulses of 10 A on the trigger electrode, the main gap will break down in less than 0.1 μsec with jitter times of about 30 nsec when 30 kV is applied to the main gap. The trigger energy required is less than 0.01 J. The main gap may be broken down with trigger voltage pulses as low as 50 V; however, longer delay times result. A detailed study of the trigger and firing characteristics of the vacuum gap at low voltages are presented in a separate paper by Farrall [1806].

The quantity of hydrogen released on firing the trigger is extremely minute. Experiments indicate that 75% of the hydrogen comes from the negative trigger electrode where intense local heating is produced by the cathode spots of the trigger discharge. On a single firing of the trigger the gas pressure has been observed to rise to about 2×10^{-6} Torr in a 700-cc vessel. Thus about 1.4×10^{-6} Torr-liters of hydrogen are released. This corresponds to the dissociation of 8×10^{-9} g of titanium hydride (TiH). Since the trigger contains a few tenths gram of titanium hydride, adequate trigger life is assured for most protective applications. Trigger life has not been determined experimentally, but little deterioration was found in more than 600 single operations.

Under normal nonrepetitive operation of the gap there is no accumulative pressure rise due to hydrogen. There are several places where the hydrogen can be gettered, at room temperature after each operation of the gap. It can be reabsorbed by the trigger electrodes or by an auxiliary getter consisting of titanium hydride granules that are present in most of the gaps. There is also the possibility that the hydrogen may be temporarily adsorbed on the fresh surface of gas-free electrode material that is deposited on the walls of the gap by erosion of the main electrodes. The sophisticated design of a heavy duty vacuum switch gap is shown in Fig. B6-1.

Small gaps will carry currents of the order of 10,000 A for a few microseconds. Larger gaps are for 60-cycle operation and will carry currents up to 40,000 A for $\frac{1}{2}$ cycle. A large demountable gap carries currents up to 100,000 A for $\frac{1}{2}$ cycle. The triggered vacuum gap is more expensive than the conventional three-electrode spark gap operation in air, but it has a number of advantages including small size, rapid deionization time, possible operation over wide range of gap voltage, possible operation in strong radiation environment, gas cleanup presents no problem, no audio noise or shock waves, and no explosion hazard. Applications of the triggered vacuum gap include its use as an overvoltage protection device and as a switch for rapidly diverting large current flow.

High-current megajoule capacitor banks are generally switched by a large

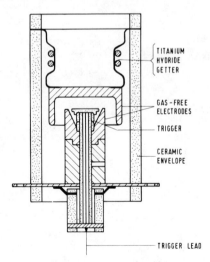

FIG. B6-1. An example of a triggered vacuum gap of ceramic–metal construction designed to carry current pulses of a few microseconds duration.

number of spark gap switches connected in parallel to a common load impedance [1627]. To reduce the complexity of such switching systems, it is desirable to minimize the number of parallel switches by using spark gaps which can each switch a peak current of ~500 kA and pass 100 C/shot for about 10,000 shots. However, at such high ratings there is a greater tendency for premature voltage breakdowns in the spark gaps to occur which could result in an unacceptable failure rate in a system having a large number of such components operating together.

Some preliminary experiments have been carried out to obtain data on the premature breakdown and electrode erosion rate of a pressurized air three-electrode spark gap. The latter has been developed for use in 50-kV low-inductance, high-current fusion research applications. Consideration has also been given to the use of a similar 200-kV spark gap in low-frequency, high-inductance circuits such as series-connected 1.0-MV capacitor banks. The electrode geometry of the spark gap under investigation is shown in Fig. B6-2. The center electrode consists of a circular disk of tungsten alloy placed between plane–parallel brass main electrodes, space 15 mm apart. The electrodes are placed in an enclosure containing air at up to 6 atm pressure.

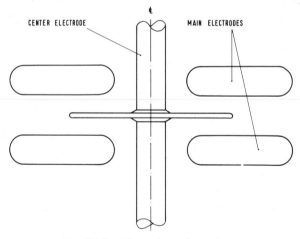

FIG. B6-2. Electrode configuration.

The DC potential of the center electrode is held at an intermediate value such that the electric field between the main electrodes is almost uniform. However, if a trigger pulse is applied to the disk appreciable field distortion occurs at the edge of the disk. For this reason this electrode arrangement has been referred to as a field distortion spark gap. Breakdown of the two parts of the gap occurs in the cascade mode in about 30 nsec with a jitter of ± 3 nsec for a 50-kV gap. The spark gap was used to discharge a 45-kV, 30-kJ capacitor bank into an inductive load at a peak current of 400 kA and with 28 C being passed through the gap per shot. The repetition rate was 1 shot/min, the air in the enclosure being changed every shot. For more details see Burden and James [1627].

7. The Lightning Protectors

Lightning protectors are gas-filled discharge tubes [1764]. Their tightly sealed electrodes protrude into a discharge chamber filled with rarefied gas at low pressure. The protectors are designed for extreme pulse power ratings reaching into the megawatt region. These devices are primarily used for protection of telecommunication equipment and other installations against overvoltages from any source. They are also used as electronic switches in sweep circuits.

Overvoltages endanger telecommunication equipment and operating personnel; may damage components and units, destroy the insulation rated for normal operation, and constitute a danger even for telephone subscribers. Such overvoltages are generated by power systems or atmospheric discharges. Power systems may become dangerous through direct contact or induction by inductive or capacitive coupling. In the case of a direct contact, the full voltage of the power system can affect the telecommunication equipment. Short-circuit currents in the power system may cause by induction overvoltages in telecommunication lines. A similar effect is brought about by capacitive coupling between the two systems.

Overvoltages due to atmospheric discharge are set up by lightning strokes in transmission lines or in their vicinity. They are also generated by the steady electric field in the air where transmission lines span large difference in altitude above sea level. Among the measures protecting telecommunication equipment against these effects, the discharge tube is an outstanding remedy both from the technical and the economical viewpoints. When the dangerous overvoltage reaches a certain magnitude, the breakdown or striking voltage, the discharge tube provides a reversible short-circuit path between its electrodes, thus bypassing dangerous currents. Thus excellently protected circuits can be designed with the aid of the discharge tube.

As long as the interfering voltage has an uncritical value, the discharge tube remains an inactive, low-capacitance device of an extremely high insulating value. In other words, it takes no part in normal operation of telecommunication equipment. Compared with other types of protecting devices, the outstanding feature of the discharge tube is the absence of any upper limit of the dangerous voltage surge. Hence, the discharge tube is a highly effective protecting device especially in areas where high-voltage surges due to lightning are to be expected. Since the discharge takes place in a tightly sealed envelope, the striking voltage is independent of ambient air pressure, air humidity, and contamination. Requiring no maintenance, the discharge tube will repeatedly handle many discharges in short succession, virtually with unchanged electric properties.

When the alternating voltage at the terminals of the discharge tube is high enough, the oscillogram of this voltage has the form shown in Fig. B7-1. The section O–A is that of a sine wave. When this voltage exceeds the value V_s (striking voltage), the curve of the voltage falls abruptly from A to B, then follows between B and C an almost constant ordinate V_r (arc residual voltage). The current passing through the discharge tube is then very large. Towards the end of the cycle, the discharge tube recovers, its terminal voltage rises quite sharply (section CD) and then resumes the value of the sine wave. The same phenomena are repeated during successive cycles.

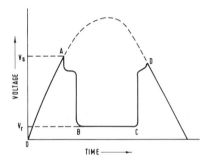

FIG. B7-1. Operation of a lightning flash protector (see text).

The extremely low arc residual voltage of about 10 V prevents excessive heating of the protector even when a high current is flowing. For the reliability of operating, it is important to know the protector endurance, that is, the duration of its operation until destroyed. The endurance characteristics of a number of neutron protectors are shown in Fig. B7-2. It may be seen that all four protector types are dimensioned to withstand their respective rated ac for 1 sec as denoted by the plus signs (+). The protector type represented by curve 1, rated for 10 A, will withstand 50 A for this time without becoming destroyed. Accordingly, the 20-A protector will withstand 90 A (curve 2), the 30 A protector, 130 A (curve 3), and the 50 A protector, 250 A for 1 sec. When the values of this diagram are exceeded, irreversible destruction takes the form of the two electrodes being welded to each other, thus maintaining safe conditions on the line.

The mode of operation of the protector is the same for dc and ac; however, a direct current causes the destruction sooner than alternating current. The terminal voltage of the protector does not change within the limits set by the rated protector current and the discharge time. The glow to arc transition is

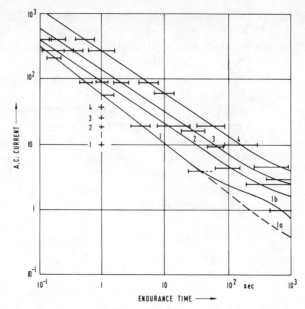

FIG. B7-2. Endurance characteristics of neutron lightning flash protectors. 1a: type B; 1b: type E, G; 2: type A, C, R, M5; 3: type M6; 4: type M8.

caused and the protecting operation is introduced at the very low arc voltage.

An arrangement for protection against overvoltage, including the current-limiting fuses, is shown in Fig. B7-3. Parallel to the switch device to be protected are gas-discharge tubes which bypass any overvoltage (atmospheric discharge, induction, or contact by high-voltage lines, etc.).

The current-limiting fuse should respond to bypass currents exceeding a certain duration, not to the vigorous, but short atmospheric discharges. A fuse of this type will generally be useless for the equipment to be protected. For this reason, the arrangement of Fig. B7-3 is now provided without current-limiting fuses in most cases. Considering the high-grade perfomance of modern gas-discharge tubes, the spark gap owes its existence to historical reasons and can be eliminated altogether. Recently, instead of gas-discharge types, a new type of lightning flash protector became available using metal–ceramic overvoltage protectors [1764]. Table B7-1 shows data for modern metal–ceramic lightning flash overvoltage protectors.

Pfeiffer [1908] deals scientifically with the voltage-dependent resistance of silicon carbide resistors. He found that they operated without delay up to 10^{11} A/sec rise speed. When using gas discharges or spark gaps as pulse voltage protectors, it is necessary to be familiar with the behavior of the spark resistance itself. Theoretical calculations can be made using the laws of Toepler, Rompe–Weizel, and Braginskii [1906]. Pfeiffer [1906] checked these laws over a wide range of gap width, pressure, and kind of gas. Only the time interval

7. THE LIGHTNING PROTECTORS

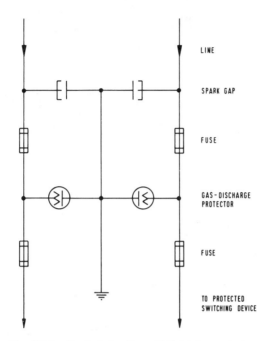

FIG. B7-3. Typical circuitry with lightning protectors.

TABLE B7-1

ELECTRIC DATA FOR METAL–CERAMIC LIGHTNING PROTECTORS[a]

Response dc voltage (V)	150	230	350	600
Response pulse voltage at 5 kV/μsec (kV)	< 1	< 2		< 3
Conducted ac current (A)			20	
Conducted pulse current (kA)			10	
Insulation resistance at 100 V (Ω)			10^8	
Capacity (pF)			ca. 3.0	

[a] See [1764].

of the voltage breakdown at the spark gap was studied. The measurements show that all spark laws are useful in these conditions. The highest accuracy and the easiest applicability are attained using the spark law of Toepler. In another paper the same author [1907] studied the occurrence of steps in current progress during spark discharge in carbon dioxide. During the investigation of spark discharges in nitrogen, carbon dioxide, and argon, aberrations from normal current progress are observed for carbon dioxide and small gap widths. The effect if accompanied by a remarkable scatter in breakdown time and even in breakdown voltage. Increasing the pressure for all gap widths causes a transition to normal current progress.

8. Laser-Controlled Spark Gaps

Pillsticker [1624] presented a summary about laser-triggered, high-voltage spark gaps. Triggering of spark gaps by a focused laser beam seems to be superior to the conventional triggering system with a high-voltage pulse: No electrical connection exists between the trigger circuit and main circuit with the spark gap. Disturbances are not transmitted and the electrodes of the spark gap may take any potential. The formative time lag and its jitter is much less than that observed in electrical triggering. From the economic point of view, triggering with a laser beam can compete with electrical triggering.

A comparison of both the possibilities of laser triggering shows that the beam should be in the direction of the electric field between the electrodes. The beam has to be admitted to the gap through a hole in one of the electrodes (see Fig. B8-1b). These are the conditions for having a high value of the first Townsend coefficient within the breakdown region.

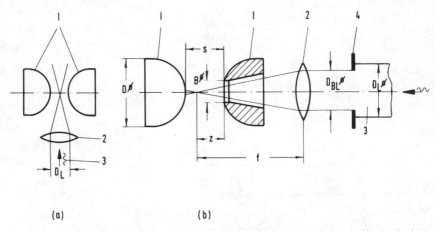

Fig. B8-1. Electrode configurations of laser-ignited gaps: (a) perpendicular ignition, (b) axial ignition. 1. Spark gap electrodes, 2. lens, 3. laser beam, 4. diaphragm.

The gas in the gap region should be a noble gas. Its ionizing voltage is low and there is a high probability of collisions between electrons and atoms. This probability increases with increasing pressure of the gas, and therefore the spark gap must be under high-pressure conditions. For the experiment, argon is used because it is not very expensive. In order to raise the static breakdown voltage, a dosed gaseous molecule has to be added.

It is pointed out that the frequency of the laser light should be as low as possible and the duration of the laser pulse has to be short to keep the formative time lag and its jitter short. Other components such as optical parameters are selected to achieve a laser power as small as possible. Two important regions must be considered in order to have a survey of the breakdown mechanism in the laser-triggered spark gap:

(1) A laser-produced plasma is induced in the gap at two points: (a) in the gas within the focus and (b) at the surface of the electrode without a hole.

(2) There is only a laser-produced plasma at the surface of the electrode without a hole.

(3) No plasma is produced by the laser beam.

A formative time lag versus gap voltage plot shows that the largest region of voltages with short formative times and jitter can be realized if the conditions of point 1 are observed. The operation of a laser-triggered spark gap is absolutely reliable. Some difficulties arise in technical research on light transmission and problems of adjusting the optical setup. A solution seems to be only a matter of time and expenditure.

A paper on the switching jitter in laser-ignited spark gaps has been published by Milam *et al.* [1637]. The first study of jitter in a laser-triggered spark gap switched by pulse trains from a mode-locked ruby laser is described. The spark gap was fired by producing gas breakdown in a high-pressure argon–nitrogen mixture between the pole pieces. Jitter was measured as a function of

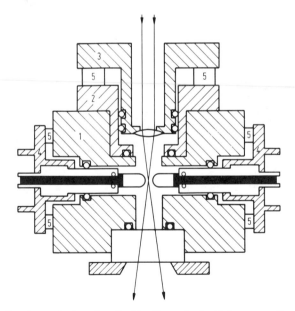

FIG. B8-2. Spark gap design: 1. cylindrical brass body; 2. lens mounting tube; 3. lens mount; 4. UG-560/U coaxial fixtures modified to be gas-tight; 5. spacers. A second window (not shown) allows observation of the gap breakdown. A small inlet across from the observation window is used for gas filling.

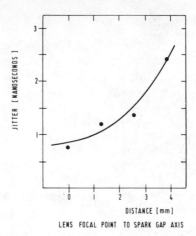

FIG. B8-3. Switching jitter as a function of the separation between the spark gap axis and the point at which the laser beam is focused.

the position of the lens used to focus the laser beam and as a function of the ratio of the applied voltage to the self-breakdown voltage of the gap. Jitter values of less than 2 nsec were obtained under optimum conditions provided that the gap was fired by the early part of the pulse train. In its experiments the group used the arrangement shown in Fig. B8-1a, the beam being provided

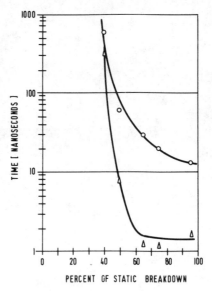

FIG. B8-4. Jitter as a function of the ratio between the voltage applied to the gap and the self-breakdown voltage. ○—delay; △—jitter.

8. LASER-CONTROLLED SPARK GAPS

by a Nd-laser operated in the straight TEM-mode. The spark gap is shown in Fig. B8-2, and the jitter measurements are plotted in Figs. B8-3 and B8-4.

In 1969 Impulsphysik GmBH (contribution of the author, unpublished) constructed a laser for the synchronous firing of two 120-kV gaps for use at CERN, Geneva [1810]. The jitter of these two gaps was so low that simply by the shortening or elongation of the light path length between a beam splitter

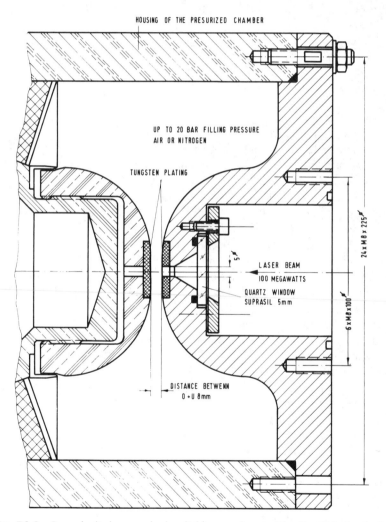

FIG. B8-5. Laser-ignited pressurized switching spark gap. The Q-switched ruby laser beam is coaxially injected. It passes a quartz window and generates a kind of an ionized hose between the electrodes of about 0.5 cm diam. Operation: 120 kV; time jitter: can be less than 0.5 nsec.

and the gaps, subnanosecond intervals could be realized. The coaxial mode of firing was used, the 120-kV gaps operating in argon at a pressure of ~ 20 bars. Figure B8-5 shows the cross section of this gap. As can be seen, the laser pulse of ~ 100 MW peak power passes a 5-mm Suprasil window and hits the tungsten plate electrodes producing between both electrodes a highly ionized plasma in which a relatively great diameter spark grows with sufficient ability to transport the coulomb amount of the capacitor discharge circuit at negligible time jitter.

9. Quenched Spark Gaps

a. *The Quenchotron*

Früngel developed the quenching spark gap principle (first published reference here) to a family of commercially available Quenchotrons. The Quenchotron interrupts the current by cooling the hydrogen ions between a series of parallel disks. After trapping the missing electrons the ions become insulating atoms, and the smaller the distance between the copper disks the shorter is the time of deionization. At 0.1-mm distance after each spark 2 μsec are sufficient to attain about 90% of the full breakdown voltage. Main feature of these tubes is an extremely high repetition rate. Well-controlled, it can switch 12 kV, 10^4 A at a rate of 10^5/sec, and with repetition rates higher than $\sim 10^3$/sec, a time jitter of less than 1 nsec has been observed. The current risetime depends on the external circuit data. For each 400 V an additional disk must be provided in the Quenchotron. A Quenchotron used in the Strobokin heavy-duty flash unit (see Volume II, Chapter K7a) has 30 tungsten–copper alloy disks of 8 cm diam spaced by mica rings. It is controlled by three parallel-fed pulses at the 8th, 16th, and 24th disks. Normally it is filled with H_2 gas at 1.2 bars. The nominal permissible thermal loss capacitance is 200 kJ/burst of discharges. If the consumer circuit has a resistance of one or more ohms, the efficiency is higher than 90%, which means that a megajoule energy content in a series of several thousand discharges can be handled in the form of a well-controlled pulse burst. The Strobokin control unit together with the built in TPA (trigger power amplifier) serves to trigger the Quenchotron with quartz stabilized or other frequencies [1073].

b. *The Blast Circuit Breaker*

Another method for interrupting the current uses a blast effect to replace the ions with fresh cool unionized gas. Inall [1909] realized a "fast circuit breaker for the discharge of a storage inductor." Very high power lasers, certain plasma machines, apparatus for the study of high power arcs, and similar equipment, often need a means of storing some tens of megajoules, to

9. QUENCHED SPARK GAPS

be discharged in ~ 1 msec. Capacitor banks of this rating are so large and expensive, that several proposals have been made for the use of rotating machines to store the energy.

The homopolar generator at the Australian National University in Canberra can supply 1.5×10^6 A at 800 V, that is 1.2 MJ msec^{-1}, and it can do this for about 0.1 sec. But, for most applications it is an advantage to increase the peak power and reduce the duration of the discharge available from a rotating machine. The report deals with tests on a model of equipment planned to take 20 MJ in 0.3 sec from the Canberra generator, and deliver 15 MJ to a load in 1 msec. The model should be able to take 400,000 J from the generator, but the present tests have been limited by the testing arrangements to 100,000 J. The equipment includes several unusual components, however, and their successful operation at this energy is notable.

The circuit of the system is shown in Fig. B9b-1. The most important and most difficult component to design was the fast circuit breaker (FCB). For this reason the tests were chiefly concerned with the FCB and the fuse connected across it. Currents of up to 76,000 A were used, with a load resistance of 15.2 mΩ; this could produce a maximum potential of 1.160 V between the opening contacts. The maximum measured was 1.120 V, when the current was 76,000 A, and 1800 V when the current was 72,000 A and the load resistance was 26 mΩ.

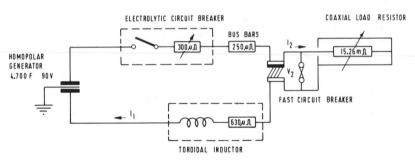

FIG. B9b-1. Circuit used to test the fast circuit breaker, the fuse, and the storage inductor.

A cycle of operation of the equipment started with the homopolar generator running at 180 rpm, thereby generating 90 V, with the FCB closed. The electrolytic circuit breaker (ECB) was closed at the gate to give a minimum resistance of 300 $\mu\Omega$ 0.4 sec after the first contact, and then opened to clear the contact at 0.75 sec from the start.

After the peak value of the current I_1 was reached, and the ECB had commenced its opening travel, the FCB was tripped. The high-pressure nitrogen drive in this breaker moved the contacts apart, and as they separated, an arc developed with a potential of about 20 V across it. The oscillogram in Fig.

B9b-2 shows the current I_1 in the inductor, the current I_2 in the fuse and load circuit, and the potential V_2, across the fuse. The 20 V across the arc caused the current to transfer to the fuse circuit at the rate of 2×10^8 A sec^{-1}, as set by the 0.1 μH inductance of the fuse circuit. The resistance of the cold fuse was 0.25 mΩ and all the current transferred to it within 250 μsec. The fuse melted and the current passed into the load within the next 20–100 μsec. The maximum current in the load was 0.9–0.99I_{10}, depending on the design of the fuse, I_{10} being the current in the inductor before the FCB operated. The FCB, the fuse, and the inductor, were designed and constructed for this project and are described briefly here.

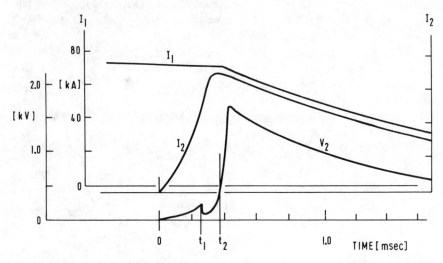

Fig. B9b-2. Oscillograms of the current in the inductor, I_1, the current in the fuse and load, I_2, and the voltage, V_2, across the load. The zero of the time scale is when the contacts separated, about one millisecond after the "open" signal. The arc between the contacts was extinguished at t_1, and the fuse melted at t_2.

The fast circuit breaker could carry a current of 150,000 A for 5 sec or 10,000 A continuously. It was designed to mount in a coaxial or interleaved bus bar. The contacts were 9 cm diam and the over-all height of the breaker, the gas chamber, and the trip mechanism was 40 cm, but the current path was between two terminal plates only 10 cm apart, as shown in Fig. B9b-3. The stationary contact was a copper block 9 cm diam and 4 cm thick, clamped to one terminal plate. The moving contact was a copper disk 0.1 cm thick, connected by 3600 copper wires 0.084 cm diam and 5 cm long, to a second disk. The wires were inclined to the plane of the disks and located in a ring to form the thick stranded wall of a hollow cylinder. The inclined wires allowed the disks to move towards or away from each other by a small amount as the wires bent and the inclination changed. Rims on the disks were fitted with O-rings and the disk-wire assembly was contained in a cylinder with one disk against the closed end, where a constant electrical contact was made. The other disk protruded beyond the open end of the cylinder and pressed against the stationary contact of the circuit breaker. The closed end of the cylinder was clamped to the second terminal plate. Nitrogen at a pressure of 28.1 kg cm^{-2} in the cylinder flattened the disk against the stationary surfaces, thereby ensuring the intimate contact needed to carry such large currents. The disks were made thin to enable them to follow the shape of the stationary surfaces, and their ability to do so is clear from the fact that no hot contact spots occurred even after the stationary contact face was eroded around the rim. A chamber (with a volume of 0.32 liter) filled with nitrogen at 141 kg cm^{-2} was mounted on the stationary contact, with a "pop" valve arranged to allow

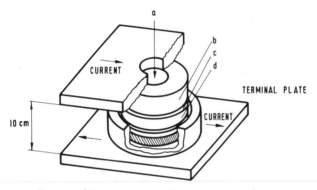

FIG. B9b-3. A diagram of the current-carrying components of the fast circuit breaker, for 100,000 A, 2000 V, pulsed. a, nitrogen at 141 kg cm^{-2} to drive electrodes apart; b, stationary electrode; c. intercontact disk; d, moving electrode; e, 3600 wires to moving electrode plus nitrogen at 28.1 kg cm^{-2} to maintain contact between electrodes.

nitrogen to discharge through the stationary contact, onto the center of the moving contact disk and radially outwards between the contact faces. The period from the first movement of the trigger to the time when the potential between the contacts began to rise was about 1 msec. This potential caused the current to transfer to the fuse, and after 250 μsec all the current was flowing in the fuse circuit. The potential rose to about 120 V before the arc was extinguished by the sheet of nitrogen flowing between the contacts. The separation of the contacts was sufficient to hold off 1800 V after a further 100 μsec, when the fuse ruptured. New breakers have been designed for 400,000 A and 2000 V. Two of these will be used in a series circuit to provide a peak power of 1600 MW.

An extra copper disk 0.08 cm thick was placed between the contact faces. It was driven by the nitrogen at the face of the moving contact. The only damage to the contact faces was produced by the arc as the faces separated, and this minor damage was confined to the outer edge of the stationary contact and the replaceable intercontact disk. A single arc caused small pits near the outer rim of the faces but no marking was visible within the central area of the contact, even after 60 operations. But for a reliable operation it was necessary to replace the intercontact disk after every ten operations. although one disk was used for 30 operations. It is expected that the number of reliable operations will be increased in future, when the inductance of the fuse circuit is reduced to a quarter of the value used.

The fuse was designed on the basis of the work reported by G. Schenk at the Fifth Symposium on Fusion Technology, but with emphasis on the lowest possible resistance and inductance, and on a simple enclosure to allow the fused foil to be replaced automatically in less time than the shortest pulse interval set by other requirements. The fuse link consisted of a copper foil 6 cm wide, 0.005 cm thick, and 4 cm long, clamped to the terminal plates and between two layers of woven asbestos cloth. A current density of 2×10^6 A cm^{-2} allowed a period of about 100 μsec between the extinction of the arc between the contacts and the rupture of the fuse, when the current transferred from the contacts into the fuse at $2-3 \times 10^8$ A sec^{-1}. These values were used when the oscillogram shown in Fig. B9b-2 was recorded. The inductor was wound as a toroid to minimize external magnetic field, the minor diameter being 57 cm and the major diameter 144.3 cm to the outer edge. This is surprisingly compact in view of its ability to store 400,000 J.

These tests have demonstrated that the circuit breaker and fuse can carry a current of 76,000 A and present a negligible impedance up to 50 to 100 μsec before diverting the current into a parallel circuit with 1800 V across it. The time required is within 1 msec from the trip command. This is the only device known to be able to manipulate currents of such a magnitude. It is possible that versions of the circuit breaker could be used to introduce a resistor into the circuit of a superconducting inductor if it started to go normal, thus limiting the energy dissipated within the cryostat. Second, if a superconducting inductor was built with sufficient insulation and pulse characteristics, a very compact energy store for pulse generation could be constructed. Two further important applications could be the switching of high-voltage, dc power transmission circuits, and the protection of rectifiers during a reverse current breakdown.

Inall designed the fast circuit breaker while on leave from the Australian National University and attached to the Culham Laboratory of the UKAEA, where the first model was made. It is now on loan to the Australian National University where it has been further developed and tested at high current. It will be used to power a high power laser, while further development is continued on a system to handle several megajoules.

10. Crowbar Switches

Crowbar switches have been developed alongside high-energy plasma generators. The typical job of a crowbar is to effect a short circuit over a low-Ohmic component or a capacitor bank, e.g., to protect a critically dimensioned capacitor against voltage reversals due to oscillation of the circuit. Another function is to cut the tail of a high-voltage, high-current pulse after a certain time, e.g., to produce a nearly rectangular pulse to feed into another piece of equipment.

The technique for high-Ohmic, high-voltage pulses is entirely different from that used for heavy currents. Feser and Rodewald [1570] reported on a triggered multiple chopping gap for high lightning and high switching surges. The trigger performance of the whole multiple chopping gap essentially depends on the trigger performance of the first stage. A trigger range of more than 15% in the negative polarity with a jitter of less than ± 50 nsec makes it possible to use this multiple chopping gap for transformer tests. The multiple gap can be operated in the voltage range 300 kV $\leq$ U $\leq$ 3600 kV without any change in the circuit parameters. For chopping lower voltages down to 20 kV, a part of the gaps can be short-circuited. Figure B10-1 shows the circuit. Without any trigger, the multiple gap shows a remarkable low-standard deviation of the flashover voltage. This allows the gap to be used as a protective device in high-voltage systems. The multiple gap can be used for chopping of lightning impulses, switching surges, and ac voltages. A further aspect of the new device is the possibility to use the capacitive voltage grading as the high-voltage arm of a voltage divider. The chopping gap used as a voltage divider has similar transfer characteristics as a low damped capacitive voltage divider. Therefore, in high-voltage test technique, the new system presents three applications in one apparatus: first as a chopping gap, second as a voltage divider, and third as a load capacitor.

For current crowbars, Kasuya and Murasaki [1636] are utilizing reflected shock waves. A new method is proposed for crowbar techniques utilizing the ionization behind reflected shock waves in air. It can easily crowbar discharged currents when the time of the current peak coincides with the arrival of the shock wave at the crowbar gap. When we use capacitor banks to produce current pulses, there are many cases in which we must crowbar them to supply loads with unidirectional currents. (A) In particular, this is required in plasma confinement experiments, where in most cases the rise times of the currents are very short. (B) On the other hand, more moderate cases are quasi-steady current sources; for example, for magnetohydrodynamic arc jets, for which we often make use of so-called LC-ladder networks. These networks require impedance matching and most of the stored energy is wasted in the ballast or matching resistors. Therefore if we can easily crowbar capacitor banks, it will be simpler to supply unidirectional currents to various loads such as (A) or

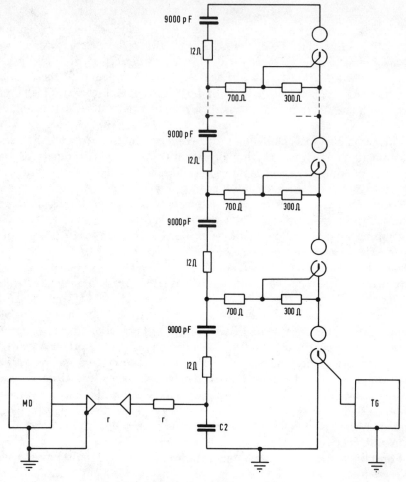

Fig. B10-1. Circuit of the multiple chopping gap.

(B) without waste of energy. The test of a new method for crowbarring, utilizing the ionization behind reflected shock waves is discussed here. Because the voltage on loads or crowbar gaps at the moment of crowbar is nearly equal to zero, it is desirable to use this method since it is capable of producing very short rise times and sufficient duration of uniform plasmas.

Figure B10-2 shows the outline of the experimental setup similar to that previously described. A crowbar gap is mounted at the end of a pressure-driven shock tube. An incident shock wave propagates from left to right and reflects at the end wall of the shock tube. The stepwise signal of the first pressure pickup from the incident shock wave is sent to an amplifier to trigger

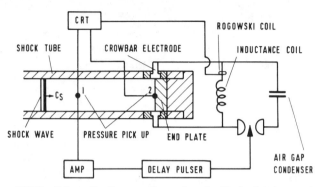

FIG. B10-2. Schematic diagram of experiment with shock tube crowbar.

the main air gap via a delayed-trigger circuit. After an appropriate delay time according to the speed of the shock wave, the capacitor (3 μF, 4 kV) begins to discharge through the load inductance (4 μH). We can then crowbar the discharge current if we make the time of the current peak coincide with the arrival of the shock wave at the crowbar gap.

As observed by the pressure pickups the incident shock wave takes about 50 μsec to travel 40 cm (from the pressure pickup 1 and 2), when the initial conditions of the shock tube are: pressure of driven gas (air) = 23 mTorr and pressure of driver gas (H_2) = 11 kg cm^{-2}. At this time the shock Mach number is about 25. A crowbar spark gap is also often used to generate well-defined pulses for feeding hydrogen streamer chambers in the field of nuclear physics (see Volume II, Chapter K6a). Schmied *et al.* [1225] show the mathematical treatment and experiments aimed at realization of nonoscillating crowbar switches, voltage area about 140 kV.

With very high currents and in complete contact between the electrodes of the crowbar, electric arcs will form in the metallic vapor of the electrodes. Reeves-Saunders at the Australian National University wrote a book [1910] about the evolution of high-current dc arcs on rotating anodes and a shorter paper [1911] about transition into a stable mode for an arc burning on a rotating anode.

11. Other Unusual Switching Gaps

a. *Megavolt Switching Spark Gaps*

There is an increasing demand for multimegavolt pulses especially to test equipment like switches, lines, protection devices, and to simulate the largest and highest pulse generators, namely thunderstorm flashes [1227]. In another field, i.e., the reliability of high-voltage equipment relative to dielectric, mechanical, and climatic conditions in service, the situation happens to be similar. The parameters involved are so numerous that, as far as we can see, full scale tests cannot be avoided. In its "Les Renardières" testing center, Electricité de France has recently built a laboratory dealing with all dielectric problems related to both high voltage and ultrahigh voltage ranges.

Ultrahigh Voltages

At the highest test voltage level under switching impulse, the test circuit must be corona-free to avoid any predischarge causing ripples on the generated wave shapes and initiating unexpected flashovers. These conditions are fullfilled when the electric field on the surface of all conductors is limited to about 20 kV/cm. This requires the addition of electrodes of various forms: Single toroid with or without upper plate, double toroid, etc. It also necessitates a correct knowledge of the electric field itself, which can be achieved as some progress has been made recently in this subject.

In 1970 a paper was presented in the name of Study Committee 33 (overvoltage and insulation coordination) by the "Les Renardières" group [1226].* This paper describes the first stage in the analysis of a series of experiments in which positive switching impulses with a time to crest between 100 and 500 μsec have been applied to 5 and 10 m rod–plane gaps in air. Three rod tips were used, hemisphere, hyperboloid, and cone. Voltage, current, and field (at up to 13 points) at the plane and current or charge at the rod were measured. Two image converter cameras, one image intensifier camera, two still cameras, and two photomultipliers were used simultaneously to measure the light output from the gap. 50% breakdown voltages, the standard deviation, times to breakdown, corona inception voltages and fields are reported. Of the 7000 photographs, about half have been analyzed so far to give values for leader length and velocity, the charge injected into the gap and the time dependence of the field at the plane. A detailed analysis is given for one shot in which the development of the leader across the gap is followed and correlated with the field at the plane.

In June 1971, a number of European scientists and engineers gathered at the EdF Les Renardières Ultrahigh Voltage Laboratory to carry out experiments on the breakdown of long gaps in air. The scientists taking part came from the CEGB, CESI, EdF, and from the universities of Braunschweig, Munich, Padua, and Stuttgart. It was the first stage in the formation of a group which will draw together the resources and expertise of electricity supply industries and universities to collaborate on a program of high-voltage

---

* The organization participating were CEGB (United Kingdom), CESI (Italy), EdF (France), Universities of Braunschweig, München and Stuttgart (Germany), and Padova (Italy).

research. The project was stimulated by the current interest in ultrahigh-voltage transmission systems, with voltages as high as 1.5 MV. The object of the group is to study the basic physics of breakdown and its engineering applications in these systems under the conditions which are likely to occur in practice.

The 50% breakdown voltage for a switching impulse on a 5-m rod–plane gap is about 1.350 kV and on 10 m, 2.000 kV. When the time to crest is varied in the range 130–500 μsec, the 50% breakdown voltage exhibits a U-curve with a maximum dip of about 13.5%, the effect being most marked for the cone tip and least for the hemisphere. In these tests, breakdown occurs after the crest of the impulse voltage if the time to crest is less than the corresponding to the minimum in the U-curve, and occurs before it if the time to crest is longer. With the cone tip, the first corona is separated from the subsequent prebreakdown events by a dark period lasting from 6 to 30 μsec. Whereas with the other tips, the first corona is followed immediately by leader development. The corona inception voltage has a large variation, from about 300 kV with a cone to 1600 kV with a hemisphere.

In the 10 m gap, the maximum charge which can be injected without producing breakdown ranges from ~100 μC for the 500 μsec wavefront with a cone to 200 μC for the 130 μsec wavefront on the hemisphere. The general trend is to increase from cone to hemisphere, with the hyperboloid in between, and to increase, with shorter wavefronts. The velocity of the leader tip in the 10 m gap is 1–5 cm/μsec but does not appear to be simply related to breakdown. Figures B11a-1 and 2 display some of the more important results based on thousands of image converter and photomultiplier measurements.

While this group operated mainly with positive high tension gaps, Levitov et al. [1594] describe the peculiarities of discharge developed in long gaps with negative voltage polarity. From the experimental results, the probabilities of gap breakdown depending on voltage amplitude and standard deviations were found. Table B11a-1 gives the values of discharge voltage, U_{br}, with 100 and 50% breakdown probability and standard deviations, σ, for the latter case.

Predischarge time duration of the gap fell into the range 7–75 μsec, 15–110 μsec, and 20–220 μsec, respectively, for impulses having fronts of 1.5, 20, and 125 μsec and decreased with increase in discharge probability. Other contributions on this field are presented by Mosch et al. [1595] about the breakdown tests on inhomogeneous air gaps with switching surges superimposed on dc voltage. The arrangement of rod-to-plane ($s = 0.2 \cdots 2m$) has been investigated with dc voltage (up to 300 kV) superimposed with switching surges of the 60/1200 and 240/1200 μsec type (up to 1.5 MV).

With unipolar voltage combinations the breakdown develops similarly to that of single switching surges. With bipolar voltage combinations (positive

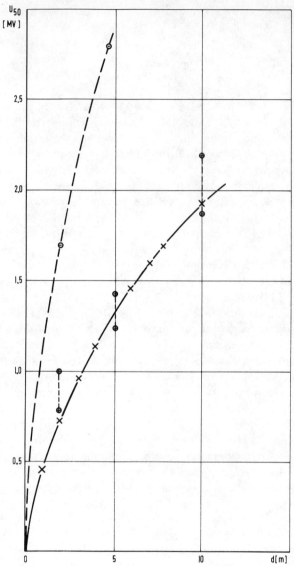

Fig. B11a-1. 50% breakdown voltages as a function of gap length.
⊕ - - - ⊕ positive polarity, range observed with different electrodes and impulse voltages; ⊖ negative polarity; × values from 2005.

dc voltage superimposed on negative switching surges or negative dc voltage superimposed on positive switching surges), the space charges, arising from the preexisting dc voltage stress, can retard or accelerate the formation of the

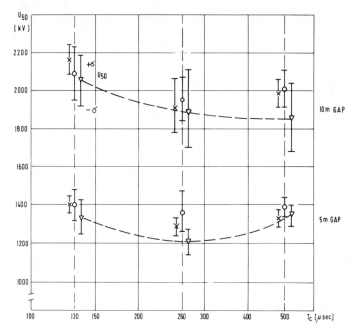

FIG. B11a-2. 50% breakdown voltages U_{50} and their standard deviation as a function of time to crest T_c.

predischarges. In the investigated gap range for unipolar as well as bipolar voltage combinations, the breakdown voltage lies between or above the values occurring when using only one of the two voltage types.

An important paper by Krasser [1592] discusses the behavior of insulators during flashovers in air on outdoor bushings and how to control these by means of potential grading. Final measurements on a 245-kV capacitor type bushing, using continuously variable parameters, indicate that they influence the voltage for 50 cycles in a manner similar to the impulse tests. Based on the

TABLE B11a-1

VALUES OF DISCHARGE VOLTAGE WITH BREAKDOWN PROBABILITIES AND STANDARD DEVIATIONS

Front/tails (μsec)	U_{br}: 100% (kV)	U_{br}: 50% (kV)	σ (%)
1.5/3500	3630	3270	5.2
20/3500	3140	2540	9.0
125/3500	2800	2640	4.2

most favorable combination of all these factors, which must evidently be adapted to each particular case, it is possible to determine the best dimensions of such insulating devices.

Hepworth *et al.* [1590] developed a quantitative theory of breakdown between concentric spheres in atmospheric air in which it is assumed that a conducting corona cloud produced at the inner sphere expands until electron avalanche growth is only just possible at its surface. Breakdown of the gap occurs if the corona sphere radius exceeds a fixed fraction of that of the outer sphere. This fraction differs for positive and negative impulses. Secondary emission processes are included where necessary.

The theory applies to rise times, such that equilibrium is established between the applied impulse voltage and avalanche growth at the corona cloud surface (1 μsec for negative impulses and 100 μsec for positive). It compares well with experiment for small concentric (hemi-) spheres where data are available. It is extended to rod–plane and sphere–plane gaps, where good agreement with the experiment is also found. The enhanced strength of sphere gaps is successfully explained and the theory extends naturally to the case of uniform field breakdown.

The effect of impulse voltage wavetail duration on the prebreakdown phenomena in long positive point–plane gaps is the subject of a paper by Aked and McAllister [1586]. One of the many factors which affect the impulse breakdown of long positive point–plane gaps in air is the form of the applied impulse voltage. In a previous paper [1807], some of the effects of variation of the impulse wavefront duration were considered. The effects of variation of the impulse voltage wavetail duration on the breakdown of gaps of 50 and 175 cm are now reported. The impulse voltage waveforms used has fronts of 1.7 μsec and wavetails ranging from 16 to 1950 μsec. The results show that, when the wavetail is short and the voltage is close to the 50% breakdown level, it is possible to have leader development, without subsequent breakdown of the gap. The wavetail is considered to be short, if it is less than four times the median time to breakdown when applying short-fronted long-tailed impulses. This leader development is characterized by the existence of a current pulse subsequent to the initial corona current pulse and also by the formation of a bright channel in the discharge. It has been shown that the maximum wavetail which will produce this leader development is dependent on the gap length, as is the median time to breakdown. For a 175 cm gap, the duration and the amplitude of the current pulse associated with the leader development can vary between very wide limits, depending upon the peak value of the applied voltage and the wavetail duration.

The physics of such streamers and leaders in long spark gaps with impulse voltage is described by Böcker and Fischer [1587]. In order to explain the development of the breakdown of long spark gaps, the model of the channel

11. OTHER UNUSUAL SWITCHING GAPS

discharge of Raether can be used for the small, single parts. The transition between the different steps of impulse corona and leader development is most important. When the velocities of both events are combined with the increasing field strength given by the voltage steepness in a certain way, this leads to most favorable conditions for their succeeding preparation. This can be seen by a minimum of charge and a minimum of the breakdown voltage for a certain voltage steepness. The optical observation of the dimensions of the different predischarge phenomena depends very strongly on the equipment used for this purpose.

The influence of the shape of the testing waveform on breakdown of long gaps is described by Jones and Whittington [1591]. Comparative breakdown tests have been made on a 0.8-m rod–plane air gap using conventional double exponential switching impulses and sinusoidal half-waves, which more closely approximate to the switching overvoltage occurring on power systems.

Novel techniques for the measurement of prebreakdown currents and time to breakdown have been developed. Digital calculations have been made of the shape of testing waves. The prospective peak voltages for breakdown and the instantaneous voltage at which breakdown occurs are shown to be influenced by quite small differences in the shape of the test wave. The paper also indicates how the measurement of prebreakdown current can assist in the interpretation of test results.

Multimegavolt switching problems can be eased by the use of SF_6 gas. Boeck and Troger [1912] describe a SF_6-insulated metal-clad switchgear for ultrahigh voltages (UHV). The heavy gas SF_6 has been employed as an arc-extinguishing medium since 1959 and for the insulation of switchgear since 1965 [1913–1915]. This relatively new practice would appear to be successful even at the very high service voltage of 1300 kV. This is proved with the aid of drafts for SF_6-insulated and conventional switchgear for 1300 kV.

Excellent experience has been gained so far with ~320 bays in the voltage range up to 145 kV, and ~60 bays from 170 to 362 kV [1916]. The problems with voltages of up to 500 kV can be mastered easily today, but considerable research and development are still necessary for the construction of metal-clad switchgear for higher voltage.

The increasing application of pure SF_6 gas or SF_6–air mixtures has been discussed in a paper by Takuma *et al.* [1604]. For the tests, SF_6 gas were taken repeatedly from a reserve tank. Analysis by use of the mass spectrometer has shown the absence of distintegrated products of SF_6 gas, but the presence of 0.47–0.48% air (by volume). The SF_6 gas pressure tested is 0, 2, 4, and 6 $kg/cm^2 \cdot g$ (rod-to-plane), and 0, 2, and 4 $kg/cm^2 \cdot g$ (sphere-to-plane). The test was made twice in the same experimental conditions (except at the maximum pressure) for the rise and fall of the pressure. The experiments were done in a pressure tank up to 1.3 MV and the waveform of applied impulses was $1\frac{9}{49}$

μsec. Time-resolved photographs of the discharge development have been taken in SF_6 gas at very large gap lengths (30 and 50) with an image-converter camera.

The discharge channels in SF_6 consist of a lot of steps (discharge repetition along the spatially identical paths) having luminous points at the tip and they move downward (toward the plane) with the discrete extension of each step. The pause time between steps decreases approximately in inverse proportion

TABLE B11a-2

COMPARISON OF REAL AND EFFECTIVE LEADER LENGTHS FOR BREAKDOWNS AND WITHSTANDS[a]

Gap (m)	Tip	Wave shape	Breakdown				Withstand				
			l_f (m)	l_{fz} (m)	$\bar{\psi}_f$	σ (%)	l_w	l_{wz}	$\bar{\psi}_w$	σ (%)	l_{wzM} (m)
5	Cone	130/3300	4.0	3.6	0.90	—	1.88	1.36	0.72	14	1.96
		260/3400	—	—	—	10	—	—	—	—	—
		500/3800	4.0	3.6	0.90	—	—	—	—	—	—
5	Hemisphere	130/3300	—	—	—	—	—	—	—	—	—
		260/3400	—	—	—	10	1.1	0.65	0.59	16	0.76
		500/3800	—	—	—	—	1.58	1.00	0.59	16	1.80
10	Cone	130/3300	8.9	7.5	—	—	3.85	2.00	0.52	23	2.70
		260/3400	8.6	7.2	0.84	10	6.37	3.20	0.50	14	3.80
		500/3800	8.5	7.1	—	—	—	—	—	—	—
10	Hemisphere	130/3300	8.0	6.7	—	—	4.15	2.80	0.67	25	4.30
		260/3400	8.4	7.0	0.84	10	3.83	2.60	0.68	13	3.76
		500/3800	8.7	7.3	—	—	4.25	2.70	0.63	24	4.40
10	Hyperboloid	130/3300	8.2	6.9	—	—	4.08	2.18	0.57	60	2.83
		260/3400	7.6	6.4	0.84	10	4.68	2.86	0.62	34	4.72
		500/3800	8.5	7.1	—	—	3.80	2.68	0.70	8	4.09

[a] List of symbols: l_f = leader length at the beginning of the final jump; l_{fz} = projection of l_f on axis z; l_w = maximum leader length when breakdown does not occur; l_{wz} = projection of l_w on axis z; $\Psi = l_z/l_f$; $\Psi_f = \bar{\psi}$ at the beginning of the final jump $= l_{fz}/l_f$; $\Psi_w = l_{wz}/l_w$.

with gas pressure. Extremely distinct steps have been photographed at near atmospheric pressure. These photographs show that the behavior of the discharge extension and the sparkover in SF_6 is quite different from that of long gaps in air, rather, it has a strong resemblance to that of a lightning discharge.

For electrical measurements of megavolt pulse behavior, a reliable multimegavolt voltage divider is needed. One has been developed by Pellinen and

Smith [1808] and is fast enough to display events down to 1.5 µsec risetime.*
The monitor is able to handle a pulse energy up to 40 kJ.

Regarding measurements of fast varying pulse voltage, see also [1311–1324]. The European Research Group "Les Renardières" has also published a paper of fundamental value [1702], a collection of highest multimegavolt spark gap measurements. It includes the recent international standards and symbols, the breakdown parameters, details of the experimental circuit, calculation of electric field strength, techniques for optical measurements in the gap, measurements at the high-voltage electrode, measurements at the plane, measurements of the circuit response, and test procedure. One set of results is shown in Table B11a-2.

b. *Multichannel and Nanosecond Gaps*

Multichannel operation of gaps has the purpose of lowering the unavoidable geometrically defined inductance of the switching spark channel itself by using a number of sparks in parallel circuit. They have to operate synchronously to fractions of a nanosecond. Martin [1028, 1029] presented a review of such switches. Transit time isolated single channel gaps have been operated in parallel for several years, but these do not come within the rather strict definition of multichannel operation used in this paper, i.e., operation when all the electrodes are continuous sheets of conductor. While transit time isolation may play a small part, the main effect is that before the voltage across the first channel can fall very much, a number of other channels have closed. Hence, it is the inductive and resistive phases which are important in allowing a very brief interval in which multichannel operation occurs. This time ΔT is given by

$$\Delta T = 0.1\tau_{\text{tot}} + 0.8\tau_{\text{trans}}$$

τ_{tot} is as defined in the previous section and τ_{trans} is the distance between channels divided by the local velocity of light. Both of these terms are functions of n, the number of channels carrying currents comparable to the first one to form. ΔT is then placed equal to 2δ, where δ is the standard deviation of the time of closure of the gap on a rising trigger pulse. This is related to the standard deviation of voltage breakdown of the gap by the relation.

$$\delta(t) = \sigma(V)V(dV/dt)^{-1}$$

where dV/dt is evaluated at the point on the rising trigger waveform at which the gap fires.

Figure B11b-1 gives the jitter of a negative point or edge–plane gap as a function of the effective rise time of the trigger pulse. For ordinary gaps (both

---

* See High Speed Pulse Technology, 1965, Volume II, Chapter M5.

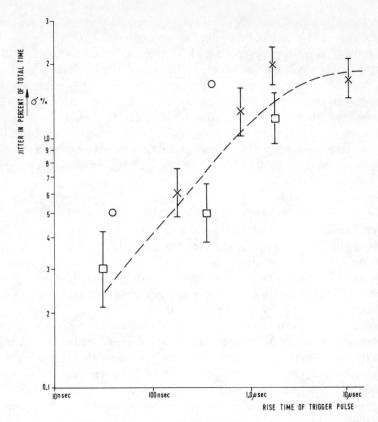

FIG. B11b-1. Rise time of trigger pulse, jitter of pulse-charged point and edge–plane gaps.

gaseous and liquid) a good jitter is 2% or so, but for edge–plane gaps charged quickly, the jitter can be down to a few tenths of 1%. Typically, ΔT is a fraction of a nanosecond whereas ordinary gaps require trigger pulses rising in 10 nsec or so. However, for edge–plane gaps, the trigger pulse can rise in 100 nsec or more and still give multichannel operation. In one experiment 140 channels were obtained from a continuous edge–plane gap. The expression has also been checked for liquid gap operation and more approximately for solid gaps, and gives answers in agreement to some 20% for ΔT. The way it is used is to define a rate of rise of trigger pulse and for such a purpose this sort of accuracy is ample. Fast trigger pulse generators are required to obtain multichannel operation and these must have impedance of 100 Ω or less. Of course the pulse length produced by the trigger generator need not be very great and the generator can be physically quite small. James [1809] has done much work on high speed systems and gaps in general and his paper gives an example of a high-current, low-inductance multichannel gap he has developed.

Multichannel spark gaps would cause no difficulty if all the gaps would have exactly the same breakdown voltage. While the inaccuracy of these values in air and other gases are well known, it was hoped to find suitable insulating liquids with higher accuracies of the breakdown voltage. Martin [1028] measured these values in carbon tetrachloride. It is necessary not only to keep the spacing of the gap constant but also to have an adequately sharp edge, so that the rather large jitter in voltage at which the streamers start out will have a negligible effect on the final jitter at closure time. This effect also applies to gas edge–plane gaps, but here the radius of the edge can be 10 mils or more and is easily obtainable. However, with liquids, working at only 400 kV or so, the edge needs to have a radius of 2 mils or less. In the jitter experiments, as indeed with all the liquid point- and edge–plane experiments, the points and edges were frequently sharpened, to ensure that the streamers started out at a low voltage. Table B11b-1 gives the breakdown voltages and jitters for an effective pulse time of 80 nsec.

As can be seen, the jitter is approximately 1% and the edge–plane holds off some 18% more volts than the point–plane. Even though the edge was only some 3 cm long, multichannel breakdowns were frequently observed. The streamer channels were not straight but were mildly zigzagged in reasonable visual agreement with the jitter of 1%. The measurements with the multichannel gap, using slightly longer pulse, also gave jitters of about 1%.

In the general field of high speed pulse generators, the breakdown strengths of gases, liquids, and solids in uniform and nonuniform fields can be estimated well enough for design purposes, using approximate relations and measurements obtained in the past few years. This suffices to do a first design for the high speed section, and the feed from this section to the load. If a single channel

TABLE B11b-1

EXPERIMENTAL RESULTS GIVING BREAKDOWN VOLTAGES AND JITTERS[a]

Experiment No.	Gap type	Voltage (kV)	Jitter (%)
1	Point–plane	428	±1.3
3		420	±1.1
2	Edge–plane	504	±1.4
4		492	±1.1

[a] Material: carbon tetrachloride; gap = 2.05 cm; t_{eff} = 80 nsec. Mean point–plane breakdown voltage = 424 kV; mean edge–plane breakdown voltage = 498 kV; ratio of point-to-edge 1:1.18.

gap is used, approximate treatments exist to enable the rise time of the pulse to be obtained to a rough but adequate accuracy. However, if this rise time is inadequate, the position becomes more complicated. Single gaps, transit time separated, can be employed in some design, but then the resulting combined pulse must have a rise time comparable with the jitter of the individual gaps. Such an approach is usually expensive and where low impedances and high voltages are required, it may be physically impossible to realise.

Several years ago, when the stabbed solid dielectric switch was developed, multichannel operation of a number of closely spaced spark channels was readily obtained. The corresponding development of liquid and gaseous multichannel gaps has only occurred in the last couple of years. The reason why multichannel operation of solid switches was obtained so easily is a combination of the relatively long resistive phase of the spark channels and the fact that trigger pulses of several hundred kilovolts rising in a few nanoseconds were easily generated by using a solid master gap.

In contrast to an array of single gaps, where transit time isolation is used, multichannel gaps have continuous electrodes. Transit time isolation may play some part in enabling a number of spark channels to take comparable currents, but this is usually of minor importance and is only included in the analysis for completion. The continuous electrodes of the gap may be serrated or have localized raised areas on them, but this is used to stabilize the number of channels formed at a number lower than that which would have been obtained with continuous electrodes.

There are several reasons why multichannel gaps may be useful, but in general they fall into two classes. First, a lower inductance is obtainable and to a lesser extent a reduced resistive phase. Thus faster rise time pulses may be generated and in high-voltage, low-impedance systems, the rise time can be made to decrease an order of magnitude. Second, electrode erosion may be very serious in a single channel gap and this is reduced drastically where the current is carried in a reasonable number of channels. A second effect of using large continuous electrodes is that such erosion as does take place causes little change in the geometry of the gap and, in addition, magnetic forces and high-current density blow up are considerably reduced.

In a multichannel gas gap as the first closing channel starts to take substantial current, the voltage on the electrodes begins to collapse. Other channels closing after the first will have less volts across them and hence will finish up carrying significantly less current when the gap voltage has fallen to a low value. After this phase is over, the currents carried by the various channels will redistribute themselves over a very much longer time scale. The current eventually finishes up flowing in one channel, after a time typically three or more orders of magnitude longer than in the first phase. The analysis given here is restricted in general to the first phase, when the gap voltage falls to a low value, although some data on the long-term current rearrangement are given towards the end of this too.

Hauer [1701] realized the construction of heavy duty EMP generators at IPC* using multichannel operation. The technique developed considerably improves the risetime capability of the switch by reducing the over-all switch inductance and improving the energy flow across it.

Figure B11b-2 illustrates the principle involved. A common trigger pulse generator is coupled symmetrically to two separate trigger pins (Trigatron mode). The criterion for obtaining multiple channels simultaneously across a switch gap is that the difference in formative times between adjacent channels shall be less than the sum of the time of the resistive phase and EMP propagation time (spacing) between channels.

Switching Systems

The duty of the switch is to isolate the energy store from the load during the charging or energy storing period and then, at the required time, close and discharge this energy to the load. The ideal switch will perform these functions without interposing an impedance discontinuity between the store and load.

A most important factor of the EMP pulse is the pulse rise time. This is typically specified to be 10 nsec. If one assumes correct geometries for the energy store, transmission system to the load and the load itself, then impedance characteristics of the switch during the first 10 nsec, or so, largely determine whether this rise time specification can be met. This switch effects may be separated as spark channel formative times and spark impedance discontinuity between store and load.

The resistive phase is the initial period of dielectric breakdown when the closure mechanism is initiated and the switch impedance collapses from a few times Z to the reciprocal of this value.

For a gaseous medium the resistive phase is given in nanoseconds by

$$t_{res} = [88/(Z+R_L)^{1/3}(E)^{4/3}](\rho/\rho_0)^{1/2}$$

where Z is the impedance of energy store (driving impedance), R_L the load impedance, E the electric field across the spark gap in MV/m, ρ the gas medium density, and ρ_0 the gas density of air at NTP. For typical operating values:

$(Z+R_L) \simeq 150 \, \Omega$
$E \simeq 30 \, \text{MV/m}$
$\rho/\rho_0 \simeq 20$ (300 psi gas pressure in switch)

It can be seen that this phase has a duration of nearly 1 nsec. The "inductive" phase follows the resistive phase as the spark channel diameter grows and the

---

* Ion Physics Corp., Inc., Burlington, Massachusetts.

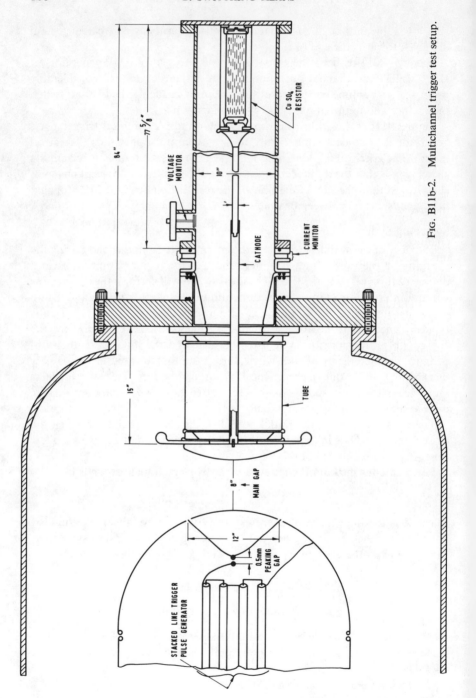

Fig. B11b-2. Multichannel trigger test setup.

11. OTHER UNUSUAL SWITCHING GAPS

switch impedance falls to the "fully closed" value. The effective duration of this phase is given by

$$t_{ind} = 2.2[Ld/(Z+R_L)]$$

where d is the gap length (electrode separation) and L the channel inductance per unit length. For a typical L of 15 nH/cm, $d = 10$ cm and $(Z+R_L) \geqslant 150\,\Omega$, the inductive phase has a duration of 1 nsec. The projected system rise time can be calculated from

$$t_{10-90\%} = 2.2(t_{res}^2 + t_{ind}^2)^{1/2}$$

which leads to a rise time around 3 nsec due to switch discharge characteristics. Estimation of system rise time due to the switch, as above, is useful but ignores the geometrical effects of the switch. In megavolt technology the switch electrodes are relatively large, maybe several feet in diameter. A switch channel initiated on the axis of such an electrode structure presents a transmission line discontinuity which is usually analyzed by means of low-voltage modeling techniques. The interactions of the switch geometry with the rest of the structure can then be optimized.

In the IPC machines described, all the switches are of trigatron type. Those which require to be accurately timed and are electrically triggered have a very small scatter or "jitter" in their closing delay times. Figure B11b-3 demonstrates this performance for a typical machine.

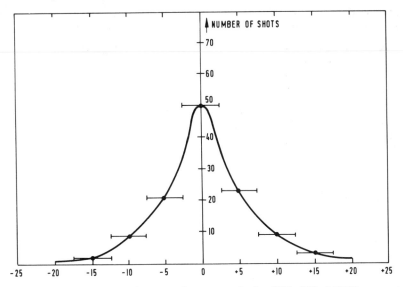

FIG. B11b-3. Electronic trigger performance of the IPC FX 1-EMP generator ($N_t = 120$ pulses, $V_0 = 3.6$ MV, the abszissa indicates the $+$ or $-$ nanosecond jitter, 0-time corresponds to the highest probability.

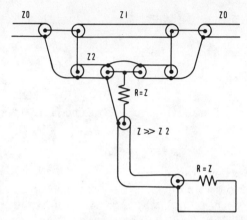

Fig. B11b-4. Diagram equivalent to high-voltage attenuators.

A high-tension generator and attenuator of pulses of 30 kV in the nanosecond and subnanosecond range is also described by Blanchet [1021]. The electronic control of biplanar image-converter tubes and electrooptical shutters (Kerr or Pockels effect) uses ultrashort electrical pulse generators and high-tension broadband attenuators. Concepts are viewed which make it possible to generate and measure the high voltage and subnanosecond impulses necessary for high speed photography and cinematography. A generator is described that produced 10 to 30 kV impulses, lasting less than 1 nsec and having a rise time of ~ 100 nsec with a delay of ~ 50 nsec and jitter below 1 nsec. Different types of attenuators are presented, specially designed for high speed radiography with exposure times of less than 1 nsec. Their working range extends from several kilovolts to several megavolts and from 10 to 10^4 A (see Figs. B11b-4 and B11b-5). Also, the conventional Marx generator can be used for nanosecond pulse generation by extreme reduction of all components.

Anderson et al. [1569] describe such a low inductance, compact, modularized Marx generator. It can be precision-switched over a 3:1 voltage range with a jitter of a few nanoseconds. The generator is built by stacking modules,

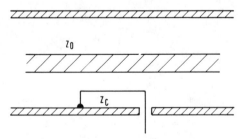

Fig. B11b-5. Principle of high-current attenuators.

each module $3\frac{1}{4}$ in. high, and containing two 50 kV Marx stages. The modules are connected in series using plug and socket connections to carry all services from stage to stage. Thus, generators having a wide range of output characteristics can be built by using different numbers of the series stacking units and adjusting the stored energy per stage. This allows a designer great flexibility in designing a generator for specific applications.

The Marx generator module designed for precision applications has been built and tested. This generator has an inductance of 1.5 μH/MV and an erection speed of 2 nsec/stage. Low-erection jitter, $\sim 1\%$, has been demonstrated using a simple resistive-coupled trigger system first developed by J. C. Martin (private communication). A unique modularized tray system used in the construction of the basic Marx generator provides unusual flexibility. Low inductance, three-electrode spark gaps designed for compact Marx generator applications are utilized. Trays are simply stacked to produce a Marx generator with 2.5 MV output. It has a peak output of 40 kA and an energy storage capacity of 25 kJ. These components and trays are housed in a graded fiberglass envelope pressurized with an electronegative gas to achieve a compact, low-inductance configuration. Higher voltages can be achieved by stacking the basic 2.5 MV modules.

A supplementary paper from the same working group is presented by Kolb *et al.* [1574] about high-voltage output switches for fast Marx generator pulse discharge systems.

Some unusual methods of switching spark gaps are reported in a volume by Rompe and Steenbeck [1163a]. A commercially available family of demountable three-electrode spark gaps in which pressure, gas, and electrode configuration can all be varied are the duratrons made by Impulsphysik GmbH, Hamburg [1964]. They cover a range of switched voltage of 5–50 kV and peak currents to 100 kA. The largest type handles, with forced air fan cooling, 4000 W average power. Because of the interchangeability of all components they have a practically unlimited life. Anode and cathode are of 15% thoriated tungsten and the ignition electrode is made of nickel, which has good resistance against the aggressive rare gas ions. (Ar^+ ions are as active chemically as chlorine.)

Deutschmann [1007] makes a low-inductance triggered discharge switch. Multichannel triggered discharge switches which are strongly constructed to withstand the high peak currents and severe impulsive forces produced during the discharge of high-energy capacitors. Planar electrodes having high thermal and electrical conductivity are incorporated in the switch design to minimize inductance and erosion.

Four triggered spark channels evenly placed along the electrode assure uniform current distribution. The inner surfaces of the insulated switch housing are ablated during discharge, resulting in a self-cleaning switch. The switch is easily disassembled for maintenance to provide extended life. All parts are removable and replaceable. The high-power trigger circuit requires a 10-kV low-current source. This may be obtained from the storage capacitor charging supply through a suitable dropping resistor. A delay or lag time of 50 nsec from the 250-V input trigger signal to main gap breakdown is typical. Total system jitter is less than 5 nsec.

When using nanosecond spark gaps as switches for capacitor discharges, it is often necessary to know more about the ion and electron mechanisms of the development of the conductive path of the spark channels. Fischer *et al.* [1221] give a survey on that theme in their paper.

C. Line conductors

1. The Influence of Impedance and Its Transformation by Connecting Lines

Transmission line dimensioning is one of the main problems of EMP generators which have to produce voltage pulses with extremely short rise times. During recent years there has been an increasing need to test equipments under strong EMP field conditions. In nature, the field close to the sharp second stroke of lightning flashes often has a rate of rise of 1 MV/10 nsec, while fast-rising fields are generated artificially by nuclear explosions. In civil defense, EMP megavolt generators are therefore needed to test vehicles, bunkers, and communications equipment. Generally, the transition of the pulse energy between the generator itself and the linear overhead antenna presents a problem. This antenna is often known as a transmission line because the produced pulse passes this expanded antenna-like line till it disappears in the Ohmic resistor. The transmission line is used to produce the EMP field in which the equipment is tested. The energy dissipation load has the character of a matching load to avoid all kinds of standing waves (see especially Fig. C1-10). Due to the very high voltages of these nanosecond pulses, special connecting members are to provide so-called wave launchers. IPC Burlington, Massachusetts is one of the leading groups working in this field. The following are extracts from Ref. [1701].

The EMP test volume is contained between the upper and lower electrodes of a parallel plate (or strip) transmission line. The lower electrode is normally the ground plane, which for practical purposes, is considered infinite. The dimensions of the transmission line vary according to the object of the test. The height, or spacing, can range from a few meters to tens of meters; the width and length can be tens of meters. The characteristic impedance for varying of height to width is shown in Fig. C1-1.

The output bushing of a typical pulse generator which is intended to supply the EMP to such a transmission line is normally coaxial with an outer diameter of one meter or less. A geometric transformation must be made between the coaxial pulse generator output and the parallel plate transmission line such that (1) minimum impedance discontinuities occur which can significantly degrade the pulse quality; and (2) a safe insulating condition is maintained throughout the transformation.

For the second condition, the stripline is tapered down as it approaches the

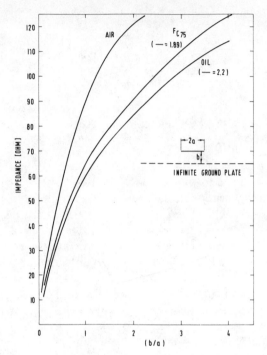

FIG. C1-1. Impedance curve for two-plate stripline [1701].

generator, keeping the height-to-width ratio, or impedance, constant. The rate of this taper is controlled such that the distance to the apex is of the same order as the line width. As the taper is continued, the height above ground is decreased to the point where it is just adequate to support the peak pulse voltage, and beyond which air breakdown would take place. At this point, a dielectric which is superior to air must be used. This can be SF_6 or one of the Freons®. Therefore, under these conditions, an insulating enclosure must be provided for the new medium and the criterion is the insulation flashover across its face in the air.

Normal design ratings across surfaces in air lie in the range 20–30 kV/cm under EMP conditions. Under these modified dielectric conditions, the tapering can continue until the dielectric strength again becomes inadequate. If this happens before the pulse output bushing is reached, the medium can be modified to pressurized electronegative gas or liquid (oil). At any interface between a liquid dielectric and SF_6 or Freon® (1 atm), a peak stress in the range 50–70 kV/cm can be obtained. Normally, two medium changes are sufficient to satisfy the electrical stress conditions at the output of the pulse generator. This stress may be as high as 500 kV/cm. A typical assembly is shown in Fig. C1-2.

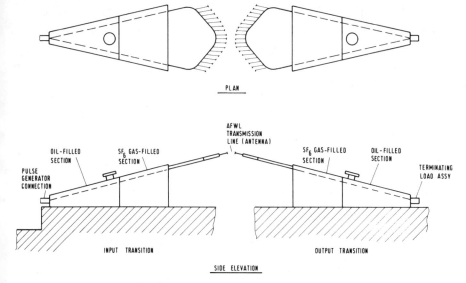

Fig. C1-2. Transition assembly, EMP-10-generator, IPC [1701].

Maintaining a close tolerance of impedance match during the physical transformation from pulser to main transmission line is a major design aim. Two impedance matching problem areas exist: tranformation from coaxial to stripline geometry; and transmission across interface between dielectric media of differing dielectric constant. The latter is by far the more significant and will be discussed further.

For a transmission which is constructed of an oil dielectric section at the pulser output, the EM wave must be propagated through an interface between media of dielectric constant ~ 2.2 and 1. Figure C1-1 shows that for a given ratio of height–width of a strip transmission line, the impedance is proportional to $1/\varepsilon^{1/2}$. (ε is the permittivity and ε_0 is the value for vacuum.) It is clear that for a specific system impedance, this ratio must be different in the oil from that in the gas. At the interface, the two geometries must blend to give the minimum impedance mismatch.

a. *Normal Incidence*

For an interface that is normal to the direction of electromagnetic energy flow, there are a number of possible approaches.

(1) Constant taper, discontinuous impedance at interface, using one stripline (Fig. C1a-1).

(2) Constant impedance, discontinuous width stripline at interface (Fig. C1a-2).

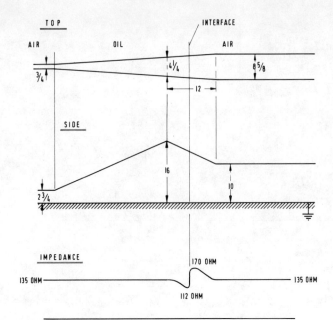

Fig. C1a-1. Constant taper-discontinuous impedance dielectric transition.

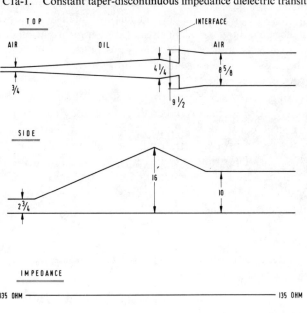

Fig. C1a-2. Constant impedance dielectric transition.

1. THE INFLUENCE OF IMPEDANCE AND ITS TRANSFORMATION

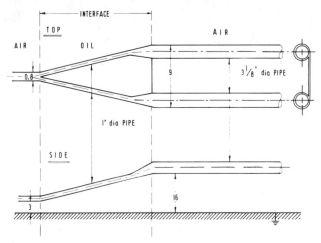

FIG. C1a-3. Constant width–constant impedance dielectric transition.

(3) Constant impedance, two parallel striplines in oil for improved geometric continuity (Fig. C1a-3).

Since each method of handling the interface problems will satisfy transmission line requirements, the choice of one method over another is determined by the combination of electrical stress and structural requirements, such as wind and weather resistance for the particular system being designed. For large interfaces, the response time across the interface is in the order of 2 nsec.

b. *Inclined Incidence*

One should be aware of the angled incidence technique at the interface although it is not used in the EMP systems under discussion. It is preferable to utilize Brewster angle techniques to maintain a constant impedance through the interface between the oil and gas, and minimize the electric gradient at the face (Fig. C1-6). The Brewster angle is given by

$$\tan \theta = x/s = 1/\varepsilon^{1/2}$$

For transformer oil into gas, $\theta = 34°$. The electric field continues normally through the system as shown, and the power flow is the same in both media. Figure C1b-1 is an example of the situation for an infinitely wide stripline (i.e., a pure TEM wave), but adapted for the line taper. Unfortunately, the system is not wide and edge effects are strong. Figure C1-1 shows that for the spacings on Fig. C1b-1, the widths at A and B, to maintain the same impedance (120 Ω), are respectively, 12 and 21.6 in. This requires a change in the width of the top plate to occur where energy is flowing through the interface (Fig. C1b-1). For the 120-Ω case, the total taper angle on the plate to accommodate the width change through the Brewster interface is 27°, which is not very different from the width taper throughout the transition (~23°), so that there is no great discontinuity in energy flow for this reason. For the 150-Ω case, the corresponding angles are 21° and 14°.

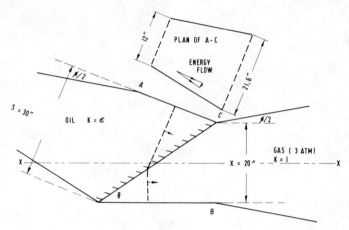

FIG. C1b-1. 5 MV Brewster transition, oil—3 atm of SF_6 (or mirror image about XX).

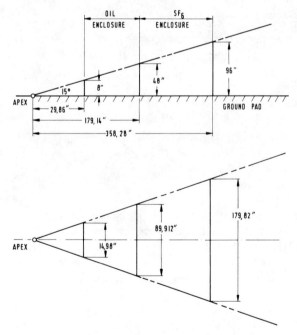

FIG. C1b-2. EMP-28 transition dimensions.

1. THE INFLUENCE OF IMPEDANCE AND ITS TRANSFORMATION

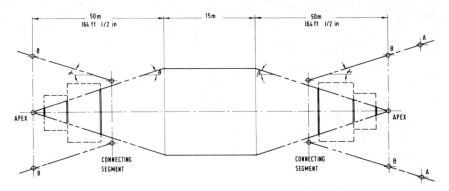

Fig. C1b-3. EMP-28 simulator layout tension cable locations.

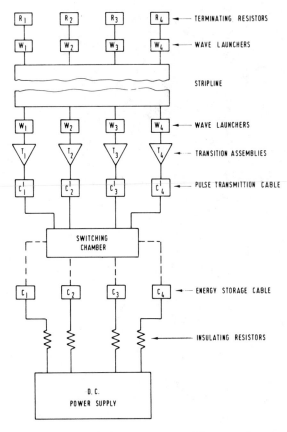

Fig. C1b-4. Block diagram of Siege 1.3 system. The wave launchers are specially designed constructions providing the required flashover protection at megavolts.

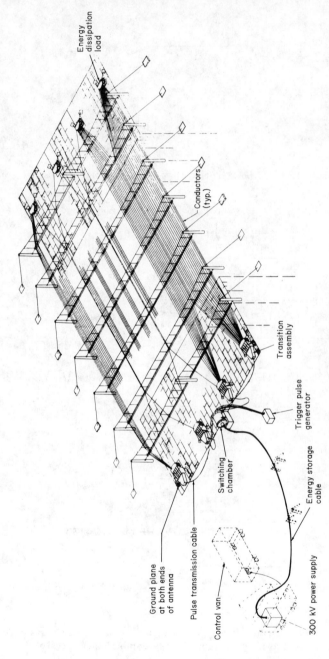

Fig. C1b-5. Siege I, Phase I system layout (not to scale).

1. THE INFLUENCE OF IMPEDANCE AND ITS TRANSFORMATION

There must be a component of energy flow across the system, as shown in Fig. C1b-1 which requires the Brewster face to be curved. This situation is further complicated because the greater part of the energy necessarily flows in the "edges" of the system. A suitable interface shape to handle the curving field lines, while presumably derivable from field analysis, would be difficult to construct.

The geometry of the transition and stripline system of a large IPC EMP generator is shown in Fig. C1b-2. This information is necessary to initiate the integration of the pulse generator into the total system. Determination of the apex is a form of benchmark for the facility setting as it does the transition lengths and the elevation of the pulse generator. Figure C1b-3 shows a plan view of the simulator layout with tentative transition position governed by the tensioning restraints of the arrangement.

A still larger assembly of an EMP transmission system has been realized in the so-called Siege-Installation (IPC). Figure C1b-4 shows the circuit and Fig. C1b-5, an artist's sketch of the layout of antennas together with ground plane and the connection cables.

c. *Impedance*

The first step in calculating the proper geometry is to refer to the graph of Fig. C1-1, which shows the impedance of a two-plate stripline plotted against the ratio of (b/a). In this graph b is the distance of separation between the conducting plane and ground, and a is the half-width of this plane. For a 95-Ω line we obtain

$$(b/a)_{\text{air/SF}_6} = 1.1 \quad \text{and} \quad (b/a)_{\text{oil}} = 2.58$$

These two factors determine the height and width of the conducting plane within the transition assembly. The problem of voltage flashover must now be considered so that the dimensions of the two stages of the transition may be established.

d. *Impedance Mismatch Considerations*

At the oil/SF$_6$ interface in the stripline, there must necessarily be an impedance mismatch due to the fact that the width of the conducting plane must be increased to maintain the 95-Ω impedance in the SF$_6$-filled portion. For a width $2a$ in these media we have

$$(2a)_{\text{oil}} = 2b/(b/a)_{\text{oil}} = 2b/2.58 \quad \text{and} \quad (2a)_{\text{SF}_6} = 2b/(b/a)_{\text{SF}_6} = 2b/1.1$$

Thus,

$$\Delta(2a) = 2b\left(\frac{1}{1.1} - \frac{1}{2.58}\right) \frac{2.58 - 1.1}{(1.1)(2.58)} \approx b$$

This increased width which takes place at the interface coupling within the oil effectively reduces the impedance of the stripline to $\sim 75\ \Omega$ in the immediate vicinity of the oil–gas interface. The effect of this mismatch on the peak voltage of the pulse launched onto the antenna is quite small as can be seen from the following analysis. The impedances of the transition and the associated voltage magnitudes are shown in Fig. C1d-1.

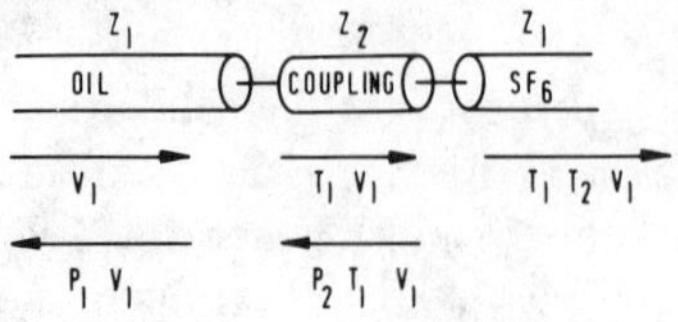

FIG. C1d-1. Transmission line model for interface mismatch.

The reflection and transmission coefficients, ρ_i and T_i are defined in terms of impedance as follows:

$$\rho_1 = (Z_2 - Z_1)/(Z_1 + Z_2), \qquad T_1 = (1 + \rho_1)$$
$$\rho_2 = -\rho_1, \qquad T_2 = (1 + \rho_2) = (1 - \rho_1)$$

and V_1 is the magnitude of pulse in oil transition.

The magnitude of the voltage transmitted to the SF_6 transition stage is

$$V_{SF_6} = T_1 T_2 V_1 = (1 - \rho_1^2) V_1$$

Since ρ_1^2 and ρ_2^2 are $\sim 10^{-2}$, the voltage pulse transmitted to the SF_6-filled section is very nearly equal to the voltage pulse in the oil-filled section. Furthermore, these small reflections indicate that a negligible amount of energy is directed back toward the pulser from the interface.

2. The Coaxial Low-Impedance Water Line and the $CuSO_4$ Output Resistor

A coaxial water-filled cable is used in IPC's 5-kJ "Neptune" electron accelerator [1710], the characteristics of which are given in Table C2-1. Because of the high dielectric constant of water, the impedance of the coaxial line is only $1.5\ \Omega$. Another advantage of using water is that repeated occasional breakdowns can occur without damaging the equipment. The data for the pulse generator using this low-impedance, high-voltage coaxial water line, are shown in Table C2-1.

A block diagram for the accelerator is shown in Fig. C2-1. The capacitors in the LC generator are charged and a command pulse to switch I array erects the generator. At the peak of the erection, switch II is triggered and the LC

2. LOW-IMPEDANCE WATER LINE AND $CuSO_4$ OUTPUT RESISTOR

TABLE C2-1

NEPTUNE C PULSE GENERATOR CHARACTERISTICS (IPC).[a,b]

Impedance	1 Ω
Pulse width 95%	40 nsec
FWHM	> 75 nsec
LC generator voltage	400–950 kV
Electron energy	100–400 keV
Beam energy	5 kJ
Fluence reproducibility	±5%
Prepulse	<1% (peak voltage)
Jitter	±5 nsec rms
Turnaround time	<15 min
Vibration	Very low
Size (over-all)	24 × 8 × 7 ft (high) (excluding control room)

[a] Ref. [1710].
[b] A water line of 1.5 Ω impedance is used in the energy transfer area.

generator transfers its charge to a pulse-forming line. The pulse-forming line is a water-insulated coaxial line whose length determines the electron beam pulse length. When this charge transfer is completed, switch III is triggered and the pulse-forming line discharges into a matching transformer which feeds the output diode.

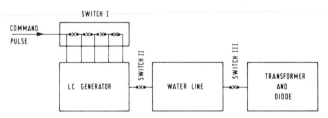

FIG. C2-1. Block diagram of Neptune system.

a. LC Generator

Figure C2a-1 shows the principle of operation of a two-stage LC generator. The capacitors, charged to a voltage V, are close packed and connected together as shown. Alternate capacitors are shunted by an inductive coil in series with a switch; when both switches close, the voltage on the two capacitors oscillates between $+V$ and $-V$. The output of the generator thus swings between 0 and 4 V. It is usual to trigger a load-connecting switch (switch II) at peak voltage. The basic difference between an LC generator and a conventional Marx generator is the lower inductance of the LC generator, since the

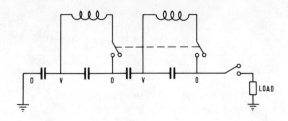

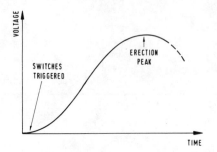

Fig. C2a-1. Principle of operation of an LC generator.

switch inductances are not a part of the generator inductance. An LC generator was chosen over a conventional Marx generator because the lower inductance allows charging of the coaxial water line faster than a high-inductance Marx generator. Since the holdoff strength of water varies approximately inversely with the third root of the charging time, the peak voltage that can be obtained on the water line is higher with an LC generator.

The loss in an LC generator is caused by damping resistors fitted to guard against fault conditions. This remains constant and the time and amplitude jitters are completely negligible.

The erection transients in an LC generator differ markedly from those of a Marx. In the LC generator the voltage on half the capacitors is reversed during erection. This stresses the dielectric. The effect of this is, however, lessened by using a comparatively slow swing time. The switch triggering transients are negligible. In contrast, the erection of a Marx generator superimposes a large oscillatory voltage across each capacitor. The voltage does not reverse but the peak voltage may be very much higher than the static charging voltage. These transients are extremely fast and poorly damped. In practice, reliability depends more on the detailed design than on the basic choice of generator.

Occasionally, the output of an equipment has to provide short-circuit operation at several megajoule energy. In such cases low Ohmic resistors must be provided which can never be destroyed. A practical solution is the $CuSO_4$ resistor [1029].

2. LOW-IMPEDANCE WATER LINE AND $CuSO_4$ OUTPUT RESISTOR

$CuSO_4$ resistors are of great use in nearly all high-voltage applications. They are very much cheaper and simpler to make than soldering together 500 1-W resistors. They can be flexible and arranged in elegant shapes. They can absorb vast amounts of energy, one cubic foot or so of solution quietly and safely taking the energy out of a 1 MJ bank. Therefore they act excellently as high-energy terminators for transmission lines. They have a resistive response into the several hundred megacycles region and apart from the electrode–liquid interfaces, obey a much better understood section of physics than carbon resistors. One disadvantage is that it is quite easy to measure their resistance incorrectly. This is because insulating films form on the electrodes rather like electrolytic condensers and hence, resistance measuring systems working at fractions of a volt see large capacitive components. However, these films can be easily stripped off with a direct current, or the voltage raised to the point at which the efficiency drops away. The fact that there is, say a 5-V drop at the electrode–liquid interface is irrelevant when a one million volt pulse is applied to the resistor.

In making $CuSO_4$ resistors, it is sensible to use deionized water and $CuSO_4$ only, and copper electrodes. The body can be either flexible PVC tubing or, if rigid resistors are required, Perspex. The resistance of ionic conductors is a function of temperature, and in long resistors, temperature gradients can exist, but this is an easy matter to measure and in most cases should cause no problems. However, when large energy densities are dumped into $CuSO_4$ resistors the temperature rises, the resistance changes, and in addition, a pressure wave develops because the liquid does not have time to expand. Consequently, for dump resistors which may experience rapid rise of 10° or 20°C, flexible-walled vessels should be employed.

In general, for low-energy applications, resistors, or resistor chains are quite adequate up to 50 or 100 kV but above this they become inconvenient, and for high-energy absorbing applications and/or high voltages, $CuSO_4$ resistors are greatly to be preferred.

D. Conversion of capacitor energy into current impulses

1. Direct Discharge through a Conductor

Now that dc capacitor discharge circuits together with power transistors and SCR's (thyristors) are available, many kinds of controlled or stabilized circuits for current pulsing are feasible. The following pages give only a survey of available equipment in the form of examples. Other industrial equipment of no less importance or interest has not been mentioned simply because of limitations of space.

A 5000-A, 450-V solid-state switch for pulsing an arc lamp has been developed by Alexander et al. [1263]. It switches direct current on and off for the controlled pulsing of a high-intensity light source. Using 400-A (continuous rating) silicon-controlled rectifiers, the switch has a designed maximum power rating of 5000 A for a 1-msec pulse. The circuit in Fig. D1-1 can switch direct currents for pulse periods ranging from 20 μsec to minutes with electronic termination of the pulse. In the present application of the switch, a 1.7-msec pulse with a 4700-A peak value is obtained through an Osram XBO1600 W xenon lamp from a capacitor-bank source having an initial voltage of 450 V. (Lamp deterioration is noticeable at this current level.) Greater pulse lengths are possible only at lower current levels because of power dissipation in the silicon-controlled rectifiers (SCR's) and the limited energy capacity of the source.

The light-generating equipment consists of a power source, the lamp with its power supply, and the switch. A bank of heavy-duty, 6-V storage batteries and a 0.02-F (24×850 μF) capacitor bank comprise the power source. (The capacitor bank permits a high rate of current buildup.) For times of 1 msec or less, the capacitor bank is charged by a separate, variable power supply and is used alone. A $7\frac{1}{2}$-hp, 4-pole, 220-V contactor connects the batteries to the capacitor bank. Pushbuttons serve to activate the contactor. The contactor, therefore, may be used as a safety device should the electronic switch fail to turn off after a long pulse operation.

The lamp is housed in a Zeiss light coagulator which has an internal dc power supply capable of delivering, in eight steps, 40–130 A to the lamp. When the lamp is to be pulsed, it is operated at ~ 8 A by means of a "simmer" resistor (5 Ω) in series with the internal power supply, which is set at its lowest step.

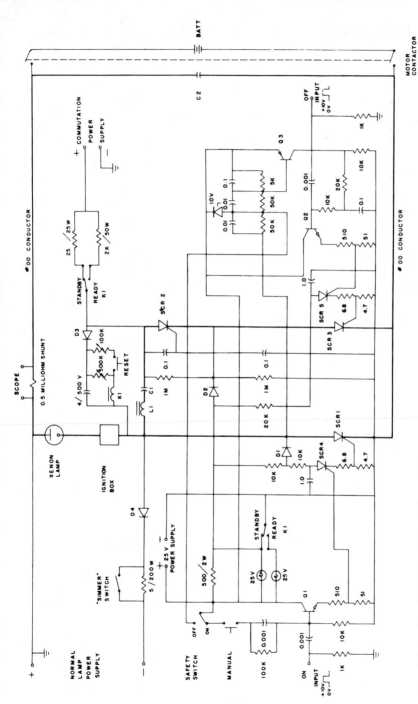

FIG. D1-1. Circuit diagram of the 5000-A, 450-V solid-state switch. Capacitance in μF and resistance in Ω, unless otherwise noted. Following are key to and data for components: D1, 2, 3, Int. rect. 1 A/1000 V; D4, In 3268; L1, ~100 μF; C1, 500 μF/450 V (nonpolarized, noninductive); C2, 20,000 μF/450 V; SCR 1, 2, 3, Westinghouse 221 K; SCR 4, 5, Motorola 1305-6; K1, Sigma 42 R 01000S-SIL; Q1, 2, 2N 697; Q3, 2N 384.

D. CONVERSION OF CAPACITOR ENERGY

The electronic switch components are separated into three groups: switch and commutation, reset and recharge, and safety. The first group consists of three high power SCR's, SCR 1–3 (400 A, 500 V); two low power SCR's, SCR 4 and 5; the commutation capacitor, C1 (5×100 μF, nonpolarized); the inductor, L1 (100 μH) and associated signal-handling circuitry. Upon receiving a positive turn-on pulse, SCR 1 begins the operation of the pulse circuit by initiating current conduction from the capacitor bank C2 through the 1600 W lamp. The usual lamp power supply is isolated from the circuit by the action of the power diode D4.

For lamp turn-off, SCR 3 is supplied a positive turn-on pulse. SCR 2 and 3, C1, and L1 turn of SCR 1 by diverting the current from it. Minimum turn-off is about 1β μsec when SCR 1 is cold and increases as the temperature of this SCR increases. SCR 2 and 3 turn off after the following operations: (1) the discharge, through the lamp, of C1 and then its recharge with the opposite polarity, and (2) the "open circuit" action of the relay K1.

The second group consists of K1 and associated components; K1 turns off SCR 2 and 3 by interrupting the current through them from the commutation power supply. Also, K1 provides a recharge path for C1 and operates indicator lights. Its action is internal and automatic and its coil is energized when the circuit is in the "ready" position.

The third group consists primarily of transistor Q3 and an RC network. The transistor and network protect the lamp and SCR 1 from prolonged overload. Upon conduction of SCR 1, the transistor current charges a capacitor and causes SCR 3 to turn on after a preset time of ~ 1 msec.

Forgacs [1229] realized a fast thermal pulser for metallic phase transformation studies. In this kind of metallurgical processing, "fast" means durations of 0.1–10 sec but in the field of capacity discharge this would be regarded as slow. A special feature is the temperature stabilization in the sample by a probe. The penetration of metallurgical techniques by modern pulse technology permitted unusual progress in accuracy and thermal calibration, especially with steel. The thermal pulser of Forgacs was developed for use in the study of the kinetics of phase transformations in metals. The design specifications were based on the desired experimental parameters for a particular study of a particular alloy; however, the apparatus is sufficiently flexible to permit a variety of conditions to be attained with a variety of metals. The specific requirement of the experiment is to thermally pulse a sample of an aluminum–4% copper alloy to a selected temperature up to 600°C. The temperature pulse should be flat within 1 or 2% and have a duration of 0.1–10.0 sec measurable and controlled to a similar accuracy. To attain such accuracy in a 100-msec pulse means that rise and fall times should preferably not exceed a few milliseconds. A sample is repeatedly pulsed to a selected temperature, cooled, and examined for phase transformations between pulses;

1. DIRECT DISCHARGE THROUGH A CONDUCTOR

thus it is desirable that the system permit many pulses to be made on an individual sample at temperatures approaching the melting point.

Karlyn [1811] has described a pulse heating system in which fast temperature rise is obtained by discharging a high-voltage capacitor through the sample. For heating pulse durations of several seconds, a battery is switched on to maintain the sample at temperature; fast cooling is obtained by water jet. The present system uses the same basic technique; however, improvements were made to increase the accuracy of temperature control, which becomes more difficult when high thermal conductivity metals are tested, particularly for pulses of long duration.

Therefore, a thermal pulser utilizing I^2R heating was developed for phase transformation studies in metals. The pulser combines fast rise and fall times and a feedback-controlled, flat-topped temperature pulse. The temperature sensor is either a three-legged thermocouple arranged to cancel out IR drops due to current in the sample, or a radiation thermometer. Sample and thermocouple fatigue life were materially improved by use of a sample-mounting arrangement which combined prestretched spring clamps to eliminate sample bowing and a symmetrical design to eliminate transverse magnetic forces on the sample. Measurements show peak temperature of consecutive pulses repeatable to $\sim \pm 1\%$, and temperature having a pulse duration from 0.1 to 10 sec constant to $\sim \pm 0.7$–1.5%, even for metals with high thermal conductivity. The power capability is adequate to melt thin samples of most metals.

A functional diagram of the basic system employed is shown in Fig. D1-2. The high-voltage supply charges a bank of high-voltage capacitors totaling

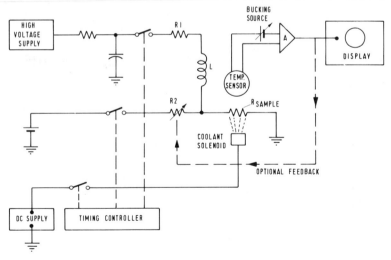

FIG. D1-2. Functional diagram of thermal pulser.

425 μF, rated at 4 kV, through a resistor. Operation has been arbitrarily restricted to 3 kV maximum since the life of these capacitors can be shortened by discharging into an underdamped circuit at high power levels. The high-voltage switch is an ignitron. Resistance R_1 in the high-voltage discharge circuit consists of the internal series resistance in the capacitor, wiring resistance, and current monitoring resistors. The series inductance L represents stray inductance in the circuit. The resonant frequency of the discharge circuit is ~ 5 kHz and peak discharge current at 3 kV is $\sim 30{,}000$ A.

The low-voltage supply is a 12-V heavy-duty truck battery, connected to be chargeable between pulses. Series resistance R_2 consists of a power transistor shunted by a group of air-cooled, high power capability resistors operated well below ratings to ensure minimum change of resistance due to self-heating. R_2 may be varied to yield direct currents from ~ 1.3 to 144 A. The sample resistance is typically from ~ 2 to 50 mΩ. The large values of battery voltage and external series resistance compared to what is required to deliver the quoted currents in the low sample resistance help to ensure constant current despite changes in sample-clamping-contact resistance, switch-contact resistance in the contactor employed as the switch in this circuit, and sample resistance (from self-heating).

Two types of temperature sensor have been employed: the thermocouple and the radiation detector. The thermocouple has the advantage of greater accuracy and faster response, but certain problems are encountered. An intrinsic junction is employed in which the wires are welded to the sample rather than to each other; this technique ensures fast response and that the measured junctions are essentially at the same temperature as the sample surface. However, if a normal intrinsic thermocouple is employed and if the two junctions are not at precisely the same potential in the high amperage battery circuit, an appreciable error may result due to IR drop. For example, with Chromel–Alumel wires, with one leg welded only 0.1 mm downstream (current flow) from the other, a 5-cm long, 10-mΩ sample would yield an IR drop error at 100 A equivalent to 50°C. This error would also vary as the current and sample resistance change during the pulse such that it cannot be corrected accurately. This problem has been solved by utilizing a three-legged thermocouple in which the two negative legs are welded slightly upstream and downstream, respectively, of the positive leg. The negative legs are then connected to the ends of a low-resistance pot whose wiper provides a voltage equivalent to that of a single thermocouple precisely positioned to cancel out IR drops. Balance is independent of current amplitude, and if the welds are made in positions symmetrical with respect to expected thermal gradients, errors due to intrapulse variations in sample resistance are minimized.

The constancy of a 6-sec pulse at 500°C was measured at $\pm 0.7\%$. The cooling rate attainable with water jet is not shown in the temperature–time

1. DIRECT DISCHARGE THROUGH A CONDUCTOR

TABLE D1-1[a]

TYPES OF HIGH POWER PULSE GENERATOR

Model No. output (either polarity)	E_{out} peak max (V)	I_{out} peak max (A)	Pulse width range (μsec)	Rise time max (μsec)	Fall time max (μsec)	Max droop at full-rated pulse width (%)	Overshoot max (%)	R_L nominal (Ω)
1761	300	100	0.1–1.0	0.08	0.08	3	10	3.0
1762	300		1–10	0.3	0.2	5	6	
1763	300		10–100	3.0	3.0	5	6	
1764	280		100–300	20	20	10	6	
1765	120	250	0.1–1.0	0.09	0.09	3	6	0.5
1766	120		1–10	0.3	0.3	5	6	
1767	120		10–100	3.5	3.2	5	6	
1768	110		100–300	25	20	10	6	
1769	60	500	1–10	0.4	0.4	5	10	0.12
1770	60		10–100	4.0	3.5	5	10	
1771	58		100–300	30	25	10	10	
1772	40	750	1–10	0.5	0.5	5	10	0.055
1773	40		10–100	4.0	3.5	5	10	
1777	38		100–300	30	25	10	10	

[a] All units rated at 1.5% duty factor into resistive load. Backswing for all units <25%.

TABLE D1-2 STANDARD OUTPUT PLUG-IN UNITS FOR VELONEX MODEL 360 HIGH POWER PULSE GENERATOR

Model No. output polarity		E_{out} peak max (kV)	I_{out} peak max (A)	Pulse width range (μsec)	Preliminary specifications				R_L nominal (kΩ)
(+)	(−)				Rise time max (μsec)	Fall time max (μsec)	Max droop at full-rated pulse width (%)	Overshoot max (%)	
1720	1721	30	1.0	10–100	6.0	7.0	5		30
1723	1724	20	1.5	1–10	1.5	1.8	5		13
1725	1726	20		10–100	4.5	5.5	5		
1727	1728	10.0		1–10	0.4	0.4	5		
1729	1730	10.0	3.0	10–100	3.0	4.0	5		3.2
1731	1732	9.5		100–300	15	20	10		
1733	1734	5.0		0.1–1.0	0.08	0.08	3		
1735	1736	5.0	6.0	1–10	0.3	0.3	5		0.800
1737	1738	5.0		10–100	2.0	2.0	5		
1739	1740	4.8		100–300	12	15	10		
1741	—	2.5		0.1–1	0.06	0.06	3	6	
1742	—	2.5	12.5	1–10	0.15	0.15	5		0.200
1743	—	2.5		10–100	1.3		5		
1744	—	2.3		100–300	10	12	10		
1745	1746	1.2		0.1–1.0	0.07	0.07	3		
1747	1748	1.2	25.0	1–10	0.1	0.1	5		0.050
1749	1750	1.2		10–100	2.0	2.0	5		
1751	1752	1.1		100–300	10	10	10		
1753	1754	0.600		0.1–1.0	0.12	0.12	3		
1755	1756	0.600	50.0	1–10	0.4	0.4	5		0.0125
1757	1758	0.600		10–100	2.2	2.4	5		
1759	1760	0.550		100–300	12	12	10		

plots; a typical figure is 8 msec for a 50% reduction in temperature with 2 atm air driving the water column. The complete circuit diagram can be seen in Ref. [1229].

Another field of application of stabilized high current pulses is the diode lasers. Velonex [1231] developed a line of high power pulse generators. Table D1-1 shows types with pulse width between 0.1 and 300 μsec and 100 to 750 A. Table D1-2 gives the data of a high power pulse and burst generator which has 22-kW peak power and can provide bursts of 3000–2,000,000 pulses/sec. Duty factor during burst intervals is up to 50% with 50,000 bursts/sec. This model is flexible and easy to operate (see Table D1-3).

TABLE D1-3

SPECIFICATIONS FOR MODEL 570[a]

Output voltage (pulse)	0–2100 V negative
Output current (peak)	0–10.5 A
Output power (peak)	0–22 kW[b]
Burst Mode:	
Burst-gate characteristics:	
Width	0.3–300 μsec
Repetition rate	3–100,000 pps (only single pulse above 50,000 pps)
Pulse characteristics:	
Rise time/fall time	30/50 nsec
Width	0.1–300 μsec
Repetition rate	$\times 10^3 - 2 \times 10^6$ pps
Duty factor[c]	50% at 22 kW peak
Amplitude variation[d]	Less than 5%
Nonburst Mode:	
Rise time/fall time	30/50 nsec
Width	0.1–300 μsec
Repetition rate	3–100,000 pps
Droop[d,e]	Less than 3 or 0.05%/μsec, whichever is greater
Overshoot[d]	Less than 5%
Ripple on pulse top	Negligible
Duty factor	1% at 22 kW peak, greater for lower output levels
Weight	140 lb (160 lb shipping)
Cabinet size	$19\frac{3}{4} \times 18\frac{1}{4} \times 18$
Power input	650 W from 115 V[f], 50–60 Hz

[a] All specifications are given when operating into 200-Ω resitive load.

[b] Peak power may be increased 20% for duty factor less than 0.1%.

[c] Over-all duty factor (burst duty factor × gate duty factor) should not exceed 1% at 22 kW peak.

[d] Measured at output levels above 400 V peak.

[e] Pulse repetition rate may be extended to single pulse. Manual single pulse switch (V-1208) available as accessory.

[f] Specify model 570-F for 230 V, 50/60 Hz operation.

a. *Pulsing Diode Laser Devices*

Semiconductor laser devices, such as diodes, normally require a high current source and a load impedance typically ranging from 0.1 to 4 Ω, depending on the configuration. Diodes are often connected in parallel to form a pulse light source. The Velonex plug-ins V-1009, V-1010, V-1140, V-1260 through V-1263, and V-1197 have an output pulse width capacity from 0.1 to 10 μsec and can drive low-impedance loads.

The low characteristic impedance requires a stripline configuration to preserve wave shape and pulse fidelity. It is especially important that the stripline configuration be maintained from the output terminals to the point of connection to the laser diode. Failure to provide a suitable stripline into Dewar flasks will reduce output power to one-fourth or one-tenth of the maximum power available. It is recommended in this application that one side or the other of the diode be connected to ground with a low-inductance ground return conductor back to the grounding terminal on the face of the plug-in.

Many semiconductor devices, such as SCRs, and laser diodes are susceptible to the rate of rise of the voltage pulse dV/dt. To test this characteristic, the rise–fall time control plug-ins V-2170 and V-1276 are recommended.

b. *Pulsing Gas Lasers*

Gas lasers sometimes exhibit a wide range of dynamic impedance when subjected to high-voltage pulses. In applying the Velonex high power pulse generator to a gas laser application, it has been found useful to insert a bias dc voltage suitable to bias the laser to a threshold mode and to use the pulse from the Velonex model 350 to drive it into lasing. In the lasing mode, the dynamic impedance of the gas laser many times will exhibit a change of 10:1. A connection as shown in Fig. D1b-1 will normally provide optimum operating conditions. Typical output plug-ins for these applications include V-1105 and V-1077. This same arrangement has useful application when the load impedance has a sharp decline after reaching a threshold voltage.

For feeding Fischer–Nanolite spark lights, which are basically two-electrode

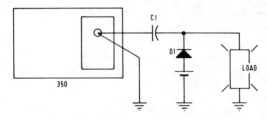

Fig. D1b-1. Circuit for driving gas laser using Velonex pulse generator model 350. C1: suitable rated dc blocking capacitor; D1: fast recovery diode with PIV rating > peak pulse amplitude.

spark gaps with a capacitance of about 5000 pF and a flashover voltage between 3 and 5 kV, Impulsphysik [1073, 1058] produces two very high peak power current pulsers, one suitable for single and repetitive flashes up to 30/sec, and a bigger one for repetitive flashes up to 10,000/sec. Both devices rapidly charge the 5000-pF capacitor to more than 3 kV that the measurable jitter time in the spark amounts to a few nanoseconds only. The circuit of the smaller Nanolite pulser uses a specially developed trigatron which charges the capacitance of the Nanolite directly while the bigger unit uses an H_2-thyratron, the deionization time of which is so short that recycling speeds of 1–10 kHz are possible with quartz-controlled frequency. On request, a second H_2-thyratron in the same equipment is available which has a delayed output between a few nanoseconds and 1 μsec. This permits the operation of a slightly delayed Kerr cell by which, for example, the light tail of the Nanolite spark can be cut. Nanolites with such a tail-cutting Kerr cell have a nearly rectangular light pulse shape and together with narrow band optical filters they permit the simulation of all kinds of laser pulse light, e.g., for testing photomultipliers or to excite fluorescence in the fields of biology or organic chemistry.

About 3 nsec pulses of high voltage and high, stable current are required for feeding N_2-lasers on the superradiant principle. Small and Ashari [1712] show a design of such a pulser using a so-called Blumlein pulse generator to limit the pulse time. N_2-superradiant lasers operate without mirrors and generate a 3371-A pulse of only 3 nsec at several megawatts. These will become important tools in the field of atmospheric measurements, especially for air pollution studies using the Raman effect.

The so-called Blumlein circuits (see Fig. D1b-2) have been successfully applied to the H_2, N_2, and Ne high-gain pulsed lasers. Here a modified electrode configuration which allows uniform excitation over a larger cross section and has other advantages is described. In order to indicate the new configuration, a design typical of the methods used previously is first considered.

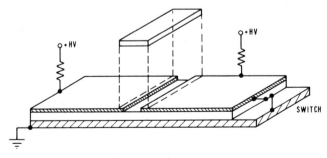

FIG. D1b-2. A typical Blumlein circuit for high current pulsed gas discharges. The discharge occurs across the gap in the upper electrode and is confined to a narrow region by dielectric walls.

A parallel plate transmission line, insulated with a solid dielectric, has a transverse gap cut from its upper conductor (see Fig. D1b-3). The gap is covered with a dielectric slab, usually glass, forming a long narrow cavity. A low-pressure gas discharge is excited across the gap with the dielectric walls serving to confine the plasma. A discharge is initiated by charging the entire transmission line and then closing a low-inductance switch at one end. A voltage step propagating away from the switch suddenly overcharges the gap. A high-current discharge, typically a kiloampere per centimeter of electrode length, is sustained until the remainder of the transmission line is discharged.

One of the first papers (but not referred to in the first two volumes of this book) about drivers for optical diodes is published by Bonin [1233]. Such drivers are used to feed laser diode arrays, which begin to lase at a certain current threshold. Since all current below this represents thermal loss, the aim is to generate 10–25 A pulses with ~50–100 nsec width and rise and decay times of only a very few nanoseconds. The required voltage is something like 100 V. Circuits with tubes are also still being used. Kubis and Strassburg [1199] describe a very simple pulse generator for rectangular $+/-$ pulses at pulse lengths between 1 μsec and 50 sec (!).

For all applications of light pulses in which the light is not permitted to be monochromatic, pulsed xenon arcs or sparks are suitable light sources. Xenon arcs burning at very low current, permit the generation of properly shaped light pulses by superposing current pulses over the holding current. Such a 35-V 180-A pulse generator with droop control for pulsing xenon arcs is described by Hviid and Nielsen [1559]. To measure the far-ultraviolet absorption spectra of hydrated electrons and hydrogen atoms dissolved in water, the technique of pulse radiolysis has been used. It was found necessary

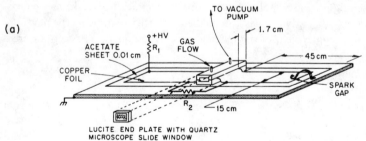

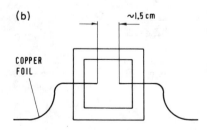

Fig. D1b-3. The modified Blumlein excitation method. (a) Over view of the transmission line and vacuum enclosure. The acetate dielectric can withstand 20,000 V. $R_1 = 22$ MΩ, 2 W; $R_2 = 10$ kΩ for pulse repetition rate of 1/sec. This laser produced 10–20 kW peak power output. (b) End view of discharge region. The discharge is pressure confined and does not touch the vacuum enclosure.

1. DIRECT DISCHARGE THROUGH A CONDUCTOR

to increase the luminance of the Osram XBO 450 W/4 xenon lamp used as analyzing light source by a factor of more than 25 at 200 nm. The increased luminance was obtained by pulsing the lamp with 180 A pulses from the pulse generator described in this note. The pulsing of xenon lamps has now become established in a number of laboratories, and Hodgson and Keene [1812] have recently described the pulsed operation of a XBO 250 xenon lamp from a 180-V 700-A source using thyristor switches. They obtained up to 700 times increase in arc luminance at 200 nm from their smaller arc and used a variable series inductance as droop control for the resulting light pulse.

The pulse generator works as a combined switch and series current regulator. The advantages of this circuit compared with that described by Hodgson and Keene is that it works with a lower supply voltage (35 V) and smaller electrolytic capacitors, and esily allows the shape of the current pulse to be adjusted at each optical wavelength to produce a flat pulse of monochromatic light. The light intensity can be further increased by operating several pulse generators in parallel.

Another feedback-controlled pulse generator for use in echo signal simulation in the field of geology has been realized by Matson [1645]. A precision current pulse generator for geologic diffusion measurements by a pulsed gradient technique is described for delivering controlled pulses of current into an inductive load. Digital logic is utilized for control of the pulse duration, and current feedback is used to control the magnitude of the pulse. Diffusion results are shown for No. 15 white mineral oil using pulse durations of 10 msec and magnitudes >25 A.

With this current pulse generator, Helmholtz coils can be fed so as to generate, for example, rectangular-shaped field pulses. The coils consist of a double layer of windings of No. 25 gauge copper wire. The coils extend from ~ 1.98 to 2.49 cm from the midplane between the magnet pole faces in a 7.6-cm magnet gap. Both the coil form and the NMR probe were constructed from nonconducting materials. Measurements on the self-diffusion of water gave a coil factor of 3.38 G/cm-A at 25°C.

The goal of current pulse generator design is to avoid tubes and to replace them by more reliable semiconductors. In the kilovolt–kiloampere area hydrogen thyratrons and ignitrons are still the devices used most often. Recently a microsecond pulse generator has been realized by Campbell and Kasper [1196] employing series-coupled SCR's. A pulse generator circuit is presented in which capacitive discharge through a high-voltage switch incorporating a chain of series-coupled SCR's produces pulses of microsecond durations with peak currents of 1 kA at 1 kV. The basic operation of the switch, limitations governing the choice and arrangement of components, extension to voltages higher than 1 kV, and the electrical performance of the circuit when used to initiate the firing of xenon flash tubes are discussed.

The circuit diagram for the SCR pulse generator is shown in Fig. D1-6. The

D. CONVERSION OF CAPACITOR ENERGY

energy storage capacitors C_4 and C_5 are charged by a full wave bridge rectifier, consisting of diodes D_1–D_4, through the current limiting resistor R_9 and the primaries of output transformers T_2 and T_3, respectively. The peak output of the full wave bridge and thus the voltage across the fully charged capacitors, is ~ 1.0 kV.

The series-coupled SCR's 1, 2, and 3 constitute the high-voltage switch. To prevent undesired breakdown of the switch, the charging voltage is divided equally across the three SCR's so that no single SCR is overcharged. To close the switch, an external trigger pulse is supplied to one end of the SCR chain. By use of appropriate coupling capacitors and resistors, this trigger pulse is

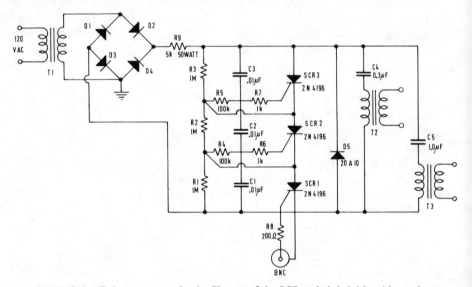

Fig. D1b-4. Pulse generator circuit. Closure of the SCR switch is initiated by an input pulse through R_8. In the circuit as constructed, T_2 and T_3 are located external to the chassis and are connected to the circuit by Amphenol type C connectors and RG-8/U coaxial cable. All resistors 1/2 W 5% except R_9. D_1–D_4-Motorola HEP-170, 1 kV PIV; $T_1 = 700$ V, 90 mA plate transformer; $T_2 =$ xenon Corp. TX-298 trigger transformer; $T_3 =$ primary, 10 turns 18 AWG, 5 kV insulation, secondary, 30 turns, 40 kV CRT wire, core, Carsted Research, Inc., CRL 14.

directed through successive stages of the chain effectively turning on all the SCR's within small fractions of a microsecond of each other. More specifically, closure of the switch is initiated by triggering into conduction SCR 1 which has the lowest anode voltage. This causes the cathode voltage of SCR 2 to drop. Capacitors C_1–C_3 hold the gate voltage of SCR 2 essentially fixed until the rapidly rising gate to cathode current triggers SCR 2 into conduction. The triggering of SCR 2 occurs ~ 40 nsec after the triggering of SCR 1. A similar sequence of events triggers the next higher SCR in the chain; in this manner,

1. DIRECT DISCHARGE THROUGH A CONDUCTOR 139

the entire chain becomes conducting. Capacitors C_4 and C_5 then discharge through the primaries of their respective transformers T_2 and T_3, thereby generating the output pulses.

In photodissociation laser work, the pulse generator initiates the firing of a flash tube filled with xenon at low pressure. The output pulse generated by T_3 reliably triggers a Westinghouse WL 8306 ignitron. However, dependable firing of a flash tube connected in series with the ignitron will occur only for flash tube voltages exceeding a threshold characteristic of the particular tube and its history. From the quoted power requirements for triggering the ignitron, a 4-μF capacitor was initially selected for C_5. However, subsequent tests demonstrated that a 1-μF capacitor will fire the ignitron with T_3 constructed as described in Fig. D1-6. Reliable firing of the ignitron at lower capacitances is desirable since the shorter charging time permits higher repetition rates. By applying the greater than 50 kV trigger pulse generated by T_2 directly to a trigger wire wrapped around the flash tube, the threshold firing voltage of the tube is markedly reduced.

Very similar techniques are applied in the pulse generators used for picture and line scanning of television picture tubes. German readers will find more information on such circuits in a paper by Walz [1197], which induces a diagram showing very fast synchronizing by phase balance as needed on videotape–recorder operation. The feedback circuit is strongly damped to avoid flickering on the upper side of the TV picture. All US makers of IC's recommended types of IC for such TV pulse techniques in their solid-state bulletins.

For purely scientific applications, pulses may be obtained by using a sophisticated pulse generator and amplifying these to the required level. Berkeley Nucleonics Corp. [1644] offers a series of pulse generators which generate precision amplitude pulses for testing linearity, stability and resolution of amplifiers, analog–digital converters, and pulse height analyzers, and also produces high speed pulse generators whose main function is to provide high repetition rates and fast rise times for testing TTL, ECL, and NIM logic (see Tables D1-4, D1-5).

If very fast pulses have to be amplified without risk of distortion, gigahertz transistors operating as on-line for microstrip-circuits can be provided. RCA [1232] shows for its 2N5921-npn overlay transistor two such typical amplifier arrangements as shown in Figs. D1b-5 and D1b-6.

Singer and Ems [1560] describe a fast, high-current pulser suitable for low temperatures which is a very sophisticated kind of current pulse generator. Many experiments require fast, high-current pulses delivered to low-impedance loads ($\lesssim 50\ \Omega$) at low temperatures. It is well known that avalanche transistor pulse generators are commonly used as fast, high-current pulsers at room temperature. But it is also true that the avalanche effect in semiconductors occurs down to liquid helium (LHe) temperatures ($\sim 4\ °K$). An avalanche

TABLE D1-4 PRECISION PULSE GENERATORS

Model	Temp. coeff. (±ppm/°C)	Linearity Integ. (%)	Linearity Diff. (%)	Frequency Max (MHz)	Frequency Min (Hz)	Rise time Min (nsec)	Rise time Max (μsec)	Fall time Max (msec)	Fall time Min (μsec)	Tail pulse	Flat top pulse	Flat top width (μsec)	Amplitude jitter (ppm, rms)	Output Z (Ω)
BH-1[a]	100–300[b]	0.1	0.25	1	10	20	50	1	0.05	Yes	No	—	20	50
RP-2	50	0.1	0.15	0.05	1	50	5	0.1	2	Yes	No	—	20	100
PB-2	33	0.1	0.25	1	1	50	2	Variable[c]	Variable[c]	Yes	Yes	0.3–100	20	100
GL-3	33	0.1	0.25	1	1	50	2	Variable[c]	Variable[c]	Yes	Yes	0.3–100	20	100
PB-3	10	0.005	0.05	0.1	1	50	2	Variable[c]		Yes	Yes	0.3–100	10	100
PB-4	5	0.005	0.03	0.25	2.5	50	10	1	0.5	Yes	Yes	0.5–25	10	50

[a] Instrument has internal ramp generator.
[b] Temperature coefficient is dependent on fall time: less than 100 ppm/°C up to 50 μsec then increasing to 300 ppm/°C at 1 msec.
[c] Fall time depends upon load resistance and rise time setting. Ranges from 0.3 μsec to 1 msec.

TABLE D1-5 HIGH SPEED PULSE GENERATORS

Model	Application	Frequency Max (MHz)	Frequency Min (Hz)	Rise time (nsec)	Fall time (nsec)	Delay	Pulse width	Output amplitude (V into 50 Ω)
8010	Low cost, general purpose for TTL, ECL, and NIM	50	1	3.5 and 5	3.5 and 5	20 nec–1 sec	20 nsec–1 sec	+5 (TTL), −0.8 (NIM), −1.5 (ECL)
8020	Fast NIM logic, 2 outputs	125	0.5	1	1.3	0–100 μsec	3 nsec–100 μsec	−0.8

1. DIRECT DISCHARGE THROUGH A CONDUCTOR

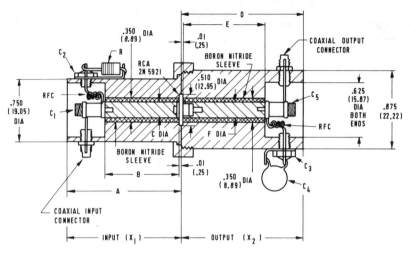

FIG. D1b-5. Constructional details of 1.2/2 GHz coaxial-line test circuits.

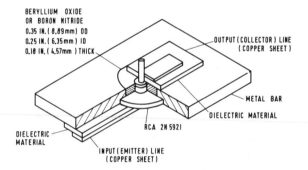

FIG. D1b-6. Suggested mounting arrangement of the 2N5921 in a microstripline circuit.

pulser, built with standard components, is found to operate reliably in LHe. Moreover, it works as well as, and in one respect even better than, the room temperature pulser. An avalanche pulsing circuit, constructed with off-the shelf components operates reliably at 4.2 °K. Nanosecond pulses with amperes of peak current are delivered to low-impedance loads. Physical proximity of generator and load eliminate cross talk and transmission line distortion.

A typical free running pulser is shown in Fig. D1-9. Capactior C_1 is charged through R_c toward V_c. However, Q_1 avalanches at some voltage $V_{B1} < V_c$ allowing C_1 to discharge through R_L to V_{B2}. [Note that blocking resistor R_B ($R_L \ll R_B \ll R_c$) must be inserted between Q_1 and the coaxial line to prevent the line from discharging through R_L.] When the current drops below the holding current of the transistor (10–100 mA), the transistor turns off and C_1 begins charging again.

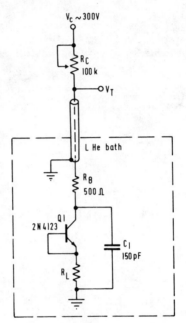

Fig. D1b-7. Liquid helium avalanche pulser circuit.

2. Transformed Discharges for Highest Currents at Low Voltage

Many applications of pulsed electric power require currents in the multimegampere range with rise times of 30 to 120 μsec. The corresponding charges of typically 40–160 C are uneconomical to store in fast capacitors and cannot be switched conveniently and reliably by means of available switch gear. To meet this need, Advanced Kinetics, Inc. [1646] has developed a line of low-inductance air-core pulse transformers. Typical parameters are: input: 50 kJ at 10 kV, 320 kA peak, 50 μsec rise time; output: 40 kJ at 1.1 kV, 2300 kA.

The main applications are high-current pulses in high-temperature plasma research, shock tubes, high-pulsed magnetic fields, magnetic metal forming, exploding wires, electrohydraulic metal forming, pulsed light sources, pulsed underwater sound sources. The features of the Adkin system are: current step-up pulse transformer (kiloamperes to megamperes), high-efficiency power transfer, 50 kJ–10 kV unit operation, coaxial cable connection to any capacitor bank, massive low-inductance transmission line to loads, specially suited for power-crowbar at high currents and availability of custom-built, larger sizes.

2. DISCHARGES FOR HIGHEST CURRENTS AT LOW VOLTAGE

The transformer is designed to transfer stored energy at 80% efficiency into inductive loads as small as 0.015 μH. These specifications call for (secondary) leakage inductances as low as 0.0015 μH, equivalent to about a centimeter of ordinary coaxial cable. The extremely restrictive condition on the leakage inductance dominates the transformer design and limits its physical size. The small size, together with the requirement of operability at high power level, makes the use of an iron core uneconomical. The secondary shunt inductance of the transformer with an effective one-turn secondary connection is required to be ~0.15 μH, which is met in the present air-core transformer. Parameters are summarized in Table D2-1 for a 10-kV, 1000-μF, 50-kJ source, a 0.015-μH optimum load, and for three typical transformers.

The input current and discharge time (quarter-cycle time) depend on whether the primary windings are all connected in parallel or in series–parallel. The connections are made so as to keep constant, under all conditions, the current flowing in each primary turn, so that the efficiency properties of the transformers are basically unaffected. Efficiency tends to drop off for the longer pulse times, due to Ohmic losses.

The transformer is constructed of massive conductor, following techniques developed in high-field magnet technology. Maximum fields appearing within the transformer are kept below 20,000 G, so that mechanical stress is negligible. Input terminals are provided for 40 low-inductance (13 mμH/ft) coaxial cables from the capacitor bank. The output is by means of 40 cables in parallel, which are collected on two parallel output strips of 16 in. width. A coaxial output can also be provided.

The transformer winding is composed of N winding stacks, each comprising a primary helix of n turns and n one-turn secondaries. The following configurations are offered: $N = 10$ and $n = 4$, or $N = 8$ and $n = 5$, or $N = 4$ and

TABLE D2-1

SPECIFICATIONS OF ADKIN PULSE TRANSFORMERS[a]

Transformer	N	n	Transformer ratio	Input current (kA)	Output current (MA)	Discharge time (μsec)	Output voltage (V)
PT-4-50-A	10	4	4:1	640	2.3	25	2200
PT-4-50-B	10	4	8:1	320	2.3	50	1100
PT-5-50-A	8	5	5:1	500	2.3	30	1800
PT-5-50-B	8	5	10:1	250	2.2	60	900
PT-10-50-A	4	10	10:1	250	2.2	60	900
PT-10-50-B	4	10	20:1	130	2.1	120	400

[a] Source: 10 kV, 1000 μF, 50 kJ; load: 0.015 μH (optimum); N: number of winding stacks; n: secondary turns in each winding stacks.

$n = 10$; and there are 40 primary and 40 secondary turns in all. A variable connector system is provided, which permits the ratio of a given transformer to be altered at the factory by a factor of 2.

The transformer is insulated for 10 kV operating voltage, and the windings are immersed in oil. For applications requiring high repetition rate, a circulatory cooling system can be provided. In the power-crowbar application, the output cables are separately connected to each crowbar ignitron, and ensure equal current-sharing.

3. Subnanosecond Pulse Generators

Pfeiffer [1918] developed a kiloampere pulse generator with rise time in the subnanosecond area. He was able to generate these current pulses with only one switching spark gap, by using high-pressure nitrogen as an additional

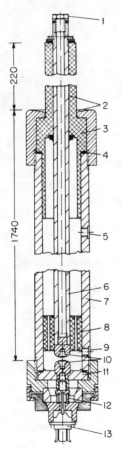

Fig. D3-1. Entire design of the subnanosecond current pulse generator. 1. High-voltage connection with screening toroid; 2. filling with silicone oil; 3. Teflon insulation; 4. washer; 5. pressurized gas inlet; 6. internal conductor (brass tube); 7. outside conductor (brass tube); 8. additional capacitor with holes for pressure equalization; 9. pressure outlet and ground connection; 10. brass electrodes with tungsten tips; 11. PVC cover; 12. resistor on a ceramic body; 13. voltage outlet via bushing (dimensions in millimeters).

pulse-forming network. His paper deals with the influence of the form of electrodes on the shape of the current pulse. The best result obtained was 1 kA pulse with 3 nsec rise time, at a pressure of ~ 20 bars. The design of this gap can be seen in Fig. D3-1.

4. Inductive Energy Storage Systems Applied to Extend the Duration of Current Pulses from Capacitor Banks

Very high-current pulses are required for high-temperature plasma experiments (see Chapter G, Section 7; Chapter H, Section 6b) and to extend the confinement of high-temperature plasmas, power-crowbar circuits are used. Salge and Braunsberger [1919] reported the application of inductively stored energy in combination with capacitor discharges for high-current purpose. A rapidly increasing magnetic field in the load produced by load-current has reached its first peak, the capacitor bank is crowbarred and the load is then supplied from an inductive storage. The function of the system, the requirements for the circuit elements, and the possibilities for a technical realization of the system are investigated. In addition the results of experiments conducted on a model circuit are discussed. In this case, the inductive storage system is built up as a transformer; its main inductance is charged by a dc power source.

Fast rising pulsed magnetic fields with long pulse duration can be generated in so-called active- or power-crowbar circuits [1848, 1849]. During the discharge of a condenser bank one or more additional power sources are switched in. For this purpose so far, capacitive energy storage systems were mainly used [1850]. This report will discuss the use of inductively stored energy in power-crowbar systems.

Figure D4-1 shows schematically a basic circuit developed during these investigations with the corresponding current shapes. The Ohmic resistance of the circuit elements are not shown in Fig. D4-1. In the starting position the condenser bank C is charged. Closing the switch S_1 at the time t_0 initiates the charging of the inductive store L_{St} by the direct current source. The current rises according to the values of the charging circuit. At time t_1 the current has approximately reached its final value. The discharge of the condenser bank C across the load is started and the direct current source connected, by closing the switch S_2. The charging voltage of C and the circuit elements are matched in such a way that the discharge current of the capacitor bank C and the current through the inductive store L_{St} are equal at t_3. Thus, a current zero in the dc power source at time t_3 can be achieved. By opening the switch S_1 at this moment, the charged inductive store is switched into the oscillating circuit and its original frequency is reduced (dashed current curve, Fig. D4-1). The crowbar switch S_3 is also closed at time t_3. Now, the discharge circuit consists of the load, the inductance L_{St} and the crowbar switch S_3. The current decays

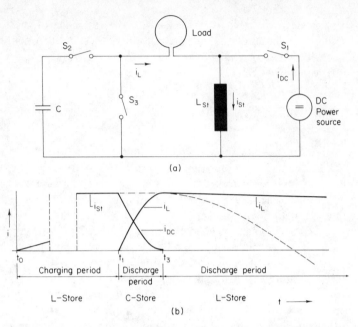

FIG. D4-1. Fundamental circuit fall time extention of current pulse duration by an inductive energy store. (a) Circuit diagram. (b) Current curves shown schematically.

according to a time constant which results from the inductances and the Ohmic resistances of the discharge circuit.

Compared to other power-crowbar circuits, one advantage of the system described here is that the current through the load is stabilized by the inductance L_{St}. This permits superposition, on the basic field, of other magnetic fields which can be used for further heating or stabilizing of a plasma. The basic field will not be essentially influenced. Another advantage of the system is that it makes the switching problem easier than in other inductive energy storage circuits. The discharge current of the condenser bank C is simultaneously utilized both for a fast rise of the magnetic field in the load and for a current zero passage in the dc power source. This reduces the stress on the switch S_1 when opened.

Many circuit elements are stressed in a similar way as in the usual power-crowbar circuits. But for the charging circuit for the inductive store, and especially the switch S_1, there are remarkable differences. The dc power source has to deliver currents of some megampere, which the switch S_1 has to carry. At the current zero passage, S_1 has to open.

To control such high currents technically, two methods are available:

(1) By using a number of parallel connected switches.

(2) By building up the inductive store as a transformer with low-leakage inductance.

Thus it is possible to match the current of one switching element to the rated values of industrially made switching devices. Achievement of a current zero passage is not very difficult. Care has only to be taken that the discharge current of the capacitor bank is higher than the current through the dc power source. The requirements regarding jitter and recovery of the dielectric strength of S_1 depend mainly on the current pulse shape and they are higher, the faster the current in the load rises.

There are several possibilities for the technical realization of the switch S_1: explosive type switches [1851, 1852], fast acting switches with mechanically moved contacts [1853] with vacuum spark gaps in parallel [1854], and also gas-filled control tubes. Investigations are being carried out regarding these possibilities. The inductive store can be charged by rotating machines, rectifier devices fed from the ac power network and lead storage batteries.

For studying the problems of the system, a model circuit shown in Fig. D4-2 was used. The circuit was built up with elements existing in the laboratory and was not optimized. The storage coil was a transformer with a conversion ratio of 1:5 and a leakage inductance of 0.3% of the main inductance. The transformer was charged by a rectifier unit supplied from the ac power network and started by the switch S_1 on the ac side. To break the charging circuit at current zero passage, a triggered exploding wire in series with a nonlinear inductance (S_1') was used. This was necessary to avoid switching voltage influence on the load current. The energy stored in the main inductance of the transformer was used to supply the energy lost during the crowbar

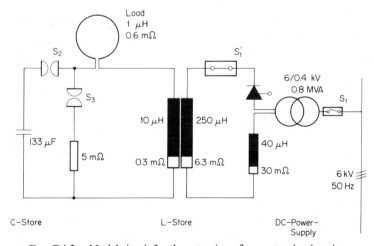

FIG. D4-2. Model circuit for the extension of current pulse duration.

period and was in this case ten times higher than the magnetic energy in the load, delivered from the condenser bank. The current rise in the load was strongly influenced by the unavoidable inductance of the rectifier unit used. Figure D4-3 compares the calculated load current for passive and active crowbar attainable in this circuit. Through the use of the simple crowbar spark gap, 85% of the energy stored by the main inductance of the transformer was absorbed by the crowbar switch S_3, 10% was dissipated in the load, and 5% remained in the store. First experiments showed that the circuit would run. Investigations just started are aimed at improving the efficiency and economy of the circuit compared to other solutions.

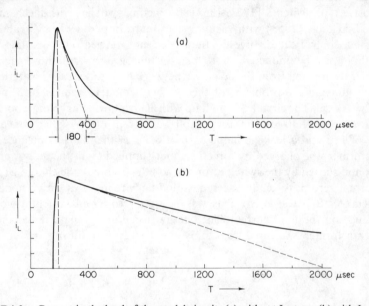

FIG. D4-3. Current in the load of the model circuit: (a) without L-store, (b) with L-store.

Finally, Fig. D4-4 shows a sketch of a power supply for a toroidal plasma experiment. Ten storage coils (4) are arranged circularly beneath the experiment (6). The inductive stores with an energy content of 6 MJ are built up as transformer units with a conversion ratio of 1:10 each. They are charged by lead storage batteries (1). The 600-kJ condenser bank (not shown in Fig. D4-4) for producing the fast rise of the magnetic field in the load is connected by cables (2). The system is able to feed a load inductance of 300 nH with a current of 2 MA. The decay time of the current is in the order of 0.1 sec. A rough calculation of costs for the inductive store and the batteries shows that this system can offer economical advantages compared to a solution using a condenser bank as a power-crowbar source.

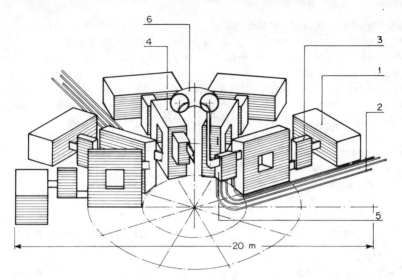

FIG. D4-4. Sketch of a toroidal plasma experiment supplied by a capacitor bank (600 kJ) and an inductive store (6 MJ). 1: Direct current power supply; 2: cables to capacitor bank; 3: switch S_1; 4: L-store; 5: switches S_2, S_3; and 6: load.

This work is being carried out in cooperation with the Institut für Plasmaphysik, Kernforschungsanlage Jülich (Euratom/KFA/Prof. Kind) and supported by the Bundesminister für Bildung und Wissenschaft, Bonn.

A proposal for the construction and operation of an inductive store for 20 MJ has been made by Inall [1920]. Briefly, the plan is to couple 15 MJ of energy initially stored in the Canberra homopolar generator, to a load at a peak power of 15,000 MW for ~ 1.0 msec. About 22 MJ would first be transferred to a coaxial inductor at a peak current and voltage of 1.5 MA and 190 V, respectively. Fast mechanical switches being developed would open to produce a potential of 1000 V and a peak power of 15 GW into the load.

Recent developments in the design of lasers has led to the need for energy storage devices with a rating of some tens of megajoules and capable of discharging in ~ 1 msec, when using a laser material having a fluorescence lifetime of about this value. Capacitor banks of this rating are so large and expensive that several proposals have been made for the use of rotating machines to store the energy. The homopolar generator at The Australian National University in Canberra could supply 1.5 MA at 800 V, that is 1.2 MJ msec^{-1}, and it could do this for more than 0.1 sec. For many important applications of these lasers, however, it is an advantage to increase the peak power and reduce the duration of the discharge.

The generator is capable of storing 560 MJ, but for the operation discussed here it would be charged to only 81 MJ (i.e., a speed of 356 rpm, giving a voltage of 190 V) before each pulse. During the discharge, 22 MJ would be transferred to an intermediate storage inductor which would supply 15 MJ to the load in 1 msec. Between 8–15 MJ, depending on the operation sequence, would be lost in the resistance of the inductor, and an equal amount in the series control resistor and busbars. A further 8–10 MJ would be lost in a surge limiting resistor across the generator and between 18–40 MJ would remain in the generator at the end of the pulse. With the present generator drive, the system could supply a pulse every 3 min. The features of the possible operating sequences will be discussed later and it will be shown that the sequence which takes 40 MJ from the generator and delivers 15 MJ to the load is possible if the switches, which are required for either sequence, can be made.

a. *Circuit Components*

The governing factors in the choice of the type of inductor to use for an energy store are:

inductance in relation to physical size: this is important because of the cost of the copper to make it and the building required to house it;
convenience of the output connections;
confinement of the associated magnetic field;
simplicity of construction;
the configuration of the electromagnetic forces on it and the convenience of the constraints to oppose them; and
the current and voltage at which it is to operate.

An inductor consisting of coaxial conductors is larger than a coil of the same inductance and resistance, but it meets the other requirements so well that it has been chosen as the most suitable for use with the high currents available from the homopolar generator.

The maximum current available from the generator for reliable regular operation is 1.5 MA. At this current an inductor of 20 μH is required to establish a magnetic field with 22 MJ of energy. About 15 MJ of this can be delivered to the load in 1 msec, and the power would be as high as one could consider using at this stage. Therefore, these values have been chosen for design purposes.

Of the possible charging and discharging circuits, Fig. D4a-1 was chosen because it best meets the requirements by providing:

the voltage increase to couple the power into arrays of xenon arc lamps which require 1000 V to operate;
protection of the generator against production of transient high voltages on the terminals under normal or fault conditions; and
for the energy to be delivered to the load in $\sim$1 msec.

4. INDUCTIVE ENERGY STORAGE SYSTEMS

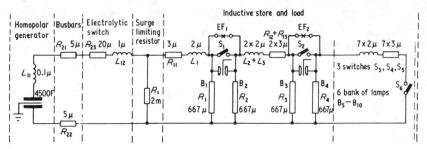

FIG. D4a-1. The charging circuit for the 20 μH coaxial inductor coupled to the homopolar generator via the electrolytic switch. The discharge into the 10 banks of lamps, B_1–B_{10}, occurs when S_1–S_5 open simultaneously as the charging current reaches 1.5×10^6 A.

The circuit (Figs. D4a-1, 4a-2) would consist of the homopolar generator, using one disk of each rotor and full-field excitation to give an equivalent capacity of 4500 F, the electrolytic control resistor with concentrated electrolyte to give a minimum resistance, $R_{23} = 20$ $\mu\Omega$, and the 10 $\mu\Omega$ resistance of the busbars from the generator. These would connect to the ten sections of the coaxial inductors, L_1–L_{10}, each with an inductance $L = 2$ μH and a resistance $R = 3$ $\mu\Omega$. In addition there would be switch S_6 to start the charging cycle, switches S_1–S_5 to control the discharge cycle, 10 banks of xenon arc lamps, B_1–B_{10} each containing 250 flash tubes, and a surge limiting shunt resistor R_s of 2 mΩ. R_s would dissipate all the energy in the generator if R_{23} failed to terminate the pulse according to its preset program. The function of S_6 could be performed by any one of S_1–S_5 if this were convenient.

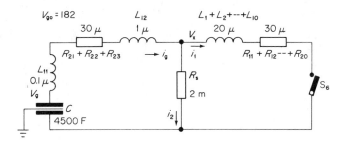

FIG. D4a-2. The circuit during the supply of current to the inductive store.

b. Peak Current Operating Cycle

One operating sequence could be to charge the generator to 195 V, raise the electrolyte to its maximum level between the electrodes of R_{23} and then, as the level starts to recede at the maximum rate of 80 cm sec^{-1}, close the switch S_6. The circuit at this stage would be that shown in Fig. D4a-2. The voltage would have fallen to 182 V due to energy being dissipated in R_s while the electrolyte was being raised.

If the effect of R_s is neglected, the current i_1 would build up according to the expression

$$i_1 = \frac{-2V_{g0}C}{(4LC-R^2C^2)^{1/2}} \exp\left(-\frac{Rt}{2L}\right) \sin\frac{(4LC-R^2C^2)^{1/2}}{2LC}t \quad (1)$$

in which

$$L = L_1 + L_2 + \ldots + L_{12} = 21.1 \times 10^{-6}$$

$$R = R_{11} + R_{12} + \ldots + R_{23} = 60 \times 10^{-6}$$

$$C = 4.5 \times 10^3$$

$$V_{g0} = 182 \text{ V}.$$

Substituting these values in (1),

$$i_1 = 2.96 \times 10^6 \exp(-1.42t)\sin(2.93t). \quad (2)$$

The maximum value of this expression is 1.55 MA at $t = 0.38$ sec. When the effects of R_s is taken into account, the maximum value of i_g would be 1.6 MA. The values of i_1, i_2, V_x, and V_g shown in Fig. D4a-2 are plotted in Fig. D4b-1. When the current reaches 1.5 MA, the switches S_1–S_5 will open and the circuits will then become as shown in Figs. D4b-2 and 4b-3. Referring to Fig. D4b-2, as the arc in S_1 becomes high in resistance, the current i_1 will flow into the exploding foil EF_1 and the arc will be extinguished. EF_1 can be designed to carry the current for ~200 μsec before it ruptures, causing the voltage across it, S_1 and B_1, to rise rapidly. The trigger pulse will be applied to B_1 at this

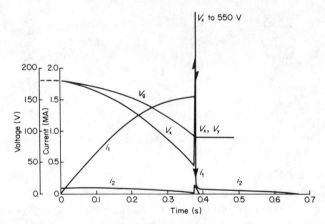

FIG. D4b-1. The current and voltage as a function of time. Switches S_1–S_5 open at $t = 0.38$ sec.

4. INDUCTIVE ENERGY STORAGE SYSTEMS

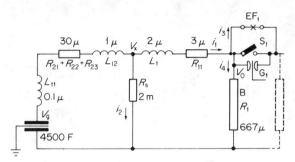

FIG. D4b-2. The circuit of the generator and the first bank of lamps, B_1, after S_1 has opened. $i_1 = 1.5 \times 10^6$ A.

instant and current will start to build up in R_1 at a rate set by the stray inductance in series with it. The voltage across the lamps will be ~ 500 V soon after the trigger and the current will rise to its full value in ~ 100 μsec as shown in Fig. D4b-3. If the current increases more rapidly than this, the lamps will be damaged. Some current will continue to flow in the disintegrated foil for ~ 100 μsec, until it is completely established in the flash tubes.

For information about the blast switch used for S_1 to S_5, see this volume, Chapter B9. The limitations on the rate of rise of the pulse current imposed by the skin effect in the rotors of the homopolar generator of the ANU are given in a separate paper by Carden [1921].

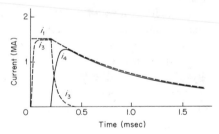

FIG. D4b-3. The current i_1 in the inductor, the current i_3 in the exploding foil, and the current i_4 in the lamps of B_1.

E. Conversion of capacitor energy into voltage impulses and its practical applications

1. Very Fast Voltage Pulse Generators for Scientific Applications

Robra [1180] prepared a general computer program for all circuit components (R, C, L) for achieving a predetermined pulse form. In particular, page 276 of the paper gives the entire program for medium sized analog computers. The need in biochemistry and biology to study such matters as fluorescence decay and relaxation kinetics at nanosecond intervals led to the development of suitable equipment. Because of the very limited field of application, such equipment is still not generally available commercially.

a. *Nanosecond Temperature-Jump Machines*

Hoffman [1277] developed a nanosecond temperature-jump apparatus that extends the applicability of the temperature-jump method to the investigation of fast chemical kinetics in solution from the microsecond to the nanosecond time range. A sample solution can be resistively heated to a maximum of 10°C within 50 nsec using the discharge of a coaxial cable capacitor. Relaxation of the reaction to a new equilibrium is followed spectrophotometrically. The design of the sample microcell (minimum volume 40 μl) is of particular interest for biochemists working with relaxation kinetics, because the volume of solution required is an order of magnitude less than for previous energy-heated temperature-jump cells.

The development of relaxation methods by Eigen and Maeyer [1818] for the study of fast reactions has led to important advances in understanding chemical processes which were previously too rapid for direct observation. Of the various relaxation techniques developed, the temperature-jump (T-jump) method with electrical energy heating [1819] has proved to be the most versatile, and hence has enjoyed the widest application.

In previous T-jump machines [1820] a high-voltage capacitor C was discharged through the sample cell (resistance R) giving an exponentially decaying current-time curve with a time constant RC (generally a few microseconds). In the case of the newly developed apparatus, a coaxial cable is used as capacitor, and by impedance matching, a rectangular current pulse can be obtained. Its duration corresponds to the time for an electromagnetic wave to traverse the length of the cable twice, in this case 50 nsec. Even shorter discharge times (5–10 nsec) are possible with shorter cables. Attempts to reach this nanosecond time range using heating by a Q-switched laser pulse [1821, 1823] have been expensive and unsuccessful due to shock effect saturation of the absorbing dye [1822], or detection system sensitivity limitations. Both the Q-switched laser and the matched coaxial discharge method involve power pulses of the order of 100 MW, but the homogeneous absorption of sufficient energy by a small sample solution can be achieved much more easily with the electrical method. The general idea of a coaxial capacitor was suggested in 1963 by Eigen and De Maeyer [1824]. This article describes the first experimental realization of the concept.

The discharge characteristics of a coaxial cable are not the same as for an ordinary capacitor because the cable has inductive as well as capacitive properties.

Energy Considerations

If a voltage V is maintained for a time interval Δt across a sample solution of mass m, specific heat C_p and resistance R, the electrical energy which is transformed into heat energy is given by

$$(V^2/R)\,\Delta t = mC_p\,\Delta T \tag{1}$$

where ΔT is the rise in temperature of the sample. Then from the relationships

$$m = pdA \tag{2}$$

$$R = d/\sigma A \tag{3}$$

$$V = Ed \tag{4}$$

where p is the sample density, σ the sample conductivity, d the electrode separation, A the sample cross-sectional area (perpendicular to field), and E the electric field, Eq. (1) can be rearranged to give

$$\Delta T = E^2\,\Delta t/pC_p \tag{5}$$

Thus for a particular sample of given conductivity subjected to a rectangular voltage pulse, the temperature rise is a function only of the electric field and the duration of the pulse. The ultimate time resolution limit is set by the dielectric breakdown strength of the sample solution. For water, this is 300 kV/cm for a static field; for a very short pulse this value can be exceeded, since breakdown is preceded by a "statistical delay" followed by a "formative delay." Tests using a cell with a 1-mm electrode separation have shown that a 50-nsec electric field pulse can exceed 500 kV/cm without breakdown occurring; even higher field strengths could be used with shorter pulses. With 500 kV/cm, and a solution whose conductivity is equal to that of a 0.1-M KCl solution, evaluation of Eq. (5) for a 5-nsec pulse gives a temperature rise of 2.5°C. Limitations in time resolution due to capacitive effects at the sample cell can easily be shown to arise only for subnanosecond times.

The apparatus is shown in Fig. E1a-1. The cable used is 5 m of a 28-Ω characteristic impedance coaxial high-frequency impulse cable supplied by Felten and Guilleaume of Cologne, Germany. The inner conductor is 24.6-mm-diam copper tubing, and the dielectric is 49.6-mm-diam polyethylene; the outer conductor is a 53.0-mm o.d. aluminum cylinder. The inner conductor is charged to a voltage of up to 100 kV. This is insulated from the sample cell by a high-pressure spark gap. The discharge is triggered by a magnetic valve

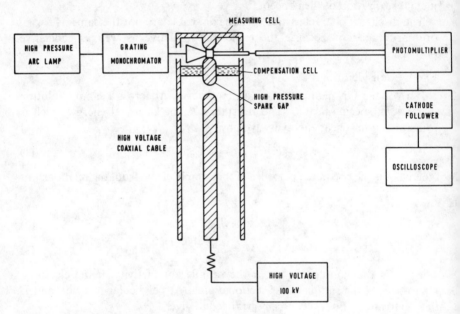

FIG. E1a-1. Block diagram of the nanosecond temperature jump apparatus.

releasing this pressure (~ 10 atm). A variety of sample cells have been tested and found satisfactory. Figure E1a-2 shows a cross section of the cylindrically shaped Plexiglas cell body. It has a cylindrical (5 mm diam) or square (3 × 3 mm) cross-sectional compartment for the sample, sealed above and below by gold or platinum electrodes. The electrode ends are flat so that there is homogeneous heating of the sample. The electrode spacing is 4–5 mm, resulting in a sample volume of 40 to 100 μl. This small volume is particularly advantageous for measurements with expensive biological substances. For this reason, it is also of interest for slow measurements. The solution cools back to the prejump

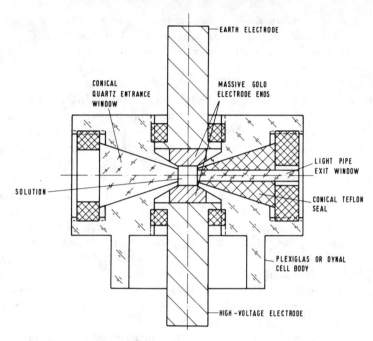

FIG. E1a-2. Section through the temperature jump microcell.

temperature monotonically and to a very good approximation, exponentially (no convection) with a time constant of 4.8 sec for the 5-mm-diam cell.

b. *Feeding of Ultrafast Image Converters*

Ultrahigh speed pulses are needed if a biplanar image converter tube used as a shutter has to be fed with an exactly synchronized pulse between anode and cathode. Marilleau *et al.* [1020] presented a paper about an ultrahigh speed, biplanar shutter with exposure times separately adjustable from 10^{-9} to 0.5×10^{-6} sec; adjustable time interval between frames; with external triggering and synchronization; and a unique system of amplification and recording of video signals. The different parts of this camera are described. Applications to the field of high explosive research are illustrated (see also [1825]).

Important work concerning the making and using of these tubes for high speed photography has been done in France since 1965 in the Laboratoires d'Electronique et de Physique Appliquées, the Commissariat à l'Energie Atomique and the Compagnie Francaise TH-CSF. A 38-mm-diam tube is used with an S20 photocathode and a P11 glass-window output screen. The use of planar tubes with glass-fiber optic output is planned later on. The exposure time is obtained by using the duration of a 10-kV pulse applied between anode and cathode. This square pulse signal is obtained by two identical step waveforms separated by a length of time equal to the chosen exposure time.

From the electrical point of view, each plane circular electrode of the tube can be considered as part of a transmission line connected up through two coaxial cable stubs terminated by high-voltage connectors. Cathode and anode structures are obviously coupled via the interelectrode space, but a time-domain study led to a satisfactory solution regarding the parasitic reflections and couplings and the rise time of the pulse differentially applied to the two electrodes.

The tube can be operated by either of the two circuits of Fig. E1b-1a and b. In the circuit of Fig. E1b-1a, the two step signals are obtained from a step function generator feeding into two lines whose different lengths determine the exposure time. Each step signal crosses the system only once, but the step function generator is loaded by $Zc/2$ (Zc: characteristic impedance of the coaxial cables). In the Fig. E1b-1b setup, the same step signal crosses the anode and cathode structures but the step function generator is loaded by Zc; the exposure time is determined by the transit time between arrival of the pulse, first at one electrode and then at the other.

The electrical response of the system (equipped with tube) to an applied step function has a rise time of about 0.7 nsec after one electrode crossing and 1 nsec after two electrode crossings. For exposure times above 3 nsec, it is possible to use the circuit of Fig. E1b-1b; for shorter exposure times it is better to use the Fig. E1b-1a circuit. Different shutters can be supplied by an equal number of independent step function generators which can be triggered at any time. They can be cascade-connected to a single step function generator; the gap between the frames is then determined by the length of the coaxial connecting cables between the shutters. For example, by adjusting the length of these coaxial connecting cables it is possible to fix the exposure times for three

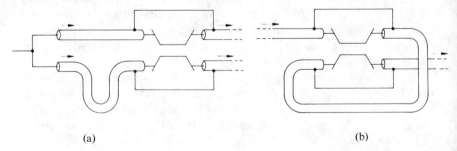

(a)   (b)

FIG. E1b-1. Two circuits for subnanosecond exposures with image converters (see text).

frames independently, using either type of system Fig. E1b-1a or 1b, with one or three generators.

Clement et al. [1019] also wrote a paper on a similar subject. In this study of high speed shutter tubes they mention various parameters that tend to limit the exposure time of shutter tubes, the conductivity of electrodes, the space charge resulting from the intense photocurrent, the electron transit time, and the deformation of the pulse as it is applied to the tube. A new tube, capable of reaching exposure times as low as a few hundred picoseconds, has its voltage impulse propagated along the stripline which is integrated in the tube itself; the input and output pulses are carried through vacuum-tight coaxial connectors with matched impedances. The electron transit time has been calculated as a function of the wave shape of the applied voltage pulse. A direct optical measurement of high speed camera exposure times has been developed using ultrashort light pulses (of a few picoseconds) delivered by a mode-locked Nd^{3+} laser.

The active element of the flat line constitutes of the photocathode; this is deposited on top of a semitransparent conductive layer with a specific resistance of 50 Ω; the photocathode is longitudinally bordered by two conductive strips, 3 mm long, which ensure the transmission of the opening signal with very little loss. These strips eliminate the edge effect of the electric field and avoid marginal distortions of the image. Commercially, Tektronix [1012] realized a 70-psec pulse generator. This Type 284 pulse generator provides the facility for verifying the performance of sampling oscilloscopes. It offers, in one small instrument, all of the signals required to check the rise time, vertical deflection factors, and horizontal sweep rates. A pretrigger output is also provided. In addition to checking the transient response of sampling oscilloscopes, the fast-rise step of the pulse output is an excellent 50-Ω signal source for TDR measurements. The 284 is available in a cabinet version, or modified for rack mounting in a standard 19-in. rack using the optional rack adapter. The pulse output has 70 psec or less rise time with a pulse width of more than 1 μsec and a repetition rate of approximately 50 kHz. Aberrations immediately following positive-going transitions are less than ±3% and after 2 nsec, less than ±2%. Pulse amplitude is more than +200 mV into 50 Ω. Source resistance is 50 Ω.

Another equipment is the Type 2101, a compact, 25-MHz, 10-V, general purpose pulse generator with simultaneous positive-going and negative-going pulse outputs. Switch positions are provided for selection of a specific pulse period, duration, and delay, within the calibrated range of the respective control. Ranges may be extended by user-supplied capacitors to reduce the repetition rate and to increase the duration and delay. The only limiting factor in extending the range is the physical size of the user-supplied timing capacitors. An external gate input permits output pulses for the duration of the gate signal, useful for all pulse modes. Independent baseline offset controls are provided for the positive-going and negative-going pulse outputs.

c. High-Voltage Pulse Generators for Large Electron Guns

A commercially available high-voltage pulse generator for driving large electron guns has been developed by IPC (Ion Physics Corporation, Burlington, Massachusetts) [1723]. Its specifications are:

Output: -100 to -300 kV, continuously variable; pulse width: 1–10 μsec ± 0.2 μsec, stepwise variable at all operating voltages; normal operating pulse width, 4 μsec.

Voltage droop: 10% at 4 μsec and 300 kV for specified load.

Rise time (10–90% of specified output) $\leqslant 0.4$ μsec. Fall time (90–10%) $\leqslant 0.2$ μsec.

Jitter: ± 0.2 μsec (between trigger input and voltage output).

Maximum overshoot: 25% at beginning of pulse; maximum undershoot: -10% at end of pulse.

A resistance of 200 Ω can be fed at 300 kV. The energy delivered to load at 300 kV, 4 μsec is 1.8 kJ. The total system is guaranteed for 10,000 shots or one year, whichever occurs first. The typical load is an emission-limited hot cathode electron gun. The electron gun shunt capacitance is 4 nF and will be operated in the temperature limited case for a constant current of 1500 A for all voltages >100 kV. The electron gun will be located not more than 50 ft from the pulser unit. The pulser is designed for oil insulation, both tank and oil being furnished by supplier. The oil tank will be for the generator only and will not include an oil reservoir or oil handling system. The tank will have a removable top with ports for output cables.

d. Nanosecond Pulsed Streamer Chambers

Rudenko [1018] discusses very fast, high-voltage pulses necessary for the so-called streamer chamber operation. At the present time the streamer chamber (spark track chamber) is finding more and more application for the detection of high-energy particles. Its advantage is the greater isotropy, compared to spark chambers, in the recording of particle tracks; its disadvantages are the low brightness of the track, which makes photography of the track difficult, and the instability of operation of the chamber as the result of variation of the pulse length from event to event.

Experimental verification of the calculations was carried out in a streamer chamber consisting of three glass cases $160 \times 100 \times 600$ mm, cemented together with epoxy resin and filled with neon to a pressure $p = 750$ Torr. The distance between the electrode plates of the chamber was 150 mm. A pulse was fed to the plates from a generator which made it possible to obtain pulses of double amplitude on the plates up to 500 kV (unmatched loading) and with a rise time of 1.5×10^{-9} sec. The generator was triggered by a coincidence circuit which received pulses from two rows of counters forming a telescope. The delay time of appearance of the first pulse on the plates after passage of a charged particle through the chamber was ~ 1 μsec.

In front of the chamber a cutoff spark gap was set up in an atmosphere of nitrogen at $p = 10$ atm. which permitted smooth control of the duration of the first pulse. Pulses of alternating polarity were obtained after cutting off the first pulse because of the oscillatory discharge of the capacity of the chamber plates through the cutoff spark gap. The period of the oscillations was determined by the capacity of the plates and the inductance of the plate and spark gap system and amounted to 35 nsec. The number of oscillations depended on the attenuation in the system and was controlled by varying the resistance of the spark in the cutoff spark gap. To obtain a pulse of definite length without succeeding oscillations, an air cutoff spark gap at $p = 1$ atm was used, whose spark resistance was greater than in the pressurized spark gap.

Results are presented of the theoretical calculations and experiments which were carried out with the aim of obtaining bright and narrow charged-particle tracks in a streamer chamber by appling a number of pulses of alternating polarity. Tracks are obtained which have almost the same size and brightness of the luminous centers ($\sigma = 1$–2 mm) in the directions along and perpendicular to the electric field. The experimental results are in satisfactory agreement with the calculations. It is shown that the new method of high-voltage power supply yields high-quality tracks more consistently than when single high-voltage pulses are applied to the chamber plates. Bayle and Schmied [1223] presented the analysis of the streamer chamber operation. For further useful papers on fast high-voltage pulse generators see: Früngel [1062]; Alcock et al. [1815]; Bacchi and Pauwels [1011]; Blanchet [1021]; Capdevielle et al. [1816]; Laviron and Delmare [1817]. All the pulse generators mentioned in these papers use more or less conventional switching means.

e. *The Single Spark Ring Transformer*

An entirely new kind of circuitry is the so-called single spark ring transmitter described by Landecker et al. [1264]. A basic paper on this technology has been published previously by the same author [1265]. It has now become possible to achieve the same result with the aid of a single spark gap either by incorporating distributed circuit elements, essentially half-wave transmission lines acting as 1-to-1 transformers, into the arrangement, or by assembling the array of tightly coupled individually tuned circuits having a common single spark gap. It is shown that the latter arrangement is superior to the use of distributed circuit elements. This results in a very great simplification of the design because the question of synchronization of a multiplicity of spark gaps does not arise.

The capacitors distributed equidistantly around the circumference of the transmitter may advantageously be replaced by one single ring-shaped capacitor.

In the original system the capacitors were charged in parallel and discharged in series through multiple spark gaps to produce an in-phase oscillatory current. The mean diameter of the circular array should generally be from 0.2 to 0.3 of one wavelength if exponentially decaying pulses of useful duration were to be produced. This is because the diameter of the current path alone determines the radiation resistance. Under these circumstances the ring radiates as a short magnetic dipole. The frequency of oscillation is determined by a capacitor composed of all capacitors along the circumference in series and by a circuit composed of the circuits associated with each capacitor. The circuits associated with the current path are the internal circuits of the capacitors, of their connecting leads, and of the spark gaps. For highest efficiency all these circuits must be made as small as possible, otherwise the energy storage of the transmitter is reduced, because in a spark transmitter all energy is initially stored in the capacitor (see Fig. E1e-1).

Apart from the simplicity of the concept of such a transmitter, two particular advantages were claimed: (1) Very high peak powers, possibly extending into the gigawatt range, could be produced; (2) no scalar potentials exceeding the charging potential of the capacitors appear anywhere in the circuit if the condition of circular symmetry of the current path is satisfied.

In the most successful design the multiple cables were replaced by a "radial transmission line." All tuning capacitors were connected between the outer peripheries of two circular copper disks, as shown in Fig. E1e-2. The capacitors were mounted slanting with respect to the plane of the disk so that the major component of the current was tangential to the disks. The inner peripheries of the disks were connected through the spark gap to a strip transmission line of adjustable length to complete the required electrical length of one-half wavelength between the outer circumference of the lines and the short-circuiting bar of the transmission line. When this arrangement was first tested, it was found that very strong undesirable eddy currents were induced in the disks by the peripheral rf current. These currents were suppressed by providing radial slots ~ 3 mm wide in the copper disks, as indicated in Fig. E1e-2, and this resulted in a large increase in radiated power. However, the tuning of the transmitter depended unpredictably on the spacing between the copper disks, whereas one would expect it to be substantially independent of this spacing until the distance between the disks amounts to an appreciable fraction of one wavelength.

From the foregoing it is clear that means must be provided to establish the

1. VERY FAST VOLTAGE PULSE GENERATORS 163

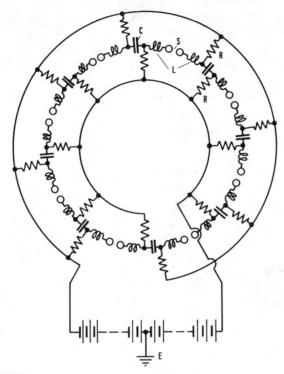

FIG. E1e-1. Circuit of multispark transmitter [1264].

dominant (circumferential) mode of oscillation as rapidly as possible. Initially, it appeared most attractive to couple a resonant circuit or a tuned stub into the half-wave transmission line in the vicinity of the spark gap, but attempts to do this were unsuccessful and will not be described.

Most of these experiments were carried out at a frequency of 70 MHz. The

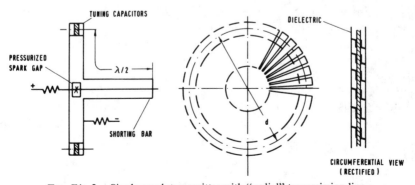

FIG. E1e-2. Single spark transmitter with "radial" transmission line.

mean diameter of the current path at this frequency is almost exactly 1 m and the number of capacitors distributed evenly around the circumference was fixed at 40. Transmitters of various configurations have also been operated at 500 MHz. For such a high frequency construction is much easier and cheaper, and free-space testing is more practicable. However, dielectric and spark losses increase, and the small energy storage limits the application of these transmitters to model experiments. In most experiments carried out in order to optimize the configuration described, a strip tuning line was joined to one of the copper disks and to the spark gap, as shown schematically in Fig. E1e-2.

The performance of the 70-MHz transmitter was completely satisfactory. The tuning capacitors of 63.5 pF may be safely used up to 40 kV potential. The calculated peak power is then ~ 70 MW. The efficiency of the transmitter defined as energy radiated to energy stored was determined from field strength measurements made at seven wavelengths from the transmitter. For these measurements a Tektronix type 585 oscilloscope was used. The voltage induced in a small pickup loop was read directly on the calibrated input of the oscilloscope. The field strength measured in this way and the power radiated by the transmitter are related by Eq. (20) [1265]. With the transmitter operating at 40 kV the transmitter radiated 14.8 MW, that is, the efficiency is $14.8/70 \cong 0.2$ or 20%.

f. *Pulse Generators for Testing Equipment*

A versatile pulse generator, especially useful for testing lightning pulse protection equipment, is the so-called Isopuls 10 [1017] made by IPH (Impulsphysik GmbH, Hamburg 56, W. Germany). The Isopuls 10 is a pulse generator producing high-voltage pulses with standardized waveform and defined energy. The pulses are defined by VDE[*] Standard 0433, §3/4.66, which conform to IEC[†] Standard. The parts are subjected to overvoltages in accordance with the VDE Standard. There is a choice of positive or negative pulses, and the impulse voltage is continuously adjustable. The instrument can also be used for the following applications, in some cases using additional plug-in units: stabilized dc voltage generator with choice of polarity up to 10 kV; impulse generator for square wave or triangular pulses up to 10 kV, for low-impedance loads; high-voltage generator up to 500 kV, using an impulse transformer. It has three ranges, 0.25–2.5, 0.5–5.0, and 1.0–10.0 kV, and there is free choice of polarity. The voltage is continuously adjustable within the ranges given below. The storage capacity is $1+3+6$ μF. Process control permits a pulse separation between 10 sec and 6 min. Plug-in units provide

---

[*] Verband Deutscher Elektrotechniker.
[†] International Electrical Commission.

standardized waves of 1.2 μsec/50 μsec, 5 μsec/∞, 50 μsec/500 μsec, and 10 μsec/1000 μsec; low rise/decay time is used for low-impedance objects, and longer time for high impedances, like transformers.

A special feature of the Isopuls is the possibility of measuring the peak voltage at which the breakdown of the tested device has occurred.

2. Kerr Cell and Pockels Cell Pulsers and Related Techniques

Kerr cells are often used with a high-voltage rectangular pulse generator to open camera lenses or spectrographs for a predetermined time or to cut the light tail of the afterglow of a spark discharge lamp. For the theory of Kerr cells, see Refs. [1282] and [1025]. The molecules of the liquid which by their very nature are electric dipoles (as for instance those of nitrobenzene, carbon disulfide, or water) assume statistically all possible directions in space so that the liquid is isotropic. As soon as an electric field is applied the molecules are aligned; the liquid becomes optically active and assumes the characteristic of a uniaxial crystal, the optical axis of which is in the direction of the electrical field. This effect is so fast that its delay is negligible down to the nanosecond range. The effect was discovered in 1875 by J. Kerr and later on named after him.

If the electric vector of linearly polarized light is at an angle of 45° to the electric field of the Kerr cell, the two components parallel and perpendicular to the electric field travel with different velocities. Therefore the light leaves the liquid in an elliptically polarized state. In order to use this phenomenon for light modulation, a Kerr cell is used. This consists of a plate capacitor placed inside a closed housing with windows, and having a Kerr liquid as dielectric (Fig. E2-1). Usually the cell is arranged between the two crossed polarizers in such a way that the polarizing planes of the latter are at a 45° angle to the electric field vector.

In the nonexcited state therefore, no light can pass through this system. If, however, a voltage is applied to the Kerr cell then elliptically polarized light will arrive at the analyzer. This transmits the component which is in its polarizing plane and the system allows some light to pass. The intensity I of

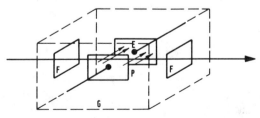

Fig. F2-1. Kerr cell.

the light beam which leaves the cell can then be calculated, if reflection and absorption losses are neglected, according to the following formula:

$$I = \tfrac{1}{2}I_0 \sin^2(\tfrac{1}{2}\phi) \tag{1}$$

where I_0 is the intensity of the unpolarized light beam which enters the polarizer and ϕ is the phase difference between the two vector components of the light parallel and perpendicular to the electric field, and is given by

$$\phi = 2\pi k l \mathbf{E}^2 \tag{2}$$

where k is the Kerr constant, l the length of light path in the Kerr cell, and $\mathbf{E}$ the electric field vector. By combining Eqs. (1) and (2), the light intensity can be written as

$$I = \tfrac{1}{2}I_0 \sin^2(\pi k l \mathbf{E}^2) \tag{3}$$

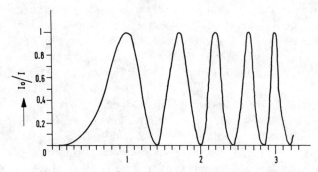

FIG. E2-2. Effect of voltage on transmission of a Kerr cell (voltage in units of first opening voltage).

Figure E2-2 shows this characteristic of the light intensity as a function of the voltage applied to the Kerr cell. The abscissa is in units of the first opening voltage. The opening voltage of the Kerr cell may be defined as that voltage which must be applied to the cell in order to achieve $I = I_0/2$ and there obtain the maximum brightness of the transmitted light beam. This is the case if

$$\sin^2(\pi k l \mathbf{E}^2) = 1 \tag{4}$$

or

$$\pi k l \mathbf{E}^2 = \pi/2, \quad 3\pi/2, \quad 5\pi/2, \quad \text{etc.} \tag{5}$$

is complied with. The total rotation of the plane of polarization during passage through the liquid must therefore, with this arrangement (i.e., crossed polarizers), amount to 90°, 270°, 450°, etc.

For a light path l and a plate spacing d, the first and subsequent opening

2. KERR CELL AND POCKELS CELL PULSERS

TABLE E2-1

CHARACTERISTICS OF SOME KERR LIQUIDS

Liquids	Kerr constant k at 538 nm V cm^{-2}	Dielectric constant ε	Short wave cutoff (nm)
Nitrobenzene	4.3×10^{-10} (546 nm)	36	440
Benzonitrile	1.3×10^{-10}	26.3	330
Water	4.4×10^{-12}	80	180
Carbon disulfide	3.7×10^{-12}	5.1	—
Chloroform	3.2×10^{-12}	2.6	—

voltages u are calculated as follows:

$$\mathbf{E}^2 = 1/2kl, \quad 3/2kl, \quad 5/2kl, \quad \text{etc.} \tag{6}$$

or

$$u = d \cdot (1/2kl)^{1/2}, \quad d \cdot (3/2kl)^{1/2}, \quad d \cdot (5/2kl)^{1/2}, \quad \text{etc.} \tag{7}$$

Characteristics of some Kerr liquids are shown in Table E2-1.

The Impulsphysics Association manufactures three basic types of Kerr cells shown in Table E2-2. All these Kerr cells are completely demountable. This allows:

(1) use of different Kerr liquids;
(2) change of electrodes for different electrode distances;
(3) replacement of windows in case of damage, adjustment of windows for high Q-switch efficiency (sapphire windows), and insertion of uv-transparent windows for use of the Kerr cells in the ultraviolet spectral range;
(4) use of different polarizers; and
(5) use of polarizing foils and/or adjustable prisms.

The Kerr cells are made of materials selected very carefully on the basis of many years' experience, so as not to cause contamination of the nitrobenzene or other Kerr liquid.

A typical circuit is shown by Preonas and Swift [1022]. The Kerr cell chosen for their task has an aperture of 1.52×3.91 cm and requires 25 kV across its electrodes for full opening. Normally, the cell is placed immediately in front of the light source spark gap as shown in Fig. E2-3. The Kerr cell may also be mounted in front of the camera whenever self-luminosity of the event presents problems.

A 25-kV square-wave pulse generator with an adjustable pulse duration and with rise and fall times of less than 10^{-7} sec has been developed to drive the Kerr cell (see Fig. E2-4). The operating sequence of the light source is initiated by triggering the light source and two time-delay generators. The

TABLE E2-2

DATA FOR KERR CELLS

	Minikerr	Standardkerr	HV–Kerr
Optical data			
Microadjustment of polarizers	×[a]	×[a]	×[a]
Prisms	×[a]	—[b]	×[a]
Polarizing foils	—[b]	×[a]	—[b]
Entrance window	10 × 10 mm	24 × 30 mm	5 mm diam
Blocking factor[c]	$1.6 \times 10_5$	10_4	$1.6 \times 10_5$
Spectral range when filled with nitrobenzene, $C_6H_5NO_2$	450–1400 nm	480–820 nm[d]	450–1400 nm
Setting accuracy of the adjustable windows BK7 or suprasil better than 30 in.	×[a]	×[a]	×[a]
Plainness of plates better than $\lambda/10$ NaD	×[a]	×[a]	×[a]
Electrical data[e]			
Electrical data			
Opening voltages: 450 nm	18	18	15
(kV) 700 nm	24	24	23.2
$\lambda/4$ voltage 694.3 nm	15.5 kV	—[b]	—[b]
Cross resistance	$5 \times 10_8$	10_8	$3 \times 10_8$
Capacitance, pF	30	63	42

[a] Available with this Kerr cell.
[b] Not available with this Kerr cell.
[c] The blocking factor refers to a beam of 1 mm diam with a deviation of up to 4° from the optical axis, or to a parallel light beam of 8 mm diam with 0° deviation from the optical axis.
[d] These low values are due to the use of polarizing foils.
[e] For HV–Kerr cell, at 10 mm electrode distance.

first time-delay generator is used to trigger spark gap 1 of the square-wave generator, opening the Kerr cell after the light source has reached sufficient intensity. The second time-delay generator triggers spark gap 2, short-circuiting the Kerr cell at the end of the desired optical output. Three resistors are placed strategically in the discharge circuit to stabilize the capacitor

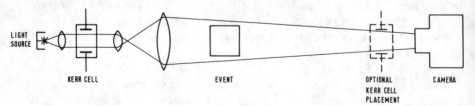

FIG. E2-3. Optical schematic showing Kerr cell placement.

2. KERR CELL AND POCKELS CELL PULSERS

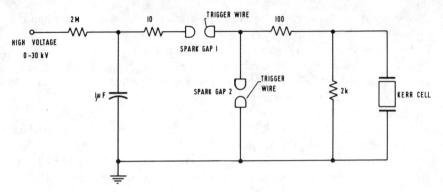

FIG. E2-4. Kerr cell pulse forming network.

discharge and to inhibit oscillations. These resistors are stacks of 70 2-W composition resistors in a seven parallel–ten series combination. This geometry is required to carry high power—the 10-Ω resistor stack dissipating ~ 60 MW peak power when the second gap is fired.

Figure E2-5 is a circuit diagram of the basic light source. Energy is stored in three 14-μF 20-kV energy storage capacitors (a total of 8.4 kJ). This energy is switched through a three-element spark switch and a tapped inductor. The pulse energy is transmitted to the remotely located light source gap through two RG8U coaxial cables wired in parallel. Double cables are required to carry the current pulse without damage; they also lower the circuit inductance. Care was taken to make solid contact with both the central and outer conductors of the cables and to restrain the cables from moving under the influence of the magnetic forces. The internal inductance of the capacitors, the spark switch, and 8 m of cables are sufficient to extend the light source output to

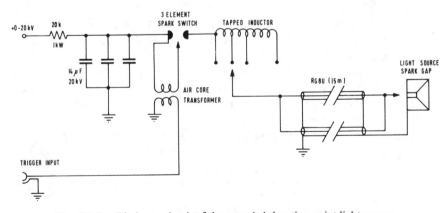

FIG. E2-5. Discharge circuit of the extended-duration point light source.

20 μsec. Various taps on the air core inductor are employed to lengthen the source duration stepwise to 1000 μsec.

The light source spark gap consists of three coaxial elements shown in Fig. E2-6. The steel ground electrode is essentially a cylinder with a small central hole to emit light from the spark channel. The high-voltage electrode is a 0.32-cm drill rod sharpened to a 90° included angle cone. The two electrodes are separated by a Plexiglass washer that restrains the lateral expansion of the arc channel and seeds the arc with organic matter which substantially increases the plasma luminosity. A similar application had already been described in 1965 by Fründel.* A multiple Kerr cell camera operated by coordinated pulsers is the subject of a paper by Persson [1826].

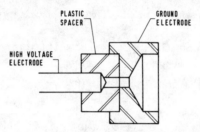

FIG. E2-6. Coaxial spark gap light source elements.

Pockels cells can be operated at lower voltages (see also this volume, Chapter B3). Some commercially available cells are shown in Table E2-3. The only disadvantage of these cells is its limitation to nearly parallel light beams. Rowlands and Wentz [1023] describe a low-voltage Pockels cell permitting high repetition rates and short exposure durations. For those who are more familiar with the conventional technique of using tubes, it may be interesting to show a simple circuit for driving Pockels cells with high frequency bursts of rectangular pulses.

The cell is ideally suited for ultrahigh speed photography (conventional and holographic) either for multiple shuttering of a steady or quasisteady light source or for sequentially Q-switching a laser light source. It permits independent control of exposure duration and framing rate. The electrooptical cell consists of two potassium dideuterium phosphate crystals, both arranged so that their optical and electrical axes are perpendicular to the light axis. Switching requires considerably less than 400 V, and the cell can be operated at a rate well beyond 5,000,000 frames/sec. Variable exposure durations as low as 5 nsec and rise and fall times of a few nanoseconds are readily attainable. While emphasis is on utilization of the cell as either a shutter or as a laser modulator, the application of crystal units for the rapid deflection of light

---

* High Speed Pulse Technology, Vol. I, pp. 277–279.

TABLE E2-3

TRANSVERSE POCKELS CELL MODULATORS FOR Q-SWITCH APPLICATIONS AND MODULATION[a]

Type	01.PC19	01.PC20	01.PC17/2.5	01.PC17/5	01.PC22
Material	KD*P	ADP	ADP	ADP	KD*P
Half-wave voltage at 0.6328 μV	1200	1000	200	400	500
Open beam diam, mm	6	6	2.5	5	3
Length of crystal, mm	36	36	80	80	36
Diam. of cell, mm	50	50	45 × 48	45 × 48	50
Length of cell, mm	87	87	127	127	87
Window material	Spectrosil	Spectrosil	Spectrosil	Spectrosil	Spectrosil
Electrodes	Gold	Gold	Gold	Gold	Gold
Highest permissible voltage in kV (peak)	5	5	0.4	0.8	2
Spectral transmission in μ	0.3–1.1	0.3–1.1	0.3–1.1	0.3–1.1	0.3–1.3
Capacity, pF	10	15	60	60	10
Dark-bright ratio at 1 mrad divergence	1000:1	100:1	20:1	100:1	50:1

[a] See Ref. [1855].

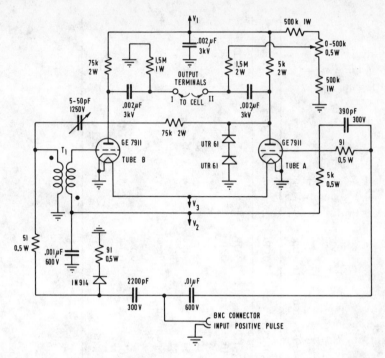

FIG. E2-7. Pockels cell driver. $V_1 = 1100$ V dc; $V_2 = 30$ V dc; $V_3 = 6.3$ V dc.

beams and the harmonic generation of light of different frequencies is also indicated.

Figure E2-7 is a schematic diagram of the circuit used to drive the electro-optical cell. Rapid rise and fall times are desirable when employing cells either as Q-switches or photographic shutters. The electronic driver of Fig. E2-7 has a rise and fall time of a few nanoseconds. The circuit depends upon the turn-on time of the respective tubes to produce the rise and fall time of the voltage pulse applied to the two crystal systems. At the driving frequencies of interest, the electrooptical crystals appear as pure electrical capacitances to the driving circuit. When operating as a laser Q-switch, both tubes are initially in the off condition and the voltage across the crystals will be the quarter-wave voltage by virtue of the biasing circuitry.

The technique of light modulation and beam deflection with potassium tantalate–niobate crystals is given by Chen *et al.* [1284]. The dielectric and electrooptic properties of $KTa_xNb_{1-x}O_3$ (KTN) are discussed from the point of view of the material's usefulness in light modulators and beam deflectors. It is shown that baseband light modulators with 200–300 MHz bandwidths and analog deflectors with 200–300 resolvable spots are within the practical capabilities of this material. A pulsed laser Kerr system polarimeter for electrooptical fringe pattern measurement of transient electrical parameters is described by Cassidy [1283].

Instead of nitrobenzene, liquid carbon disulphide can be used [1244] if a high nonlinear optical performance is needed. Cuchy and Landovsky [1176] describe an electrooptical amplitude modulator using the linear electrooptic effect on ADP, KDP, RDP, DKDP, and LiNbO$_3$ crystals. The modulator of the Billings cell type needs a modulation voltage of some kilovolts. Its great advantage is its simple construction with a large aperture. For the modulation of a laser beam, MLZ and MLY-45 modulators are recommended. For a 70% modulation, they need a voltage of only 180 or 100 V, respectively. Table E2-4 gives in addition to the optical data, the amplitude needed from the pulse generator to ensure proper modulation with each type of crystal. Instead of using Kerr cells as shutters, they can be used for the stepwise deflection of a parallel laser beam for sampling areas. The technique is described in detail by Hill [1175].

A paper by Hepner [1171] also deals with digital light deflection, using prisms and a polarization switch based on the Pockels effect with transverse field. The rapid deflection of light can be applied to the reading and writing of optical memories, problems of visualization, and optical printers. Digital deflection has the advantage of accuracy of addressing and simplicity of command.

The Kerr effect permits a large angular field but, for a great number of beam positions, the voltage is very high. Cubic crystals are interesting for this purpose but are difficult to obtain in good optical quality. So, KDP with transverse electrical field in a birefringent compensated mounting was preferred. The switching voltage is less than 1000 V, and there are no electrodes in the optical path. The angular field in this mounting is 10° for a polarization switching of better than 90%. The angular field typical for switching with KDP in a longitudinal mounting is 1°.

The design and development of an electrically tunable birefringent filter using dihydrogen phosphate type crystals is described by Ahmed and Ley [1169]. It is the first practical realization of an old idea, which the French astronomer Bernard Lyot proposed in 1933 [1827].

Repetitive pulsing of a Kerr cell modulator operating with a laser is the basis of stroboscopic holography using an electrooptical modulator. An example of this unusual kind of system is described by Kennedy et al. [1170]. Holographic interferometry depends on the exact superposition of the object under investigation and a three-dimensional reconstruction (hologram) of the object. Any subsequent displacement of the actual object is then registered as a series of interference fringes, the plane of localization of which is dependent upon the form of displacement. The number of interference fringes is indicative of the magnitude of the displacement of the object expressed in terms of the

TABLE E2-4
DATA FOR POCKELS CELL CRYSTALS

Crystal	Index of refraction n_o/n_e at 632.8 nm	Cut	Electrooptic constant (10^{-12} m/V)	Voltage for $\lambda/2$ retardation (kV) at 632.8 nm	Crystallographic system	Maximal dimensions of the crystals $X \times Y \times Z$ (mm)
ADP	1.5246	Z 0°	$r_{41} = 21$	10.9	tetragonal	$50 \times 50 \times 80$
$NH_4H_2PO_4$	1.4792	Y 45°	$r_{63} = 8.3$			
KDP	1.5095	Z 0°	$r_{41} = 9$	8.6	tetragonal	$40 \times 40 \times 60$
KH_2PO_4	1.4684	Z 45°	$r_{63} = 10.7$			
RDP	1.479[a]	Z 0°	$r_{63} = -11$	5.6	tetragonal	$20 \times 20 \times 30$
RbH_2PO_4	1.508[a]					
DKDP	1.503	Z 0°	$r_{41} = 8.8$	3.6	tetragonal	$20 \times 20 \times 30$
KD_2PO_4	1.464	Z 45°	$r_{63} = 26.4$			
$LiNbO_3$	2.163	Z 0°	$r_{23} = r_{13} = 8$	2.8	trigonal	$10 \times 10 \times 10$
	2.277		$r_{33} = r_{26}$			
			$r_{51} = r_{42} = 28$			
			$r_{12} = r_{22} = 3.4$			

[a] At wavelength 546 nm.

wavelength of the laser light for illumination and the angles of viewing and illumination within the interferometer. Figure E2-8 illustrates the layout of the various components of the two-beam interferometer used in this project.

The difficulties associated with the movement of the object during periods of illumination—manifested as a general "jitter" of the interference fringe pattern—were reduced to an acceptable level by utilizing the minimum possible light pulse duration obtainable from the system. Ideally, light pulses of 20 nsec duration would be necessary to reduce the fringe jitter to an imperceptible level; however, such narrow pulses were entirely beyond the scope of the Pockels cell/switch combination. A as direct consequence of the modulation

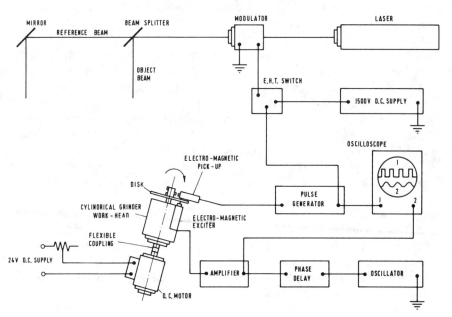

FIG. E2-8. General layout of interferometer.

of the cw laser beam to produce a series of extremely short duration pulses of light, the general intensity of light reflected from the surface of the object was considerable reduced, e.g., for a rotational speed of 2500 rpm and pulse widths of 10 μsec, the intensity of the light reflected from the object was found to be 0.05 mW/cm^2.

The sudden pulsing of optically responsive crystals for shutter purposes is the principle applied in a patent by Marks *et al.* [1085] on an electrooptically responsive flashblindness controlling device. This invention relates to solid-state devices and specifically, such as are capable of controlling the passage of light through a partially transparent assembly of electrooptically responsive crystals and light polarizers in response to the application of an electrical charge.

A Russian paper by Marugin and Ovchinnikov [1647] deals with the effect of electrode arrangement and z-cut KDP crystal configuration on the control voltage for switching an electrooptical shutter. It shows the dependence of the pulse voltage at the electrodes of a simple KDP crystal on the crystal length. The influence of the length is small and amounts between 10 and 14 kV.

A Pockels effect electrooptic image modulator, which combines properties that make it uniquely applicable in many areas of optical image processing, has recently been developed at Itek Corporation [2006]. The device combines high sensitivity, resolution, and optical efficiency with high speed of operation and apparently unlimited recyclability. The active material consists of single-crystal $Bi_{12}SiO_{20}$ (bismuth silicon oxide) which has a large electrooptic effect at relatively low voltage, a sensitive photoconductive effect which modulates the electrooptic voltage in an imagewise pattern, and high resistivity with long trapping times, providing optical memory effects over many hours. Device applications currently being investigated include incoherent-to-coherent converter—an incoherently illuminated scene read into the device is read out with coherent light for coherent optical processing; block data composer—serial data read into and stored on the device are read out coherently into a holographic memory; on-line Fourier-plane filter—Vander Lugt filters are composed on the device for real-time Fourier-plane cross correlation.

An unusual way of obtaining picosecond light pulses was described in 1970 by Duguay and Hansen [1027] in their account of how pulses of light were photographed in flight for the first time. Picosecond green light pulses passing through a light scattering medium were photographed by a camera positioned behind a shutter of $\geqslant 10$ psec framing time. Color photographs show a packet of green light suspended within the scattering medium. The experiment demonstrates a fundamentally new and yet simple way of visualizing light pulses, viz., to look at them from behind a fast shutter. The method can also be used for the photographic measurement of ultrafast relaxation times in dielectrics and fluorescent dyes.

Green light pulses at 0.53 μm are obtained by second harmonic generation from the 1.06 μm pulses generated by a mode-locked Nd:glass laser. The laser generates infrared pulses ~ 8 psec duration. The harmonically derived green ones are ~ 6 psec in duration. The green pulses are sent through a cell containing water to which has been added a small amount of powdered milk in order to make the liquid an efficient scatterer. A camera is positioned behind an ultrafast shutter. This shutter or light gate is a new type of Kerr cell optically rather than electrically driven. The 1.06 μm pulses are used to drive the ultrafast Kerr cell shutter. When CS_2 is used in the shutter, the opening time or framing

2. KERR CELL AND POCKELS CELL PULSERS

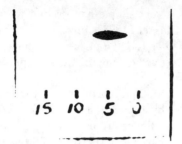

FIG. E2-9. A "flying" parcel of green light photographed by an optically generated vector of an infrared picosecond pulse.

time of the shutter is ≥ 10 psec. The infrared and green pulses are generated in synchronism: Using variable optical delay lines, it is easy to arrange for the shutter to open at any chosen time during the passage of the green pulse through the water cell.

The type of photograph obtained in this fashion is shown in Fig. E2-9. When the shutter is properly synchronized, a packet of green light is seen as if suspended in the middle of the cell. By changing the time at which the shutter opens, the pulse could be seen at the entrance or at the end of the cell. The pulse appears as a bright spot on a very dark background.

German readers are recommended to read "Kurzzeitphysik" by Vollrath and Thomer [1703], in which Müller [1703 f] wrote an excellent chapter about all types of electrooptical shutters. Pages 253 through 258 contain a further 118 references. Two of the many circuits are shown in Figs. E2-10 and 11 as examples.

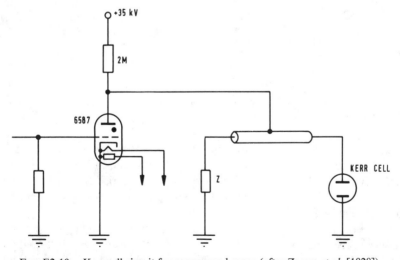

FIG. E2-10. Kerr cell circuit for nanosecond range (after Zarem *et al.* [1828]).

178   E. CONVERSION OF CAPACITOR ENERGY AND ITS APPLICATIONS

Figure E2-10 shows a very simple nanosecond Kerr cell pulser, given by Zarem, Marshall, and Hauser [1828]. Figure E2-11 shows a subnanosecond pulser, given by Hull [1829]. It is like a Lecher system with a crowbar, the pulse voltage being on the Kerr cell during the time the pulse voltage takes to pass the crowbarring cable.

Turning now to picosecond pulses for auxiliary purposes, it is also possible to step up voltage pulses by application of so-called snap-off diodes as shown by Pfeiffer [1856]. These diodes are used to modify the output pulse of an avalanche pulse generator to give a rise time of 70 psec only. Unfortunately,

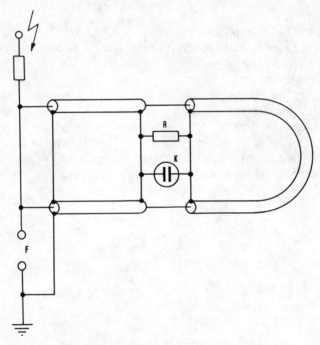

FIG. E2-11. Pulse generator (after Hull [1829]). F = switching spark gap, R = resistor to be fed, K = Kerr cell.

this technique is limited to pulse voltages of 15 V although it appears possible to amplify such pulses. van de Veeke [1814] describes a high-voltage pulse amplifier which drives capacitive loads with short rise times. For such a purpose, current must be delivered to and withdrawn from the load only during the pulse fronts.

Figure E2-12 shows a practical circuit designed for supplying steep positive 'ses between grid and cathode (in particular for testing the spot quality of tubes). Coupling capacitor C_2 must be chosen for optimum results,

2. KERR CELL AND POCKELS CELL PULSERS

as the charge on this capacitor delivers the base current of T_{2a} during the charge time of C_{load}.

When the T_1 transistors are turned on, part of V_{cc} appears across T_{2C} for a fraction of a microsecond. Thus T_{2C} breaks down briefly before the divided resistors have time to recharge the base capacitors. The energy stored in the (18 pF) base capacitors, however, is insufficient to destroy the junctions; optimal rise time of the leading edge requires only small capacitances.

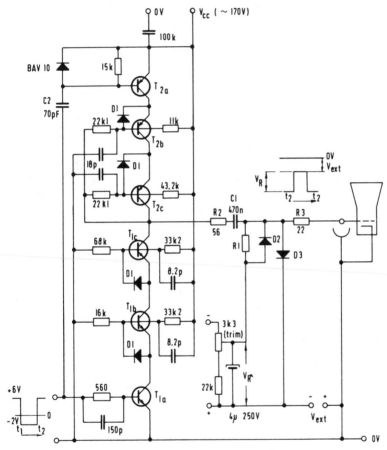

FIG. E2-12. Practical circuit for steep rectangular high voltage pulses to open grids of tubes. T_{1a-c} = BSX 60; T_{2a-c} = 2N2905 A; D1 = BAV 10; D2 = D3 = BA 145.

The same remarks apply to the base capacitors in T_1 transistors. The output pulse is capacitively coupled to the load via C_1, in order to add to the pulse voltage an external dc voltage V_{ext} for correct adjustment of the picture tube current. Diode D_2 and resistor R_1 serve as dc restorer, while the clipping diode

D_3 gives the pulse a flat top. With $V_{ext} = 0$ V, the pulse must never exceed the zero volt line, which is accomplished by making V_R somewhat smaller than the pulse amplitude (about V_{cc}). Finally, resistors R_2 and R_3 have been added to avoid damage to the circuit by current surges during short-circuiting of the output terminals. The specifications of the pulse amplifier with an amplitude ~ 170 V are shown in the following table.

C_{load} (pF)	t_{rise} (nsec)	t_{fall} (nsec)
20	20	15
100	30	25
300	60	50

3. Pulsers for Laser Excitation and Flash Photolysis

a. *Laser Excitation*

The usual laser pumping lamps need conventional capacitors as described in Chapter A6. Because of their millisecond operation, there is no special need to develop sophisticated circuits. Organic dye lasers and the superradiant nitrogen laser operate in a different way that is similar to the technique of flash photolysis. In a few microseconds, energies up to 100 J must be fed into the flash lamp. Ewanizky and Wright, Jr. [1859] developed a coaxial Marx–bank driver and flash lamp for optical excitation of organic dye lasers.

Since organic dye lasers typically have high power thresholds, an efficient excitation source should produce fast rise, high-intensity optical pulses. The design of flash lamp pumps has, correspondingly, evolved to coaxial flash lamps with driver circuits of small capacitance and low inductance. However, practical difficulties may arise at even moderate levels of input energy since the small values of capacitance generally used to produce microsecond duration pulses require that relatively high dc charging voltages be applied. Meticulous care with insulation and large components with high dc voltage ratings must be used to avoid frequent breakdown. In an effort to alleviate these problems, but retain compact, low-inductance component packaging, a coaxial Marx–bank type driver was developed. The essential property of the Marx–bank circuit is that a number of storage capacitors may be charged in parallel, then discharged in series connection. An advantage immediately realized in practical circuit application is that smaller components need only be charged to a relatively moderate dc voltage to obtain high pulse voltage and input energy to drive the flash lamp.

A schematic of the two-element circuit developed here is shown in Fig. Initially, the storage capacitors are all charged in parallel by the dc power supply. When the first spark gap is triggered, a negative voltage appears at the ground leg of the center spark gap which causes it to break down rapidly. This process would continue at an accelerated rate in a circuit with a larger number of elements since the spark gap overvoltage increases at each stage. When all spark gaps are conducting, the storage capacitors are effectively connected in series with the flash lamp. The number of stages employed would be limited when the total spark gap and circuit resistance are appreciable compared to that of the flash lamp or if pulse shape is degraded due to large total inductance. This latter factor may be minimized by coaxial construction with low-inductance components.

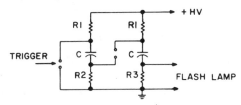

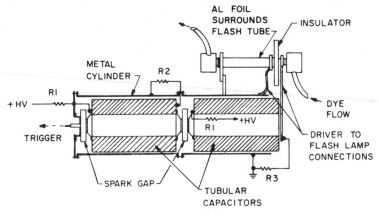

FIG. E3a-1. Coaxial Marx–bank driver and flash lamp. (a) Circuit schematic: $R1 = 4$ MΩ, $R2 = 160$ kΩ, $R3 = 6.2$ kΩ. (b) General arrangement of components. Other components are described in the text.

Fig. E3a-1 shows a sketch of the finished unit with a commercial flash lamp connected. Cylindrical, low-inductance, 0.3-μF capacitors and pancake-type spark gaps were installed in a close fitting metal cylinder lined with a layer of Mylar foil. Simple, copper metal sheet straps connect the capacitors and spark gaps. Insulated leads were brought out either directly through the cylinder wall or through the central hole in the tubular capacitors to external charging resistors.

TABLE E3a-1

LASER SYSTEM PERFORMANCE[a]

Experimental conditions	dc voltage (kV)	Input energy (J)	Laser output energy (mJ)	Total efficiency (%)
(a)	12.5	47	200	0.42
(a)	15	67.5	295	0.44
(a)	17.5	92	380	0.41
(b)	17.5	92	320	0.35

[a] Output measurements were made with an ITL laser calorimeter placed near the front of the laser. Experimental conditions: (a) Phase-R flash lamp with 140-Torr xenon gas-filled pressure, close-spaced plane-parallel cavity with 42% reflectivity (at 5900 Å) dielectric front mirror, 5×10^{-5} M solution of rhodamine 6G in ethanol; (b) lab constructed coaxial flash lamp with 11-cm anode separation and 8-mm annular diameter, 50-Torr xenon gas filled, no front mirror, and 1×10^{-4} M rhodamine 6G solution.

Both commercial and laboratory constructed coaxial flash lamps were used with the driver circuit. The lamps were all similar in design and were generally fitted with an external xenon gas supply to give some flexibility in matching impedances with the driver. An external nitrogen gas supply was also fitted to both spark gaps for more reliable operation over a larger voltage holdoff range than could be obtained from the standard sealed units. Nitrogen under pressure of 6 psi was sufficient to give an operating range of 10-~20-kV dc voltage, corresponding to an input energy of 100 J. Table E3a-1 shows some results obtained with an organic dye laser and two different flash lamps.

Otis [1813], in his work on double discharge lasers of the type first described by Laflamme [1860], has proven that the use of resistive plastic materials in the trigger circuit is excellent in producing increased trigger current values, and in achieving successful operation of lasers with large (up to 7.6 cm) electrode separations.

The laser is a three-electrode system, shown in Fig. E3-2. The anode and

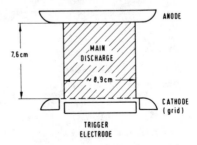

FIG. E3a-2. Sectional view (not to scale) of the double-discharge laser.

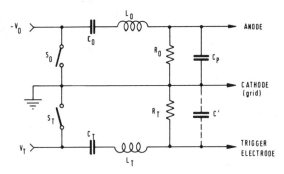

FIG. E3a-3. Electrical discharge circuits. V_O and V_T, charging voltages; C_O and C_T, storage capacitors; S_O and S_T, spark gaps; L_O and L_T, inductances; C_p, peaking capacitor; C', trigger stray capacitance; R_O and R_T, charging resistors.

cathode of the main discharge are Rogowski-shaped electrodes, separated by a 7.62-cm spacing. The trigger electrode assembly is inserted in a 10.16 × 17.78 cm hole in the cathode, over which the grid (14-mesh copper-wire cloth screening) is stretched. This ensures that the main discharge is triggered over about 8.9 × 16.5 cm of the central plane portion of the main electrodes. Each discharge is driven in sequence by a separate discharge circuit (Fig. E3-3) using a delay generator between the two spark gaps (S_O and S_T) for proper timing. This system yields reproducible main discharges under the experimental conditions listed in Table E3-2. Small signal gain values of $\alpha_0 = 0.04$ cm^{-1} have been measured by the threshold excitation method, in low He-content gas mixtures such as $He:N_2:CO_2 = 50:20:30$, at $\frac{1}{3}$ atm total pressure. Successful operation at higher pressures (up to 500 Torr), and at larger electrode spacing (up to 10 cm) has been obtained, through a proportional increase in the charging voltage.

In this geometry of the double-discharge laser, it is very important that the

TABLE E3a-2

TYPICAL OPERATING CONDITIONS OF THE DOUBLE-DISCHARGE LASER

Main discharge	Trigger discharge
$V_0 = 32$ kV	$V_T = 20$ kV
$C_0 = 0.1$ μF $= 3C_p$	$C_T = 0.01$ μF
$L_0 \simeq 20$ μH	$L_T = 5$ μH
Gas pressure—$\frac{1}{3}$ atm	
Optimum gas mixture—$He:N_2:CO_2 = 50:20:30$	
Laser small signal gain—$\alpha_0 \simeq 0.04$ cm^{-1}	
Resistive plastic material—Eccosorb HF 1000 (0.635 cm thick)	
$\varepsilon \simeq 100$ (at 1 MHz), $\rho = 10^4 - 10^5$ Ωcm	

trigger discharge injects a large number of electrons in the main discharge, thus producing an electron sheath suitable for initiating the main discharge. The use of Mylar foils and other dielectrics for electrically insulating the grid from the trigger electrode yielded satisfactory results [1860]. But the main drawbacks in using thin Mylar foils (about 0.15 mm thick) are the low trigger current values achieved and the low perforating strength under local heating due to accidental bright arcs in the main discharge volume. The author found that the use of resistive plastic materials [1861] (consisting essentially of graphite particles suspended in a polymer), between the trigger electrode and the grid, offers interesting new possibilities. This material is available in a wide range of dielectric constants ($e = 10$–2000) and resistivities ($p = 10$–10^5 Ω-cm). In fact, it is important that the grid-trigger assembly has a nonnegligible stray capacitance C^1 (see Fig. E3a-3), which should be well selected for energy storing and for pulse shaping in the trigger discharge circuit. We have achieved a trigger current density of ~ 6.1 A/cm^2, which represents a tenfold increase in the trigger current, as compared to that obtained with previously used insulating materials.

In view of the large dielectric constant of this resistive plastic material, thick slabs can be used instead of thin foils for a given value of the stray capacitance C^1. One then avoids the problems associated with the dielectric strength of the insulating material. Strength under accidental bright arcs proved excellent.

As a further advantage, the finite resistivity of the material makes it act as a resistor in the circuit, thus damping the trigger current and stabilizing the trigger discharge. Because the resistance depends on the current path, the resistive plastic material also provides excellent transverse insulation, thus limiting any surface current and preventing any arc production in the trigger discharge. When used in the double-discharge device, this resistive plastic material is capable of triggering a 7.6×8.9 cm cross-section main discharge with a length of 16.5 cm along the laser axis.

A compact high speed, low impedance Blumlein line for high-voltage pulse shaping is described by Crouch and Risk [1198]. Glycerol was used as a dielectric to obtain a 14.5-Ω line which in conjunction with a Marx generator driver could produce 5-nsec-long 240-kV pulses. The line, which is small enough to be housed in a Lucite box 30 cm wide $\times$ 30 cm long $\times$ 13 cm thick, was used to drive a bank of four 2.2 m long $\times$ 30 cm wide 7.62-cm gap streamer chambers. Figure E3a-4 shows a Blumlein line consisting of three parallel plates, equally spaced with the center plate shorter than the outer plates. The ignition by overvoltage on the spark gap G switches the center plate to ground after the plate has been charged by the high-voltage source HV. If V_0 is the breakdown voltage of G, then a ground going pulse of amplitude V_0 begins to ᴨgate down the lower line. When the pulse reaches the end of the center ᴄees a total impedance of $\frac{3}{2}Z_0$ due to contributions $\frac{1}{2}Z_0$ from the

3. PULSERS FOR LASER EXCITATION AND FLASH PHOTOLYSIS

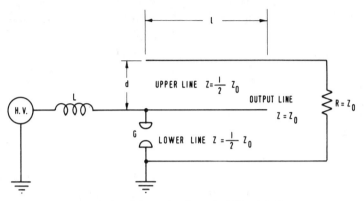

FIG. E3a-4. Schematic of matched Blumlein line.

upper line and Z_0 from the output line. Hence, there are reflected and transmitted current pulses given by

$$I_r = [(\tfrac{3}{2}Z_0 - \tfrac{1}{2}Z_0)/(\tfrac{3}{2}Z_0 + \tfrac{1}{2}Z_0)] I_i = \tfrac{1}{2}I_i, \qquad I_t = I_i - I_r = \tfrac{1}{2}I_i$$

The voltages at the junction are thus

$$V_r = I_r(\tfrac{1}{2}Z_0) = \tfrac{1}{2}V_i, \qquad V_t = I_t(\tfrac{3}{2}Z_0) = \tfrac{3}{2}V_i$$

Of the total $V_t = \tfrac{3}{2}V_i$, there is $\tfrac{1}{2}V_i$ across the upper line and V_i across the output line, i.e., the output voltage equals the input voltage.

As the output pulse V_0 propagates down the output line, the reflected pulse and the pulse in the upper line, which are of the same polarity, propagate to the end of their respective lines and reflect. The pulse in the lower line reverses its polarity since it is terminated in a shorted spark gap. Therefore, when the two pulses arrive back at the juncture with the output line, they are of opposite polarity and cancel each other, thereby driving the output line to ground. All succeeding reflections for this matched line continue to cancel in a similar manner. Thus the output pulse length is equal to twice the transit time of the lower (and upper) lines.

For driving streamer chambers, the chambers themselves have an impedance Z_0 and serve as the output line. They can then be terminated in a resistance Z_0 to eliminate reflections.

It has already been established by others that a key element in the successful operation of streamer chambers is a proper shaping of the high-voltage pulse. As has been pointed out and verified in our own work, the reason for this is the fact that the usual voltage source for this kind of work is a Marx generator, and Marx generators have too slow a rise time (~ 10–20 nsec at best) to produce the short pulses ($\lesssim 10$ nsec) needed to produce short bright streamers ($\lesssim 1$ cm long). After studying pulse shaping with capacitor discharges and air

Blumlein lines, we concluded that the Blumlein line was a simpler and more reliable method and so we proceeded to concentrate on that approach. Air Blumlein lines, however, while giving excellent pulse shapes were far too bulky for the limited space defined by the experimental apparatus. Furthermore, at the voltages required to drive the chamber ($\gtrsim 100$ kV for a 7.6-cm gap) air was not a satisfactory insulator. Since the area of the line decreases linearly with the dielectric constant of the liquid, there was good reason for trying a high dielectric constant material such as glycerol ($\varepsilon = 44$) rather than transformer oil ($\varepsilon = 2$). Water, with $\varepsilon = 80$, was not used initially because of uncertainty about its long-term stability as an insulating medium. After the glycerol line was operating satisfactorily, the glycerol was replaced by water.

The authors built a 14.5-Ω Blumlein with 3.18-cm gaps, 20.3-cm wide electrodes and a nominal 7.62-cm long central electrode. This line, with a 15.2-cm-diam nylon wall pressurized spark gap and two 15.2-cm-diam spherical section spark gap electrodes produced pulses 240 kV high and 5 nsec wide, and could be housed in a 30 cm^2 × 13 cm thick Lucite box containing the glycerol. A line this small can be attached so close ($\lesssim 15$ cm) to the streamer chambers that it is unnecessary in most cases to take special care to maintain a matched connection between the chambers and line. The resulting combined system of Marx generator, Blumlein line, and connecting electrodes considerably simplified the use of streamer chambers for the experiment. In this same paper, the problems of charging a Blumlein line, its pulse width and related matters are also treated.

For the excitation of superradiant nitrogen lasers, 3 nsec pulses of 20–60 kV are required. The usual practice is to provide a Blumlein line which is fed by a Marx circuit. This technique has been reported in Chapter D1 of this volume. See especially Fig. D1b-2 [1712].

b. *Flash Photolysis*

Flash photolysis is a process of studying fast reactions, especially the excited-state reactions in organic molecules using all the features of discharge technology. Bailey and Hercules [1877] explain that flash photolysis produces a high power of radiant energy as compared with steady-state sources, by dissipating a moderate amount of energy during a short period of time. Powers of 50×10^6 W for a few microseconds are not uncommon with flash equipment. Such high powers are capable of producing high concentrations of short-lived intermediates in chemical reactions.

The technology used in the process is strongly related to modern organic dye laser technology. Willets' [1878] paper shows the historical development 'he evolution of flash and laser photolysis techniques.

ise the electronic processes that occur within a molecule are inde-
ᶜ the mode of excitation, flash photolysis can be used to study the

mechanisms of photochemical reactions. Papers of fundamental importance have been written by Rabinowitch [1543] and Bridge and Porter [1540–1542]. Figure E3b-1 shows the electronic energy level diagram of a typical organic molecule containing a π-electron system. S_0 is the ground state (a singlet state), S^+ is the first excited singlet state, and T_1 and T_2 are the first and second triplet states, respectively. Photon absorption (process 1) raises the molecule to an excited singlet state, in which state it can undergo one of several processes. The most rapid process ($\sim 10^{-12}$ to 10^{-13} sec) is the radiationless loss of excess vibrational energy called internal conversion (process 2). This puts the molecule in the lowest vibrational level of the first excited singlet state where it remains during the lifetime of the state ($\sim 10^{-8}$ sec). A molecule in an excited singlet state may undergo radiationless deactivation to the ground state by internal conversion (process 3); it may undergo a radiative transition to the ground state called fluorescence (process 4); or it may cross to a triplet state via intersystem crossing (process 5), finally reaching the lowest vibrational level of the first triplet state. Because spectroscopic selection rules formally forbid transitions between states of unlike multiplicity, a triplet state has a long intrinsic lifetime ($\sim 10^{-3}$ sec or longer) relative to that of a singlet state ($\sim 10^{-8}$ sec). Deactivation of the triplet state may be accomplished by internal conversion (process 6) or by the emission of a photon giving a long-lived luminescence called phosphorescence (process 7). Because of the long lifetime of the triplet state, it has been possible to observe it has been possible to observe triplet–triplet absorption (process 8) for a number of molecules by flash photolysis.

Besides losing its excitation energy by the processes described above, a molecule in an excited state may enter into a photochemical reaction from either singlet or triplet levels. A valuable feature of flash photolysis is that it

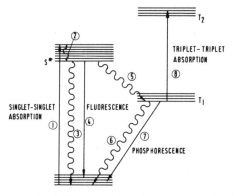

FIG. E3b-1. The energy level diagram of a typical organic molecule. Singlet states, S_0 and S^+; triplet states, T_1 and T_2; absorption processes, 1 and 8; internal conversion, 2, 3, and 6; fluorescence, 4; intersystem crossing, 5; phosphorescence, 7.

can be used to detect and monitor triplets and/or free radicals as intermediates in photochemical reactions. Often such observations are impossible by other techniques because generally, the concentrations of intermediate species are too low to allow detection under even high-intensity steady-state illumination.

c. *Experimental Design*

Apparatus for flash work ranges from simple designs, such as setting up a flash tube and reflector next to a reaction vessel, to highly complex, well-baffled systems, such as those used for obtaining spectra of intermediates. Porter has described the latter in considerable detail.

Figure E3c-1 shows a generalized design of the apparatus used for spectroscopic flash studies. A is the sample cell; B is a water jacket for thermostatting the cell and/or filtering the output of the flash tube(s); C is the flash tube(s); D is a reflector; E is a spectroscopic source, either a flash tube or a steady-state lamp; F is a monochromator for selecting radiation of a particular wavelength from the spectroscopic source; G is another monochromator for analyzing the light emerging from the cell; and H is a detector, the exact nature of which depends on the experiment. By rearranging the various components and choosing the proper detector, this basic design can be modified for use in a wide variety of experiments.

To determine the complete absorption spectrum of an intermediate, one would use a flash tube at E, no monochromator at F, and a spectrograph with a photographic plate as a detector (combining G and H). A suitable period of time (~ 10–1000 μsec) after the intermediate is produced by the main flash (C), a delay circuit would fire a flash tube at E and the spectrum would be recorded. This may be done as many times as necessary to obtain the proper density on the photographic plate.

For spectrophotometrically monitoring the concentration of an intermediate with a known absorption spectrum, one would use a tungsten or hydrogen lamp as the spectroscopic source at E and a photomultiplier tube with an oscilloscope readout at H. The monochromator at F would be set to the

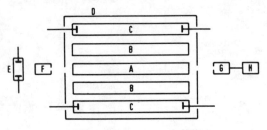

FIG. E3c-1. An experimental design useful for spectroscopic monitoring of a flash ...sis experiment. A, Sample cell; B, water jacket and/or filter; C, flash tubes; D, ... spectroscopic source; F, monochromator; G, monochromator; H, detector.

proper monitoring wavelength while the monochromator at G would be omitted. The study of duroquinone by Bridge and Porter provides an excellent example of this type of experiment.

Radiative triplet-state lifetimes (phosphorescence lifetimes) may be determined by setting the analyzing monochromator (G) at the phosphorescence maximum and eliminating the spectral flash (E) and preliminary monochromator (F). Also, excited singlet-state lifetimes (fluorescence lifetimes) have been determined using a flash of extremely short duration (~ 10 nsec) and electronics which respond rapidly. The oscilloscope trace obtained for either of these experiments would be similar to those except the ordinate would record phosphorescence or fluorescence intensity rather than light absorption.

Figure E3c-2 shows a block diagram of a flash-type circuit. The high-voltage power supply must provide from 200 to ~ 20 kV dc depending on the particular application. The storage section is usually a capacitor bank but may also contain inductors to control the flash duration. The energy and duration at a flash are given by $E = \frac{1}{2}CV$ and $t \simeq (LC^{1/2})$ where V is the voltage, and C and L are the total capacitance and inductance, respectively, of the discharge circuit. This indicates that the condition of high voltage and low capacitance and inductance produces the highest power. The charging resistor isolates the storage section from the power supply in order to prevent the capacitor bank from discharging through the power supply when the flash is triggered, and to prevent the flash tube from continuing to fire after the capacitor bank has been discharged. The flash tube is triggered by a pulse of ~ 5–25 kV from the trigger section. This pulse is applied either to a trigger wire wrapped around the flash tube or to a spark gap connected in series with the flash tube.

The detailed setup of the flash equipment is shown in Fig. E3c-3. Specifically, this circuit is designed to fire two 500-J flash tubes producing a flash of 20 μsec duration. The high-voltage dc supply charges each of the 10-μF 10-kV capacitors to 10 kV; a voltage at which the lamps would fire spontaneously without the spark gap is a type of high-voltage, high-current switch which is activated by a triggering pulse (~ 5–25 kV) obtained by discharging a small capacitor through the primary of the trigger transformer. When the spark gap

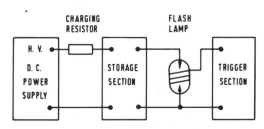

FIG. E3c-2. A block diagram of a basic flash lamp circuit.

is triggered, the full 10 kV of the main storage capacitors appear across the lamps and they fire. Then the storage capacitors automatically begin to charge for the next flash. The high-frequency choke coils in the circuit permit the use of one power supply to charge both capacitors while isolating them from one another during the discharge.

Also included in the circuit of Fig. 3c-3 are safety features which are strongly recommended due to the combination of high voltages and large capacitances encountered in flash circuits. When the power is off, both the main capacitor bank and the trigger capacitor are shorted through appropriate resistances, by use of relays. Another relay may be inserted io the primary of the trigger transformer to prevent accidental firing of the lamps. All safety features should be of the type which will automatically return to, and remain in, the safety position when the power is shut off.

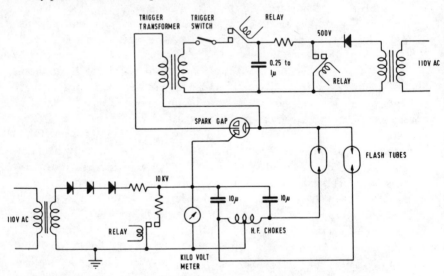

FIG. E3c-3. A detailed circuit for the operation of two flash tubes.

Should a flash of even shorter duration be desired, special components and circuits must be used. For example, a very short flash has been reported, producing only ~ 40 μJ/flash and having a full width at half-height flash duration of less than 10 nsec. A special flash tube was made by opening a General Electric NE-2 neon bulb and bending the electrodes until they were parallel and within 1/64 in. of one another. The bulb was then connected to a diffusion pump and gently flamed. Pure hydrogen gas was admitted to a ˙essure of ~ 100 mm and the bulb was reasealed.

˙ure E3c-4 shows the circuit used with this flash tube. Only the distributed ˙ce (C^1) of the circuit was used, unless higher energy per flash was desired, in which case a special low-inductance ceramic capacitor (C) was used. In place of a spark gap, a 2D21 thyratron tube was used to trigger the flash. The extremely high plate voltage used required a high variable bias to prevent the thyratron from firing because of the plate voltage alone. Although these conditions exceeded the recommended operating conditions for the 2D21, the authors reported that most 2D21 thyratron tubes performed satisfactorily. The 20 MΩ resistor between the power supply and the flash tube was large enough to cut off both the flash tube and the thyratron after firing. The 500 kΩ resistor across the lamp prevented the flash tube from breaking down prematurely under the high-voltage supply aod prevented it from firing at all unless it fired within a few microseconds of the thyratron. Such a circuit must be wired carefully to eliminate all stray inductance which would prolong the duration of the flash.

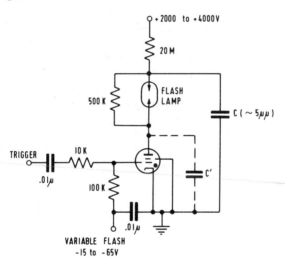

FIG. E3c-4. A circuit designed to produce flashes in the nanosecond region.

An ultraviolet lamp for generating weak subnanosecond flashes with 4.10^5 photons only, but of constant shape has been reported by D'Alessio [1762]. This lamp was made from a commercially available mercury contact relay, using the spark emission from the contact as the source of radiation. The rise time of the pulse was 0.4 nsec and the half-width of the pulse was 0.5 nsec. An interesting feature is that the lamp can be operated between 3 and 5×10^4 pulses/sec. This mercury lamp has a basic hydrogen filling pressure of 18 atm. The energy of the flash comes from its own capacitance.

Currently there are several commercially available complete flash units for flash photolysis studies. Some of these units were developed for high speed photographic use and provide ~ 1.5 J/flash with a duration of ~ 0.5 μsec.

Some units have high repeatability rates (up to ~6000/sec) while others may be flashed only once in 5 sec. Other units were developed for exciting laser action in crystals. Because they provide powers up to 20,000 J/flash, these units should have application in flash photolysis studies requiring moderate to high energies.

The construction of flash lamps has been described in detail by Porter [1542]. Electrodes, spaced appropriately, are sealed into each end of a quartz tube of the proper size and shape. The tube is flamed gently while under vacuum and then filled to a pressure of ~100 mm with an inert gas (argon, xenon, or krypton are usually used) and fired several times to outgas the electrodes. It is evacuated again and refilled, then sealed off. Such a tube will have a satisfactory life if the outgassing of the electrodes has been complete and the operating conditions are moderate. It cannot be emphasized too strongly that the quartz to metal seals must be constructed to withstand the violent electrical and thermal shock of firing the lamp. It is recommended that sturdy safety shields be used around any flash apparatus, especially if high energies are being dissipated in a shorter time, because lamp failures are often quite violent.

A large variety of flash tubes are available from several manufacturers. Normally, the tubes are made of quartz, although some are made of Pyrex. Xenon seems to be preferred although on special order the flash tube may be filled with another gas or vapor. There are two basic designs in use, the straight tube and the helix. Table E3c-1 summarizes representative flash tubes and their operating characteristics under "normal" conditions.

Spark gaps consist of two heavy tungsten electrodes separated by a few millimeters with a sharply pointed trigger wire ending a short distance from one of them. The electrodes are enclosed in a container to reduce noise. The trigger probe provides the first spark which breaks down the gap to allow the main current surge to pass. Table E3c-2 gives the operating characteristics of a large number of triggered spark gaps currently available.

For providing the triggering pulse, ordinary automobile spark coils or tesla coils have been used, but now, trigger transformers are commonly used. A wide range of trigger transformers is available for most applications.

Except for flash circuits designed for very short duration flashes, any commercially available capacitor having the appropriate capacitance and voltage ratings may be used in the energy storage section. For flashes of short duration it may be necessary to use special low-inductance capacitors. If it is necessary to prolong the flash duration to prevent failure of the flash tube, special inductors may be used in the circuit. These inductors must be capable of ~dling high currents and must be mounted securely and away from metal which may be attracted to the inductor by the high magnetic field during the discharge.

TABLE E3c-1

SUMMARY OF OPERATING CHARACTERISTICS OF SOME TYPICAL COMMERCIALLY AVAILABLE FLASH TUBES

Mfr	Model	Shape	Energy per flash (J)	Flash duration (μsec)	Flash rate	Arc Length (in.)	Envelope
EG&G[a]	FX-1	Straight	400	250	1/10 sec	6	Quartz
	FX-3	Straight	1.25	2	800/sec	$3\frac{5}{8}$	Quartz-inner Corning#7740-outer
	FX-12	Straight	5	6	6000/sec	$\frac{1}{4}$	Quartz
	FX-38A	Straight	400	1000	1/20 sec	3	Quartz
	FX-38A	Straight	200	400	1/10 sec	3	Quartz
	FX-38A	Straight	100	80	1/5 sec	3	Quartz
	FX-42	Straight	600	600	1/10 sec	3	Quartz
	FX-47	Straight	10,000	2200	1/4 min	$6\frac{1}{2}$	Quartz
	FX-51	U	600	600	1/10 sec	$3\frac{1}{2}$	Quartz
	FX-100	U	100	150	1/10 sec	$2\frac{3}{8}$	Quartz
GE[b]	FT-91	Straight	125	—	30/min	3	Quartz
	FT-151	Helix	125	—	2.5/sec	—	—
	FT-218	Helix	200	—	—	—	—
EPP[c]	S-13-138	Straight	600	—	6/min	3	Quartz
	S-13-139	Straight	2000	—	2/min	6	Quartz
	S-13-140	Straight	10,000	—	1/4 min	$6\frac{1}{2}$	Quartz

[a] Edgerton, Germeshausen & Grier, Inc., 160 Brookline Av., Boston, Massachusetts 02114.
[b] General Electric, Photo Lamp Department #281, Nela Park, Cleveland 12, Ohio.
[c] Electro-Power-Pacs, Inc., 5 Hadley St., Cambridge, Massachusetts 02140.

The high-voltage power supply needs to provide only the appropriate voltage and current necessary to charge the storage capacitors in a reasonable time. It need not be filtered since it is effectively disconnected from the circuit during discharge.

An apparatus for flash photolysis kinetic spectroscopy in the vacuum ultraviolet has been developed by Goodfriend and Woods [1900]. This apparatus allows spectra of transient species to be observed in the vacuum ultraviolet down to the fluorite cutoff and allows photolysis radiation with wavelengths as low as 1650 Å to be utilized.

In order to make radiation of wavelength less than 2000 Å available for photolysis, all air must be removed from the space between the flash tube and the sample cell. This was done by using a concentric flash-tube sample-cell configuration. The sample cell lies along the axis of a 7.6-cm-diam section of

TABLE E31c-2

SUMMARY OF CHARACTERISTICS OF SOME COMMERCIALLY AVAILABLE SPARK GAPS

EG&G Model No.	Operating range (kV)		Static break down (kV)	Peak current (A)	Peak current duration (μsec)	Energy discharge (J)	Trigger potential needed (kV)	Typical delay time (μsec)
	Min	Max						
GP 11	1.8	3.5	4.2	5000	20	25	5.5	0.10
GP 12	10.0	24	30	100,000	10	2500	15	0.05
GP 14	12	36.5	42	100,000	10	2500	20	0.05
GP 15	25	70	86	100,000	10	4000	25	0.10
GP 16	1	2	2.6	5000	20	25	5	0.20
GP 17	4.4	10	12.5	5000	20	25	7	0.02
GP 19	2.5	5	7	75,000	10	150	5	2.0
GP 20	3.5	11	14	15,000	20	200	10	0.06
GP 22	6	16	19	100,000	10	2500	15	0.04
GP 26	2	3.7	4.8	5000	20	25	6	0.10
GP 27	2	3.7	4.8	5000	20	25	6	0.10
GP 30	2	5	7.5	100,000	10	2500	15	0.08
GP 31	2	6	7.5	15,000	20	200	10	0.10
GP 32	20	50	70	100,000	10	4000	25	0.10

double-tough Pyrex pipe. The two end plates are clamped onto the flanges of the Pyrex pipe by means of glass-pipe clamps. Coupling to the sample cell is by means of Veeco type O-ring vacuum couples. Mallory 1000 (tungsten–copper) electrodes in the end plates then make the space between the two concentric cylinders effectively into a flash tube.

Source flash tubes are constructed of quartz, using an "end-on" configuration with electrodes inserted perpendicular to the light path. The middle part of the flash tube is constructed of a 2- to 3-mm quartz capillary which in the assembled apparatus lies along the optical axis.

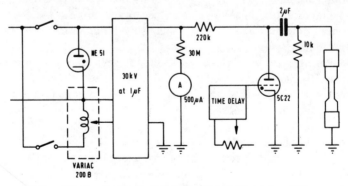

FIG. E3c-5. Source flash charging circuit for vacuum UV.

The capillary serves several very important purposes. The photolysis flash is rich in lines as well as continuum. The source flash on the other hand must be almost entirely continuum. In the capillary, not only in line emission under great pressure, but continua due to silica eroded from the walls contribute significantly to the emission. The capillary makes the source flash essentially into a pulsed-Lyman source. Of course, concentrating the emission into a small space also aids in increasing the intensity of radiation focussed onto the slit of the spectrograph.

Although erosion of capillary material is essential to the operation of the flash tube, it also produces certain difficulties. The explosive discharge through the capillary produces sputtering which both pits and deposits material on optical surfaces in front of it. Occasionally the shock produced can shatter the optics. Figure E3c-5 shows the circuit used.

A number of heavy-duty flash photolysis units have been constructed at Uppsala University, Sweden, by Claesson, Lindqvist, and Strong [1545, 1546, 1548, 1549], with emphasis on combined high-light output and short flash duration time. Capacitor discharge across flash lamps of suitable design was used, and the required characteristics were obtained by the use of high-energy capacitors, and by reducing the self-inductance in the discharge circuit to a minimum. As a recent addition to the previous units, a very fast apparatus with a maximum energy of 8 kJ has been constructed. It is called Apparatus V.

The particular properties which are most desirable in a flash photolysis apparatus depend very strongly on the problem under study and the kinetics of the reactions subsequent to the flash. However, critical damping of the circuit is advantageous from several points of view: It minimizes the flash duration time; it permits measurements at short time after flash for a monitoring system of given sensitivity; and it facilitates a numerical integration of the kinetic differential equation since the light intensity does not oscillate.

The choice of operating voltage is somewhat critical. For a given energy, a higher voltage means a smaller capacity and therefore also (at least in principle), a shorter flash duration time. On the other hand, a higher voltage means more insulation difficulties and greater voltage for triggering. After some consideration we have chosen 50 kV as operating voltage and ~ 100 kV for triggering. Higher voltages might have been more advantageous but ease of operation and design were the ultimate reason for this choice.

Introducing carbon resistance to accomplish proper damping improves the "$1/e$ time" for a 3-kJ flash from 13 to 8 μsec, without seriously affecting the peak intensity, and with almost constant "$\frac{1}{2}$ time," 6 μsec. For the critically damped circuit, the flash has approximately an exponential decay (with a half-life of about 4–5 μsec) down to about 1% of maximum light intensity. Increasing the discharged energy above a certain limit (about 2 kJ for the underdamped and 3 kJ for the critically damped circuit) leads to only small increase in peak light intensity. For very fast reactions, the optimum working condition is at the onset of this "saturation."

Straight heavy-discharge tubes of the type described by Claesson and Lindqvist were used as flash lamps. They were made from quartz tubing (17 mm i.d., 3 mm wall thickness) provided with tungsten electrodes at the ends. The electrode distance was 20 cm. Two lamps were required to withstand the full energy (8 kJ) and they were connected io series to facilitate triggering. The lamps were mounted parallel to each other at a distance of 8.5 cm between their axes. They were connected in common to a 2-liter ballast flask filled with oxygen to a pressure slightly higher than the lamp breakdown pressure (40 Torr). In order to avoid breakdown through air between the lamps, a layer of epoxy resin was cast around the high-voltage lamp electrode.

The capacitor bank consisted of 16 low-inductance capacitors (0.4 μF each, 50 kV dc, ringing frequency 1.1 MHz, Type CTU 50, ASEA, Sweden), storing a highest energy of 8 kJ. This type of capacitor has one isolated terminal and the casing is provided with a flange to make possible a coaxial attachment of the output leads to the capacitor (Fig. E3c-6). The capacitors were connected in parallel with 1.4-m long coaxial cables made from 6-mm copper wire isolated from an outer copper braid by means of PVC tubing of 3-mm wall thickness. The cables were brought together at a connector consisting essentially of two conducting plates spaced a distance of 8 cm from each other. The detailed construction of the connector is shown in Fig. E3c-6. The lamps were attached directly to the connector, at right angles to the plates. The plates were connected to a variable high voltage dc power supply. The lamps were connected to each other directly by means of a short copper bar, or, alternatively, via a damping resistor discussed in the following paragraph.

The discharge was initiated by apply a 100-kV pulse at the interconnected lamp electrodes. These electrodes were biased at half the charging voltage by means of a voltage divider. Figure E3c-7 shows a diagram of the trigger circuit. The flash unit was self-contained, and the optical bench supporting the reaction cell was mounted on a separate table to reduce vibrations caused by the discharge.

A typical light intensity oscillogram for a 2.9-kJ (6.4 μF, 30 kV) flash is shown in Fig. E3c-8. The measurements were taken at 420 mμ using as light detector a special multiplier unit, and the time resolution was 0.2 μsec. Peak light intensity is reached in 3 μsec. There is a shoulder on the flash profile at about half of peak light intensity; consequently, the width of the flash profile at $1/e$ of the maximum (13 μsec) is abnormally much higher than the width at $\frac{1}{2}$ of the maximum (6 μsec), commonly given as the "flash duration time." The shoulder indicates that the electrical discharge circuit is underdamped. Assuming that the time between the maximum and the shoulder (6.5 μsec)

3. PULSERS FOR LASER EXCITATION AND FLASH PHOTOLYSIS 197

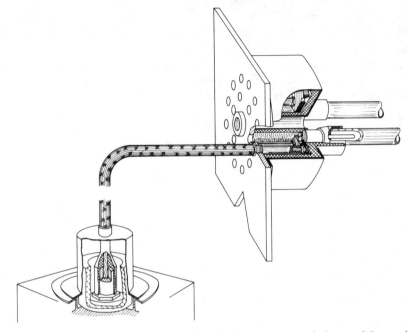

FIG. E3c-6. Details of the construction of the connector and the coaxial capacitor terminals.

corresponds to a half-cycle in a conventional Ohmic system, this gives a self-inductance of 0.65 μH and a total circuit resistance of approximately 0.1 Ω. The total number of quanta observed per flash by a uranyl oxalate actinometer solution in a 200-mm long cylindrical quartz cell of 25 mm i.d. positioned between the two flash lamps was 5×10^{19} quanta [1546].

For very fast reactions (i.e., those with half-lives of the order of that of the light intensity) following flash initiation, energy dissipated as light during

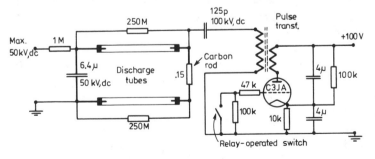

FIG. E3c-7. The discharge and trigger circuits of the flash photolysis apparatus (capacity, farads; resistance, ohms).

subsequent oscillations following the peak current is not effective in producing further measurable photochemical change. It is therefore desirable to reduce the "tail" of the flash as much as possible, consistent with maintaining a high peak light intensity. Gover and Porter have shown that the addition of 6 mm nitrogen to 60 mm krypton significantly reduces tailing as compared to that with pure krypton. However, it is difficult to see how the addition of nitrogen could have a significant effect when oxygen is used as the primary filling gas, and indeed it has been found with other flash units at this Institute and elsewhere that the discharge characteristics are identical when pure oxygen or air is used. The optimum condition under which to operate would seem to be when the electrical circuit is critically damped. Figure E3c-8b shows an oscillogram of a 2.9-kJ (6.4 μF, 30 kV) flash with a 0.15-Ω resistor (10 mm diam, 127 mm long cylindrical carbon arc electrode) added in series between the two lamps. It has been shown—by measuring the current derivative as a function of time with a coil placed near the lamps—that the circuit was very slightly underdamped even with this added resistance. The width of the profile at 1 e of maximum is appreciably reduced (8 μsec), although the width at $\frac{1}{2}$ of maximum is about the same as before; this shows the disadvantage of using the $\frac{1}{2}$ time as "flash duration time." The peak intensity is only slightly reduced by adding the resistance. The total integrated intensity is decreased by this technique, however; under identical conditions with those given above, uranyl oxalate actinometry gave only 2.3×10^{19} quanta absorbed per flash, showing that 54% of the energy is dissipated across the carbon resistor.

In Fig. E3c-9, the logarithm of the light intensity is plotted as a function of time for a flash with the 0.15-Ω resistance added. The discharge is almost exponential with a half-life of 4.5 μsec down to 1% of the peak light intensity (35 μsec), after which time the light continues to tail off—either from continued current oscillations in the slightly underdamped circuit, or from an "afterglow" in the quartz flash lamps. Without added damping resistance, the flash intensity decays to 1% of its maximum value in ~ 50 μsec.

The reason that the peak light intensity is only slightly affected by the reduced energy output of the lamp when the resistance is added is that there is a "saturation effect" in peak light intensity at the higher energies, whereas at low energies the peak light intensity is a linear function of electrical energy. This is shown in Fig. E3c-10 where the peak light intensity is plotted as a function of energy in the capacitor bank with and without added damping resistance. (A further increase in energy from 3.9 to 6.5 kJ—6.4 μF, 45 kV— results in only a 5% increase in peak light intensity for the undamped discharge. This effect has been observed with other flash units, but the reason for it is not known. Since the total number of quanta emitted by the lamps per flash is proportional to the energy discharged (except for very low energy outputs), it is obvious that the saturation of the peak light intensity must

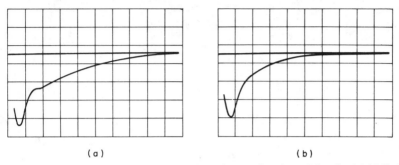

Fig. E3c-8. Oscillograms showing light intensity as a function of time for 2.9 kJ flashes (6.4 μF, 30 kV). Sweep times = 5 μsec per major horizontal division. (a) No added resistance; (b) 0.15-Ω resistance added in series with the flash lamps.

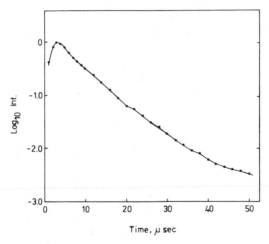

Fig. E3c-9. Plot of the logarithm of the light intensity divided by the peak intensity, as a function of time, for a 2.9-kJ flash (6.4-μF, 30-kV), with 0.15-Ω resistance added in series with the flash lamps.

result in longer flash duration times and longer times for the intensity to decrease to 1% of its maximum value. In the region of lower energy, where the plots are roughly linear, the flash discharge times are approximately constant for each case. It would seem, then, that for very fast photochemical reactions the optimum condition under which to work is in the region of maximum curvature in the peak light intensity versus energy plots. This corresponds for this flash apparatus to a voltage of ~23 kV in the undamped discharge, or to 30 kV when the 0.15 Ω resistance has been added. Willets [1942] of the Kodak Ltd. Harrow (Middlesex, England) has described the evolution of flash photolysis and laser photolysis techniques.

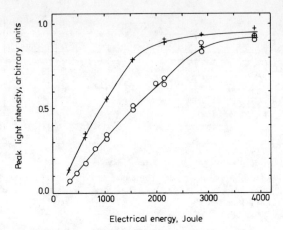

FIG. E3c-10. Plot of peak light intensity v^8 discharged electrical energy, ○ with 0.15 Ω in series with the flash lamps, + no resistance added.

4. High Voltage Pulse Transformers and their Application to Spark Tracing (Aerodynamics) and to X Ray, Electron, and Light Pulses

a. *High-Voltage Pulse Transformers*

High-voltage pulses are generated mainly by direct capacitor discharges produced in various ways. Pulses from a few kilovolts up to ~2 MV are also obtainable with much less effort using pulse transformers. These transformers have long open cores of laminated iron. The primary winding is fed by a capacitor discharge of 2 to 30 kV pulses/turn. The secondary is either wound on a long cylindrical core, or, in the case of unsymmatrical grounding, a conical core with a winding diameter that increases with increasing voltage to ground. During the growth of magnetic flux, a voltage is generated in the secondary that in no-load operation is almost constant, as long as dJ/dt is constant in the primary.

Früngel and Ebeling [1572] presented a survey of this last mentioned technique at the International Symposium Hochspannungstechnik, TU München, March 1972.

Pulse generators based on this principle are generally very useful and have advantages over bulky Marx generators where only the peak voltage of a certain duration is of interest. Typical pulse width on the secondary is in the microsecond range at 100 to 1000 kV. It depends on the load but can be designed by properly selected primary discharge circuit members. Examples of applications are EMP (electromagnetic pulse) simulation for investigation of electronic equipment, generation of x-ray flashes up to frequencies of 100 kHz and generation of high-frequency spark trains—the Strobokin "spark tracing method"—for visualization of laminar or turbulent air flow patterns.

For simple pulse generation with continuously variable output voltage between 0 and 500 kV at 5 nF, Impulsphysik GmbH, Hamburg [1013] manufactures voltage surge equipment called "Isopuls I."

At this equipment the primary winding consists of 8 turns only and is designed for 30 kV, the secondary winding for 600 kV. An output voltage of 500 kV is reached with a load of 5 nF and a parallel resistor of 1500 Ω. The peak output amounts to $\sim$180 MW and the no-load voltage to $\sim$600 kV. The rise time is then $\sim$2 μsec, decay to zero, 25 μsec.

To produce a fast rise time, the secondary feeds into a low inductance, 5-nF capacitor, which can be discharged very rapidly into the line by a fast switching spark gap, for example with the help of a Quenchotron, so that 500 kV pulses with 10 nsec rise time are available. For some small equipments, these pulses are sufficient to simulate the required EMP voltages on an antenna array of about five 10-m lines parallel to the surface of the earth at a height of 3–4 m. As a result a small vehicle can be tested with pulses of field strength 120 kV/m.

The high-voltage pole is insulated with SF_6 gas and has a Plexiglass lead-out bushing for connection of an ordinary high-voltage measuring cable. A built in capacitive measuring voltage divider permits connection of oscilloscopes. The weight of the gas filled high-voltage unit is 400 kg, and its height 170 cm, plus 70 cm for the high-voltage output terminal. If the transformer primary is fed by a series of pulses at more than 5 kHz, the so-called spark tracing technique for three dimensional aerodynamic flow research can be realized.* New results at low and high pressures have been reported by Früngel and Thorwart (1015).

b. *Spark Tracing*

In studying gas flow, the physical motion of the medium is normally made visible by blowing in smoke or lightweight particles. These methods are limited to the observation of more complicated phenomena like turbulence, boundary layer flow or three-dimensional flow. The well-known interference method does not show the flow but only the pressure pattern. The Schlieren and shadowgraphy methods show density variations over the cross section but both are limited in application to thin cross sections because of the impossibility of taking high speed stereo pictures with these techniques.

The spark tracing technique is the method of choice [1863]. The main applications are: three-dimensional flow, boundary layer flow and all other aerodynamic studies up to Mach 20 and down to as low as 10 m/sec. The

---

* High Speed Pulse Technology, Volume II, Chapter K8.

method can be applied up to pressures of 50 atm and, under proper conditions, down to 10 Torr.

In spark tracing, the luminous ionized plasma channel of an electric spark discharge is the indicator. This offers the particular advantage that it is comparatively mass-free, because the luminous ions of a gas discharge consist of the material of the flowing medium. This method has special advantages with three-dimensional or nonstationary flows [1864]. The first discharge generates a spark path between two points or wires. If high-voltage pulses are now fed to the same electrodes at such high repetition rates that during the interval between two pulses the plasma of the spark path has not yet completely deionized, then every subsequent voltage pulse will trace the path of the first spark, which in the meantime has been moved away by the air current.

Figure E4b-1 shows how the first spark over occurs between two point electrodes and the subsequent spark shape depends on whether the air flow is parallel, convergent, or divergent or whether or not there is an obstacle [1864]. The block diagram, Fig. E4b-2, shows a spark-triggered heavy-duty pulse transformer operating on the differential principle. This design is particularly suitable where rather large spark-over distances must be achieved. The triggering spark gap is a three-electrode spark gap (Strobokin) system which can be controlled up to 100,000 fps. Free running operation is possible up to 300,00 fps.

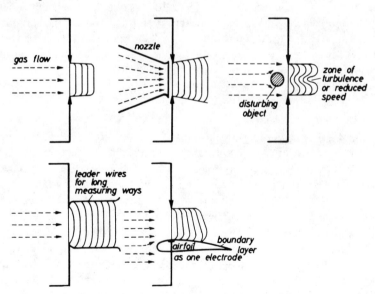

FIG. E4b-1. Spark tracing method. The first spark over occurs between two electrode points and the subsequent spark shape depends upon whether the air flow is parallel, convergent, or divergent, or whether or not there is an obstacle.

The plasma channel can thus be made to light up periodically at the preset pulsing rate and during a preset burst time [1865]. Since the pulse power system (Strobokin) provides an electric pulsing rate generated with extreme accuracy—e.g., by quartz-control of the oscillator—it is possible to obtain, with the same accuracy, velocity or acceleration vectors or isochrones in a representative section of air flow by photographic methods. A simple camera with open shutter is the only recording instrument required. Often for three-dimensional flow patterns three-dimensional photographs are necessary, calling for two cameras or a stereoscopic camera [1864].

As material for the guide wires any bare metal wire can be used. Manganin wire is particularly suitable because of its low electronic function [1866]. The first spark over occurs at the point where the gap is smallest, and can be made

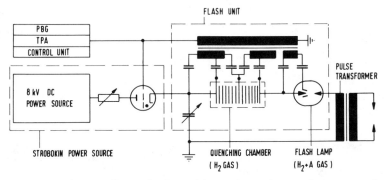

FIG. E4b-2. Diagram of a spark triggered heavy-duty pulse transformer operating on the differential principle.

linear by an electric arrangement which destroys the homogeneity of electric fields, e.g., by inserting insulated needles into the first spark path. Other spark path shapes, such as sharply bent primary spark configurations, can also be obtained in this way.

Japanese quantitative measurements and considerations regarding accuracy of spark tracing show that at air speeds above 2 to 4 m/sec, the disturbance of the flow by the spark energy can be neglected [1867]. The spark tracing method has proved to be suitable for investigating the flow conditions in high speed turbines. It yields a qualitative picture of the air current flow, and also permits quantitative analyses. Since the spark over always starts at the narrowest part of the electrode gap, the method is also suitable for air flow investigations in tunnels with divergent walls which at the same time serve as electrode supports, i.e., for decelerating flow patterns which also occur in such equipment as radial compressors [1868].

It is quite easy to investigate several events simultaneously with different spark gaps. The utilization of this method in compressor design, for example, is comparatively easy, even at high rotation speeds, since these do not constitute a major difficulty for the mounting of the wire electrode on the rotor [1869]. Likewise, there is no difficulty in transferring high voltage to the rotor. No conductor rings are necessary, since simple electrodes which are mounted at a short distance from rotating metal rings will suffice to transfer the energy in the form of short sparks. High flow velocities between the turbine blades are desirable in order to keep the relation of the disturbing energy to the kinetic energy as small as possible. However, excellent insulation is an indispensable prerequisite, to make the channels under investigation suitable for photographic recording.

Among the first industrial applications of the spark tracing method was the Institute of Aerodynamics at Bochum University dealing with special forms of radial compressors. Figure E4b-3 shows how the sparks reveal the air flow between four pairs of shovels in a radial compressor. Additional problems arise in this application because the sparks between the shovel blades not only move radially outwards but also rotate with the rotor. According to Fister, this additional angular velocity can be compensated for by combining the taking camera with a rotating Dove prism which rotates in the reverse direction to the rotor and with half its speed. With this compensation, the shovel wheel

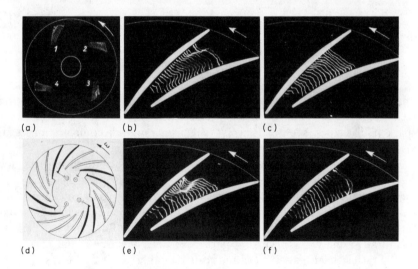

FIG. E4b-3. Application of the spark tracing method to determine the air flow between the shovels of a compressor. (Photographs by Professor Fister, Bochum University.) (a) Complete picture, (b) channel 1, (c) channel 2, (d) arrangement of electrode, (e) channel 3, (f) channel 4.

4. HIGH VOLTAGE PULSE TRANSFORMERS AND THEIR APPLICATIONS   205

is optically stationary and a series of photographs can be taken with any ordinary camera. With this technique Fister was able to make decided improvements in the efficiency of compressors of various types and to show the limits of validity of the calculating formulas being used.

Nagao and Ikegami, Kyoto University, Japan, applied the spark tracing method to the investigation of the rotary flow inside a Diesel engine cylinder during the compression phases. Here the aim was to create effective turbulence in order to improve the combustion in the cylinder. Figure E4b-4 shows an example of this work and Nagao and Ikegami confirmed that reproducibility of tests had been checked with good results. Apart from the fact that Professor Nagao's investigation was unique and novel, this was the first time that the spark tracing method had ever been applied to the investigation of flow

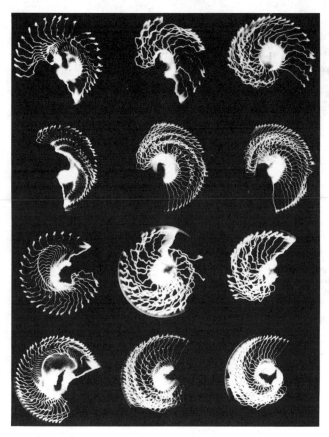

Fig. E4b-4. Rotary flow inside a Diesel engine cylinder during the compression phase made visible by the spark tracing method. (Photographs by Professor Nagao, Kyoto University.)

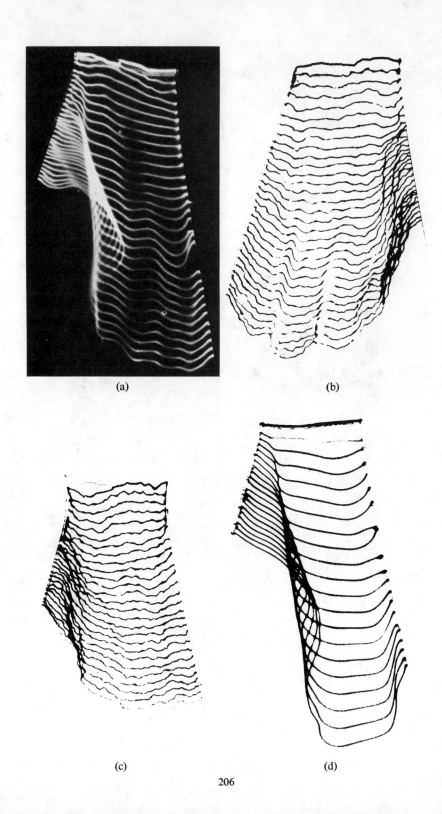

4. HIGH VOLTAGE PULSE TRANSFORMERS AND THEIR APPLICATIONS 207

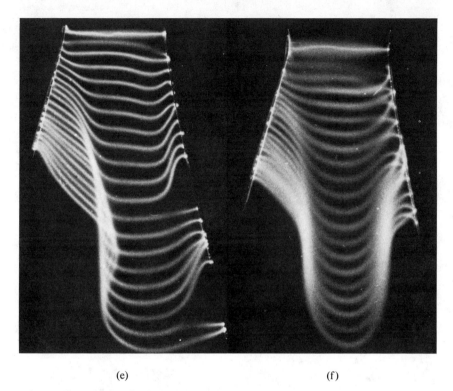

(e) (f)

Fig. E4b-5a–f. Application of the spark tracing method in overpressure and underpressure conditions. (Photographs by W. Thorwart, International Impulsphysics Association, Hamburg.)

phenomena in an overpressure atmosphere. Consequently, two questions arose simultaneously:

(a) If the spark tracing method is applicable to overpressure, is it also applicable to underpressure, because this would be of interest in hypersonic wind tunnel work?

(b) What are the limits in both directions? An investigation was instigated in order to answer these two questions.

Figure E4b-5 shows three photographs taken at various overpressures, namely (a) at atmospheric pressure, (b) at a pressure of 6 atm, and (c) at a pressure of 26 atm. Though particularly the last picture shows that the higher the pressure, the more irregular becomes the trace of the individual spark, it has nevertheless been established that the spark tracing method will certainly work up to overpressures of 60 times atmospheric pressure. The remaining pictures in Fig. E4b-5 show spark tracings at low pressures, namely, (d) at $\frac{1}{3}$ of normal atmospheric pressure, (e) at $\frac{1}{8}$ of normal pressure, and here the sparks are no longer as needle sharp as they were at overpressures but become slightly blurred due to plasma expansion at reduced pressure, and (f) at $\frac{1}{12}$ of normal atmospheric pressure (60 Torr) showing plasma expansion already at a disturbing magnitude, including that practical applications will not be possible at much lower than 60 Torr, though the method as such operates down to still lower pressures.

Another problem of great interest is the flow in hot gaseous media like flames, for instance, the flame of a Bunsen burner which is normally not visible. Here again experiments with the spark tracing method were made and Fig. E4b-6 shows the flow in a Bunsen burner flame. At the right-hand side we have one electrode which is glowing hot. At the left-hand side there is a needle electrode and the Bunsen burner is at the bottom center so that the

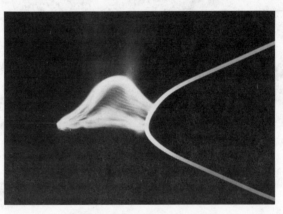

FIG. E4b-6. Spark tracing inside a Bunsen burner flame. (Photographs by W. Thorwart, International Impulsphysics Associations, Hamburg.)

flame is vertical. The needle-sharp spark lines in the flame are clearly visible so that quantitative calculation of flow pattern and flow velocities within the flame is easy. After it has been established that the spark tracing method is also applicable to investigation of the flow conditions of hot gaseous media, practical applications follow immediately. It is of great economic interest to achieve the most effective burning of materials in any type of flame, whether from gas, oil, or coal dust burner, since in order to obtain the most heating calories from the material burned, it is of vital importance to know everything about the flow pattern, the velocity of the flow, and the environmental conditions around the flames of the burners and alongside the walls of the combusion chamber.

Bernotat and Umhauer [1862] found another application of the spark tracing method, namely, in conjunction with high speed photography, to investigate the flow conditions inside a cross stream air classifier. The photographs obtained clearly demonstrated the increasing mutual interplay between gas flow and particle stream with increasing concentration of the solids. The results show that in fluid–solids systems the method can generally be used up to volume concentrations of $\sim 1 \times 10^{-2}$.

The study was carried out on a new classifier, working on the cross stream principle, that had been constructed at the Karlsruhe Institute of Mechanical Process Engineering. Previously, a so-called black box method had been used for testing the new classifier, i.e., the input and output streams of solids were analyzed for their particle size distributions. But with this method no detailed information about the process inside the classifier could be obtained. The results showed that the existing theory can explain the phenomena only for

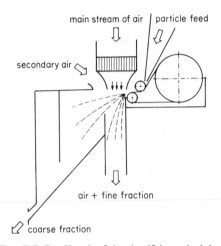

FIG. E4b-7. Sketch of the classifying principle.

low particle concentrations. At higher concentrations, direct observation and measurement of the motions of solid particles and gas became essential. Therefore, the "spark tracing method," according to Weske and Früngel, and high speed photography using a camera with rotating prism have been applied to the problem.

As indicated in Fig. E4b-7, the feed is accelerated by a belt and thrown as a thin layer of particles into an air stream, leaving a nozzle with constant velocity. In this stream the particles are spread into fan-formed trajectories for the different particle sizes x. A movable blade divides the particle stream into two fractions. The trajectory which hits the tip of the blade is that of particles of cut-size x_T. The coarse fraction, i.e., all particles with $x > x_L$, crosses the separation zone and leaves the air stream above the blade while the fine fraction with $x < x_T$ is carried out by the air. The coarse material is caught in a wide collecting space whereas the fine particles are precipitated on a cloth filter.

Single particles of distinct size x move along determinate trajectories as indicated in Fig. E4b-7 which can be precalculated from the interaction between particle and fluid. Therefore, in the ideal case of classification all particles with $x > x_T$ should be found in the coarse fraction. Real classifiers, however, can be operated neither on laboratory nor on plant scale in such a manner that all particles of exactly the same size move along exactly the same trajectories, even at very low particle concentrations. In every case fine particles will be found in the coarse fraction and vice versa. There will always be an overlapping of the two fractions.

At higher particle concentrations in the separating zone of the classifier, in addition to the particle–fluid interactions, interactions between particle clouds and the fluid and particle–particle interactions, mainly particle collisions, become more and more effective. These influences can be described theoretically by a stochastic transport process where a random movement is superposed on the determinate particle movement. The additional interactions lead to a change in cut size and to a wider range of overlapping of the two fractions, i.e., the sharpness of cut decreases. A measure for the sharpness of cut for instance is the gradient of the grade efficiency curve in its middle region.

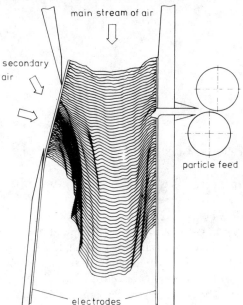

FIG. E4b-8. Example of a spark tracing photograph out of the experiments with clean air; frequency of the spark series $f = 10$ kHz. Courtesy of Dr. Umhauer, Karlsruhe Inst. of Mechanical Process Engineering.

4. HIGH VOLTAGE PULSE TRANSFORMERS AND THEIR APPLICATIONS 211

The question was whether the spark tracing method could be applied to two phase flow in the presence of solid particles and if so, under what conditions and up to what particle concentrations usable photographs could be taken. About a year ago, Muschelknautz and Rink succeeded in studying solids–gas flow in a jet mill with the help of spark tracing. Encouraged by these results, we started to apply the spark tracing method to several two phase problems, especially to the investigation of the air classifier described. The first results of these investigations are reported here.

Two different electrode positions were used. To visualize the movement of the main stream of air, two spark guide wires of nickel were placed in the central plane of the flow channel (compare, for example, Fig. E4b-8). The wire on the right-hand side is interrupted at the solids entrance. Both parts however are connected to the same electrical potential. The guide wire on the left-hand side is fixed to the movable blade and rises above the edge of the blade into the mouth of the nozzle. Small disturbances of the secondary air flow caused by the wire are accepted. The wires diverge in the flow direction causing the primary spark to flash over at a distinct point, which means the spark series always starts in the flow direction at the same point. The distance between the electrodes at the level of the particle entrance is 85 mm. Discharge between two point electrodes was used to investigate the inflow of the secondary air. The lower electrode is mounted on the blade and ends at its tip. The upper electrode is placed at the edge of the nozzle. The spark over now occurs exactly in the entrance region of the secondary air.

The equipment required to generate the spark series consists essentially of a power supply, a frequency generator with a control unit, and a pulse transformer. The pulse transformer is charged on-line by a quenched spark gap and produces high-voltage pulses of more than 150 kV of a few microseconds duration. Sparks with a length of 10–15 cm can easily be realized. This equipment has been developed by Impulsphysik GmbH, Hamburg.

For an accurate measurement of velocity as well as for a relatively simple evaluation, it is desirable to be able to choose a constant operating frequency. The equipment made it possible to use an accurate frequency generated by a crystal oscillator which remained constant at 10 kHz. This frequency is suitable for the measurement of velocities between 20 and 40 m/sec. In most of the photographs a series of 60–70 flashes was produced and the flash duration was of the order of a few microseconds. The discharge energy of the flashes may be controlled by limiting the current by means of a resistance placed on the secondary arm of the pulse transformer. One must be careful to avoid disturbance of the flow by the imparting of too much thermal energy to the sparks, while ensuring that their luminous density is high enough to allow photography. The photographs were taken using a 6 × 6 cm plate camera. The first experiments concerned themselves with the determination of the gas flow patterns by means of spark tracing, for different combinations of the main and secondary air flows, when particles were absent. Figure E4b-8 shows an example of the typical appearance of a series of spark paths, which was obtained using a flash frequency of 10 kHz in a flow with an air velocity of 25 m/sec at the exit of the nozzle. These operating conditions with $\sim 10\%$ secondary air are the same for all further examples. The jet aperture dimensions were 8×10 cm^2 with consequent turbulent flow ($Re \geqslant 1.5 \times 10^5$).

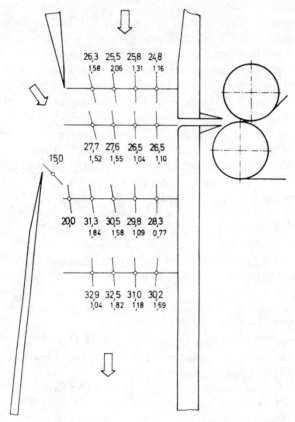

FIG. E4b-9. Average air velocities and standard deviations in the case of clean gas flow Units: m/sec.

In order to provide a better view of the system, the outline of the electrodes, the wall on the right-hand side of the flow channel, the opening through which the particles enter the system, the outline of the nozzle and the separator blade were superimposed on the photographs. The movement of the ionized gas marks the movement of the flow itself. It must, however, be made clear that the spark paths, themselves, do not provide a picture of the velocity profile. It is necessary rather to compare successive paths to obtain the required information. It must also be noted that the photographic method used only provides information on the velocities and movements in the plane in which the photographs were taken. The direction of the flow is indicated by the propagation of irregularities in the spark path, so that in such places the flow velocity can be measured according to direction and magnitude. For stationary flow behavior, one has to take several photographs, through which average values may be calculated.

The normally irregular shape of the ignition spark path smooths itself after five or six flashes and it is only then that an evaluation is meaningful. Even then, in the vicinity of the wall and of the electrodes no definite conclusions may be made, due to the irregular distances between sparks, caused by small velocities and local differences in the work function. A special advantage of the method lies in the graphic nature of the flow visualization. Thus, in Fig. E4b-8 one sees clearly how the main air flow is compressed and accelerated by the secondary air which flows around a separation vortex behind the separator blade. From other photographs it was observed that the sparks did not penetrate vortices, consequently making their visualization possible. Figure E4b-9 shows the quantitative evaluation of a spark tracing picture with air velocities and standard deviations.

Figure E4b-10 is an example of the results of experiments in which particles were injected into the gas flow by means of the belt accelerator. The operating conditions of the classifier remained the same for all these photographs. The average entry velocity of the particles was about 12 m/sec. The photograph differs as regard the material of the particles, their size, and their concentration. The following values quoted for concentrations relate to those at the entry point, i.e., before the particle stream is spread by the flow.

For Fig. E4b-10, a narrow size fraction of limestone (~ 500 μm) was used as feed. The concentration was ~ 100 particles/cm^3 corresponding to a volume concentration of 5.3×10^{-3}. The most striking fact about this photograph is the appearance of strong luminescent "tails" in the flow direction. They are caused by particles which find themselves in the spark path and may partly undergo size reduction. Obviously the smallest particles created thus, and which follow the flow exactly, are stimulated to luminescence. In the case of limestone particles red light is produced (the Ca flame). This light is however undesirable, since with high concentrations the light nearly drowns the spark paths. Because the electric sparks radiate short wavelength light, it is possible to eliminate the stimulated red light of the tails using a filter. On the other hand this luminescence may be used with small concentrations of particles to determine the local flow directions. The agreement between the observed luminescent particle paths and the direction of local turbulent velocities can be demonstrated by many examples.

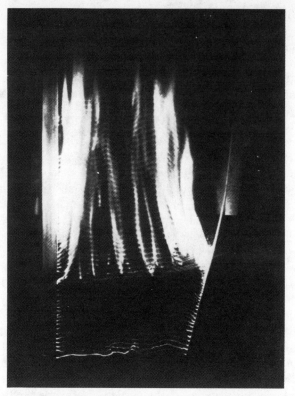

FIG. E4b-10. Spark tracing photograph, limestone particles being injected; $x = 520\ \mu\text{m}$, $C_v = 5.3 \times 10^{-3}$, $f = 10$ kHz.

It was obvious to try avoiding the tails by choosing another material. Accordingly, glass beads and quartz sand were tried. As assumed, these did not show the luminescence observed before.

c. *Generation of Pulses of X Rays, Electrons, and Light*

At the 10th Congress on High Speed Photography,* Mattsson [1641] reported the design and performance of a high-voltage pulse generator and vacuum-diode system. The pulse generator delivers 250–600 kV, 10 kA, 20 nsec pulses with positive or negative polarity. It is used to pulse a continuously pumped vacuum-diode having an exchangeable electrode assembly. Short and intense bursts of x rays, electrons, and light can be obtained by pulse. The diode can be used, with a reflection anode, as a fine-focus flash x-ray tube having a source size of the order of 1 mm. Using a transmission anode, a broad and intense x-ray beam can be generated suitable for radiation effect

---

* Further information will be available after issue of this particular congress volume.

studies. High-intensity electron pulses can be extracted through a thin metal anode for external use. The electron pulses can be converted into short, superradiant-light pulses using a semiconductor target. The emitted light is nearly monochromatic and different wavelengths are obtained for different target materials. In several applications a combination of more than one type of radiation may be used to increase the information.

5. Megavolt Pulse Techniques: Lightning Flash and EMP Simulation

a. *General*

Multimegavolt switches have already been dealt with in Chapter B11a. Figure B1a-1 shows a circuit of a multiple chopping gap which has three applications, namely as a chopping gap, as a voltage divider, and as a load capacitor.

b. *Simulation of Lightning Flashes*

Wiesinger [1858] is one of the leading workers in the field of lightning flash simulation. In this paper he gives examples of how, with modest means, the electrical effects of lightning flashes can be simulated. Such equipment is important for testing the reliability of all lightning flash protectors and related lightning conductors, protective housings, etc.

A typical circuit operates a 80-μF 50-kV battery at levels given in Table E5b-1. This table shows that it is relatively easy to simulate the rate of rise of the current and its maximum value; but extremely difficult to simulate the charge. Figure E5b-1 shows a proposal for the next step. The circuit would provide, at 100 kA, a charge of 100 C and an I^2-time integral of 3.5 kA^2 sec. If the ten parts of the circuit were triggered sequentially, 10 discharges of $\sim$10 C could

TABLE E5b 1

Lightning Flash Values Obtained in Laboratory Simulations Compared with 10% Values of an Average Lightning Flash

	Max current i (kA)	Charge Q (C)	Max rate of rise (kA/μsec)	Time integral of square of current $\delta i^2 \cdot dt$ (kA$^2 \cdot$sec)
Laboratory value	145	4	25	0.46
10% value of a typical lightning flash	0.5	17	3	4

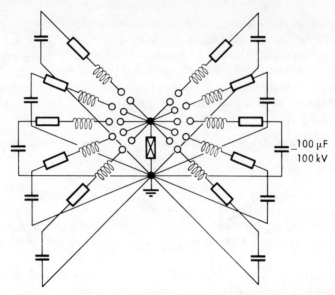

FIG. E5b-1. Purposed circuit for a 5-MJ 100-kV lightning flash simulator.

be obtained. With the circuit divided into two parts, a 50-C main pulse and a subsequent 50-C tail would be available.

The choice of circuit constants for a long-duration current generator for surge diverter testing has been discussed by Guraraj [1573]. Until recently, surge diverters were applied for protection of power apparatus against lightning overvoltages only. As economic incentive point towards lower insulation levels for EHV power apparatus, much effort has been directed towards reduction of the switching impulse protective level afforded by the diverters. During the past decade, surge diverters are increasingly being applied for limiting switching overvoltages as well.

The energy stored in the capacitance of the line discharges into the surge diverter after spark over. This energy is proportional to the length of the line, the line capacitance per unit length, and the square of the charge voltage on the line. As diverter rating and the charge voltage increase directly as the system voltage, the energy discharged in the diverter per kilovolt rating of the diverter increases more than linearly with system voltage. This is due to the fact that higher voltage lines have increased capacitance due to conductor bundling and are necessarily longer.

In order to assess the performance of the surge diverter for this duty, a long-duration current impulse test has been specified in almost all the national and International Electrotechnical Commission (IEC) standards. Guraraj's paper considers the choice of the circuit constant for the generation of such long-duration currents. The following references also deal with the

5. LIGHTNING FLASH AND EMP SIMULATION

generation and propagation of lightning flash surges: Heymann [1582]; Blasius et al. [1579]; Müller [1575]; and Pflanz [1576].

Breakdown in sulphur hexafluoride and nitrogen under direct and impulse voltages has been investigated by Binns et al. [1278]. Because of its experimental character, this work is reported here in some detail.

Several studies have been made of the breakdown voltage of SF_6, and of mixtures of SF_6 and air or nitrogen, at a range of pressures, and for different field configurations. The time lag to breakdown in SF_6 has also been measured, and observations have been made of the growth of current pulses in SF_6. Previous time-lag measurements have been confined to atmospheric pressure and below, even though SF_6 is mainly of value as an insulant at higher pressures. At these lower pressures, only the measurements of Efendiev appear to be for sufficiently intense irradiation (provided by ultraviolet light from a spark gap) to allow formative time lag to be measured separately from statistical lag. The time lags reported by Efendiev are of the order of 10^{-7} sec for gap lengths of a few millimeters and overvoltages of $\sim 10\%$. This contrasts with values of 10^{-5} sec or so at 10% overvoltage, reported elsewhere for similar gap lengths. It is likely that these longer time lags result from using comparatively weak continuous irradiation, and represent an appreciable statistical waiting time for appearance of an electron that produces a successful avalanche.

Time lags are reported here for SF_6 and nitrogen, at pressures from 200 to 2830 Torr under irradiation by a synchronized pulse of ultraviolet light (UV) from a spark gap. The fields are uniform, and gap lengths vary from 0.15 to 1.5 cm with overvoltages between 1 and 20%. The purpose of this study is to examine the time-lag characteristics of SF_6, and to compare them with those of nitrogen.

The high-voltage impulse circuit, used for the studies, is shown in Fig. E5b-2. Negative impulse voltages were derived from a single low-inductance storage capacitor of 930 pF, charged from a positive direct voltage supply, and switched by a trigatron operating in compressed nitrogen. The direct voltage was provided by a high-frequency voltage–multiplier circuit. The impulse waveform was critically damped, using a series resistor R_D, and a 0.1/320 μsec waveform was produced. This was effectively a step function of voltage, giving a fall of 2% in 10 μsec, compared with time lags which were considerably shorter than 1 μsec. The rise time of the wavefront was quite acceptable, even for very short time lags of the order of 10 nsec, since breakdown of the gap was initiated just after the voltage peak, by a pulse of ultraviolet light. High direct approach voltages were not used in these tests, but a potential of -100 V was maintained on the high-voltage electrode, to assist in dispersing any remanent ionization which might be left in the gap after breakdown or might build up from background irradiation. It might also be

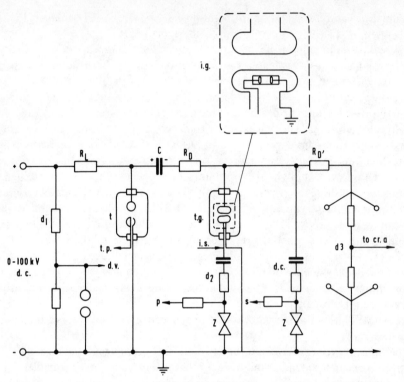

FIG. E5b-2. High-voltage-impulse circuit. d.c., differentiating circuit for deriving "stop" pulse for timer; d.v., connection to digital voltmeter; d_1, direct voltage resistance divider; d_2, voltage divider of start pulse; d_3, capacitance-compensated resistance divider for observing test waveshape; R_L, 100 MΩ resistor; R_D, main damping resistor (400 Ω); R_D', second 400 Ω damping resistor in series with divider capacitor; C, main storage capacitor (930 pF, low inductance); T, trigatron, working in compressed nitrogen; Z, Zener diode; p, "start" pulse; s, "stop" pulse; i.s., impulse voltage to supply irradiating gap; t.g., test gap with earthed anode having irradiation window; t.p., 10 kV impulse to trigger electrode; i.g., irradiating spark gap.

effective in reducing positive charges that can build up on oxide deposits on the cathode.

The test gap consisted of uniform-field brass electrodes of 5.5 cm over-all diameter, set at spacings varying from 0.1 to 1.5 cm. The electrodes were highly polished before the commencement of each test, using metal polish or a very fine abrasive paper, and were then cleaned with carbon tetrachloride and acetone. The test gap was enclosed in a Pyrex vessel (with brass end flanges and rubber gaskets) which was substantially free from leaks. It could be evacuated with a rotary pump, and subsequently filled from a cylinder of nitrogen or SF_6. After the required pressure had been obtained, the gap length was set, by first bringing the electrodes together and then separating them by a measured distance. Irradiation was provided by a spark–discharge gap consisting of $\frac{1}{8}$-in.-diam tungsten electrodes, with a separation of about $\frac{1}{16}$ in., sealed in an air-filled quartz bulb, which was contained inside the brass anode. This discharge gap had a breakdown direct voltage of about 5 kV, and was subjected to a 12-kV impulse, obtained from a 0.001-μF capacitor discharged through a thyratron. This caused rapid and consistent breakdown of the discharge gap, with very small jitter, and an intense irradiation pulse was produced. The peak photoelectric emission currents were of the order of 1 μA with a duration of the order of 10 nsec; i.e., perhaps 10^5 photoelectrons would be released. The time lag to breakdown in the main test gap was then defined as the interval between the application of the light pulse and the instant of voltage collapse, and was measured using a specially constructed digital timer working at 110 MHz. The digital timer is started and stopped by voltage pulses of ~ 10 nsec rise time produced by differentiating circuits giving an output limited by Zener diodes to ~ 15 V. The first pulse is derived from the decline of the surge voltage applied to the irradiating discharge gap when this fires, and the second from the collapse of voltage across the test gap. The resolution of the timer is ~ 10 nsec, being determined by a basic clock frequency of about 110 MHz, and, with 14 binary digits used, the full range of the timer is about 150 μsec. However, intervals greater than 10 μsec, i.e., 1000 times the timer resoluation, were not of interest in the present study. An output from the timer is provided, to operate a paper-tape punch which records time lags in a form permitting direct processing by a digital computer. The results are given in Table E5-2. Binne *et al.* also discuss the form of voltage collapse in SF_6 and draws the essential conclusions. Twenty references are given.

Uman *et al.* [1279] made experiments on four-meter sparks in air using high speed image-converter photography, current and voltage measurements, absolute measurements of radiated light intensity, and high speed image-converter spectroscopy.

The spark-gap geometry and electrical circuits for producing the breakdown voltage and for measuring the gap voltage and current are shown in Fig. E5b-3. The voltage applied across the rod–plane gap would have been a standard 1.5×40 wave (1.5 μsec to peak and 40 μsec to half-value) with a crest value of approximately 3.3×10^6 V in the absence of gap breakdown and corona load. The charging voltage and thereby the stored energy were kept constant for both positive and negative rod polarity. The critical breakdown voltages for a 4-m rod–plane gap are given by Uman *et al.* as 1.9×10^6 V for positive rod polarity and 2.8×10^6 V for negative rod polarity. Consequently, the applied overvoltages differed considerably for the two polarities.

The peak power input to the discharge channel from the impulse generator

TABLE E5b-2

DIRECT AND IMPULSE BREAKDOWN VOLTAGES IN SF_6

Pressure p (Torr)	Gap length d (cm)	pd (Torr-cm)	Mean direct breakdown voltage at 24°C with UV (kV)	Mean direct breakdown voltage at 24°C no UV (kV)	Reduction in breakdown voltage due to irradiation %	Direct breakdown voltage gradient, E no UV (kV/cm)	E/P direct voltage breakdown no UV (kV/Torr-cm)	Minimum impulse breakdown voltage (kV)	Temperature (°C)	Impulse breakdown voltage gradient (kV/cm)
200	1.5	300	—	—	—	—	—	38.3	25	255
1030	0.2	206	—	—	—	—	—	24.8	26	124
1280	0.15	192	22.1	22.8	3.07	152	119	22.8	20	152
1790	0.15	268	31.8	32.5	2.16	217	121	32.2	23	215
2310	0.15	347	42.2	42.5	0.72	283	123	42.5	23	278
2830	0.15	423	variable up to 53 kV	—	—	≃350	121	51.2	22	341

occurs when the voltage is collapsing and the current is rising to peak. From the measured V–I characteristics, the peak power was found to be about 6×10^9 W for the negative rod and about 5×10^9 W for the positive rod. Errors of $\pm 20\%$ are present in these values, due to the inaccuracy of the V–I characteristics and due to the uncertainty in the actual value of current in the gap.

The total energy input to the gap has been calculated by two independent methods. The first is by integrating the power curves over time; the second is by subtracting from the initial stored energy of the impulse generator the energy dissipated as $I^2 R$ across the 960-Ω series output resistors shown in Fig. E5b-3. The total energy input to the gap so determined for both the positive and the negative rod cases was $2 \pm 1 \times 10^4$ J. It was not possible within

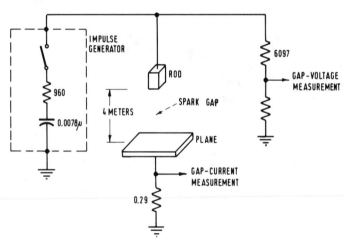

FIG. E5b-3. Schematic diagram showing circuit for producing 4-m rod–plane sparks and circuits for measuring spark gap voltage and current.

the accuracy of the measurements to determine which discharge polarity absorbed more input energy.

If the spark is assumed to be straight and vertical, rough values for the input energy and peak power per unit length can be derived. For the positive rod these rough values are 1.3×10^9 W/m and 5×10^3 J/m; for the negative rod, 1.5×10^9 W/m and 5×10^3 J/m. In addition to the uncertainties in these values discussed, the actual values per unit length should be lower than those given, because the spark channel is tortuous and the sparks are often not vertical. The paper also discusses the energy balance of input against the heat which produces the sound shock wave and the optical radiation measured by image converter technique and photography.

Triggered spark gaps for very short rise time high-voltage pulses have been developed by Blanc [1640]. Very short pulse techniques are called for, more and more, in a variety of investigation methods in various fields of research such as: Storage of energy in coils, capacitor banks, lines, etc., and its release in a very short time; study of electrical circuit parameters; production of ultrafast phenomena (electrooptical shutters, behavior of materials in pulse operation, igniters, etc.). There is also a growing need for a precise knowledge of the parameters that characterize the relevant pulses. The author gives the various parameters of such pulses: maximum amplitude, duration, rise time, waveform. He also lists the short pulse generators that are employed in each particular application.

Besides the conventional circuits of pulse generators and their control devices, he describes in detail the main component of these systems, i.e., the triggered spark gap by which it is possible to obtain pulses with the required characteristics of rise time, amplitude, duration, waveform, and power. The principle of these spark gaps is the use of high pressure gas tubes with three electrodes, one being a triggering electrode to initiate ionization between the two main electrodes.

The technological problems associated with the design of these spark gaps have been cleared up and their performance can be reproduced with a high degree of reliability.

The behavior of electromagnetic waves generated by high-voltage pulses on lines and radiated into space is the subject of a paper by Böcker [1563] it deals with the ratio between the charging energy and the energy radiated.

The potential gradient on the surface of switching electrodes for ultrahigh voltages has been measured and reported on by Moeller et al. [1564].

c. *EMP Simulation*

A good introduction to the subject of intense electromagnetic pulse (EMP) simulation is provided by Ion Physics [1701].

An important effect of a nuclear explosion is the generation and propagation of a strong, impulsive, electromagnetic field capable of impairing the performance of critical electrical and electronic systems. The engineering solutions seem straightforward; all critical circuits must be shielded, arranged for minimum coupling or be fitted with transient damping networks. The route to providing all necessary solutions on existing systems and developing new design techniques for the future, however, lies in comprehensive system testing programs using accurate simulations of the threatening environments.

The systems under test are subjected to uniform EM fields of peak amplitudes near "threat level." Since the systems are varied, the uniform field volumes are required to be large. This implies that the peak voltages and energies of the simulating EM pulses are also large. In practice, the "working test volume" is often provided between a stripline formed between an upper structure and the ground plane. An appropriate pulse generator discharges to this line structure via matching transitions; the EMP is propagated in the fundamental TEM mode so long as the frequency components and structure dimensions are consistent with transmission line operation through the line, which is fitted with a terminating load.

The peak electric fields desired are in the 10^5 V/m range. Since the working line spacings are a few to tens of meters, pulse generator peak voltages within the range 10^5–10^7 are demanded. Generators capable of these peak voltages have been available for many years, but the application of these generators to the EMP task has necessitated the development of special techniques. In particular, the rise time to peak pulse voltage is specified as <10 nsec. Since the pulse must be obtained by transmission from the dielectric environment of the pulse generator to an atmospheric test environment, the combination of the peak voltage and rise time specifications present high-voltage design problems which are peculiar to the EMP field.

The basic concept of the Ion Physics approach to the production of high energy pulsed radiation sources is outlined in the following:

A coaxial gas capacitor, which is essentially the elongated terminal of a Van de Graaff generator and the surrounding pressure tank, is charged to several megavolts then discharged at the open end into the load by a fast spark gap switch. The combination of a fixed voltage and a single output switch make the output pulse highly reproducible.

There are four major elements in the system. One is the voltage generator, which has been developed to beyond 10 MV in the form of the Van de Graaff machine. Its characteristics are not significant to the discharging (transient) properties of the system, except that the stored energy in the column adds slightly to the tail of the radiation pulse and that it represents a high impedance power supply. The other major elements are the coaxial gas capacitor, the triggered megavolt switch, and the load, which is a field emission accelerator tube.

The coaxial gas line is an elongated terminal for the Van de Graaff generator. The dielectric is high pressure gas, usually a mixture of nitrogen with 20% sulfur hexaflouride. Electrical breakdowns of the line are self-healing, and breakdowns result in increased voltage hold-off strength. The length of coaxial line determines the pulse length, in as much as the double transit time of the pulse on the line is the time for the stored energy to flow. To a first approximation, one may assume a velocity of the disturbance to be 1 ft/nsec with the result that a 15-ft long line will yield a 30-nsec pulse.

The trigger electronics and the trigatron switch are mounted within the coaxial line. The single output switch is electronically triggered, effecting a spark discharge across the gap between the front of the coaxial line and the field emission tube housing. The switch triggering system is, in summary, a 100-kV generator firing a small spark gap on the terminal surface. For the use of this equipment as on a flash x-ray machine, a single spark channel suffices to meet the rise time needs of the field emission tube. However, when these systems are used in the EMP simulation application, multiple spark channels materially reduce switch inductance permitting the attainment of 6–10 nsec pulse rise times in the multimegavolt regime. The basic IPC trigger system design will permit the operation of a multichannel switch. Jitter on these systems is typically ± 5 nsec rms.

An important consideration in the operation of the IPC FX series of machines is the relationship of the line impedance and the impedance of the load (field emission tube or EMP antenna).

First, it is necessary to note the output characteristics of a transmission line charged to a voltage V_0 which is discharged into a load R (assumed resistive). A primary pulse of voltage amplitude $V = \alpha V$, current amplitude $I = V/R$ and duration 2τ is produced where

$$\alpha = R/(R+Z_0) \tag{1}$$

where Z_0 equals characteristic impedance of the line and τ equals the electrical length of the line. For a coaxial line with a dielectric constant of l, the electrical length is ~ 1 nsec/ft of length, and:

$$Z_0 = 60 \ln(r_2/r_1) = (L/C)^{1/2} \tag{2}$$

where r_2, r_1 are, respectively, the outer and inner radii of the coaxial cylinders and L, C are the inductance and capacitance per unit length of the line. Figure E5c-1 taken from a paper by Milde [1224] may explain the values under consideration.

Maximum power is transferred into the load when $R = Z_0$, but from expression (1) it can be seen that this is when $V = V_0/2$. The requirement is not for maximum power transfer, but for the maximum radiation intensity produced by an accelerated electron beam, which varies roughly with the third power of the voltage. At this point, it is necessary to examine how a transmission line will perform when discharged into a field emission load.

The upper plot of Fig. E5c-1 shows a typical current–voltage relationship for a field emission gap. An excellent review of the field emission process and its potentialities is given by Dyke. The minimum field geometry in a coaxial gas capacitor is when $\ln(r_2/r_1) = 1$ or $Z_0 = 60 \, \Omega$, and the lower plot on Fig. E5c-1 shows for a specific charging voltage the load characteristic of a transmission line with that impedance. A replot of the field emission charactersitic is also given. The intersection of the two curves is the operating point, with the transmission line developing a pulsed voltage corresponding to a load $R = V/I$ at the operating point. There is no time delay between field

5. LIGHTNING FLASH AND EMP SIMULATION

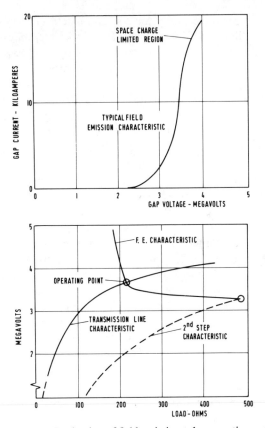

FIG. E5c-1. Derivation of field emission tube operating point.

emission and applied field, at least in the time regime of interest here, which with the above argument implies that the field emission tube can be treated as a resistive load R on the line as determined above. Actually, during the pulse when there are high currents flowing, the tube impedance will fall somewhat because of the changes in the field emission characteristic.

Finally, the installation at the Sandia Base, Albuquerque may be mentioned [1030] because it is available for classified users to test equipments. The pulse generator at the ARES facility is basically a low-impedance coaxial gasline which may be charged to peak amplitude in 45 sec by a Van de Graaff generator and discharged directly into the transmission line load. This gasline concept achieves high repeatability. A pressure vessel (see Fig. E5c-2) containing pressurized gas encloses the generator, terminal assembly, and output to the assembly. Access to the internal components is provided by a hinged closure 9 ft diam at the output end and by ports in the vessel near both ends.

Viewing ports are provided along the top and sides. The vessel is mounted on heavy-duty dollies and supported by four jack assemblies which permit height and angular adjustment. The trackmounted vessel is positioned by a hydraulic actuator whose 12-ft stroke length, combined with a 15-ft long removable coaxial section, provides ample space for installation of the dummy load or low-voltage adapter, and removal of major units from the vessel. An output switch consisting of six electronically triggered channels initiates breakdown of the gap between the charged terminal assembly and the output bushing. There is also a standby switch that is triggered mechanically. The gap width is adjustable from 3.5 to 15 in. by means of an external gap control mechanism. The oil-filled output bushing couples with the oil-filled 125-Ω coaxial line, which contains three assemblies: (1) The energy diverter, which can be switched in parallel with the peak of the pulse waveform, (2) the safety shorting mechanism, and (3) the adjustable peaking gap. A flexible outer conductor section at the interface with the input transition assembly facilitates mating alignment. A dummy load with impedance approximating that of the antenna is provided for calibration and testing of the generator. A low-voltage adapter is provided for testing the transitions and antenna. The input transition assembly consists of an insulating oil-filled enclosure (oil box), a 1-atm SF_6 gas-filled enclosure (gas box), a ground plane, and a top plate electrode. Within the oil box, the tubular coaxial geometry changes to diverging stripline geometry. Permanent E-field and B-field probes to monitor the voltage waveforms, are located in the gas box. Access ports are provided in the oil and gas boxes for inspection and maintenance.

The over-all length of the transmission line is 189 m from apex to apex. The top plate of the transmission line consists of 75 equally spaced wires. Each wire consists of six strands of aluminum around a core of steel and is secured inside the connecting segment. The ground plane is constructed partially of wire (under the working volume), mesh, and perforated plate. The width of the ground plane is 264 ft at the working volume, tapering to 50 ft at each apex. The ground plane in the working volume area is embedded 2 in. below the concrete surface. The transmission line segment of the terminator begins approximately 15 ft above the ground plane and continues for a horizontal distance of 25 ft. The transmission line segment is then connected into the vertical line-to-ground load. This resistive liquid-type termination dissipates the energy through heat with less than 10% reflection.

Both a positive and a negative pulse are possible with 10-min downtime necessitated for reversing polarity. Maximum rise time, measured between 10 and 90% of peak amplitude, is less than 10.0 nsec measured at the gas section sensor (see Fig. E5c-3). Peak amplitude is 3.9 MV and is variable from 700 kV to this peak amplitude. The pulse decay time from peak amplitude to 10% of peak amplitude is approximately 300 nsec. When not charged to

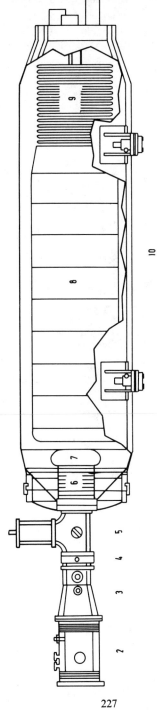

FIG. E5c-2. Cross section of the ARES megavolt Van de Graaff machine feeding a discharge capacitor to produce x-ray flashes or megavolt pulses, see text. 1. Input transition; 2. tube extension; 3. peaking gap; 4. safety short; 5. energy diverter; 6. output bushing; 7. variable gap; 8. terminal assembly; 9. Van de Graaff generator; 10. pressure vessel.

full voltage, the pulse generator is capable of firing up to 1 shot per 3 min with a waveform repeatability within ±10% of the peak amplitude and the decay time.

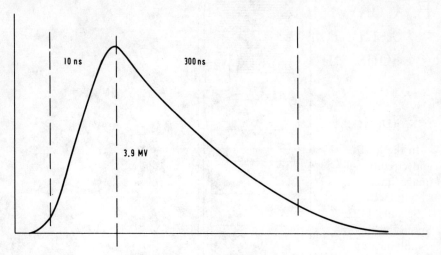

FIG. E5c-3. Specifications for waveform of voltage pulse in transmission line.

F. Conversion of capacitor energy into x-ray flashes and beams of electrons, ions, and neutrons

1. New Methods for Generating X-Ray Flashes[*]
(Debye–Scherrer and Laue Patterns, Plasmas and Lasers)

In the field of physics, very essential progress has been made in the design and circuitry of x-ray pulsers since 1965. By combining x-ray flashers with image intensifiers, very weak monochromatic x-ray flashes can be used to obtain Debye–Scherrer patterns at very short exposures.

Schaafs [1922] presented at the 10th International Congress on High Speed Photography (1972) a paper showing how to produce K-radiation using a molybdenum anode. Debye–Scherrer patterns can be obtained either with simple flash operation and an image intensifier or by summing the effect of a large number of repetitive weak flashes. It may be mentioned that with high capacitor energy more or less "white" polychromatic radiation will, in any case, be generated. To produce monochromatic radiation a well-defined voltage pulse, just a little above the excitation voltage of the particular anode metal, must be applied to the anode.

Jamet [1923] produced Kα-radiation flashes of 100 nsec duration with a Cu anode fed with 25–30 kV pulses. This tube still operated properly after 100 pulses. The experiments discussed at the same conference showed that for pulsing copper anodes 120 keV and for molybdenum 250 keV, are necessary to obtain 6–12 keV monochromatic Kα-radiation. Beryllium exit windows on the tube are useful as they transmit soft x rays down to 4 keV. An attempt to use soft electrons of the same energy had negative results.

Charbonnier [1924] had success with Debye–Scherrer patterns using a three-step image intensifier. He obtained Laue patterns with Al$_2$O$_3$ crystals at 180 keV with good intensity, and he found 300 keV suitable for molybdenum targets, 150–200 keV for Cu-targets (anodes). An empirical rule is to apply 15 times the voltage of the resonance radiation of the anode material used. Cu Kα-radiation with 15 nsec pulse time (6–8 keV quality) therefore needs ~ 120 keV.

---

[*] See [1703]. Basic recommended literature on x-ray flash techniques: G. Thomer [1703g]. Sixty-six other references included.

Stenerhag et al. [1925] discovered x-ray flashes in exploding wire experiments. It seems that the voltage remaining on the electrodes between which the wire exploded may accelerate the generated electrons in an axial direction. To obtain this result, the current in the exploding wire must have more than 10^9 A/sec rise speed. At 30 kV, a tungsten wire 0.1 mm diam and 60–80 mm long delivered a 200-nsec x-ray flash through a 0.1-mm Al filter. The shortest flash was observed with 30 kV on a 0.05-mm tungsten wire 20 mm long, the duration being 170 nsec. The pressure was 100 μTorr. In one case two subsequent flashes were observed at 70 nsec intervals. Perhaps, with detachable tubes this technique could be developed into a simple method for fast x-ray double flashes for motion analysis.

Mattsson [1926] used a pinhole camera for imaging the x-ray spot on the anode and fed the anode with a 10-kA 20-nsec pulse from a Marx generator. The pressure in the vacuum diode was 10^{-4} Torr obtained with the help of a small ion pump. The observed anode spot on the tantalum anode had a diameter of 0.5–1 mm. The cathode was of V2 A stainless steel. With a CdS crystal behind the anode, he obtained a strong monochromatic superradiant flash of 5350 Å with 60 Å bandwidth. He applied 250 kV, 5 kA, 20 nsec pulses to the tube and the tantalum anode and the crystal both had a diameter of $\sim$3 cm.

Rapp [1927] presented another unusual method for producing polychromatic x-ray flashes, namely, the so-called plasma focus technique. The plasma focus tube is 66 mm diam and 100 mm long and when it is fed with 30 kJ from 150 μF, a pinching discharge at 500 kA and 30 nH entire inductance, takes place. The pinching discharge generates neutron flashes according to the reaction $D + D = He_3 + n$. The maximum neutron generation was available at 100 kJ 100 nsec at 20 cm Hg pressure of D. Jitter time was less than 100 nsec and normally, two subsequent neutron pulses, caused by additional plasma compressions, were emitted. The highest efficiency of neutron emission begins at 20 kJ.

With either D or He, the x-ray flashes have a very broad spectrum. For example, in an x-ray photograph of a mouse, the soft rays bring out all the details right down to the hairs, while the stronger rays show the skeleton. The first experiments aimed at producing x-rays by laser action have been successful. Lax [1254] gave the necessary conditions for soft x-ray lasers. In his paper he considers the criteria for the excitation of an x-ray laser by a high-energy electron beam or a high-power mode-locked glass laser or a CO_2 laser. At the present time the electron beam is incapable of adequate excitation. The lasers, however, are able to produce stimulated emission of soft x rays. Consistent with the present state of the art, this requires power levels of the order of 10^{12} W and pulse durations of several picoseconds.

The first consideration for the pump is to provide sufficient power per unit volume to competely ionize the atoms within a penetration depth in a time duration shorter than the recombination time. For a lifetime of 10^{-11} sec, the minimum power required for a 1s–2p transition in beryllium is $\sim 10^{10}$ W. This corresponds to a focal spot size of 100 µm and a penetration depth of 1 µm.

The second consideration is that the time for complete ionization be shorter than the recombination time. This is necessary in order to achieve inversion. The mechanism involved in the ionization is an avalanche process. The total time calculated for this process is 10^{-12} sec for the power density of $10^{16}/cm^2$ that is available. This easily fulfils the second criterion.

The third requirement is that the rate of stimulated emission be faster than the rate of spontaneous emission. The former can be obtained from the gain expression of

$$g = (\Delta N/8\pi) - (\lambda^2/\Delta v)\tau_{spont}^{-1}$$

This gives $g = 10^6$ cm^{-1}, $\tau_{stim}^{-1} \simeq 10^{16}$ sec$^{-1} \gg \tau_{spont}^{-1}$. The large gain thus obtained will produce a superradiant beam of 100-Å wavelength in beryllium, which can be made to be directional in a suitable geometry.

Details for detecting the presence of laser radiation by measuring directional effects, the gain, nonlinear increase in the x-ray laser output, and narrowing of the line are described in Lax's paper. Plasma effects and their expansion during the laser pulse are considered in the design of the experiment. Calculations for the operation of the soft x-ray laser and the quantitative criteria for achieving laser action in the short x-ray region are presented. Based on this approach, it is estimated that peak powers of $\sim 10^{14}$ and 10^{16} W and pulse durations in the subpicosecond region are indicated for obtaining laser action in the 10- and 1-Å regions, respectively. Calculations show that spontaneous and stimulated lifetimes at 1 Å become comparable. This and the fact that the time required for complete ionization is longer than the spontaneous lifetime, preclude the possibility of a strong x-ray laser by the techniques presented in the paper. Similar work was done by Zarowin [1255].

Since the energy separation between equivalent levels of the optical electron (s) increases with degree of ionization, there appears good reason to believe that multi-ionized atoms, isoelectronic with known ion laser species, can be expected to produce stimulated emission in the vuv/soft x-ray regime under the appropriate conditions. For example, singly ionized neon (Ne II), is isoelectronic with nine times ionized argon (Ar X), so that the laser transitions of Ne II (3p → 3s) at ~ 3300 Å are expected to be found at ~ 400 Å in Ar X. More highly ionized atoms, isoelectronic with other known ionic laser species, can be expected to produce even shorter wavelength stimulated emission. The production of such highly ionized atoms and the population of their excited levels in gas discharges require higher electron temperatures and densities than those presently attained in gas lasers. The required discharge conditions can be inferred from the experimental circumstances under which known ion lasers operate, at present up to four times ionized atoms (S V) and down to wavelengths of 2358 Å (Ne IV). These have been found to occur in moderately high current density gaseous discharges ($10^3 \to 10^4$ A/cm^2) at low pressure (<1 Torr). Furthermore, it is well known that "vacuum arc" discharges in which the vaporized electrode material forms the low-pressure gas produce highly ionized atoms in substantial densities, if sufficiently high power densities are delivered to the plasma.

This approach to the production of highly excited multi-ionized atoms has been limited by the contrary requirements of high power density and sufficiently high plasma resistance to permit efficient delivery of this power density. The high power density generates the high electron temperatures required for collisional excitation of the energetic levels of interest, but also decreases the plasma resistance so that delivery of the power becomes progressively less efficient.

Efficient delivery of this power requires that the discharge of the equivalent series RLC circuit be damped, i.e., such that $0 < 4L/R^2C \leqslant C$, where the peak current, I_p, is $V_0/R > I_p \geqslant 0.74V_0/R$ for an initial capacitor voltage V_0. Thus the inductance–resistance square ratio intrinsic to the discharge and associated circuitry determines the minimum capacitance required for damping, increasing as the resistance decreases. To restrict the capacitance to reasonable magnitude (~ 10 μF), a coaxial n turn primary transformer will be shown to allow the plasma resistance R_p to appear as $n^2 R_p$ to the capacitor, without a similar increase in leakage inductance common to the more usual transformer. In this manner, the discharge may be restricted to damped operation even with the low plasma resistance of high power density low-pressure arcs.

Such a device has been constructed and initial current densities of 10^6 A/cm^2 have been achieved in an optical interaction region 1-cm diam by 20-cm length at filled pressures less than 1 Torr. The discharge is internally concentric with the multiturn coaxial primary ($n = 6$ in this case) and forms a 1-turn secondary. The construction of the plasma tube, the differentially pumped port tube connecting it to a grazing incidence spectrometer, will be described. The procedure for detecting gain with the available low-reflectance mirrors at these wavelengths and up to date experimental results were further subjects of Zarowin's paper.

Experimental results have been presented by Tonon et al. [1928] describing a new detecting device which allows one to obtain frequency, time and space resolved pictures of uv and soft x-ray emission of a laser-created plasma in a single shot. X-ray pictures of such a plasma are shown. After these preliminary results, it is possible to set up readily an x-ray framing camera. A laser-created plasma is an x-ray source of special interest. The emitted power can be 10% of the laser intensity and the emitted spectrum is centered around 1-Å wavelength.

Another source for the generation of x-ray pulses is a very hot vacuum spark in which the vaporized electrode material is the x-ray source. Cohen *et al.* [1291] published a paper giving the results of experiments on such x rays. X-ray spectroscopic studies show, in addition to the characteristic x-ray lines of the electrode metal, lines originating from very high stages of ionization of the element. In particular, lines of the $1s^2$–$1snp$ sequence in the helium spectra Ti XXI through Ni XXVII, and the $1s$–$2p$ transition in Fe XXVI,

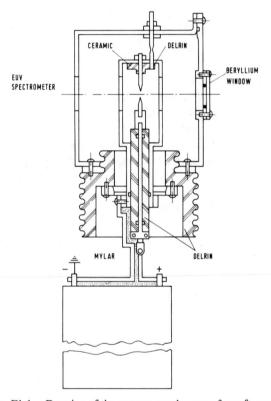

FIG. F1-1. Drawing of the vacuum spark source for soft x rays.

were observed. Pinhole photography and time-resolved studies of the total x-ray emission indicate that these highly ionized species originate in a small, hot region between the two electrodes, which is produced by a pinch in a plasma of evaporated anode metal.

Figure F1-1 shows schematically the internal structure of the spark source. Two rods of the element under investigation, of 0.2-cm diam and separated by a gap of ~ 0.3 cm, form the spark gap. At a pressure $\sim 10^{-4}$ Torr, a 13-μF capacitor is discharged across this gap. The main spark is triggered by a sliding spark across the ceramic insulator between the ground electrode (cathode) and a third electrode; this spark is in turn ignited by a tesla coil. The source has so far been operated at voltages up to 19 kV, giving 2.3 kJ/discharge. It is this comparatively large energy, combined with the low (160 nH) inductance of the capacitor and associated transmission line, which makes this source so efficient for exciting the spectra of highly ionized atoms. The maximum current in the discharge was 176 kA, as given by

$$I_{max} = V_0 (C/L)^{1/2}$$

where V_0 is the starting voltage.

The spatial distribution of the x-ray emission in the spark was determined by pinhole photography. Pinholes of various sizes were pierced in a sheet of tantalum or lead, and placed in contact with the beryllium exit window. A piece of x-ray film was positioned outside the vacuum chamber at a distance such that a 1:1 image was obtained.

The spectrum of the x rays was investigated with a crystal spectrograph of the type shown in Fig. F1-2. In this arrangement, the x rays are allowed to fall upon the crystal with a range of Bragg angles between θ_1 and θ_2. Radiation of wavelength λ will be reflected if the corresponding Bragg angle θ, given by the relation $n\lambda = 2d \sin \theta$, lies between θ_1 and θ_2. A spectral line will produce a line on the film whose width is determined by the effective size of the source, unless the source is very small, when the perfection of the analyzing crystal will be the determining factor. It was found that about 100 sparks were necessary to produce a spectrum of reasonable photographic density. As each spark occurred in a slightly different position, the lines were smeared out to a width of about 0.2 cm. In fact, for lines originating in a very small region of the source, many fine lines were seen on the plate, each corresponding to a different position of the spark. This smearing of the lines limited the precision with

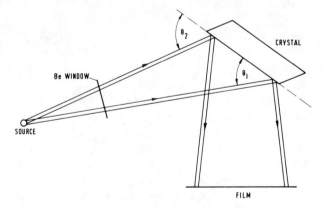

FIG. F1-2. Schematic diagram of the Bragg crystal spectrograph.

which wavelength measurements could be made to about ± 0.01 Å. The crystal used was an excellent calcite, kindly lent to us by Dr. Deslattes of the National Bureau of Standards. The (104) reflection, with $2d = 6.070$ Å, was used throughout.

In addition to the helium-like spectra, it was also possible to observe the 1s–2p Ly–α transition in Fe XXVI. The wavelength was found to be 1.79 Å. A broad feature with a short-wavelength limit at 1.63 Å was also observed.

All stages of ionization of the anode metal except the very lowest have been observed to occur in this spark. The possibilities for x-ray transitions in multi-ionized atoms, and in excited states of these ions are therefore very numerous. It is therefore very likely that the satellite structure observed by Schörling is due to effects of multiple ionization.

The time histories of the various spectral lines produced in this source, both in the x-ray region and in the extreme uv, combined with pinhole photography and other techniques, should provide useful diagnostic tools for the study of this type of discharge. Use of this discharge as a spectroscopic source should open up an interesting new region for the atomic spectroscopist.

Finally, it proved possible to generate nearly monochromatic x-ray pulses by using a high-current monochromatic electron beam pulse. The generation of such beams is the subject of a paper by Friedman and Ury [1768].

A preliminary study was performed to ascertain the feasibility of delivering large quantities of electrical energy in relativistic electron beams of long pulse duration. An electron beam of 8 kA peak current and 250 kV peak voltage was produced with a duration of $\gtrsim 1$ μsec. The electron beam was transported through a 1-m long drift tube with little energy loss. The generators producing these beams are capable of delivering from 10^9 to as much as 10^{13} W peak electrical power for durations of 50–100 nsec.

Some applications require longer duration electron beam pulses at comparable electrical powers. Moreover, present high-power short-duration relativistic electron beam generators have disadvantages which can limit the amount of electrical energy they can deliver. These limitations include the electrical inefficiency from having to pulse-charge an intermediate energy store, the increased likelihood of voltage breakdown across insulators in very low-inductance systems such as are needed in short pulse duration, and the impedance collapse of diodes when current densities exceed 1 kA/cm^2.

Some of these limitations can be reduced in a long pulse generator. Thus, the electrical efficiency can be improved by removing the pulse-forming line (e.g., Blumlein transmission line) from the system and connecting a diode directly to a capacitor bank or to an artificial delay line. The second limitation can be overcome by enlarging the dimensions of the insulators. This is possible if their inductances are not factors in the generator performance. Quantitatively, it has been found empirically that the breakdown voltage V_c across an insulator is given by

$$V_c \propto d/t^{0.17}$$

where d is the insulator dimension, and t is the time over which the voltage is maintained. The third limitation forms the most severe problem. Only two basically different designs of diodes have been employed in relativistic electron beam accelerators: planar field emission diodes and foilless diodes.

Much of the present study was devoted to exploring the impedance collapse of these diodes under long voltage pulse conditions. One experiment is shown schematically in Fig. F1-3. A swinging L–C type generator of 63 nF effective capacitance is used to produce a peak voltage of 250 kV on a 7-Ω transmission line connected directly to the diode. The diode was immersed in an 8-kG magnetic field and the entire system was evacuated to $\sim 1.0 \times 10^{-4}$ Torr.

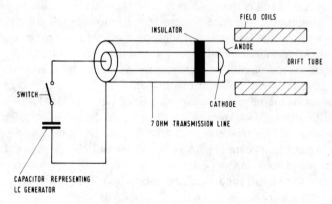

FIG. F1-3. Schematic view of the experiment with a foilless diode.

The beam produced by the foilless diode in the long pulse mode of operation was allowed to propagate in an evacuated 1-m-long drift tube. The beam propagated with little energy loss demonstrated the same qualitative features that have been observed for short pulse duration beams. In particular, long duration pulses of x-ray radiation were observed indicating little degradation of the electron energies during beam propagation. Moreover, long duration microwave pulses were obtained by passing the electron beam through a spatially modulated magnetic field as has been done with 50 nsec duration beams. Only 750 J of electrical energy were produced and transported in the form of a relativistic electron beam. However, our results show that higher currents and higher voltages can be used, and hence, higher electrical energy can be delivered.

Another paper on the behavior of the plasma focus was presented by Fischer and Bostick [1766] at a special AFCRL seminar. High resolution (0.1 mm),

5–10 nsec image-converter photographs demonstrate the existence of strong directed fields extending into a small energy, 1-kJ plasma focus, resulting from filaments which can be traced back to the early stages of the predischarge. Neutron emission appears to originate from areas of minute size, hot x-ray spots which can be observed within the filament structures. Such microscopic details will be obscured by background radiation if the energy applied is raised to the 10–1000 kJ range.

Mead *et al.* [1525] did preliminary measurements on high-temperature plasmas produced with two Nd-glass laser systems.

As produced here and elsewhere, these miniature plasmas have peak temperatures in the kilovolt range at densities of $\sim 10^{21}$ cm^{-3}. In addition to their fundamental interest, such plasmas are recognized to be of potential importance in connection with efforts to achieve controlled thermonuclear fusion.

In searching for diagnostic techniques that are suitable for exploring laser-produced plasmas, it is interesting to note that the collective and random motions of the plasma constituents give rise to characteristic emissions that can often be selectively measured. The random motion of the free electrons, for example, may yield soft x-ray bremsstrahlung and, in some cases, may lead to x-ray line emission by the bound electrons. The corresponding motion of deuterium ions can produce fusion neutrons. Collective electron motions, on the other hand, which are caused by the incident laser light, may lead to informative optical radiation. In attempting to examine all these emissions, one hopes to obtain a more complete picture of the evolving plasma and its interaction with the incident laser radiation.

Its principal components, as shown in Fig. F1-4, consist of a mode-locked oscillator, an electrooptic single-pulse selector, and a series of rod amplifiers. The system delivers subnanosecond pulses of about 5 J. A special feature of the oscillator is the inclusion within the acvity of a pair of mode-selecting splitters (tilted glass plates), which result in an extended oscillator pulse of 50–100 psec and appear to facilitate the production of clean single pulses. The pulse train emitted by the oscillator passes through a single amplifier before entering the electrooptic shutter which selects a single pulse of a few millijoules. The unwanted pulses are reduced by the shutter system to $\sim 0.01\%$ of the selected pulse. However, noise pulses that may occur within a couple of nanoseconds of the elected pulse are not significantly attenuated. The final amplifier chain provides a net gain of ~ 1000.

The spectrum above 2 keV is being explored using Bragg reflection from a cylindrically curved graphite crystal. Line radiation appears to be present in the spectra measured for laser-produced plasmas of lead and gold. Plasma temperatures implied by x-ray transmission measurements are found to vary with absorber thickness. Temperatures of a few tens of kilo-electron volts, inferred from thick-absorber measurements of low-Z plasmas, suggest the operation of anomalous heating mechanisms.

A second glass laser system, involving face-pumped disks 14 cm diam, is also in use for plasma-heating experiments and has led to the production of D–D neutrons. This system has been used to produce pulses of several tens of joules in a few nanoseconds for recent experiments. A calibrated plastic scintillator with two photomultipliers is used for neutron detection. Targets

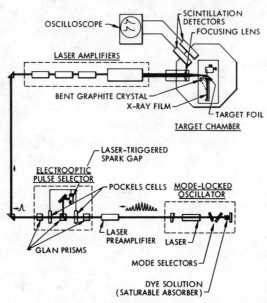

FIG. F1-4. The plasma-x laser system as used for x-ray studies. A subnanosecond pulse of about 5 J can be focused on the target. All laser rods are Brewster-ended Nd glass and, with one exception, are 1.3 cm in diameter and 24 cm in length. The final amplifier is 2.5×48 cm. Laser-beam diagnostic equipment, which consists primarily of calibrated photodiodes, is not shown.

of CD_2 have produced total yields of more than 10^4 neutrons. Effective neutron production appears to require laser pulses longer than 2 nsec and to be accompanied by a large reflected pulse. While x ray and neutron generation in plasma focus tubes is a matter of temperature and particle speed, conventional x-ray tubes use electrons emitted from a hot filament or from a cold metal surface as a result either of local heating by a vacuum spark or of high field strength.

Passner [1767] describes an entirely new method, using a laser beam which strikes the metal surface and generates an appreciable number of electrons to start an x-ray flash of well-defined duration. His x-ray tube delivers 10 A at

100 keV into a 1.5-mm-diam spot in a 10^{-8} sec pulse at 500 pps using the photoelectric effect emitted by the photocathode for the electron source. Employing the light from a nitrogen laser, the device produces a 10-nsec pulse of x rays. The tube was built for an experiment to frequently mix x-ray photons with light photons. The experiment consists of mixing a pulse of 3371-Å light from the nitrogen laser with the copper Kα x-ray emission line in diamond. Owing to the small angular region of efficient mixing, a high brightness (photons/cm^2 sr) beam of x rays is required for the experiment.

There are several ways of producing a high brightness x-ray pulse. One would be to grid modulate a thermonic cathode x-ray tube. Another might be to use a spark x-ray tube; however, high power spark tubes have only a limited lifetime to date, and do not produce subnanosecond output pulses.

It was decided to build a photoelectric x-ray tube because it simplified the problem of making the light and x-ray pulses coincide while also permitting the production of subnanosecond x-ray pulses, should the experiment require them. When the equipment was designed, it was possible to buy a nitrogen laser having a power output of 10^5 W/pulse.

The high intensity x-ray pulses are produced in a 25.4-cm diam, 30.5-cm long vacuum chamber (Fig. F1-5), which is mounted on a 10.2-cm aperture liquid nitrogen trapped diffusion pump using polyphenyl ether pumping fluid, with a foreline oil trap. The LN$_2$ trapped chamber pressure after two days of pumping is $1-2 \times 10^{-8}$ Torr. The cathode C is a quartz disk 1.3-cm thick at the edge, flat on one side, with the surface facing the anode ground and polished to a 5.1-cm radius concave surface. The concave surface is Nichrome (or gold) plated to make electrical contact with the cup D. A 4×2 mm hole in the center of the disk is provided for an x-ray output. This hole shape was chosen to match the output of the x-ray tube to the acceptance angles of the first Bragg diffraction crystal of the light x-ray mixing experiment. The quartz disk is supported by a stainless steel structure, part of which is a hemispherical cup D. The entire cathode structure is attached to an insulator and operates at negative 100 kV. The anode A normally operates at ground potential and is a copper disk mounted on a goniometer head G. The anode is electrically insulated from ground so that the current from the anode can be brought out through the feedthrough M. A hemispherical mesh E mounted on a tube is between cathode and anode. A disk F is between the insulator and cathode support structure. Its function is to shield the insulator L from contamination from the sodium gun I. H is a sliding seal to allow movement of the anode and mesh under vacuum. I is a movable sodium evaporation gun. J is a port for the laser light to enter. The light beam diameter at the cathode is determined by lenses and mirrors exterior to the vacuum chamber.

With a photoemitter deposited on the cathode, photoelectrons are emitted when light illuminates the photocathode. The electric field accelerates the

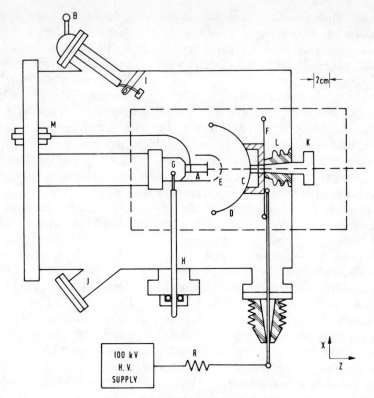

FIG. F1-5. Top view of x-ray tube. The portion inside the dotted lines is drawn to the indicated scale. The tube diameter and length are also to scale. Details are in the text.

electrons and they form a waist near the anode. The x rays travel from the anode through the hole in the quartz and through a Kapton* window in port K.

The effect of space charge is to blow the beam up radially. Once the beam has reached 100 keV, space charge effects are small in the distances used. It is the function of the mesh to accelerate the electrons to 100 keV in as short a distance as possible and also shape the field in the vicinity of the cathode. Without the mesh the beam blows up, but with it a 10-A beam focuses into a 1.5-mm spot on the anode. The spot can be made asymmetric by focusing the laser into a horizontal line on the cathode. The space charge forces blow the beam up vertically and one obtains a vertical line focus on the anode of 0.5×2 mm.

Passner also discusses the selection of the best material for the photocathode and the dimensioning of the electron optics.

---

* Trademark of DuPont.

A comprehensive survey of plasma behavior leading to the production of x rays and neutrons is given in the famous German book by Vollrath and Thomer [1703] (see especially p. 1023, subchapter 7.2, with 114 other references).

2. New Designs and Applications of X-Ray Flash Equipment for Motion Analysis

a. *Equipment*

For all cases of shadowgraph photographic motion analysis in which the subject is obscured from observation with visible light, x-ray flash penetration must be used. Each improvement in x-ray flash technique and equipment broadens the field of application. At the 9th International Congress on High Speed Photography, 1970, Ebeling [1929] presented a survey of available aids to research (or R & D) work based on x-ray flashes.

The generation of x-ray flashes for submicrosecond x-radiography requires improvements in pulse generation and tube design. X-ray flash equipment has been produced that is capable either of delivering, with one tube, one to eight flashes with at least 5 μsec flash separation or flash series up to 40,000/sec which are limited only by the thermal capacity of the anode. High-voltage pulses, variable from 30 to 300 kV, are generated by switching 2–50 J, 10 kV pulses into the primary of a specially designed ferrite core HV pulse transformer. All switching elements are normal "low voltage" thyratrons or H_2-pressurized spark gaps. Together with direct drum–camera recording or an additive image converter rotating prism camera system, the equipment allows the investigation either of very fast events with eight flashes up to 200 kHz or of long duration phenomena by long burst x-ray flashes up to 20 kHz for film lengths of ~30 m. Simple modifications permit the generation of electrons and neutrons.

The operational principle of flash x-ray tubes is based on a vacuum breakdown process between a conically shaped anode and a surrounding circular ring cathode.

As illustrated in Fig. F2a-1, the breakdown is initiated either by switching the capacitors C, charged to a voltage U_c exceeding the breakdown voltage, by means of a spark gap switch to the anode or by generating an initial plasma between the cathode and an additional trigger electrode T. In both cases the number of electrons needed to get a reasonable amount of x rays after falling through the cathode–anode potential is generated by a field emission–photoionization process. It has been shown that currents of the order of 10^3 A and more are obtainable. Calculation of the formulas of Fig. F2a-1 for typical pulse durations of about 100 nsec yields currents in the range 10–10,000 A

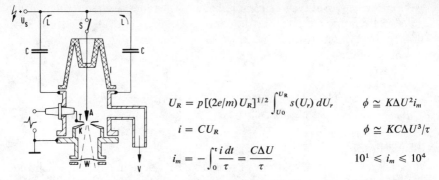

$$U_R = p[(2e/m)U_R]^{1/2} \int_{U_0}^{U_R} s(U_r)\,dU_r \qquad \phi \cong K\Delta U^2 i_m$$

$$i = CU_R \qquad \phi \cong KC\Delta U^3/\tau$$

$$i_m = -\int_0^\tau \frac{i\,dt}{\tau} = \frac{C\Delta U}{\tau} \qquad 10^1 \leqslant i_m \leqslant 10^4$$

FIG. F2a-1. Schematic diagram of the x-ray flash. U_s = max voltage; L = self-inductance; C = storage capacitor; A = anode; K = cathode; T = trigger electrode; W = exit window; S = switching element; I = insulator; V = vacuum.

for single flash exposure, which is very large compared with those obtainable with cw-operated tubes.

For time sequential flash generation with a single tube, the problem arises of switching the high-voltage pulses to the tube in the short time interval required for time resolved fast movement investigation, say, down to 10 μsec between flashes.

The principle of double flash generation, for example, is outlined schematically in Fig. F2a-2, where (a) is the conventional operation of switching to the tube, at times t_1 and t_2, and (b) two capacitors are charged to a dc voltage U_S. In order to prevent flashover of the t_2 switch at time t_1 or the t_1 switch at time t_2, both switches must have a diode characteristic at a low impedance, otherwise an unreasonable amount of energy is lost in the switch itself. For voltages exceeding 40 kV such fast diodes are not available. To overcome this difficulty, operation as outlined under (b) was chosen. All switching elements are com-

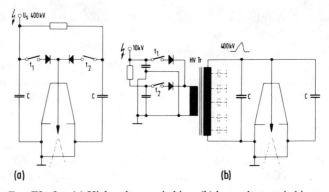

FIG. F2a-2. (a) High-voltage switching; (b) low-voltage switching.

paratively low voltage devices like thyratrons or special spark gaps feeding high current pulses into the primary of a suitable pulse transformer. Development of a new pulse transformer for submicrosecond pulses up to 400 kV was undertaken using ferrite material in a chilled core shape in order to take advantage of the highest possible magnetic coupling between the windings. The primary consists of a single one-turn layer of Mylar-insulated copper foil connected by a sandwich line to a series of extremely low-inductance capacitor spark gap systems marked by $SpG_1 \cdots SpG_n$ in Fig. F2a-3.

The capacitors C_{pr} are charged by a dc HV generator to a voltage between 10 and 20 kV. The mechanism of x-ray flash generation is as follows: At the time t_1 the first thyratron fires spark gap 1 thereby feeding a 10-kV pulse into the primary of the pulse transformer. The secondary steps up the pulse, charging the main discharge capacitor battery C_s to about 300 kV within 600 nsec. Simultaneously at t_1, the first thyratron has generated a low-voltage pulse which is delayed by a variable network exactly the 600 nsec which the 300-kV pulse needs to reach its peak. At this time the second thyratron fires the x-ray tube at T Tr by means of a 30-nsec trigger pulse, thereby discharging the extremely low-inductance capacitor battery C_s into the tube. At a preselected time t_2 the event is repeated and a new flash is generated.

In the foreground of Fig. F2a-4 the tube head mounted on a vacuum pump stand may be seen. The insulator, turned out of a polyethylene cylinder, is

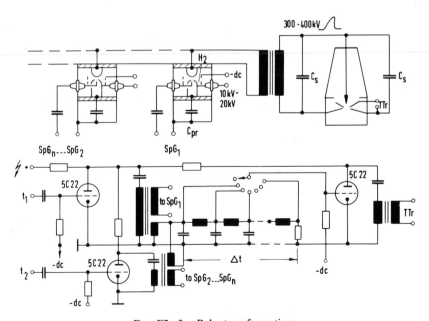

FIG. F2a-3. Pulse transformation.

Fig. F2a-4. Pulse transformer with tube head.

surrounded coaxially by six capacitor batteries of maximum 600 pF capacitance, thus delivering a discharge energy of 12 J at 200 kV anode voltage, for example. The corresponding dose rate for a single flash is 0.4 R at 20 cm focal distance. The tube is operable up to 400 kV without self breakdown and has a practically unlimited lifetime because the parts which suffer evaporation damage, like cathode rings and the anode top, are easily replaced. Further, the construction of the tube does not allow the evaporated material to reach the inner wall of the HV insulator.

For short-time operation the tube can be sealed off from the vacuum pump if the arrangement of the experimental setup makes it necessary. The HV pulse transformer in the background delivers up to 400 kV pulses into 600 pF, generated by 20 kV, 0.5 μF pulse discharge into the primary, thus yielding an energy efficiency of about 0.5. For variation of x-ray quality there are two more tapping points for 120 and 220 kV at maximum rating.

As examples three recordings of a 3-mm copper test screen with 0.5-mm holes are given in Fig. F2a-5. The distance from anode to screen was 1 m and from screen to film or image converter cathode screen was 3 cm. For the image

Fig. F2a-5. Test screen, direct recording, and image intensifier recording.

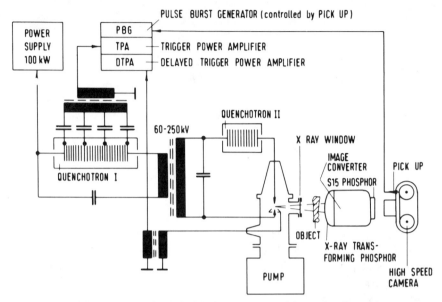

FIG. F2a-6. Operation principle for long burst high speed radiography.

converter recording, the dose rate was lowered fiftyfold in order to avoid severe overexposure of the luminescent screen. As can be seen, the 0.5-mm holes are just within the overall limiting resolution of the image converter tube, whilst direct film recording is limited in resolution only by the film itself and the size of the anode emitting spot, calculated at about 1 mm to half-power points.

For long burst high speed radiography in the medium range of 500–40,000 frames/sec with a total frame number of about 2000—decreasing to about 300 at highest speed—there are additional problems to be solved.

If a flash rate of, say 2000/sec is assumed with 8 J flash energy lasting for one second, the power supply has to deliver a total energy of 16 kJ into all switching elements. That means an energy consumption of about 25 kW out of the ac line. The principle of operation sketched out in Fig. F2a-6 is the same as for single or double pulse generation as outlined earlier. The difference for repetitive pulsing is that, instead of certain number of dc-charged capacitors being discharged into the primary of the HV transformer, the same capacitor is charged and discharged with a preselected frequency generated by a "pulse burst generator." The switching element in this case is a heavy duty quenching spark gap fired at four stages by a "trigger power amplifier" thyratron device up to 100,000 flashes/sec. The delayed output of this TPA triggers the x-ray tube itself as was outlined earlier. The deionization time of the x-ray tube, reduced by a second Quenchotron, limits the highest possible flash rate to

Fig. F2a-7. X-ray anode.

~40,000/sec. In the example shown in Fig. F2a-6 the flash rate is synchronized with a rotating prism camera, thus generating an x-ray flash through the object at the image converter each time a picture is ready to be taken. The fairly slow decay time of recently available x-ray image converters limits this operation to ~6000 frames/sec. At higher frequencies a severe reduction in information must be accepted because of "overwriting effects." Also at these higher frequencies drum cameras must be used.

In earlier cinematographic experiments the anode consisted of a steel needle with a tungsten cone on top. This type of anode will not dissipate the heat generated by a 400 J burst and has to be replaced after a short time. With the new type of anode shown in Fig. 2a-7, however, it was possible to dissipate the heat generated by 10,000 J bursts. It consists of a heavy tungsten cone shrunk on to a beryllium–copper rod with cooling disks. It is possible to drive the anode up to the red glow point for some time, taking advantage of radiational cooling, which is proportional to T^4.

Great care must be taken with lead shielding of the exit cone to avoid severe

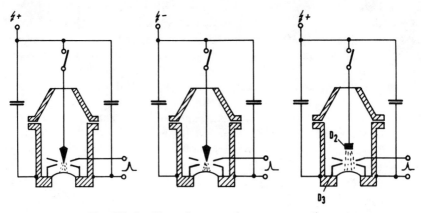

Fig. F2a-8. X-ray electron and neutron generation.

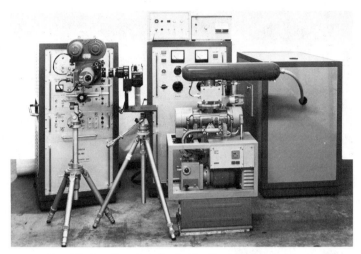

Fig. F2a-9. Complete Strobokin–Strobo x-ray flash unit.

radiation damage to the operator. A 2.5-kHz flash burst of 1 sec duration generates up to 250 R in a cone of 40° at 20 cm distance. For the whole body, that is about one-half of the lethal dose.

Simply by changing certain electrode parts and the transformer polarity the tube can be made to deliver sequences of 400 keV electron flashes of

Fig. F2a-10. X-ray flash lamp.

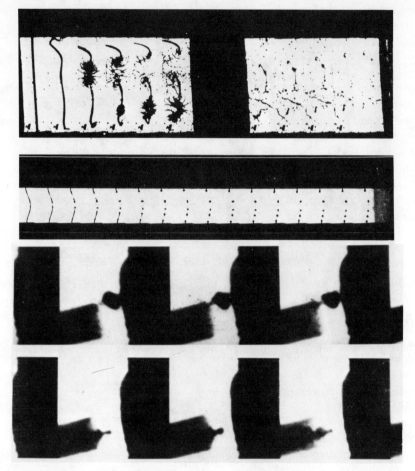

Fig. F2a-11a,b. Two samples of operation.

2 W sec/cm² flux at 20 cm distance, after focusing, thus giving the possibility of generating x radiation outside the tube or of taking e^- shadowgraphs of very low density, low-contrast objects. Neutron generation by means of deuteriated tungsten electrodes in a tritium low-pressure atmosphere is a third possibility of the tube, which will be further investigated. (See Figs. F2a-8 to F2a-11.)

b. *Practical Techniques*

In Sweden, Mattsson [1031] developed an x-ray single flasher for 1200 kV, consisting of detachable components only.

The pulsers are modified Marx-surge generators, with output voltages from 100 to 1200 kV and peak currents close to 10 kA. The capacitors act as pulse-forming transmission lines yielding a fast rise output pulse. They can be arranged to provide a high stage-to-stage capacitive coupling, ensuring proper and rapid breakdown of untriggered spark gaps; hence, no ultraviolet coupling of the spark gaps is necessary. Special care has been taken to build reliable systems of low weight and compact design, in the interest of mobility and easy operation. The x-ray flash tubes have exchangeable electrode systems which permit the electrode configuration to be easily adapted for optimum contrast and resolution in a given application. The tubes are continuously fed with small ion-supply pumps. Thus far, dose rates of 4×10^9 R/sec at 1200 kV and pulse lengths of 10 nsec have been obtained. With certain fast-film/screen combinations, stop–motion radiographs can be taken at 1 m through, for example, 5 cm of steel. Further details are given below.

The performance of the flash x-ray machine has to satisfy the three following basic conditions.

First, the radiation output must be of a sufficiently short duration to avoid motion blur. To resolve small particles moving at hypervelocities a short exposure time is essential; if the allowable blur is one tenth of the particle diameter, a 1-mm particle moving at a velocity of 5000 m/sec requires an exposure time of $2 \cdot 10^{-8}$ sec.

The second condition affecting radiographic quality is radiation output (beam intensity and energy). When radiographing small or thin objects, these objects must absorb enough radiation to cause a detectable change in film density. The minimum particle thickness, which can be resolved depends on two factors: the characteristic absorption factor of the material and the beam energy. Normally a 10% change in x-ray intensity is just detectable using commercial film screens; this means that a copper foil 0.1-mm thick cannot be resolved at beam voltages higher than 200 kV. On the other hand x rays generated at higher voltages (1 MW or more) will make it possible, with fast films and phosphor screens, to obtain radiographs through relatively thick objects (e.g., 50-mm steel or more), or, when using slower films and lead screens, to extend this high picture quality technique to objects of medium size moving at high velocities. Hence, to cope with the divergent radiographic demands formed by today's high speed phenomena, a wide voltage operating range of the flash x-ray machine is essential.

The third factor influencing resolution is the x-ray source size. The resolution (R) indicating the smallest spacing two objects can have without overlapping penumbra is defined by

$$R = (b/a)f$$

where f is the source size, a the film-to-source distance, and b the film-to-objects distance. The quotient b/a is normally of the order of 0.1–0.01. Thus, with $f \leqslant 1$ mm, R becomes small enough (0.1–0.01 mm) for all practical radiographic purposes).

A design to meet these conditions led to the development of:

(1) a compact and flexible set of pulse generators covering the voltage range 100–1200 kV, with a 10,000-A peak current and a 20-nsec pulse width;

(2) a fine-focus flash x-ray tube having an exchangeable or replaceable electrode system.

The present pulse generators are based upon the Marx-circuit; voltage multiplication is accomplished by charging capacitors in parallel and discharging them in series. Spark gaps are used to switch from parallel to series arrangement. One or two of the gaps are triggered and the remainder are overcharged by the transient voltages within the generator.

The capacitors act as pulse-forming transmission lines yielding a fast rise output pulse. They can be arranged to provide a high stage-to-stage capacitive coupling, insuring proper and rapid breakdown of untriggered spark gaps. Hence, no ultraviolet coupling of the spark gaps is necessary, a considerable simplification of the generator design.

Efficient electromagnetic shielding is achieved by making the electrical discharge coaxial. The radiated noise is further reduced to a very low level by housing capacitors, charging network, and spark gaps in an electrically grounded steel tank. In this way the pulse generator performance is also isolated from such atmospheric variables as dust, humidity, pressure, etc. The steel tank is pressurized to improve insulation and to control the breakdown voltage of the fixed spark gaps. Only one pressure system is incorporated. Dry air is used as insulating medium. The output voltage is varied by simultaneously adjusting the charging voltage and the pressure of the spark gaps.

A block diagram of the pulse generator system is shown in Fig. F2b-1. The main units are: pulse former and delay unit, pulse amplifier, trigger unit, timer unit, high voltage power supply, pressure regulator, and pulse generator. The output voltage depends on charging voltage, number of stages, and load impedance. Several generators with 10, 20, and 40 stages have been built and tested under field and laboratory conditions. Electrical and physical data of

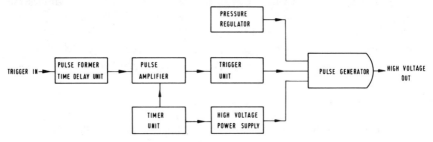

FIG. F2b-1. Block diagram indicating functional relation of major components in Scandiflash pulse generators.

these generators are compiled in Table F2b-1. Given voltages refer to measurements with leads matched to the internal impedance of each generator. The time jitter of the generators including triggering circuits is less than ±50 nsec. This figure can be reduced to ±10 nsec by careful adjustment of the spark gaps. Elevation of the generator is controlled by a hydraulic system. The very compact design (tank diameter 0.8 m, length 1.2 m) allows it to be easily moved and positioned in any direction.

Operating a flash x-ray tube in the megavolt region with nanosecond pulses requires a tube current of the order of 10^3–10^4 A. At the same time a small x-ray source size requires an electron beam cross section of the order of 1 mm at the anode. Hence, an electron current density of about 10^6 A/cm^2 is necessary. These current requirements are met by a field emission type electron source. The electric field strength necessary in field emission is high ($\sim 10^7$ V/cm). However, the corresponding voltages are moderate, depending on the choice of electrode material and electrode configuration. Thus, a tube design utilizing a field emission source, with exchangeable or replaceable electrode system and exit window, was considered the best way to ensure maximum versatility and economical operation.

TABLE F2b-1

ELECTRICAL AND PHYSICAL DATA OF SCANDIFLASH PULSE GENERATORS

Pulse generator model	300	600	1200
Output voltage (kV)	100–300	250–600	500–1200
Peak current (A)	10,000	10,000	10,000
Internal impedance (Ω)	30	60	120
Dimensions:			
diameter (mm)	700	800	800
length (mm)	600	885	1205
Weight (kg)	150	300	450

The tube as seen from the schematic Fig. F2b-2 comprises a geometrical arrangement of the anode and cathode derived from the well-known "Siemens-tube" configuration namely, a conical anode 3 and a disk-shaped hollow field-emitter cathode 4. The x rays are extracted along the tube axis and, as a result of the conical anode, the apparent emitting area is small. Operating the tube at 1200 kV, and 10^4 A, the peak power is about 10^{10} W. The generated energy is dissipated in the small volume around the electrodes. To cool the anode a small amount of anode material is allowed to evaporate. This is the only practical way to dissipate the heat generated by the discharge. To prevent the metal vapor formed from contaminating insulating surfaces of the tube structure, a cylindrical metal shield 5 surrounds the discharge volume. Erosion

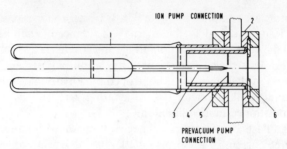

Fig. F2b-2. Cross-sectional view of flash x-ray tube with modified Siemens electrode arrangement. 1. Tube frame; 2. pump section; 3. anode; 4. cathode; 5. sputtering shield; 6. exit window.

effects are significant only when the tube is operated at high voltages (~ 1 MV). Thus, the effective lifetime of the anode is limited, a minor drawback as the anode is easily replaceable.

Erosion effects are reduced by increasing the anode area, i.e., decreasing the energy density at the anode. This is practically solved with an electrode configuration (Fig. F2b-3) featuring a cathode consisting of four edges mounted parallel to the conical anode. The drawback of this arrangement is a somewhat larger source size. Thus, in radiographic applications where a small source size is required, the modified Siemens geometry is preferred.

The risk of unwanted voltage break down in the small tube structure is controlled by critical material choice, smooth and clean surfaces, and ultra-high vacuum techniques. The tube is continuously pumped by a 1 liter/sec ion pump with the pump also functioning as an accurate vacuum gauge. Table F2b-2 shows the values obtained under routine test conditions.

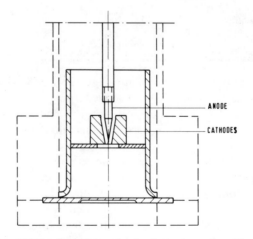

Fig. F2b-3. Low power density electrode configuration.

2. DESIGNS AND APPLICATIONS OF X-RAY FLASH EQUIPMENT

TABLE F2b-2

VALUES OBTAINED WITH SCANDIFLASH PULSE GENERATORS UNDER ROUTINE CONDITIONS

Pulse generator model	300	600	1200
Dose/pulse at 0.5 m from exit port (mR)	35	75	180
Radiation pulse width (FWHM) (nsec)	20	15	10
Dose rate at exit port (R/sec)	$3 \cdot 10^8$	$1 \cdot 10^9$	$4 \cdot 10^9$
Typical source diameter (mm)	1	1	2

Fexitron [1290] manufactures a series of different x-ray flashers for hypervelocity impact studies, chronography research, exploding wire, and plasma research, pulsed radiation testing, shock wave studies in dense solids, and metal casting research.

The Fexitron x-ray system is designed for flash radiography of high speed events in opaque media at velocities up to 20,000 ft/sec (6×10^6 mm/sec). Film density of 0.7 is provided through 3 in. of aluminum at a film-to-source distance of 4 ft and shadowgraphs in air at distances up to 19 ft. The pulsers generate square wave voltage pulses that provide the system with maximum x-ray yield at a minimum of stored energy, size, and cost. Spectrum and resolution are improved since the square waveform provides optimum anode power for a given x-ray yield. The x-ray tubes use the newly developed T-F emission electron source and not the vacuum are source employed by earlier flash x-ray tubes. The T-F emission cathode is claimed to have higher reliability and reproducibility on a pulse-to-pulse basis and relatively long life. In contrast to the Impulsphysik equipment [1036, 1868], in which the same anode flashes with a high repetition rate, the Fexitron single-channel system can be expanded in the field for multichannel application by the addition of pulsers, x-ray tubes, and an accessory attachment for sequential timing and trigger amplification. Specifications are given in Table F2b-3 for the whole system and in Tables F2b-4 and F2b-5 for the flash tube and pulser, respectively.

In all cases where the workpiece or event under examination is too thick to be penetrated by the x-ray flashes with sufficient intensity to blacken the x-ray film, x-ray image intensifiers msut be used. Their main limitation is the reduction of the number of line pairs per millimeter with increasing number of amplifier stages, caused by the inadequate resolution provided by the imaging lenses between the stages. Modern intensifiers employ, instead, glass fiber bundles, usually ground and polished to spherical form at both the cathode and the luminescent screen end. Modern types deliver a properly exposed photographic grain for each x-ray quantum received, a yield which seems to be the limit of physics.

Recent developments in the use of high-gain optical image intensifiers for

TABLE F2b-3

PERFORMANCE SPECIFICATION OF THE FEXITRON 300 MW FLASH X-RAY SYSTEM, TYPE PS-300–1000-0.1

X-ray tube	The T-300–1000-0.2 x-ray tube receives a square wave from the model 215 pulser of tolerances prescribed herein.
Resolution	Velocity: up to 20,000 ft/sec; source size: $2\frac{1}{2} \times 7$ mm; object thickness: down to 0.001 in. of copper at 100 kV.
X-ray dosage rate	10^7 R/sec at tube envelope.
Pulse width	Type PS-300–1000-0.1-0.1 μsec.
Multiflash sequence	10^6 frames/sec maximum; 2–5 pulsers may be used for sequential radiographs with interpulse delay from 1 μsec to any length delay determined by delay generator characteristics.
Transmission lines	300 Ω transmission lines may be used between pulser and tube; lengths up to several hundred feet are satisfactory.
Output voltage	Continuously variable from 150 to 300 kV at fixed impedance of 300 Ω.

TABLE F2b-4

PERFORMANCE SPECIFICATION OF THE FEXITRON 300 MW FLASH X-RAY TUBE, TYPE T-300–1000-0.2

Peak power	300 MW at 0.2 or 0.1 μsec
Cathode	T–F emission type
Cathode voltage	9.8 V ac, 50–60 Hz
Cathode current	85 A
Anode	Tungsten
Anode temperature	3000 °K max
Anode voltage	300 kV max
Beam current	1000 A max
Pulse duration	0.2 μsec
Pulse overshoot	Max of 5%
Pulse undershoot	Max of 20%
X-ray source size	$2\frac{1}{2} \times 7$ mm
X-ray dosage rate	10^7 R/sec at 3 in.
Physical size	12 in. long $\times$ $3\frac{1}{2}$ in. diam
Weight	2 lbs

high speed photography of dynamic x-ray diffraction patterns are dealt with in a paper by Reifsnider [1287].* In view of its fundamental importance, it is reported in detail.

---

* Presented on October 7, 1970, at a Technical Conference in New York by Kenneth Reifsnider, Dept. of Engineering Mechanics, College of Engineering, Virginia Polytechnic Institute and State University, Blacksburg, Virginia 24061.

2. DESIGNS AND APPLICATIONS OF X-RAY FLASH EQUIPMENT 255

With advancing technology, the possibility of monitoring in real time the continuous changes in an x-ray diffraction pattern produced by dynamic events has drawn the interest of many investigators. Progress has been slow, largely because of strict limitations on available intensity and required resolution. Also, not only the apparatus but the technique of application to research must be developed in order to make the equipment compatible with rather specific scientific interests. One new system uses a very fine grained fluorescent screen to display the x-ray patterns as optical light images. Those images are transferred via standard lenses to a low-noise, high-gain optical image intensifier. The intensified image is then viewed from the rear of the image tube with a high-resolution vidicon and the final image displayed on a TV monitor. Time exposure photographs, commonly 12 sec in duration, were made of the TV screen while the x-ray device scanned a 5-mm specimen.

TABLE F2b-5

PERFORMANCE SPECIFICATION OF THE FEXITRON 300 MW FLASH X-RAY PULSER, MODEL 215

Power input	Charging voltage for maximum output, 30 kV, charging current not critical but supply capability should be 5 mA or more.
Output voltage	Continuously variable from 150 to 300 kV; voltage exceeds 80% of maximum output for at least pulse length of plug-in storage rating; i.e., 0.1 or 0.2 μsec.
Current	Pulser capable of delivering 1000 A at 300 kV.
Impedance	Constant 300 Ω output; connections provided for attaching cable to x-ray tube or 300 Ω transmission line.
Pulse width	0.1 or 0.2 μsec maximum for plug-in storage ratings of 0.1 or 0.2. Tolerance of $+0.05 - 0.02$ μsec as measured from 80% of leading edge to 80% of trailing edge of the wave.
Pulse rise time	20% *of pulse width* 0.02 μsec maximum as measured from 10% of leading edge to 80% of leading edge of the wave.
Pulse fall time	50% *of pulse width* 0.1 μsec maximum as measured to 10% of trailing edge of the wave.

The time-resolved x-ray diffraction system described here differs from those previously reported in that a fluorescent screen which converts x rays to visible light is coupled to an electrostatic optical image intensifier by contact with a fiber-optic input rather than by standard lenses. The elimination of the front coupling lens increases the output intensity for a given input by a factor of about ten. The use of an electrostatic tube provides the additional advantage of requiring no magnetic field to focus, thereby eliminating the interference of that field with closed-circuit television systems which can be used to display the image tube output. The present system can be quickly changed to optimize its spectral response to x rays, its sensitivity to x rays, its resolution and its intensification, all within certain limits, according to the type of x-ray test being monitored. The theoretical resolution of this system is 35 line pairs/mm. Cine-recording of changing x-ray diffraction patterns during deformation has been achieved with this system. The system is quite inexpensive on a comparative basis, is nondeteriorating, and can be constructed from components commonly available. Finally, the system is by far the simplest and most versatile time-resolved system yet reported. (For an excellent review of other systems, see [1933].)

In order to understand the necessity and importance of the use of electrooptical devices in connection with x-ray diffraction equipment, it is well to examine the investigational objective of such equipment and the subsequent technological requirements. The principal contribution of electrooptical devices is their sensitivity to low-energy-level light images. In the case of x-ray diffraction, this sensitivity allows instantaneous viewing or rapid recording of diffraction patterns which otherwise could be recovered only by time exposure film techniques. However, if one wants to monitor the changes in an x-ray diffraction pattern produced by a single crystal which is being continuously deformed, the "rapid" recording mentioned must be instantaneous, i.e., the system must be "time-resolved." If we use a standard tensile test as an example, the time resolution required to resolve the formation of deformation bands and inhomogeneous plastic slip is of the order of a small fraction of a second. A reasonable time resolution for such tests, then, can be taken as the framing rate necessary for cine recording.

A second objective is to obtain meaningful information which can be interpreted in terms of standard parameters. If one wishes to examine a Laue pattern, then the electrooptical system must have a large format, i.e., a large field of view. In the investigation of deformation processes, one must be able to examine a significant portion of a specimen whose dimensions are large enough so that surface effects or size effects do not alter the normal material behavior. In some cases, it is desirable to monitor the entire specimen during continuous testing. For topographic examination of material defects (and recently the examination of integrated circuit diffusion zones), the "field of view" may be quite small, typically a few square millimeters. A versatile system, then, should have a field of view which can be changed rather substantially in size. It should also be mentioned here that different applications involve different x-ray wavelengths so that the system in use should have an adjustable spectral response which can be optimized according to the job at hand.

Also involved in this second objective of "meaningful information" is the tradeoff between a reasonable level of resolution and such things as sensitivity and cost. Looking at individual material defects at the atomic level certainly requires the best resolution obtainable. However, if one is trying to characterize deformation processes which have macroscopic significance, an effective magnification of less than 100× is frequently sufficient, especially in view of recent generalized continuum mechanics theories which simulate "farthest neighbor effects."

Finally, a reasonable objective is to construct a nondeteriorating system at minimal expense whose basic components are of standard type and are commonly available.

The following system described is time-resolved, i.e., complete x-ray diffraction patterns of all types can be recorded directly from the rear of the image intensifier tube cinematographically. A variety of x-ray fluorescent screens (including single crystal types) can be used in the system and are used in direct contact with a fiber-optic coupling rather than with a front coupling lens as in other reported systems of similar type. The system is quite inexpensive to build from parts easily obtained and is not progressively destroyed by use. Four applications of the system will be described to characterize the equipment and its capabilities. The applications are topography, Laue patterns, continuous deformation of crystals, and radiography of composites during deformation.

In order to record x-ray diffraction patterns with exposure times of a fraction of a second, the image must be operated a certain way to increase its intensity. In particular, the required intensity gain for direct viewing of such images is $\sim 10^6$ greater than that required for industrial fluoroscopy and $\sim 10^3$ greater than that required for medical fluoroscopy. The intensity of medical fluoroscopic images is low by choice in the interest of minimizing the radiation dosage of the patients being examined. However, the x-ray beams commonly used for diffraction studies must be highly collimated and originate from a small area in order to satisfy very strict resolution requirements. Hence, their absolute intensity is limited by the specific intensity of x rays which can be generated by the incidence of electrons on a target material. The latter limitation is quite severe. A standard x-ray source operating at 50 kV and 50 mA, for example, incident on a 0.8-mm thick aluminum crystal will produce a transmitted diffraction pattern which, if displayed on a standard CB-2 fluorescent screen, will generate an optical image with a brightness of only 10^{-6} fc.

In the present system, image intensification was achieved through the use of a modified RCA 8606 three-stage cascaded optical image intensifier tube. Figure F2b-4 is a schematic of this tube. The tube is electrostatically focused and has a magnification of 1:1. The input and output faceplates of each stage of the tube consist of fiber-optic bundles which are flat on the exterior ends to allow for flat-field optical system coupling and are curved on their inside faces in deference to electrostatic focus requirements. A multialkali photocathode

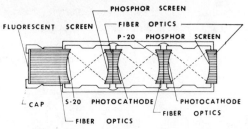

Fig. F2b-4. Electrostatic image tube.

with S-20 spectral response is deposited on the interior surface of each fiber-optic input, and an aluminized fluorescent phosphor with P-20 response is deposited on the interior surface of each fiber-optic output. Each stage of the tube is operated at a bias of 15 kV. A (permanent) focus is achieved for any operating bias voltage greater than 8 kV per stage. Our particular image tube operates with a measured luminous flux gain of 77,200. It has an on-center measured resolution of 36 line pairs/mm. The tube has a minimum useful cathode diameter of 37.5 mm.

In order to make our image tube sensitive to x rays, a fluorescent screen (of various types) is mounted flush with the input fiber-optic bundle, held in place by an annular cap (see Fig. F2b-4). The fluorescent material of the screen is in direct contact with the faceplate, while the plastic backing of the screen faces the x-ray beam. Hence, the fluorescent screen converts the x-ray image into visible light and the photocathode converts that light pattern into an electron image. The electron image is then "intensified" by acceleration in the electrostatic field and reconverted into a visible light image by the fluorescent phosphor on the output fiber optic whereupon the second and third stages of the tube perform identical functions. The final output fiber-optic faceplate displays a visible light image sufficiently bright for direct viewing, for recording by standard film techniques, or for viewing by closed-circuit vidicon television.

The annular cap which holds the fluorescent screens on the image tube input faceplate is easily removed allowing quick exchange of x-ray fluorescent screens to optimize the resolution-sensitivity trade off. Three types of polycrystalline screens were used for the present tests, a CB-2, a "Lightning Special," and a "Detail" screen, all Cronex screens made by Du Pont Photo Products Dept. The greatest resolution (and least sensitivity) was achieved with the Detail screen. The threshold resolution of that screen on the image tube is indicated by the fact that a 140-mesh of 74 μm copper wire with 105 μm nominal openings can be resolved. The CB-2 was far more sensitive but the resolution was nearly an order of magnitude worse.

Topographic studies are commonly carried out with a monochromatic x-ray beam which is highly collimated to ensure sufficient resolution. Because of these features, the resulting beam and diffraction pattern have very low intensity, especially in the case of transmission topography. A series of investigations was made with the present apparatus to determine if the sensitivity was sufficient to detect such low-intensity patterns.

A fine-focus Siemens x-ray tube with a Cu target measuring 0.4×4 mm was operated at 40 kV and 25 mA to provide an x-ray source. The first few tests were made with the target at a 3° takeoff angle (looking end-on) so that the projected target size was 400×400 μm. A 40-cm collimating tube was used between the x-ray source and the specimen (Fig. F2b-5). A silicon crystal 2 cm diam and 1 mm thick was used for all tests described below. With the arrangement for reflection topography, it was possible to find the (331), (311), and (440) reflections without difficulty. In fact, a distinct image of the reflections was displayed by the image tube at 20 kV, 20 mA, and 70% maximum intensification. A scintillation tube reading of the intensity of the latter beam indicated a flux of the order of 10 photons/cm^{2}sec. This high sensitivity is characteristic of high-gain optical image intensifier systems and is a distinct advantage over other reported types. (For further discussion of this point, see [1933].) A Ni filter for β radiation was used and the Cu Kα_1 and Cu Kα_2 images were clearly separated for the (440) reflection.

As mentioned above, the intensity is lowest for transmission topography. The same Si crystal was set up for a symmetric Laue reflection from the (220) planes with the transmitted diffraction beam incident on our image tube (see Fig. F2b-5). Operating at 45 kV and 20 mA, the intensified image was very bright at 75% maximum intensification. Hence, with the present system, intensity limitations do not seem to be a problem, at least for standard topographic test arrangements.

The question of resolution is still unsettled. The best polycrystalline screen used in the present system had a resolution (on the intensifier tube) of about 10 line pairs/mm, i.e., could resolve detail spaced apart by about 100 μm. This is certainly not sufficient for most topographic work. However, a single crystal x-ray fluorescent screen could be used with the system and would bring the resolution of the system close to the 36 line pairs/mm measured

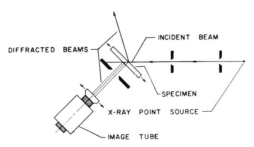

FIG. F2b-5. Topographic test arrangement of tube.

value for the image tube alone. Such screens must be specially made and are difficult to obtain. However, one such screen is presently being prepared in the author's laboratory.

Instantaneous viewing of Laue patterns has become somewhat of a comparative test and source of application of time-resolved x-ray diffraction system.

One of the more active areas of dynamic x-ray diffraction involves the study of the changes in large-beam Laue spots during continuous deformation. As a test for the usefulness of the present system for such studies, several aluminum single crystal plate specimens were extended in tension and exposed successfully to the 4-mm line source of x rays. A schematic of the arrangement is shown in Fig. F2b-6. The specimen had a gauge length of 2.54 cm × 1.27 cm × 1.59 mm thick. A notch 1.6 mm deep was cut into one edge (see Fig. F2b-6) before the specimen was extended. The x-ray beam was collimated by a 40-cm tube terminated by a set of slits exposing the total width of the specimen over

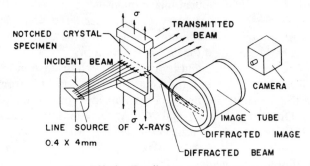

FIG. F2b-6. Tensile test arrangement.

a length of about 2 mm which included the notch. A single Laue spot was monitored by the image tube which was positioned about 7.6 cm from the specimen. The changes in the Laue spot caused by continuous extension of the specimen at a rate of 5 mm/min were recorded from the rear of the image intensifier tube by a 16-mm cine camera. Tri-X (ASA 160) film was used with a framing rate of 10 frames/sec at 40 kV, 20 mA, and 76% maximum intensification. The inhomogeneous deformation around the notch was monitored until the specimen ruptured.

The final application attempted with the present system was high-resolution radiography. Although a great majority of radiographic investigation requries very-high-intensity x-ray sources and involves resolution equivalent only to the human eye, there are several areas of investigation where additional resolution is required. One of these areas of investigation of deformation is the metal–metal composites. In addition, low-energy x-radiation common to diffraction sources is more desirable for selective absorption effects in materials with similar absorption or for materials such as boron which have very low absorption coefficients.

In order to demonstrate the feasibility of using the present system for such tests, several commercial pure aluminum matrix composite specimens were made with stainless-steel wire fibers cast in the matrix as reinforcing elements. An attempt to monitor fiber separation as a result of tensile deformation was partially successful.

c. *Applications to Fast Motion Analysis*

A paper by Bergon and Constant [1032] deals with the application of electrooptical recording equipment to short-pulse-length x-ray flash investigation of explosive processes.

The study by flash x ray of explosive processes require the recovery of a large film whose protection, in the presence of destructive charges, raises difficulties. Where the explosive charge is located on the axis of the emitted x-ray beam, and thus in direct view of the film-fluorescent screen receptor, effective protection is impossible without impairing the picture quality. This difficulty is overcome by the use of an electrooptical arrangement allowing distant recording of the fluorescent screen image before its destruction.

The equipment consists of two distinct parts, the x-ray generator with its fluorescent screen, and the electrooptical system for taking the picture of the screen. The x-ray source is a 2.3-MeV generator. The flash has a 30-nsec duration and the dosage furnished is 130 mR at 2 m. At this distance the penetration is more than 280 mm of aluminum or 80 mm of steel. The fluorescent screen is a rapid salt screen which gives a maximum luminance (Radelin "T 12" or Ilford "Fast Tungstate"). A wide aperture objective, of focal length adapted to the desired magnification, projects the image from the screen to the photocathode of a light-intensifier tube operating as a high-speed shutter.

The English Electric tube (model P829A) has the following characteristics:

Photocathode: Trialkaline S20
Output screen: P11
Image magnification: 1:1, useful screen diameter 25 mm
Minimum sensitivity: 80 μA/lm (white light from a tungsten ribbon source at 2870° K)
Minimum photon gain: 10^5 (measured at the emission peak of the P11 curve—near 4300 Å)
Resolution: better than 20 line pairs/mm in continuous operation
"Dark emission": maximum 600 scintillations/cm^2 sec

These characteristics are given for a high tension supply of 36 kV. Under these conditions a magnetic field of 260 G insures focusing. The tube comprises five amplifying stages and the distribution of voltages on the several dynodes is obtained from an external voltage divider.

The optics for taking the image from the output screen consists of two objectives mounted in tandem (Boyer objective f50–1/1.4). These two allow taking a picture of 200 mm maximum diameter. Considering the use of Polaroid film, we have chosen a magnification of 1.3/1 to conserve the definition in the output screen image. A shutter placed between the objectives and the film suppresses interfering scintillations outside of the exposure time. To protect from the destructive effects of the explosive charges, the equipment is divided into two distinct packages, the one containing the electrooptical system being buried.

Figure F2c-1 shows a section of the installation which allows the firing of a maximum explosive charge of 20 kg at 1 m distance.

Schaaffs and Krehl [1033] used x-ray flash radiography to photograph shock wave processes generated by double discharges from series and parallel-connected spark gaps. Natural electric double discharges, separated from each other by only 1 mm or so, which sometimes develop when the discharge occurs

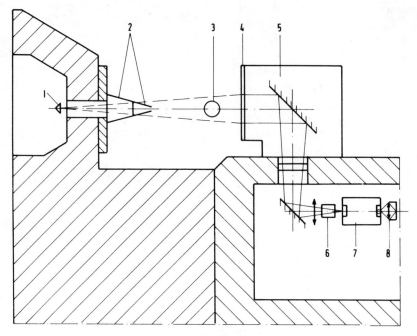

FIG. F2c-1. 1. X-ray source; 2. x-ray output and protection; 3. object radiographed; 4. fluorescent screen; 5. light-tight box; 6. electromagnetic shutter; 7. intensifier tube; 8. optical system to take picture of output screen image.

in a thin layer of a dielectric liquid, cannot, unfortunately, be reproduced at will. However, by appropriate series connection of the spark gaps, it is possible to generate, with good repeatability, well-defined discharge pairs with separations of some 20 to 40 mm. These have been observed by x-ray flash photography for the first 200 μsec. Unusual phenomena, which it was possible to record, include violent interaction of the compression rings with laterial ejection of matter; shockwave collisions preceding the formation of the high-mass density compression rings; and special effects at boundaries of media having different shockfront velocities.

Figure F2c-2 demonstrates schematically the x-ray flash arrangement used for the experiments. The electric discharge caused by the discharge of the pulse capacitor C_2 over the air switch gap S occurs in the liquid layer Fl between the two electrodes e_1 and e_2. The steep rise of current in the liquid spark triggers the delay circuit E which ignites the x-ray flashtube XT. From the anode, an x-ray flash is emitted which records the discharge in Fl in axial direction on the photographic plate P.

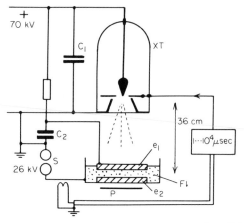

FIG. F2c-2. Schematic of x-ray flash equipment for recording dielectric discharges in liquids.

Figure F2c-3a shows a schematic sketch of a dielectric discharge obtained by the experimental arrangement of Fig. F2c-2. In the center there is a hot gas plasma G, which is surrounded by a compression ring K with the radius r_a of highly-compressed matter. The matter in K can be compressed approximately by the factor two. The compression ring is surrounded by a broad belt B of cavitation bubbles produced by transverse vibrations of the electrodes. Similar observations were also done from the other side on thin layers of nitroglycerine ignited by an electric spark. Examinations of the expanding $r_a(t)$ and the expanding velocity $\dot{r}_a$ of K have given the diagram shown in

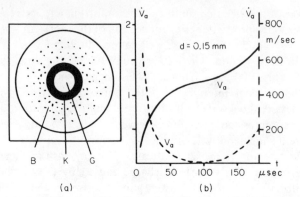

FIG. F2c-3. (a) Sketch of a radiograph of the discharge; (b) radius r_a and expansion velocity $\dot{r}_a$ of the compression ring K.

Fig. F2c-3b. From this, it follows that the expanding velocity becomes very small for a time period of about 50 μsec because for a time, the liquid undergoes solidification under the action of shockwave processes.

The experiments described here were carried out in bromobenzene C_6H_5Br, because this compound supplies radiographs of high contrast. Any other dielectric liquid, however, can be used.

Artificial electric double discharges were also studied and an experimental arrangement for the production of simulated double discharges in series and parallel connection is shown in this reference. (Eight other references of the same authors are given in this paper.)

In his paper Krehl [1725] shows the arrangement in more detail, with excellent photographs of Laue x-ray interference patterns. In his flash tube he used 0.05 mm Beryllium windows.

Schaaffs [1724] describes in detail the construction of a high-precision short-focus x-ray flash tube for 100 kV operation. Excellent pictures on p. 250 show shockwave behavior around electric sparks in liquids. Another paper by Schaaffs and Krehl [1726] refers to x-ray flash investigations into the formation of a shockwave from a liquid, solidified by shockwave processes.

A hot gas plasma in a thin dielectric layer of liquid bromobenzene is produced by an electric discharge. This plasma is surrounded by a compression ring of highly compressed matter built up by shockwave processes. The initial high-expansion velocity of the compression ring decreases with increasing compression to nearly zero. This phenomenon lasts about 50 μsec and is explained by the freezing of the original liquid matter. The frozen-in compression ring is, however, a thermodynamically unstable formation and dissolves by producing a quick-melting shockwave. These phenomena were photographed by x-ray flashes and carefully described.

2. DESIGNS AND APPLICATIONS OF X-RAY FLASH EQUIPMENT 265

A further paper on this subject by the same authors [1727] describes similar experiments using organic liquids and water. Their reference [1728] should also be mentioned. Based on two decades of personal experience Schaaffs [1730] published an educational paper on the whole field of x-ray flash applications.

After an introduction to the problem, a survey is given of the development of x-ray flash tubes. Then an x-ray flash apparatus is described which serves to illuminate high-voltage power switches during operation. Concerning the construction of the apparatus, emphasis on obtaining a great part of soft components in the spectrum of the x-ray flash was of great importance in order to attain pictures of high contrast, since it is much more difficult, in a thick object, to reveal small cavitation and vapor bubbles than metallic parts. Details are given of the penetration capacity of the x-ray flashes and of the picture quality. As an example of applications, a series of x-ray flash pictures for the illumination of a 10-kV high-power switch under electric load is given and explained.

As a "joint venture" using both x-ray and laser flashes, Lankford [1034] gave a paper on the application of these two techniques in hypervelocity ablation–erosion investigations in hyperballistics range.

Conventional capacitor–spark light sources combined with shadowgraph or Schlieren optical systems have proved inadequate to cope with the demand for sharp, unblurred photographs of models in hyperballistic flight simulation. Two basic types of light sources have been proposed for the solution of these problems: short-duration nanosecond spark sources and giant pulse lasers. Where flow field data and extreme resolution are not prime requirements, the pulsed flash x ray can also be utilized.

Approximately seven years ago, the Naval Ordance Laboratory, White Oak, Maryland [NOL (WO)], undertook feasibility studies to determine the most practical method of instrumenting its 1000-ft Hyperballistics Range Facility for the observation of ablation, erosion, and model shape characteristics at velocities up to 21,000 ft/sec (6.4 km/sec). A successful combination proved to be the use of laser frontlighting, shadowgraph, and silhouette techniques, in conjunction with flash x-ray photography, which permitted the detection of surface profile changes of only a few mils in contour. Previously unrevealed characteristics of the mechanisms of tip melting and hypervelocity particle impact have also been observed and recorded. The paper describes a few problems and some of their solutions as applied to the NOL (WO) 1000-ft Hyperballistics Range Facility.

In order to incorporate suitable power for frontlighted applications, the configurations shown in Figs. F2c-4 and F2c-5 were designed. The pulsed flash x ray was installed as indicated schematically in Fig. F2c-6. The characteristics of the laser and x-ray equipment are shown in Table F2c-1.

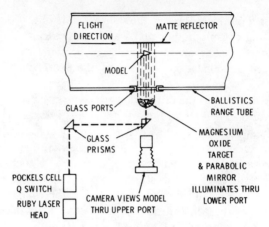

Fig. F2c-4. Front-lighted laser station.

The paper gives excellent flash x-ray photographs together with laser shadowgraphs of bullets and impact events during penetration of rainfall.

At the 10th International Congress on High Speed Photography (1972), Jamet and Thomer, ISL [1923] presented a paper on the recording of x-ray diffraction patterns for the investigation of transient changes in the crystalline structure of materials subjected to the action of shockwaves.

An arrangement including a flash x-ray tube and an image intensifier has been designed and built in order to record x-ray diffraction patterns with exposure times of the order of 100 nsec. This arrangement allows Laue

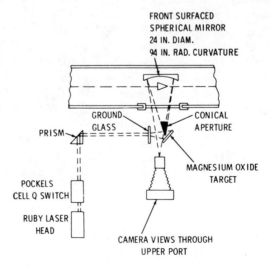

Fig. F2c-5. Laser station for silhouette or shadowgraph.

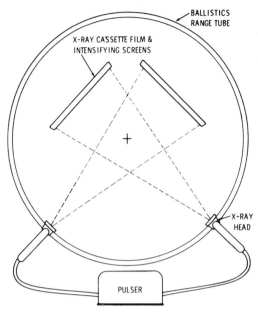

Fig. F2c-6. Flash x-ray installation.

patterns (polychromatic radiation) as well as powder patterns (Cu Kα radiation) to be recorded. Examples of records are given in their paper. As an application to the investigation of transient changes in crystalline structures, the Debye–Scherrer patterns of potassium chloride undergoing the dynamic action of shockwaves were recorded. The first results obtained are discussed.

TABLE F2c-1

CHARACTERISTICS OF LASER AND X-RAY EQUIPMENTS

(a) *Laser equipment*

Giant pulse ruby	Wavelength: 6943 Å
Pulse width. Q-switched, 15×10^{-9} sec	Line width: 0.06 Å
Peak power: 75–100 MW	Q-switch: Pockels cell
Energy: 1.0–1.2 J	Crystal: $\frac{3}{8} \times 4$ in. flat–flat ruby
Brightness: 5×10 in. W/cm²/sr	Flash lamp: Helical xenon
Beam divergence: 10 mrad	

(b) *X-ray equipment*

Output voltage: 180 kV	Pulse width: 30×10^{-9} sec
Output current: 3600 A	Kodak Blue Band x-ray film
Effective source size: 1.8 mm	Du Pont Industrial Combination Screens

Again, Krehl and Schaaffs [1922] reported on investigations concerning the applicability of x-ray flash interference and laser technology to shock-induced solidification of organic liquids. This is an extension of their studies in which they used x-ray flashes alone to investigate the expanding compression ring generated by a dielectric discharge. For investigation of the fine structure, an x-ray flash machine was developed which makes it possible to obtain, with a single flash of less than 1 μsec, Laue patterns of monocrystals, and Debye–Scherrer patterns of polycrystalline substances. The investigation of the compression rings in diverse substances by laser light revealed that the optical transparency and solidification are different in different regions of the compression ring. This provided important indications for experiments with x-ray flash interference.

Charbonnier et al. [1924] of Field Emission Corp. presented a paper on new tubes and techniques for flash x-ray diffraction and high-contrast radiography. High-energy electrons are particularly efficient in producing characteristic x rays and soft bremsstrahlung. A line of wide spectrum beryllium window flash x-ray tubes, from 150 to 600 kVp, has been developed to exploit this property. Laue and Debye–Scherrer flash x-ray diffraction patterns have been obtained using a single 30-nsec pulse exposure. X-ray diffraction tests obtained by Professor Green at The Johns Hopkins University under support from the Ballistic Research Laboratories are shown in the paper. Extremely high-contrast flash radiography of small, low-density objects has been obtained using industrial film without screen. Alternatively, particularly at high voltages and for subjects which include a broad range of materials and thicknesses, special film techniques can be used to produce extremely wide latitudes.

At the 9th International Congress on High Speed Photography (1970), a paper by Baykov et al. [1035] described a pulse x-ray system with a television method of image visualization.

For solving a wide range of experimental problems relating to high speed field investigations, the pulse x-ray technique is extremely effective. Several methods of generating powerful high-voltage pulses for feeding x-ray tubes, have been used as well as systems of visualization, electrical recording, and storing of television images received in x rays. To generate powerful high-voltage nanosecond pulses, cable generators with spark gap commutators and Lewis transformers were used. For x-ray image visualization, a television system using vidicon tubes sensitive to x-ray emission of various energy was used. The main characteristics of experimental x-ray vidicons with a beryllium input window are given in the paper. Along with standard vidicons used in commercially available TV instrumentation, experimental specimens of special x-ray vidicons with information accumulation and electric memory were used. For pulse x-ray image recording and storage, video signal re-recording from x-ray vidicon targets onto the control memory vidicon has

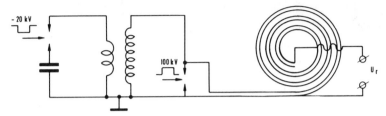

FIG. F2c-7. Circuit diagram of a pulse spiral generator.

been used. One of the main features of the system used is the generation of high-voltage pulses by means of a new spiral generator, the circuit of which is shown in Fig. F2c-7.

The generator is charged with a pulse transformer, the commutation being carried out with a film switch, triggering at the maximum charging voltage. Two models of spiral generators have been assembled and tested. In the first model, the line wave impedance $Z_0 = 4\,\Omega$, the number of turns $n = 10$, the stripline width $l_1 = 6.6$ cm, the generator capacitance $C = 1.2 \times 10^{-8}$ F. The winding was done on a frame with the diameter $D = 9.5$ cm. The insulation is ten polyethylene layers with $S = 90\,\mu$m. With charging voltage $U_{ch} = 50$ kV, the generator gave a peak output voltage $U_0 = 400$ kV with a pulse duration $t = 70$ nsec. The voltage received can be written as $U_0 = 2n\beta U_{ch}$, where β is a factor characterizing the line losses, being in this case 0.45. As the coefficient was thought to be unsatisfactory, the generator construction has been changed in the following way: $l_1 = 10$ cm, $Z_0 = 2.50\,\Omega$, $n = 5$, $D = 17$ cm, and $C = 1.2 \times 10^{-8}$ F.

With these changes, end losses have been decreased, the time constant of passive line discharge over a single turn has increased by five times and the pulse duration has been shortened to $t = 50$ nsec. The amplitude was measured with calibrated capacitive probes. It proved to be $U_0 = 350$ kV, and $\beta = 0.7$.

d. Example of a Large Field Installation for Ballistic X-Ray Studies at High Energy

At the same 9th International Congress on High Speed Photography, Viguier and Bourdarot [1039] presented a survey of a large French installation for ballistic studies at high energy. In view of the importance of this paper it is reproduced here in full.

A high speed, x-ray flash motion-picture installation, intended for ballistic studies, was designed to operate either as a framing unit, with a maximum rate of 2 million frames/sec and a maximum observation time of 30 μsec; a streak recording system, with a scanning rate variable from 1 to 15 mm/μsec and a maximum observation time of 30 μsec; or as a standard x-ray flash recording unit. A description is given of the linear accelerator which provides the supply of x-ray photons, the recording system, consisting of organic scintillator, camera with its deflection and amplifying functions, together with the results of some experiments performed with the aid of this apparatus.

A high speed cineradiography installation designed to study the effects of rapid explosives on related surroundings has been developed.

The choice of a linear electron accelerator as a source of x-ray photons was dictated by the need to obtain high–energy x-photons which would permit the radiography of large thicknesses; in addition, these photons must be emitted for a relatively long time, either continuously in order to permit radiographic observation through a slit, or by brief successive impulses for a cineradiographic observation. The images are recorded by means of an electronic deflection-type camera which provides for sweeping the time axis or for separating the images.

This installation, shown in Fig. 1 [Fig. F2d-1] consists of an emission subassembly and a reception subassembly positioned on either side of an experimentation area to facilitate obtaining, for a high speed event, either a radiographic slit image whose timewise sweep can extend from 2 to 30 μsec, or nine integral radiographic images at a maximum rate of two million images per second, or a single radiographic image.

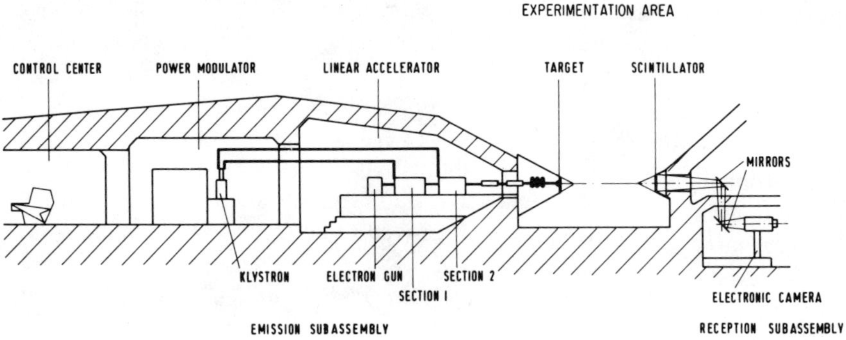

FIG. F2d-1. High speed cinematography installation.

The emission subassembly

This is a linear electron accelerator which makes it possible, after impact on a target of electrons accelerated to an energy on the order of 15 MeV, to obtain the x-photon impulse or impulses needed for the type of cineradiography chosen.

The linear accelerator built by Thomson–C.S.F., is an S band progressive wave accelerator (3000 MHz); its essential components are the following:

A triode-type electron gun, with an interchangeable cathode, controlled by an injection modulator.

Two constant-field accelerator sections; section I initially contains the collector; each section is equally under high power frequency through two waveguides in SF_6 at high pressure, by means of a two-output klystron; the filling time for these sections is on the order of 300 msec.

An amplifier klystron (Thomson–C.S.F. F 2042) which delivers, on the one hand, a peak power of 7 MW for 31 μsec (this unusual functioning duration is the principal difficulty which the accelerator builder had to solve; it is also, of course, the chief original feature of the accelerator), and on the other, a peak power of 15 MW for 2 μsec.

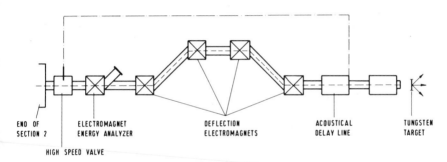

Fig. F2d-2. Diagram of course of electrons between section 2 and target.

Electronic optics, shown in Fig. 2 [Fig. F2d-2] in diagram form, providing for the focusing of the electrons onto a tungsten target and the protection of the accelerator sections against the possible destructive effects of the experiments: deviation of the electrons preventing the arrival of a burst, an acoustical delay line, and a high speed valve limiting any reentry of air.

It is understood that many other components, which need not be described here, are important for proper functioning: master oscillator, klystron power modulator, hf heads, pumping of sections, etc.

Depending on the type of cineradiography chosen, the functioning points of the accelerator on the various loading lines of the latter, as shown in Fig. 3 [Fig. F2d-3] are as follows:

Slit cineradiography: the accelerated peak current, with energy of 12.5 MeV and for a time varying from 2 to 30 μsec, is 220 mA.

Integral-image cineradiography: the accelerated peak current, with energy of 14 MeV for 100 nsec, is 220 mA. The maximum repetition frequency of such electronic impulses is 2 MHz for 30 μsec, such that it is possible to obtain a train of 60 identical impulses lasting 100 nsec, each 500 nsec from the other (a programmer selects a certain number of these impulses in the ordinary use of the accelerator).

Radiography: the accelerated peak current is 440 mA, the current impulse varies in duration from 0.1 to 1.0 μsec (the energy of the electrons varying, therefore, from 21 to 16 MeV).

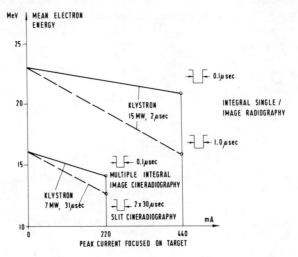

FIG. F2d-3. Load lines of linear accelerator according to mode of operation.

TABLE F2d-1

PEAK DOSES OBTAINED ONE METER FROM TARGET[a]

Accelerated electron impulses		Mean energy of accelerated electron (MeV)	Peak doses (mR)
duration (μ)	peak current (mA)		
0.1	440	21	250
0.1	220	14	50
30	220	12.5	10,000

[a] Doses were measured using ionization chambers (Massiot–Philips; Victorian) and semiconductor detectors ("lithium ion drift" solid state radiation) for a tungsten target thickness of 1.0 mm and for a 2-mm-diam spot on the target. The three accelerator operating points chosen are those giving the most important peak doses.

2. DESIGNS AND APPLICATIONS OF X-RAY FLASH EQUIPMENT 273

The peak doses obtained at 1 m from the target toward the front, on the axis of revolution of the x-ray beam, are given in Table 1 [Table F2d-1]. The accelerator operating points selected for this table are those which, after impact of the electrons on the target with a focal spot 2 mm diam, give the maximum peak x dose. The angular distribution of the doses has also been measured. Knowledge of the x-photon source has been completed by a measurement of the spectrum of the x-photons by means of a Compton electron-type spectrometer.

An electromagnet was installed after the tungsten target so as to eliminate all the electrons from the x-photon beam.

The reception subassembly

This consists of a scintillator transforming the amplitude modulation of the x-photons, produced by the object to be radiographed, into an amplitude modulation of visible photons, and a camera recording the evolutions of the latter modulation.

In the case of conventional flash-radiography operation (one integral image radiography, or two if a superposition technique is used), the "subassembly" is reduced to a radiographic film (Ferrania N; Regulix HS-Kodak) placed between two reinforcer screens (Micron R-Cawo).

1. The scintillator is organic (SPE-Radio technique) with a 4-nsec fluorescence time constant; its emission spectrum is centered on a wavelength of 4400 Å; its geometric forms are either a cylinder, for integral image cineradiography-type operation; or an oblong rectangle, metal plated on its two largest faces, for slit cineradiography operation. The radiographic analysis slit is therefore materialized by this type of scintillator; sometimes two half-slits are used, forming any angle, transferred optically onto a single slit by optical fibers.

2. The camera recording the ultrafast evolutions of the image created by the scintillator, is an electronic camera equipped with an image converter tube, with high brightness gain (high gain made necessary by the low brightness of the event to be photographed) and a single-axis deflection (slit cineradiography) or two-axis deflection (integral-image cineradiography); the tube used is the THX 423 Thomson–C.S.F. 1 tube.

Its principal characteristics are as follows:

Three-stage cascade tube supplying a photon gain at 4500 Å, on the order of $2 \cdot 10^4$; the first deflector stage has electrostatic focusing, the last stages have magnetic focusing.

Diameter of the input photocathode (S 20): 15 mm; enlargement equal to 1; output screen (P 22B) can be used over a 60-mm diam.

Dynamic resolution on the order of 12 line pairs/mm.

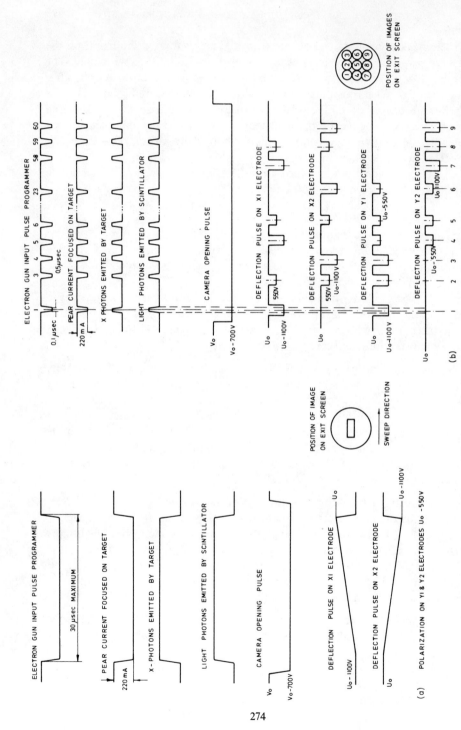

FIG. F2d-4. (a) Electronic control of the camera operating as a streak camera. (b) Electronic control of camera functioning as integral image camera.

2. DESIGNS AND APPLICATIONS OF X-RAY FLASH EQUIPMENT

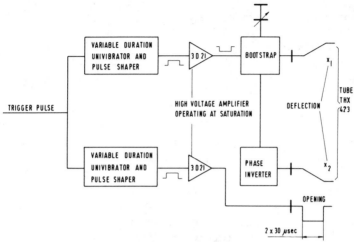

FIG. F2d-5. Electronic streak camera.

In addition to this tube, the camera includes:

A lens (Rayxar–De Oude Delft), with a 0.75 numerical aperture forming the image of the scintillator on the tube photocathode.

Standard photographic equipment (Steinheil) recording, with unit enlargement, on film (Royal X Pan or Tri X Pan–Kodak) the images formed on the screen of the output tube.

A first-tube stage control electronic system. The control needed for operation as a streak camera (slit cineradiography), summed up in Fig. 4a

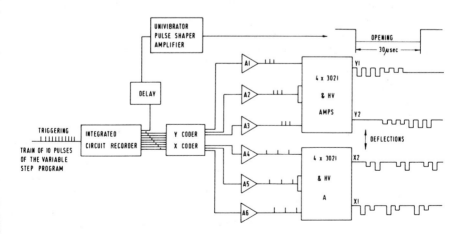

FIG. F2d-6. Electronic integral image camera.

[Fig. F2d-4a] consists of an opening impulse (amplitude: -700 V; flat stability better than 1%), with a duration varying from 2 to 30 μsec, applied to the photocathode, and two symmetrical sawteeth (amplitude: ± 1100 V), whose duration varies from 2 to 30 μsec, applied to the vertical electrodes of the deflector. One obtains on the screen of the output tube a slit image, 15 mm long and with a sweep of more than 30 mm, so that the screen sweep speed can vary between 15 and 1 mm/μsec.

The recording of nine integral images having been considered as a first stage of implementation, the electronic control needed for operation as an integral image camera (integral image cineradiography), summed up in Fig. 4b [Fig. F2d-4b], breaks down into an opening impulse (-700 V; flat stabilization better than 1%; duration 30 μsec), and a sequence of 9×4 synchronous deflection impulses applied to the four electrodes of the deflector (amplitude: 1100 V, -550 V, -0 V; flat stability better than 1%; duration of flat: 200 nsec; the four trains of nine impulses have the same time breakdown as the train of triggering impulses).

Thus, we obtain 3×3 images with 15 mm diam, 15 mm apart in each line, at a maximum frequency of 2,000,000 images/sec (this is the maximum frequency of emission of the x-ray photons). The shaping principle of all these impulses is described in Figs. 5 and 6 [Fig. F2d-5 and F2d-6].

Synchronization electronics

Simultaneity of operation of the two subassemblies together with the phenomenon to be photographed is assured by synchronization electronics. The principle is presented in Fig. 7 [Fig. F2d-7].

A 2-MHz impulse programmer [200 data generator–data pulse generating two synchronous trains of 100 programmable impulses (two successive impulses are 500 nsec apart)] provides for electron gun high frequency power supply triggering synchronism, and working in combination with a 10–1000 nsec delay generator, for synchronism of the scintillator image on the tube photocathode with the tube control electronics; a switch S, in the "Ext" position, cuts the programmer control and allows the klystron to function at a frequency of 3 Hz (S in the "3 Hz" position during adjusting of the linear acclerator); a 1–11 μsec delay generator fixes the chronometry of the observation into the chronometry of the event to be radiographed (measurement of the time intervals is made by means of the chronotron and the photomultiplier); the stroke by stroke triggering pair controls a relay which then provides for single-stroke functioning of the 2 MHz impulse programmer, of the high-voltage generator, and of the shutter of the photographic equipment (the relay operating time is compensated for by the 1–50 msec delay generator).

2. DESIGNS AND APPLICATIONS OF X-RAY FLASH EQUIPMENT

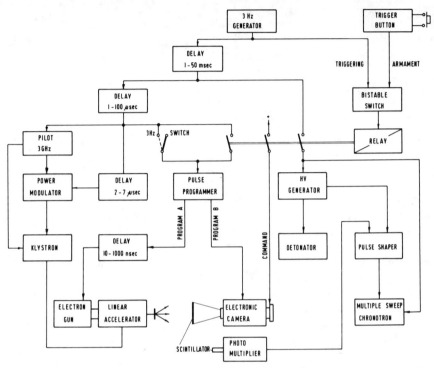

Fig. F2d-7. Synchronization electronics.

Installation

One of the two concrete housings shown in Fig. 1 [Fig. F2d-1] positioned face to face on the experimentation floor, shelters a linear accelerator, and the other the camera; a transfer by two mirrors at 45° makes it possible to protect the camera. Each terminal part is covered with aluminum protecting cones.

Results of experiments

Examples of recordings. Figure 8 [Fig. F2d-8] shows a sequence of nine radiographic images, recorded at the time of the implosion of a hollow metal cylinder by means of a convergent cylindrical detonation wave issuing from a hollow explosive cylinder surrounding the radiographed cylinder.

Fig. F2d-8. Exposure time of each image: 100 nsec, 3 μsec between two successive images.

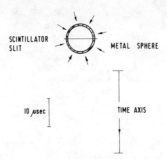

Fig. F2d-9. Displacement axis.

Figure 9 [Fig. F2d-9] is a slit radiography image recorded at the moment of implosion of a hollow metal sphere by a convergent spherical detonation wave proceeding from a hollow explosive sphere surrounding it.

The recording, lasting 28 μsec, starts shortly before the instant of arrival of the centripetal detonation wave on the outside surface of the metal sphere; the analysis slit is diametrical.

Evaluation. Eye-examination of the plate of Fig. 9 [Fig. F2d-9] for example, makes it possible to measure the outside diameter of the sphere during implosion with a precision of 5%; as to the measurement of the inside diameter, this is less precise—a precision on the order of 10% is obtained; these low precisions arise from the fact that the plates are spotted with considerable noise and the eye finds it difficult to apprehend the characteristic points corresponding to the outlines of the diameters.

Accordingly, in order to avoid this imprecision, another method of evaluation is used. Evaluation of only the slit radiographic recordings (first stage of the exploitation of all types of plates obtained) is done with the help of Laboratoire Central de l'Armement, by means of a rapid microdensitometer combined with a computer which transfers the image onto a magnetic tape. The computer processing is for the purpose of identifying the characteristic points on the densitometric profile perpendicular to the sweep. These points can be characterized simply by tracing the curve of the thickness of material traversed, projected onto the scintillator [Fig. 10 (Fig. F2d-10)].

On the densitometric curve, these points are poorly defined because of the granular noise and a permanent distortion. This distortion, due to defects in the acquisition chain, is eliminated by subtracting point by point, from a

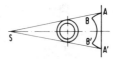

Fig. F2d-10. Curve of the material traversed as projected onto the scintillator.

photograph including an object, a photograph containing no object. The granular noise is removed by means of digital surface filters. Thus, points B and B^1 are found. Points A and A^1 are obtained by causing ellipses to pass through any two points on arcs AB and A^1B^1. The ellipse of minimum diameter is chosen. The results obtained give a precision, on the inside and outside diameters, between 1 and 2% depending on the length of the diameter.

A dynamic film marking device under development by the Laboratoire Central de l'Armement, allowing for simultaneous recording of the high speed event studied and the time bases (proceeding from laser diodes), will enable us to read the time axis at nearly 100 nsec.

Reference. L. Guyot: Reports of the 7th and 8th International Congresses on High Speed Photography.

Acknowledgments. The authors thank Messrs. Laharrague, Durand and Beaudouin, of the C.E.A., who set forth, together with Thomson–C.S.F., the operating principles of the emission and reception subassemblies; as well as Messrs. Guix and Hauducoeur, of the C.E.A., for their assistance in operating the installation described.

3. Hard X-Ray (Gamma-Ray) Flashes and Electron Beam Pulses of High Energy

There has been heavy investment in R and D work, aimed mainly at simulation of the effects of nuclear weapons on material and components. In the future the field of application may extend to more normal economic activities. This section will try to cover realized or available equipment. In contrast to applications of photography to event or motion analysis, here only the entire energy of the flash is of interest. (Where photographic analysis of the emitting zone or the subject under observation is necessary, reference may be made to a paper by Moody [1288] which gives a very good survey of gamma-ray cameras, especially those having medical applications.)

A number of high-energy flash x ray and electron pulse machines manufactured by Ion Physics Corp. are described in Ref. [1294a]. The basic component of all of the present higher energy flash x-ray systems is a multi-megavolt Van de Graaff generator. This high-impedance supply charges a gas-insulated coaxial line housed in the same pressure vessel as the supply. A triggered pressurized gap delivers the energy stored in the coaxial line into an evacuated field emission diode. The grounded anode of the diode is normally made up of a high atomic number x-ray target and an aluminum backing. The configuration of the diode allows irradiations to be performed over almost a complete hemisphere right up to the external face of the anode. Currently, the terminal or line lengths of these systems range from 3 to 11 m allowing pulse durations from 5 to 30 nsec to be obtained. Recharge time, coupled with the self-healing property of the gas trigger, allows the systems to be cycled with a repetition rate of several pulses per minute.

The system is normally supplied with a high atomic number target which, when bombarded by electrons from the field emission cathode, produces intense bursts of megavolt x rays. The systems can also be supplied with a thin window and focusing device which allows the electrons to leave the system and impinge directly on an external target or test package. Intense photoneutron bursts can be produced by the addition of a thick beryllium converter to the normal x-ray target configuration or, using the external electron beam, to an external target located at some distance from the system.

The tube anode extends beyond the end of the machine, permitting full use of the radiation field over an entire hemisphere. Packages can be placed against the anode face, roughly $\frac{1}{4}$ in. from the x-ray target. If the external electron beam is used, an even greater capability is possible, allowing under some circumstances irradiations to be carried out almost *in vivo*. The coaxial capacitor can be charged and triggered directly from the console over a wide voltage range. The charging voltage is monitored directly with a generating voltmeter and can be controlled precisely, allowing radiation output to be determined within close limits. Radiation output has been reproduced over extended periods on existing systems with an rms deviation of 2%.

A controlled external electron beam from the machine can be made available by the addition of a simple supplemental package which has been developed. Using this system, studies on material damage, plasma heating, radiography, radiation dosimetry, EMP phenomena to mention a few, can be extended under precisely controlled conditions at peak power densities of 10^{10} W/cm^2 and above. As part of its continuing development program in this area, IPC has developed differentially pumped drift and focusing structures as well as associated monitoring equipment for the determination of beam characteristics.

The field emission tubes developed by IPC feature multishot capability (guarantee 500 shots/tube) with high shot reproducibility at maximum rates exceeding 1 per minute. Typical radiation outputs exceed 10,000 R per shot at the external anode face (model FX-35). Specially focused configurations increase the radiation output but reduce the life of the anode, which can be replaced in about 15 min. Transmission windows and external targets make possible tube envelope protection at very high levels of anode peak power. One version of the tube is a composite structure which allows damaged portions to be readily replaced or reworked if anode vaporization causes envelope contamination. For more restricted use but longer time between rework, a bonded glass insulated tube structure can be provided. The unique housing/tube construction employed in the systems permits easy tube replacement within 30 min. The tube anode is detachable and can be changed or replaced with an external beam window independently.

Shielding properties of the steel envelope and the coaxial nature of the discharge lead to unusually low external rf radiation levels. As a result, study of electronic transient effects induced by the radiation pulse is possible under relatively quiet rf conditions. Screen room shielding has not been generally necessary. The system's inherent simplicity and long-life individual components minimize maintenance and downtime. The use of gas and solid insulation throughout removes the messy and time-consuming operations associated with the use of liquid dielectrics. The machine operator can service the system with little assistance, assuring low maintenance costs.

Due to the radiation intensity produced by each shot, it is necessary to erect protective shielding around the facility, predominantly in the forward direction of the x-ray beam. Primary radiation intensity in the backward direction is inherently much less than in the forward direction at megavolt energies. In addition, the backscattered component of the radiation field, together with the backdirected primary radiation, is significantly suppressed by the pressure vessel so that the shielding requirements in the backward direction are considerably reduced. The flash x-ray facility can be operated and monitored by one person at a compact console where all of the controls and monitoring equipment are conveniently located. Operation can be further simplified if desired to automatically program operation at predetermined levels. IPC offers a warranty against defective parts and workmanship on all system components and support equipment.

IPC presently offers six flash x-ray or pulsed electron systems which have a peak electron voltage range of 1.5 MeV–10 MeV shown in the following table.

Model	Peak electron voltage (MeV)	On-axis dose at 1 m (R)	Pulse width (nsec)
FX-15	1.5	0.1	
FX-25	2.5	>1.5	
FX-35	3.5	7	20–35
FX-45	5	25	
FX-75	6	100	
FX-100	10	2500	30–70

Most of these IPC systems can be constructed to allow for subsequent modifications which would increase radiation output. Such increases can be achieved either by extension of the pulse width (incorporation of a longer terminal and pressure vessel), or by the addition of various percentages of

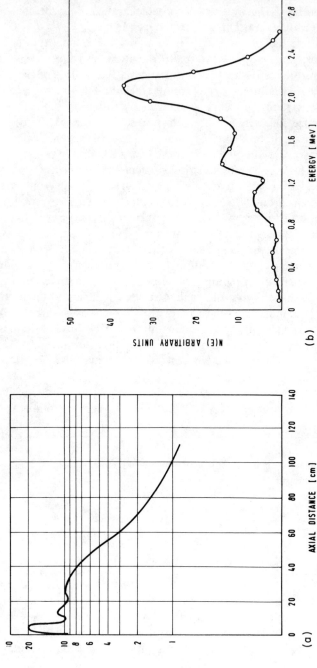

FIG. F3-1a, b. Output characteristics of an IPC flash x ray and pulsed electron system. (a) Pulsed electrons from model FX-1; dose on axis at atmospheric pressure. V_e max = 2.3 MeV, V_e av. = 2.0 MeV. (b) FX-1 external beam energy spectrum, 2.5 MeV beam, standard entrance conditions.

3. HARD X-RAY FLASHES AND ELECTRON BEAM PULSES

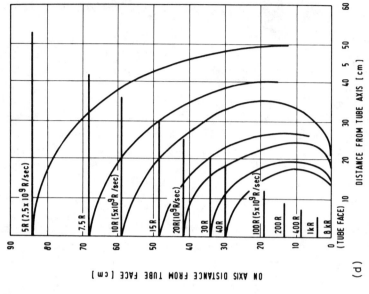

(d)

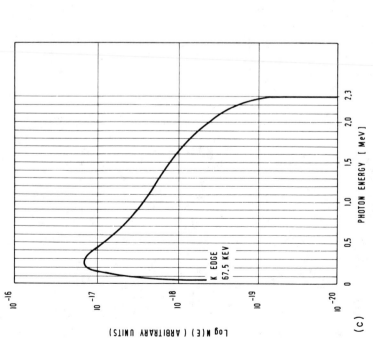

(c)

FIG. F3-1c,d. (c) Tantalum: $p = 16.6$; $E_{max} = 2.3$ MeV; $t_{Ta} = 0.038$ cm $= 0.63$ gm/cm^2; $t_{Al} = 0.95$ cm $= 2.56$ gm/cm^2; $\theta = 0°$. (d) Gamma flux map FX-1; charging voltage $= 3.6$ MV; pulse length $= 20$ nsec.

sulfur hexafluoride (SF_6), whereby greater voltage standoff capabilities may be attained.

Figure F3-1 shows (a) the typical character of the quantity (dosage) of the pulsed electrons as a function of axial distance; (b) the external x-ray flash beam spectrum; (c) the photon energy; (d) the gamma-radiation flux in the axis of the tube face.

The x-ray or gamma-ray flashes are triggered using a Lucite rod and associated photodetector and amplifier [1294 ff] as shown in Fig. F3-2. The new trigger system now allows the experimentalist to synchronize various functions of the system to be irradiated with the incoming pulse, or to initiate a pulse from the machine at a specific time upon command with a given signal from the experimental system.

This system, utilizes a high-voltage pulse generator in the terminal triggered by a spark gap light source located external to the machine.

A pulse from the light source and the pulse from the output of an x-ray scintillator located in the radiation field is fed to the input of an oscilloscope. Measurements of these waveforms give the following results:

A histogram of delay times was constructed, and from this a plot of the jitter was made using 5 nsec windows and counting the number of shots which occurred in these limits. 95% of the shots were within ±15 nsec and the remainder lie within ±20 nsec. There is an inherent nominal delay in the machine, from light pulse input to radiation output, of 900 nsec.

The light source has been tested for delay and jitter and gives results of ±10 nsec jitter and 100 nsec delay. Thus the over-all jitter would be expected to be ±18 nsec and delay 1 μsec, and the effective delay can be varied down to zero by the incorporation of suitable delay elements in the synchronization output. One large machine, IPC's model FX-75 x-ray flash system, may be described more in detail because, although still larger machines exist, the ideas

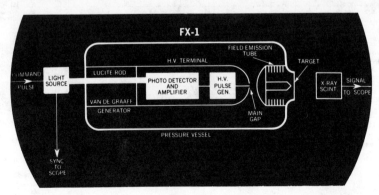

FIG. F3-2. Apparatus for triggering x-ray or gamma-ray flashes.

3. HARD X-RAY FLASHES AND ELECTRON BEAM PULSES

TABLE F3-1

FLASH X-RAY/PULSED ELECTRON SYSTEMS MANUFACTURED BY IPC

Model	FX-25	FX-45	FX-75
Line charging voltage (MV)	3.5	6.5	7.5
Electron beam energy (MeV)	2.0	4.0	5.0
Tube current (A)	20,000	32,000	60,000
Pulse length (nsec)	25 ± 5	30	35
Total energy/pulse (J)	1400	4300	10,500
Electron fluence at 5 cm (J/cm^2)	500	1000	1500
Reproducibility (%)	3	3	3
X-ray dose (rad) at 30 cm from anode	60	400	1400
X-ray intensity (rad/sec) at 30 cm from anode	3×10^9	2×10^{10}	6.7×10^{10}

embodied in this particular machine represent the state of the art attainable today without difficulty.

First, Table F3-1 gives the performance characteristics of the FX-75 machine and two smaller "sisters."* Table F3-2 and Fig. F3-3 show the physical characteristics of the FX-75. The following performance of the FX-75 machine will be demonstrated, and used as a basis for system acceptance:

(a) For five consecutive shots in 15 min a centerline dose of at least 1000 R/shot at 30 cm from the anode will be produced. Dose will be measured using Controls for Radiation, Inc. LiF thermoluminescent dosimeters which have been calibrated against a Victoreen Condenser roentgen meter.

(b) For all five consecutive shots in 15 min at a 1000-R level at 30 cm from the anode, an x-ray pulse of 40 ± 5 nsec full-width at half-maximum will be produced as measured by a fast scintillator—IT&T type FW 114 planar photo diode detector coupled to a Tektronix type 519 oscilloscope.

(c) For ten consecutive shots at the 500-R level using precision monitoring a dose consistency of 3% rms will be demonstrated. Precision monitoring involves digital voltmeter setting of the line charging voltage (column current) and dose monitoring using the scintillator–photodiode system of (b), coupled to a passive integrating circuit and peak reading transient voltmeter.

After installation at the user's site, test (a) will be made as described; for test (b), measurement will be done by a fast-response IPC monitor which utilizes a Pilot B scintillator–IT&T type FW 114 oscilloscope. In this case the

---

* A similar large machine is made by the Field Emission Corp., Inc. but only one machine can be described in detail here because of space limitations.

customer must make available a Tektronix 519 CRO or the equivalent. Test (c) will be performed only if the customer has purchased the IPC dose-monitoring peak-reading voltmeter package and digital voltmeter system. Precision monitoring involves digital voltmeter setting of the line-charging voltage (column current) and dose monitoring using the scintillator–photodiode system of (b) coupled to a passive integrating circuit and peak-reading transient voltmeter.

TABLE F3-2

MISCELLANEOUS PHYSICAL DATA OF THE IPC FX-75 MEGAVOLT FLASH X-RAY SYSTEM[a]

Pressure tank	
External length (ft)	28
Internal diameter (ft)	9
Volume (ft^3)	1780
Total weight of standard FX-system and mount (lbs)	56,000
Operating pressure of insulating gas (psig)	380
Total insulating length of generator column (in.)	108
Coaxial gas capacitor	
Length (in.)	186
Diameter (in.)	63
Time needed to discharge pressure tank (h)	4
Time needed to recharge pressure tank (h)	8
Time needed to replace field emission tube (h)	1
Electric power	
30 kVA, 380 V, 3 phase 50/60 Hz. A transformer will be supplied to convert the customer's voltage to the 208/120 V requirement of the machine.	
Insulating gas	
The pressure tank has an approximate volume of 350 ft^3 which is pressurized for normal operating conditions to 380 psig with N_2 (50%) and CO_2 (50%) insulating gas. The gas and its associated storage and transfer lines are not supplied.	

[a] Over-all dimensions of the flash x-ray system are shown in Fig. F3-3.

If instead of a high-energy x-ray output only the electron beam energy is required, a thinner tantalum anode can be provided together with a large crystal, e.g., for generation of superradiant optical energy of extreme brightness and peak power. Finally, regarding high-energy electron beam pulses, the following specifications may be given for a high-voltage pulse generator designed by IPC to drive a large electron gun [1723]:

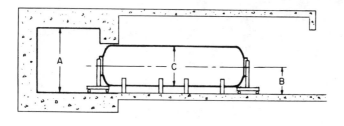

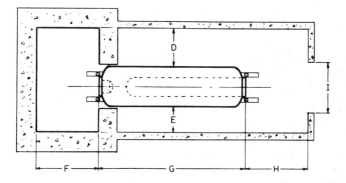

IF GENERATOR/TERMINAL ASSY CANNOT BE RETRACTED THROUGH DOORWAY, DIMENSION "H" WILL CHANGE.

	A	B	C	D	E	F	G	H	I
FX-15	10'-0"	3'-0"	2'-6"	2'-0"	4'-0"	7'-0"	16'-9"	15'-0"	4'-0"
FX-25	10'-0"	4'-0"	3'-7"	6'-0"	4'-0"	6'-0"	17'-2"	18'-0"	6'-0"
FX-45	10'-0"	4'-3"	6'-4"	6'-0"	4'-0"	10'-0"	23'-10"	16'-0"	8'-0"
FX-75	12'-0"	5'-4"	9'-2"	6'-0"	10'-0"	12'-0"	28'-0"	10'-0"	11'-0"

FIG. F3-3. Typical installation of flash x-ray system.

Output: −100 to −300 kV, continuously variable;
Pulse width: 1 to 10 ±0.2 μsec, stepwise variable at all operating voltages; normal operating pulse width will be 4 μsec;
Voltage droop: 10% at 4 μsec and 300 kV for specified load;
Rise time: (10–90% of specified output) ≤ 0.4 μsec;
Fall time: (90–10%) ≤ 0.2 μsec;
Jitter: ±0.2 μsec (between trigger input and voltage output);
Maximum overshoot: 25% at beginning of pulse;
Maximum undershoot: 10% at end of pulse.

The typical load (electron gun) has a capacitance ≈4 nF, resistance ≈200 Ω at 300 kV. The energy delivered to load at 300 kV, 4 μsec is 1.8 kJ.
The load is an emission-limited hot cathode electron gun. The electron gun

shunt capacitance is 4 nF and will be operated in the temperature-limited case for a constant current of 1500 A for all voltages above 100 kV. The electron gun will be located not more than 50 ft from the pulser unit.

4. High-Energy Ion Generation

The availability of electron beams of high energy and flux density, as described at the end of Section F3, makes possible a sophisticated technique for the generation of high-energy ions also, which may be extremely useful for ion implantation in semiconductors. Graybill and Uglum [1038] wrote the following paper of IPC about a new source of highly stripped energetic heavy ions using collective acceleration.

Experiments are described which show that pulsed high-intensity electron beams can accelerate light ions to high energies. A set of time-of-flight measurements shows proton and deuteron energies of 5 MeV, helium ion energies of 9 MeV, and nitrogen ion energies of ~ 20 MeV, produced by an electron beam with a maximum electron energy of 1.6 MeV. These measurements are confirmed by $^9\text{Be}(x, n)$ reactions for protons, deuterons, and helium ions. Further confirmation is made with an absorber measurement for the protons.

The authors [1038] present the results of some preliminary experiments which show that one can use pulsed, high-intensity electron beams to accelerate plasma ions to a high energy (a 1.3-MeV electron beam producing 5.0 MeV hydrogen ions). A great deal of interest has been shown recently in the possibility of using beam plasma systems as accelerators since they offer an extremely attractive scheme for producing high-energy ion beams.

The history of utilizing beam plasma interactions to accelerate ions goes back to the early works of Veksler [1935, 1936] who showed that there are a variety of interactions possible which can deliver a large fraction of the electron beam energy to plasma ions. Collective beam plasma interactions, such as an inverse Cerenkov process, appear to be one possible mechanism. Recently, Eastlund [1937] has shown that the inverse Cerenkov process may be a very important loss mechanism for an intense relativistic beam propagating in a plasma. Wachtel and Eastlund suggest that extremely high ion energies are possible from a beam–plasma interaction.

An alternate accelerating mechanism is the deep potential well which is produced by an intense electron beam. If the drift velocity of the beam is kept small, then ions trapped in the potential well attain the same velocity as the beam which means that their energy will be greater than the electron energy by their relativistic mass ratio. This is the underlying principle behind the recent efforts to produce an electron ring accelerator.

The experiments to be described were performed using the pulsed accelerator

4. HIGH-ENERGY ION GENERATION

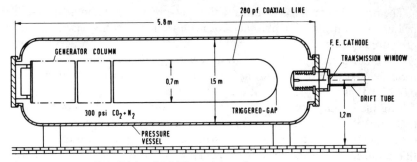

FIG. F4-1. 4.0 MV coaxial accelerator geometry.

shown in Fig. F4-1. The accelerator consists of a 280-pF coaxial capacitor which is decharged to 4.0 MV by a Van de Graaff generator. The coaxial line, which serves as a pulse forming network with a 10-nsec one-way electrical length, is command switched onto a capacitively graded vacuum tube. The vacuum tube houses a field-emission cathode and a thin (0.002-in. titanium) transmission anode.

Figure F4-2 shows, schematically, the experimental configuration used for observing the energetic ions. The electron beam produced by the field-emission diode passes through the thin anode into a 0.5-m-long drift-tube region. The conducting walls of the drift tube serve as the return current path for the beam. The beam injection current is measured by a 15-mΩ resistive shunt placed in the return current path inside the graded vacuum tube. The measured response of the shunt is less than 1 nsec. A Tektronic 519 oscilloscope was used to record the output.

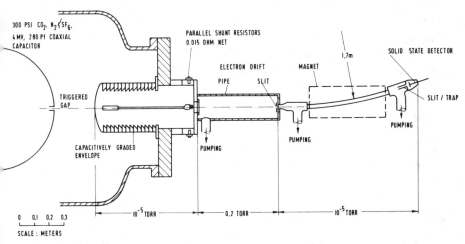

FIG. F4-2. Experimental arrangement for collective accelerator-momentum analysis.

F. CONVERSION OF CAPACITOR ENERGY

The drift-tube chamber is filled with a low-pressure gas and as the electron beam passes through it, a plasma is formed by ionizing collisions. The ambient pressure is adjusted so that force neutralization of the beam occurs quickly and the beam then pinches due to its self-magnetic field (i.e., the conditions for a Bennett pinch). The optimum pressure for beam pinching is in the range 0.05–0.3 Torr, since below that the plasma does not have enough time to form during the beam pulse (~ 40 nsec) and above that the plasma conductivity is high enough to produce volume currents which cancel the beam current. The experiments reported here were carried out in this pressure range. Prior to filling the chamber with the desired gas species, the system was pumped down to the 10^{-4} Torr range to remove residual gas. After gas filling the ambient pressure was measured with a McLeod gauge.

The cathode used was a multipin type with 20 sharp needles placed inside a 2.5-cm-diam circle. The cathode–anode gap spacing used was 2.0 cm. The energy profile was obtained by using a capacitive voltage probe which was calibrated by threshold measurements for the $Be(\gamma, n)$ reaction and then cross checked with magnetic spectrometer measurements of the beam energy spectrum. The results show a peak energy of 1.6 MeV and a mean energy of ~ 1.3 MeV. The current pulse was obtained from the resistive shunt output and indicates a peak current of 40,000 A. The peak value of v/γ for this beam is

$$v/\gamma = I/17{,}000\beta\gamma = 0.6$$

which approaches the Lawson critical current criteria. Although it might be expected that the beam would not propagate, it does in fact traverse the 0.5 m of drift region with little energy loss.

The far end of the drift tube is terminated by a blank-off plate, in the center of which a 2.5-cm-diam hole is drilled. Ions and electrons in the central portion of the beam pass through this hole into a smaller drift tube. The length of the smaller tube is modular in length and can be varied from 10 to 70 cm. The 3.0-kG magnet placed behind the blank-off plate is used to sweep out electrons entering the small drift tube. The deflection of the ion trajectories is small but does affect the amount of current reaching the ion current probe situated at the end of the small drift tube. The probe consists of a copper disk dielectrically isolated from the drift tube. The disk is connected directly to a 50-Ω coaxial cable which goes to a Tektronix 519 oscilloscope. The response of the ion current probe was measured to be less than 1 nsec.

The determination of ion energy was made by time-of-flight measurements down to the small drift tube. The trigger for all oscilloscopes was taken from the resistive current shunt so that all times were referenced to the leading edge of the injected electron beam current pulse. The reference, or "zero," detector position was fixed to be on the far side of the sweeping magnet. The other two detection points were located 30 and 60 cm further downstream. There was no error in position reproducibility since each location was defined by attaching fixed lengths of tubing to the small drift tube. The use of a time-of-flight measurement scheme depends critically on the reproducibility of the system since we could only measure at one position per acclerator pulse. This reproducibility was checked at each point and found to be excellent. This is reflected in the data given in this section.

Checks were made to ensure that we were, in fact, detecting ions. The obvious check is the polarity of the detected current pulse, which is always positive. Another verification, which was done only in the hydrogen runs, was to make an absorber measurement of the extrapolated range of the particles. Variations in the ion current made the scatter in the data large but theresults did indicate an extrapolated range of 50 mg/cm^2 in beryllium. This corresponds to a proton energy of roughly 5.0 MeV which agrees with the time-of-flight data. Finally, the Be(γ, n) reactions described further on confirmed that we were observing energetic ions.

Table F4-1 shows the results of the first measurements.

TABLE F4-1

MEAN VELOCITY AND ENERGY OF COLLECTIVELY ACCELERATED IONS[a]

Ion species	Velocity (cm/sec)	Mean energy (MeV)	Peak current (A)
Hydrogen	3.2×10^9	5.3	250
Deuterium	2.4×10^9	6.0	100
Helium	2.2×10^9	10	30
Nitrogen	1.5×10^9	17	10
Argon	$\sim 1.0 \times 10^9$	~ 25	~ 1

[a] Corrected value, March 1970.

Momentum analysis experiments have shown that the following ion species have been accelerated:

H^+ (5.0 MeV/nucleon) N^{6+} (1.6 MeV/nucleon)
He^{2+} (2.5 MeV/nucleon) Ar^{14+} (0.75 MeV/nucleon)

This unusual source has already produced particles having a charge-to-mass ratio greater than 0.3, wtih the property that these particles are supplied at high energy. Yields per pulse have reached 10^{11}–10^{13} particles.

Advances in high-voltage pulsed power technology make a pulse rate of 10 pulses/sec at appreciably higher electron currents eminently realizable; a source of energetic argon ions (> 1 MeV/amu) highly stripped ($> 14^+$), with an average ion current in excess of 0.1 μA would then be available. Further work needs to be done to determine which charge states are attainable from heavier ions (for example, xenon), and also how the energy-nucleon can be increased.

5. Neutron Flashes for Material Testing

Pulsed beams of heavy ions being obtainable, it is in principle feasible to shoot them against suitable targets to produce neutrons. What can be done with dc-neutron exposure at present is shown in Fig. F5-1.

In bullets, the metal shell is more or less transparent but the hydrogen content permits a brilliant shadow-graphic analysis. The following paragraphs present certain views regarding the feasibility of producing neutron beams [1723]. As can be seen, the present state of the art is far from permitting, for example, high speed neutron flash cinematography.

FIG. F5-1. Direct current neutron beam radiography. Metal absorbs the neutrons only a little, but much of the hydrogen in the filling.

a. *Ion Beam and Target Selection*

Having ion beams available, the following three target reactions might be employed:

(1) $T(d, n)^4He$ (neutron energy 14 MeV)
(2) $^7Li(p, n)^7Be$ (neutron energy 0.2–0.8 MeV)
(3) $^9Be(d, n)^{10}B$ (neutron energy 2–6 MeV)

The choice of the best target reaction is not simple, since the neutron yield depends upon the acceleration potential, the target composition, and the amount and kind of ion beam used. The choice is further complicated by the problems of target fabrication and limitations of useful life with ion beam currents sufficient for practical radiography. With recognition of this, the choice of the reaction for our experimental purposes was based on the following qualitative considerations.

1. *T(d, n) reaction.* The T(d, n) reaction is of interest for use with low voltage accelerators (300 to 500 kV) since it provides a high neutron yield per microampere of ion current. There are, however, several serious disadvantages associated with the use of the reaction, particularly, when considered for neutron radiography. First, the target is made with radioactive tritium diffused on a titanium substrate, therefore necessitating the user to be properly licensed for possession of radioactive material. Second, the ultimate migration of tritium throughout the entire acceleration tube line and vacuum pumping system poses a continual contamination hazard. Third, the neutron flux from the reaction is not constant, but as tritium losses occur by displacement, the neutron yield is considerably reduced. (Reduction by a factor of 10 within 300 min of operation at 4.0 mA has been reported.) Finally, the high-energy neutrons (14 MeV) are a serious disadvantage because they may travel a considerable distance away from the target, on the average, before they become properly thermalized. This means that the source of gamma generation is more generalized and difficult to control. It may also mean that, while the net total generation of fast neutrons is high, the concentration of thermal neutrons available at the entrance of a collimator will actually be lower than desired for efficient radiography.

2. $^7Li(p, n)$ *reaction.* The ^{7}Li(p, n) reaction is attractive because it has a good yield potential of low-energy neutrons which may be more easily moderated with less background gamma than from other reactions. To obtain optimum yields, it is necessary to use thick targets with as high a concentration of pure ^{7}Li as possible which brings about problems in fabrication and handling. Pure lithium is an alkaline metal that is quite chemically reactive under standard conditions. Thus, a lithium target must be made and maintained under vacuum conditions at all times if it is to remain pure. Otherwise, it will rapidly oxidize in normal atmosphere with the result that its neutron yield will be drastically reduced. Also, since it has a melting point of 180°C, it is extremely difficult to obtain extended target life with the ion beam currents required for radiography. Experimental systems have been operated wherein the target was continuously replenished by the localized condensation of lithium vapor. However, for our purpose, such an arrangement would be cumbersome in that it would displace some of the water around the target, thus reducing the possibility of optimizing the concentration of thermal neutrons in the moderator.

3. $^9Be(d, n)$ *reaction.* The ^{9}Be(d, n) reaction is of particular interest, since targets of beryllium, with its high melting temperature (1300°C) are easily fabricated and show good durability against ion beams of the intensities required for radiography. It has been shown that the neutron yield is reasonably high and increases rapidly with beam energy in the region above 2.0 MeV. Below this point, using conventional ion beam currents of 100–200 μA, the

yield falls off rapidly to levels that are impractical for radiographic purposes. With special mounting and cooling arrangements, beryllium targets have been used extensively for our experimental work to date with no evidence of yield loss with operating time. Target cost is reasonable and the replacement method is extremely simple with very little downtime.

b. *Experimental Procedure*

During the course of the experimental work, several collimators were fabricated and used in conjunction with a water moderator tank. Each unit was constructed of aluminum with welded seams for watertight integrity and, while any location would suffice, a mounting structure was positioned above the open water tank so that adjustments and collimator changes could be easily accomplished without modification of the tank itself.

The initial collimating unit used was a round aluminum tube ~ 2 in. (5 cm) in diameter by 24 in. (61 cm) long with a flat aluminum bottom. This was soon discarded because of its limited exposure field, and two collimators were subsequently made with rectangular cross sections as follows:

Input	Length	Output
2.0 in. (5.08 cm) square	15 in. (38 cm)	6 in. (15.2 cm) square
2.0 in. (5.08 cm) square	30 in. (76 cm)	4×5 in. (10.1×12.7 cm)

As a result of a series of radiographic exposures using these units, several general observations regarding design and collimator input location with respect to the thermal neutron volume in the moderating water were made.

(a) The entrance to the collimator must be sufficiently large to admit enough neutrons for reasonably short exposures. If too large, however, the collimator tube itself will displace sufficient water to increase or distort the moderating volume, thus resulting in a reduction of available thermal flux density at the collimator entrance.

(b) The ratio of the collimator length to its entrance size determines the inherent fuzziness of the projected image. While long lengths and small entrances contribute to picture quality, both also contribute to lower output flux, at which a longer exposure time for proper film density is necessary.

(c) In addition to depressing the useful thermal flux, placement of a colimator entrance too deeply in the moderator volume may result in the selection of epithermal or fast neutrons which may prove useless for the intended radiographic purpose.

(d) In the case of direct film imaging, gamma flux will also affect the film density unless properly controlled. This leads to a general densification of the film with a corresponding loss of resolution and contrast. This problem may be reduced by the use of gamma filters and location of the collimator input so that its axis is tangent rather than radial to the reaction target and the immediate moderating region.

In the initial phases of the work, use of the direct exposure technique was encumbered by a serious gamma flux in neutron beam. With the longest collimator described above, a test made without the use of gadolinium foil in the cassette resulted in a gamma-produced film density of 0.70 during a 20-min exposure using a 2.5-MeV deuteron beam of 75 μA on a beryllium target. During a second similar exposure, 0.43 in. (1.1 cm) of lead was used in front of the film cassette, and the gamma-produced density was reduced to 0.35. Subsequently, use of lead gamma filters before the subject and cassette resulted in substantial improvement of *neutron-produced* film contrast, with only slight extension of exposure time.

c. *Thermal Neutron Beam Characteristics*

Characteristics of the neutron generating and collimating system were determined by the conventional method of foil activation using both gold and dysprosium. Flux levels at the entrance and exit of the 30-in. (76 cm) long collimator were measured, and the output cadmium ratio was determined during several runs. In addition, the associated gamma intensity was determined by simultaneous exposure of type AA film in a cassette at the imaging plane during each flux measurement exposure. During a typical run using a 60-μA ion current at 2.4 MeV, a number of activation foils and a sheet of AA film in a regular x-ray cassette were simultaneously exposed to enable measurements of the ^{9}Be(d, n)^{10}B neutron beam.

The cadmium ratio was measured using 0.040 in. (1 mm) cadmium with gold foils and the neutron flux at the input to the collimator was determined by means of a dysprosium foil encased in a small thin aluminum watertight cassette. The following figures were calculated from data taken using multi-channel counting equipment and a standard photographic film densitometer:

(1) collimator input flux (dysprosium foil) 3.6×10^7 n/cm^2-sec
(2) collimator output flux (dysprosium foil) 4.3×10^4 n/cm^2-sec
(3) gamma intensity (by AA film density)   0.16 mR/sec
(4) collimator output flux (bare gold foil) 5.9×10^5 n/cm^2-sec
(5) collimator output flux (Cd-covered gold foil) 7.8×10^4 n/cm^2-sec

From these data, the following characteristics were computed for comparison with the recommendations of Berger:

       gold cadmium ratio   7.6
       therm. neutron/gamma intensity 2.6×10^5 n/cm^2-mR

Actual radiographic experience, using a cadmium image quality indicator with type AA film and 0.001 in. (0.025 mm) gadolinium foil, has shown resolution of 0.005 in. (0.127 mm) with film-indicator separation of 0.5 in. (12.7 mm). Exposure time in the order of 20 min is typical. However, larger production models of the Van de Graaff accelerator are available with ion beam capability of up to 400 μA at 3.0 and 4.0 MeV. With proper extrapolation, it is seen that good quality accelerator-produced neutron radiographs can be produced in less than 3 min.

Figure F5c-1 shows the yield of neutron versus the bombarding energy (MV) of the ions.

To close this chapter, a recent dissertation of Neugebauer [1876] may be mentioned in connection with the hope of generating more powerful neutron

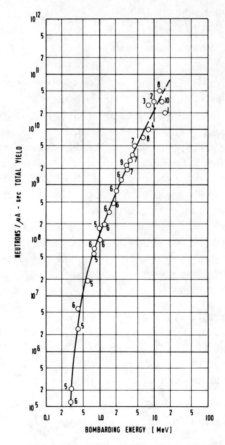

FIG. F5c-1. Neutron yield ^{9}Be(d,n)^{10}B thick target. 1. Al 51; 2. Sm 51; 3. Ac 46; 4. Fe 46; 5. Am 37; 6. Hi 60; 7. Hi 61; 8. Go 56; 9. Sh 63; 10. Li 62.

5. NEUTRON FLASHES FOR MATERIAL TESTING

flashes. He found that a hollow cylindrical beam of nitrogen clusters can be produced by expansion of nitrogen gas out of a ring-shaped nozzle cooled by liquid nitrogen. An electrical discharge along this beam produces a plasma of corresponding shape. Since plasmas of the heavy hydrogen isotopes in configurations of this kind may be of interest to produce intense neutron flashes by nuclear fusion, it was decided to build a device for production of hollow cylindrical cluster beams of hydrogen. The paper describes the design, construction and test of a liquid hydrogen-cooled cluster generator that produces a pulsed cluster beam of 40 mm i.d. and a pulse length of several milliseconds. In the discharge region the particle density in the beam is of the order of 10^{10} H_2-molecules/cm^3, whereas the interior contains only $\sim 1\%$ of the total amount of material concentrated in the beam. This should be an advantage for the generation of intense neutron flashes.

G. Conversion of capacitor energy into heat

1. Impulse Welding, Direct Capacitor Discharge

a. *Stud Welding*

In Volume I, Chapter G1, of "High Speed Pulse Technology," the reader can find some examples of pulse-welded small objects for which the welding heat has been produced by the flow of a direct (nontransformed) capacitor discharge current. Another method of world-wide importance is the so-called stud or OMARK welding (OMARK stands for "ohmically heating arc"). Here, a large electrolytic capacitor battery of 20,000, 40,000, or 60,000 μF is charged to a voltage between 50 and 200 V. The stud and the metal sheet which are to be welded together are connected with the poles of this battery. To start the weld, the pieces are pressed together with a special solenoid gun to cause a good electrical contact and are separated again immediately to a preadjusted distance after the solenoid current is cut, so that a high current

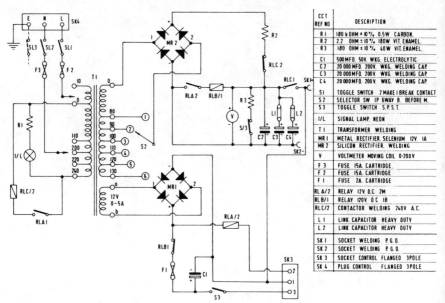

Fig. G1a-1. Circuit diagram CD 550.

1. IMPULSE WELDING, DIRECT CAPACITOR DISCHARGE

arc burns for a small part of a second. During this arc phase the surfaces of both pieces become molten and, again fully automatically, the pieces are forced together so that a molten metal connection takes place and the weld is made.

Figure G1a-1 shows the complete circuitry of Kerry's [1857] CD 550 portable studwelder. Tappings at 80–130 V on the power transformer T1 can be selected for charging, via the rectifier MR2, the three capacitors C2, C3, and C4, of which C3 and C4 are optional. The use of three equal capacitors and the availability of voltages between 80 and 130 permit the charging energy of the entire machine to be varied over a range of ~ 7.5 to 1 to suit the size of the stud. The maximum charging current is limited to a safe value by resistor R2. The voltage developed on these main capacitors is indicated by voltmeter (V) and may be varied from ~ 112 to 185 V by means of selector switch S2. The welding contactor RLC disconnects the charging current (RLC2) and connects the discharge current to the stud or the sheet (RLC1).

The maximum number of studs which may be welded per minute is governed by the time required to reload and set the gun, not by the electrical reset time of the equipment, which is not greater than 5 sec.

The stud welding system has a wide range of application. Any low carbon ferrous and many nonferrous metals may be welded including: steel to steel, steel to lead-free brass, steel to stainless steel or plated steel, steel to copper, aluminum to aluminum, aluminum to zinc base materials, copper, titanium, etc. Studs over a wide range of metals and sizes, threaded or unthreaded are suitable. In addition, insulation pins (nails or pointed studs) are available for pinning insulating materials used in the building construction and shipbuilding industries.

Indicated below are uses of this system:

(1) Domestic appliance manufacturers: including refrigerators, washing machines, dish washers, cookers, heaters, fires, fittings, etc.

(2) Motor vehicle manufacturers: for decorative trim, panels, structural, labels, locks, handles, etc.

(3) Manufacturing jewellers: for attachment of pins and clips

(4) Construction industry: for attachment of insulation and acoustic materials.

(5) Shipbuilding: for installing insulation to steel and aluminum bulk heads and panels.

(6) General fabrication: of panels, sink units, sheet metal work, duct work, etc.

(7) Electrical and electronics industry: lighting fittings, transformers, cases, vacuum tubes, semiconductors.

(8) Aircraft manufacturers: for ducting, sealing, and studding of non-stress assemblies.

(9) Chemical plant manufacturers: fabrication of pressure vessels, tanks, heat exchangers, duct work, etc.

(10) Domestic utensil manufacturers: for studding kettles, coffee pots, saucepans, pans, etc.

Apart from its low cost, the equipment has such advantages as: little or no surface marking; fast repetition welds (nominally 4–6 sec); ability to weld thin materials, including aluminum; suitability for use with small diameter studs; no need for weld electrodes or ferrules; and may be used even by an unskilled operator after initial training.

b. *Fine Wire Welding*

Direct capacitor discharge currents are often used for fine wire welding because in this application, there is no questioning the efficiency of electric power [1301].

The production of thermocouple and other junctions in wires of various materials and over a wide range of diameters is one of the main problems encountered in thermal analysis work. A junction may be formed by twisting the pair of wires together, by soldering, brazing or burying the wire ends in a small block. Each of these methods has disadvantages.

Ideally, the junction should be a point at which the thermocouple wires meet and form a strong union between the ends. The apparatus now offered is capable of producing such welds and can also be used for many other purposes. Thermocouples are often required for surface temperature measurement. Attachment of the junction is readily accomplished with this apparatus. Investigations have been made into the possibility of welding unusual pairs of metals. Platinum and uranium, for example, have been joined with this apparatus and it appears that wires of any composition may be spark welded under the correct conditions.

The standard unit houses a number of electrolytic capacitors which are charged from the output of a main transformer. Rotary switches on the power pack give a wide range of capacitance and charging voltage. There are 12 capacitance values up to 8100 μF and seven voltage settings with a maximum of 120 V. The charging time on the maximum voltage and capacity settings is $\sim$10 sec. During charging, the voltmeter rises steadily from zero to the fully charged value, while the milliammeter reading drops to zero when charging is complete. The wide range of voltage and capacitance provided enables suitable settings to be found for the welding of wires from 0.005 to 0.024 in. diam. The higher energy settings may be successfully used to weld wire to larger masses of metal. Light spot welding of foil and sheet is also possible.

A complete unit has been constructed for speedy, reliable thermocouple manufacture. The wires run from their reels to variable spring-loaded

1. IMPULSE WELDING, DIRECT CAPACITOR DISCHARGE

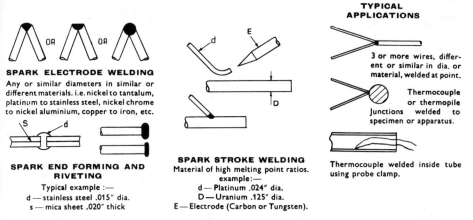

FIG. G1b-1. Welding methods obtainable with the apparatus.

manipulator clamps, which are ball-mounted to facilitate adjustment. Manipulation of the clamps easily brings the ends of the wires together and contact is maintained in V form, at almost any angle. The sliding housing carries a removable carbon (or tungsten) electrode which is connected to one discharge terminal of the power pack. The other discharge terminal is connected to the clamp frame. The lever mechanism is operated to contact the electrode with the point of the V formed by the wires and the discharge circuit is thereby completed. The resulting spark forms a weld between the thermocouple wires. Provision is made on the electrode assembly for an inert gas supply to be connected, if required. (See Fig. G1b-1.)

c. *Welds by Combined Heating and Electromagnetic Pressure*

A related process of welding is known as "electromagnetic solid-state joining" [1769]. This process uses electromagnetic induction heating and the attracted force between conductors through which currents are induced in parallel directions. Heat and pressure combinations can be precisely controlled to produce bonds of high integrity between similar and dissimilar metals and alloys, in times up to 10 sec.

A typical arrangement for welding a sleeve coupling to butting shafts is shown in Fig. G1c-1. Parts to be joined are first heated by electromagnetic induction to a temperature below their melting points. Then, for a fraction of a second, they are subjected to an electromagnetic pressure of up to 50,000 psi. The result of the combination of heat and pressure is the solid-state migration of atoms across the interface. Because the parts are heated simultaneously and uniformly, residual stresses are minimal and the heat-affected zone is virtually undetectable. The high pressure helps to destroy surface oxides at the interface

G. CONVERSION OF CAPACITOR ENERGY INTO HEAT

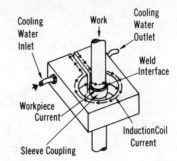

FIG. G1c-1. Work zone of process.

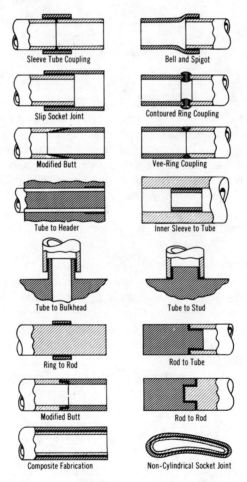

FIG. G1c-2. Examples of "electromagnetic solid-state joining."

mechanically and chemically. Also, because there is no physical contact with the workpiece, no special holding jigs or fixtures are required, and no additional restraint is placed upon the parts.

In essence, time, temperature, and pressure are the three critical elements of the process. Coalescence, or the complete growing together of the parts into a solid unit, characterizes the final condition of the joint, which has a strength equal to that of the parent metals or alloys in the annealed condition. The process in no way impairs the chemical, mechanical, or physical properties of the base metal. Metallurgically, the joints are completely sound.

Some combinations of metals and shapes of parts that were difficult or impossible to join by other methods can now be joined successfully. The method is suitable for application in the automotive, appliance, nuclear, and military fields. Figure G1c-2 shows typical examples of parts which are suitable for this kind of welding. The materials that can be joined by this method are shown in Table G1c-1.

TABLE G1c-1

MATERIALS THAT CAN BE JOINED BY COMBINED HEATING AND ELECTROMAGNETIC PRESSURE[a]

What the process joins	
Dissimilar materials	Similar materials
Aluminum to copper	Carbon steels
Aluminum-killed steel 1015 to 1243 steel	Alloy steels
Gray cast iron to carbon steel	Stainless steels
Hastelloy to stainless steel	Titanium alloys
Malleable cast iron to carbon steel	Nickel-base alloys
Molybdenum to alloy steel	Copper-base alloys
Nodular cast iron to carbon steel	Aluminum alloys
Sintered iron to alloy steel	
Stainless steel to carbon steel	

2. Welding by Transformed Capacitor Energy

a. *Ultrapulse Welding in General*

Pulse welding by transformed capacitor discharges has been described in detail in "High Speed Pulse Technology," Volume I, Chapter G2, 3. Recently, this principle for welding at all energies with a welding time below 10 msec, has reappeared in the USA under the name "ultrapulse welding" [1713, 1732]. The new name is appropriate because it emphasizes the ultrashort welding time, but the principle is unchanged. There are two ways of welding by transformed capacitor discharges. The conventional and inexpensive way uses increased capacitance for greater energy at the same voltage (normally a few hundred volts), and the discharge feeds into an increasingly heavy iron core transformer (see Refs. [1300, 1302]). The result is to increase the welding time as the square root of the energy. At 10,000 J the weld takes place very slowly in rather large fractions of a second, annealing of the environmental zone of the weld is unavoidable and the structure of the weld shows large crystals. Applying the ultrapulse welding principle, the welding time is, by definition, shorter than about 10 msec. To obtain increased energy, the capacitor voltage must increase with the square root of the energy. The transformer must have an air core and its geometrical design must ensure tight coupling with low scatter. If an iron core is used, its only purpose is to soften the first rise of the welding current at about the first millisecond until it is saturated.

b. *Electrode Holders*

The secondary circuit including the electrode holder must have a very low inductance so that a negligible part of the very high secondary current is stored and lost in the inductance. At 250,000 A and 0.1 μH, $\sim$3000 J would be lost in the circuit inductance, since $E = \frac{1}{2} L \cdot I^2$.

The transformer design is shown in Volume I of "High Speed Pulse Technology," Fig. D2-2, p. 192 and on p. 346, Fig. G2-6 shows an extremely low inductance electrode holder having a rubber backing to move the electrodes fast enough during the beginning of the melting phase, to maintain full pressure. A more versatile electrode arrangement has been shown by Kollmann [1298].

For a satisfactory and homogeneous weld, it is necessary that during the actual welding operation pressure is applied continuously to the articles to be welded, while the welding material is still in a semisolid pasty state.

In many welding operations and particularly with impulse welding machines, it is important that the movable electrode follows the welding material, as this recedes due to softening or melting in the pressure area, without any delay although the available time may amount to only a fraction of a millisecond. This ensures that, during the welding operation, the required pressure between the parts is maintained. In order to achieve this rapid follow-up action, the mass of the moving part and the stiffness of the spring element are selected to give the necessary short period of oscillation.

Figure G2b-1 is a sectional view of such an electrode holder. The electrode 10 is held by a chuck in the holder 22, the stem of which slides in the polytrafluoerothylene guide 16 and is pressed downward by the follow-up spring 19. This acts through the collar 17 which is screwed onto the stem 15. The pressure of the spring can be adjusted using the screw cap 20, and the downward movement of the collar 17 is limited by a shoulder in the vertical housing 18. This housing is rigidly attached to the horizontal piece 18', which is itself attached to a member of the machine providing the necessary vertical motion. The electrode holder 22 is connected to the input terminals by the flexible flat conductors 34. When the entire assembly is moved down to bring the electrode into contact with the workpieces to be welded, the electrode is forced upwards against the pressure of the spring. Then, as the current is applied and the material of the workpieces begins to melt, the spring causes the electrode to remain in contact with them. However, there are still cases where even this arrangement does not result in fast enough follow up to ensure entirely satisfactory welds. This is particularly so when difficult welds are to be made, e.g., copper/copper or copper/silver, and where welding times are extremely short, as with closely coupled welding transformers and low-inductance sandwich lines to the electrodes.

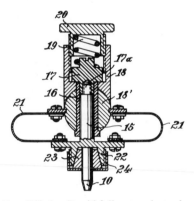

FIG. G2b-1. Rapid follow-up electrode.

Nevertheless, this kind of pulse welding with low voltage, very high current pulses of a few milliseconds duration has in the last eight years found a broad field of applications.

c. *Pressure and Energy Requirements**

In carrying out experiments, a design must be chosen (i.e., a statistical pattern) such that when starting from a base point on a pressure–energy diagram, the procedure will result in a devious route toward the maximum response area. The idea of the design is to relate a pattern of experimental response in such a way that each sequential point on the diagram is determined from knowledge gained from the previous design. In practice, the various experimental specimens are tested one after another (or at least in small

---

* See Refs. [1040, 1041, 1713].

G. CONVERSION OF CAPACITOR ENERGY INTO HEAT

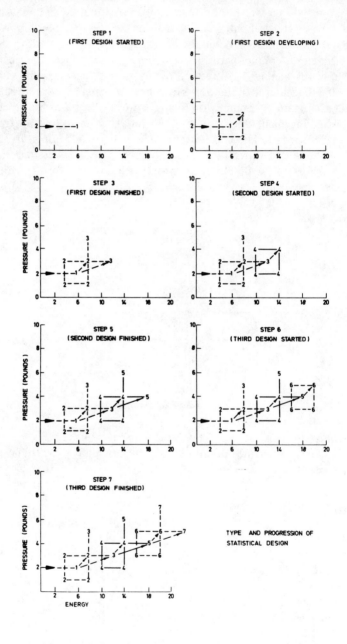

FIG. G2c-1. Finding of pressure and energy by statistical operation.

groups) so that at any stage of the investigation, the results for all earlier tests moves is determined from strength and strength consistency data, and it will be available to determine the most effective way of continuing the investigation. The key to the whole experimental process lies in making use of the sequential nature of the test procedure. The movement of the design is influenced by weld deformation, splitting, and electrode sticking. Figure G2c-1 illustrates the type and progression of the first design. This particular geometric design contains seven points of investigation and is operated in two steps. The first step involves moving into the diagram from an arbitrarily selected base point (illustrated in Fig. G2c-1 as 2-lb pressure) until a first substantially good weld is achieved and can be tested. The procedure then involves constructing and carrying out a group of experiments around the initial test to estimate the main effects, or at least to show in which direction the optimum conditions will be found to confirm the region of further experiments. The design can be repeated as many times as needed to establish the area of optimum welding conditions.

d. *Ring Projection Welding*

A small indentation or projection is made in one of the pieces to be welded. The welding current flows through this single point of contact (the projection) and the high current density at that point produces excellent local heating. With this method, two or more points can be welded at the same time and thick plates which otherwise could not be welded with a small capacity welder can be welded. An extension of this method called "ring projection welding," has been widely used in the semiconductor industry. This type of welding produces an airtight seal and is performed by simply making a ring projection instead of a spot projection.

The airtight seal welding method—i.e., hermetic seal welding using ring projections—has become indispensable to such semiconductor industries as manufacturing transistors, rectifiers, diodes, thyristors, integrated circuits, crystal oscillators, and other devices. Of major importance in hermetic seal welding is the design of the container to be sealed.

There are various types of silicon or germanium diodes and transistors, some of which are shown in Fig. G2d-1.

The welding is done around the circumference of the diode base where the upper cap and the lower base come into contact with each other. In order to facilitate this welding, a ring projection is formed on either the cap or the base by a mechanical finishing or pressing method. Since the outer diameter of the diode at the contact point naturally becomes larger with the formation of a ring projection, the welding is sometimes done without projection as in cases 2 and 3 of Fig. G2d-1. However, the projection type weld is, of course, much stronger than the nonprojection type.

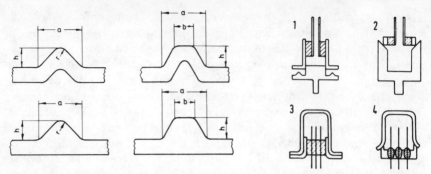

Fig. G2d-1. (a) Optimizing shapes of projections for ultrapulse welds. (b) Arrangement for vacuum-tight welds of small housings.

Projections made in soft materials such as aluminum, brass, copper, etc., are often crushed when the weld force is applied by the electrodes. For this and other reasons, other metals such as iron, Kovar, nickel or alloys of these are used. Also, it is often necessary to plate the parts thinly (2–10 μm) with gold, copper, nickel, tin or other metal, depending on the materials of the parts. An important factor in this type of welding is the cross-sectional area of the projection. It has been determined that the ideal cross section is an equilateral triangle with a slightly rounded top and a height of 0.3 to 0.5 mm depending on the size of the diode or transistor. The diameter of the projection ring has an effect on the strength of the weld also. Past experience indicates that this diameter should be from 5 to 30 mm. Projection shapes other than the circle can be welded, and it is expected that with the development of electronics, the configuration, material, and sizes of the projection will vary in many ways.

The actual method of welding diodes or transistors is shown in Fig. G2d-1. The cap and the base are tightly held by the carefully centered upper and lower

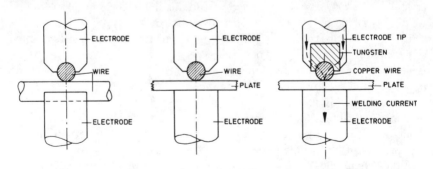

Fig. G2d-2. Welding of wires together or to a plate.

2. WELDING BY TRANSFORMED CAPACITOR ENERGY

electrodes and then pushed together. When contact is made, welding current is allowed to flow. When a small number of different types of diodes are required, the electrode tips have to be changed for each type. Even with this method, however, the operator, with sufficient training, can produce several thousand pieces a day.

The following are some practical applications of projection welding:

(1) Welding wires. The electrodes are made with grooves to fit the wires being welded. This increases the contact area between the wires and the electrodes and the point of contact between the two wires acts as the projection in this case (Fig. G2d-2a).

(2) Welding wire to sheet. In this method, the electrode bearing on the sheet is flat, whereas the electrode bearing on the wire has a groove (Fig. G2d-2b). For copper wires, the electrode tip is made of tungsten (Fig. G2d-2c).

e. *Characteristic Features of Ultrapulse Welding*

Ultrapulse welding is distinguished by the following features:

Low line load, and hence small cable cross sections between machine and power supply line.

Because of the extremely short welding time, concentration of the energy to the actual welding zone and thereby lowest heat dissipation into the areas surrounding the welding zone.

After welding the work pieces come out of the machine practically cold, keep their form and show no signs of warping or change of shape so that the work pieces to be bonded can be pretreated according to ISA quality standards 8–10, i.e., ready for assembly, without any further reworking or finishing.

By ultrapulse welding it is even possible to bond work pieces close to thermally sensitive zones.

Voltage stabilized charging of the capacitor bank ensures that line voltage fluctuations have no effect on the welding results and the machine settings once found are genuinely reproducible.

Because of the extremely short welding time the welding electrodes carry such a low thermal load that electrode cooling can be dispensed with.

It is possible to weld two punched parts to distance bolts with a diameter length ratio of 1:10 without deformation of the work pieces and without reworking.

The extremely short welding time in the millisecond range permits the proper bonding of metals with high electric and thermal conductivity, including steels with a carbon content of up to 0.3%.

Finally, Table G2e-1 shows physical and operating data for some fast pulse welding machines.

TABLE G2e-1

SPECIFICATIONS OF SOME INDUSTRIAL PULSE WELDING MACHINES

Manufacturer	Impulsphysik				Vitessa 12000[a]		Hughes	
Type	M 100	M 500	Impulsa 2000	Impulsa 4000	Table	2×Cabinet	HRW100B	HRW500
Height (cm)	27	32 28[b]	185	185	195	125	23	52
Width (cm)	28.5	35 26[b]	79	79	75	70	38	56
Depth (cm)	39	47 35[b]	130	130	171	81	35	40
Capacitor energy (W-sec)	100	500	2000	4000	12,000		100	500
Peak current (A)	7500	12,000	90,000	110,000	200,000		6000[d]	11,500[d]
Welding discharge-time (msec)	3	6	1.5	2	1.5		4.0	5.0
Capacitor capacitance (μF)	4000	20,000	400	800	2400		1550	400
Charging voltage, adjustable (V)	18–225	18–225	100–3200	100–3200	100–3200		20–410	100–1580
Number of welds per minute at max energy	120[c]	80[c]	30[c]	15[c] 100% duty	30[c]		60[c]	60[c] 100% duty
Welding head	FF102	FF101	Impulsa	Impulsa	Vitessa		VTA60	VTA70
Type working pressure (kp)	0.8–10	1.5–25	50–750	50–750	100–2100		0.2–8	1.0–22.5
The machine equals in its output a half-cycle machine of about kilovoltamp	—	—	100–500	100–500	300–2000		—	—
Welding voltage (V)	10–12	10–12	3–12	3–12	3–12		1–20	1–20

2. WELDING BY TRANSFORMED CAPACITOR ENERGY

Manufacturer	Unitek				Temi		Peco, Munich		Raytheon
Type	1-128-01	1-048-03	1-132-02	1-133-02	DCT100	DCT500	FP2K	FP3K	225C
Height (cm)	16	35	33	31 22[b]	220	450	107	46	60 25[b]
Width (cm)	31	27	25	38 25[b]	450	450	90	29	28 27[b]
Depth (cm)	38	38	41	51 23[b]	325	325	55	56	35 41[b]
Capacitor energy (W-sec)	50	100	250	500	100	500	300	1000	225
Peak current (A)	—	—	8000[d]	10,000[d]	4600	11,800	10,000	30,000	10,000
Welding discharge-time (msec)	2.6 (at 25 Wsec)	21 (at 50 Wsec)	6	11	3.6	4.4	2.5–5	2	2.0
Capacitor capacitance (μF)	755	1510	3060	6040	1410	2115	336–500	80	350
Charging voltage, adjustable (V)	13–408	20–408	20–405	20–408	0–400	0–770	300–1100	1000–5000	675–1500
Number of welds per minute at max energy	60	80 (intermittent)	100 (at 80 Wsec) 50% duty	50	60[c]	10[c]	60	20	60[c] 50% duty
Welding head	2-032-03	2-032-03	2-037-02	2-037-02	TO30 100% duty	TO70	—	—	—
Type working pressure (kp)	0.3–10	0.3–10	0.7–22.5	0.7–22.5	—	—	20–250	20–250	0.3–5.0
The machine equals in its output a half-cycle machine of about kilovoltamp	1–18	1–18	1–18	1–18	—	—	60–120	500	—
Welding voltage (V)	1–18	1–18	1–18	1–18	—	—	5–20	5–20	—

[a] There is also a 24,000 W-sec model available having a peak current of 400,000 A, and working pressure of 4,500 kp.
[b] External pulse transformer.
[c] At maximum energy.
[d] At shortest discharge time.

3. Fast Metallic Phase Transformations by Current Pulses

Research on the crystalline behavior of metal alloys with sudden jumps of temperature needs sources of energy for which electric capacitors are very suitable. A fast thermal pulser for metallic phase transformation studies is described by Forgacs [1229].

The most important operating characteristics of the thermal pulser are the repeatability of the average temperature attained during a thermal pulse, repeatability of time at temperature, and the constancy of temperature during the pulse. Three different modes of operation are discussed in the paper: capacitor discharge followed by negligible direct current, capacitor discharge followed by appreciable preselected direct current, and capacitor discharge followed by feedback controlled direct current.

The method employed to ensure synchronization of the capacitor discharge and contactor closure requires that some small direct current flow to trigger the ignitron, but the minimum selectable current of 1.3 A dc produces only 3.4×10^{-3} W in a 2 mΩ load and may be ignored compared to $\frac{1}{2} CV^2$ (850 J for 425 μF at 2000 V). The observed repeatability of the peak temperature attained by capacitor discharge in a series of successive discharges is ± 2.2°C at 220°C or $\pm 1\%$. These variations are not due to variations in capacitor voltage prior to discharge since this is measured and controlled to approximately $\pm 0.01\%$ and should therefore yield temperature variations of no more than approximately $\pm 0.02\%$. The temperature coefficient of capacitance of the capacitors averages 0.25%/°C over the 0–40°C range; however, since the temperature variations appear to be random in a time short compared to the thermal time constant of the capacitors, thermally induced capacity variations are relatively insignificant. Observation of the discharge current waveform indicates that the ignitron does not cease conduction at precisely the same point in a cycle on successive discharges; it appears likely that this is the main reason for the observed temperature variations. A spark gap in air or mechanical switch might be expected to be more erratic than an ignitron due to contact erosion and in the case of the latter, contact bounce, but alternative switches have not been tested.

The quoted repeatability was measured for a particular sample; if a second sample is to be tested which has the same nominal dimensions as a sample whose curve of V_c (capactior voltage) versus peak temperature is known, variations in sample dimensions (thickness in particular) will make pulsing to exact preselected temperatures very difficult. The best procedure is to make the sample as similar as possible and to obtain V_c versus peak temperature data at temperatures below the point where transformations are active, and extrapolate the curve, when it is necessary to avoid entering the high temperature regions prior to actual testing of a new sample.

To test the constancy of temperature during a pulse, an Al–4% Cu sample was employed with no water jet cooling. A three-legged Chromel–Alumel thermocouple, with 0.05 mm wire diam, was employed. Indicated 10–90% rise time was ~ 4 msec; however, the temperature continued to raise the last few degrees for ~ 50 msec.

4. Pulse Hardening of Carbon Steel*

a. *Modern Techniques*

Under the name "micro-induction hardening," "High Speed Pulse Technology," Volume I, Chapter G4, pp. 362 ff. dealt with the problem of hardening very small parts by high frequency power pulses fed from capacitor discharges. Since 1964 researchers in Germany and Austria [1880–1887], have worked as a team to promote this particular field, which was called "pulse hardening."

Special advantages of pulse hardening are: short time of hardening, exact determination of the hardening zone, no detrimental heating outside the hardening zone, no deformation, and the adequacy of internal chilling without additional quenching. Automatic assembly line operation can be carried out using optronic control. Capacitor storage of energy avoids current surges on the supply line, and the capacitor charging technique ensures stabilized setting values irrespective of line voltage fluctuation. The machine operates in the international industrial frequency band of 27.12 MHz.

Pulse hardening can be applied to both unalloyed and alloyed steels. Only with high chromium steel, which has low thermal conductivity, are the conditions unsuitable for pulse hardening. By pulse hardening, an extremely high degree of hardness is achieved and thereby a correspondingly high resistance to wear and tear. The grain of the steel after pulse hardening is extremely fine and because of this the hardened metal is not brittle. By pulse hardening, plain steel is considerably improved in its technological qualities and in certain circumstances can be employed instead of alloyed steels. For the larger work pieces a special method is available.

The energy needed to heat the surface of a steel tool to the required temperature can be supplied by an inductively generated heating pulse as described later, by sudden injection of frictional energy into the surface by high pressure contact against a rotor with high peripheral speed, by electron beam energy, under high vacuum, by a plasma, or finally by laser beam energy. It was found that it is the sudden application and cutoff of energy which is important and not the kind of energy used. The energy pulse should have a rectangular shape and a duration between 0.5 and 100 msec depending on the thermal

---

* See Ref. [1517].

conductivity and the alloy of the treated steel. It was found that the best quality is obtained by pulse hardening a steel that has already been hardened by conventional means. The pulse-hardened zone has a martensitic quality, but the grain becomes visible only at 5000 times magnification in an electron microscope. The main advantage of cutting edges hardened in this way is the exceptional hardness numbers of 850 to 1100 Vickers ($H_U 0.5$) combined with low brittleness. However, because of the high energy density and rapid internal chilling required, pulse hardening can be applied successfully only to small objects, such as cutting and turning tools, drills, dentists burrs, wood and metal saws, milling cutters, etc.

The time periods required for hardening and chilling are in general far below—by a power of 10 or more—the shortest used hitherto in ordinary induction hardening with continuous feed. The chilling from the maximum surface temperature, which is comparatively high and far above normal values, is effected principally by conduction of heat to the cold parts of the work piece, and a quenching agent is not required. Water is however sometimes used for insulation to prevent sparkovers (see below).

Compared to ordinary hardened steel this pulse-hardened structure has a considerably higher resistance to etching. Whilst the usually metallographic etching agents act rapidly on untreated steel, the micrographs of the pulse hardened areas, remain bright and show no structure after, for example, 3 min exposure to 3% alcoholic nitric acid. Corrosion resistance is also considerably increased.

Figure G4a-1 shows the commercially available Impulsa-H high-frequency pulse hardening machine.

Pulse hardening is achieved by feeding capacitor-stored energy into a high efficiency and constant frequency triode oscillator system. By induction from the low-resistance inductor, the work piece to be hardened is heated up within milliseconds almost to melting point in a thin surface layer and is immediately quenched by internal chilling. The oscillator unit is fitted with a small feeding mechanism for the hardening of saws. Positioning against the inductor loop to the necessary close tolerances is effected with a three-coordinate micrometer adjustment. The triggering light barrier can be adjusted in two planes.

The fine grain and elasticity of the hardened steel results primarily from sudden termination of the heating phase. Therefore, considerable effort has been directed toward giving the heating energy pulses an extremely steep trailing edge, with a decay time less than 1% of the total pulse duration. To generate the required rectangular pulse shape, the circuit, Fig. G4a-2, provides two control thyratrons, one of which starts the current flow to the pulse generator and the second cuts it off.

The heating energy E of a rectangular pulse generated by the circuit shown in Fig. G4a-2 is given by the average emission current of the tube in i by the voltage fed into i and the time difference between firing of d^1 and e^1. In case of 10 A, 8 kV, and 10 msec, 800 W-sec would be the input.

FIG. G4a-1. Impulsa-H pulse hardening machine with jigs and fixtures for hardening a band saw. The pulse is generated in the unit on the right. The unit on the left is the HF-oscillator giving up to 30 kW effective peak power for pulse durations of 0.5 to 20 msec ± 0.5%. A coupling line feeds the hardening inductor, which is provided with 3-coordinate micrometer adjustment with relation to the saw teeth.

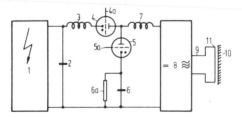

FIG. G4a-2. Block diagram for a high frequency pulse hardening unit using switching thyratrons to give a rectangular energy profile. 1 Stabilized high voltage source; 2 first capacitor; 3 HF choke; 4 high-voltage switching thyratron which receives the switch-on pulse at grid d^1; 5 switch-off thyratron, fired at grid e^1 to make the path to capacitor f conductive and to quench the discharge in d; 6 capacitor; 6a discharge resistor for re-establishing zero potential across; 7 MF protective choke, through which the current pulse feeds the 30-kW HF generator; 8 HF generator; 9 output; 10 work piece; 11 inductor.

b. *Inductors for Pulse Hardening*

An indispensable prerequisite for proper pulse hardening is the use of suitable inductors. Figure G4b-1 shows some typical inductors designed for the following applications (the plugs are 5 mm in diameter):

Fig. G4b-1. Typical hardening inductors.

(I) Hardening of slitting cutters and small circular saw blades up to 35 mm diam. In this design the hardening is achieved with one single pulse, that is, without work piece movement.

(II) Hardening of round bars up to 3 mm diam and 30 mm in length.

(III) Hardening of circular surfaces up to 10 mm diam. The inductor is a spiral.

(IV) Hardening of single saw teeth with continuous feed. The inductor must be adapted to the shape of the saw teeth. The position of the tooth in relation to the inductor determines the hardening zone.

(V) Hardening of a jigsaw blade. The feed can be continuous in this case also.

(VI) Hardening of rectangular surfaces up to $\sim 20 \times 3$ mm.

In practice, inductors cannot be calculated but must be formed for the particular hardening application by trial and error. A suitable material for the inductors is silver wire, either round or flat depending on the hardening application, and of cross section between 0.2 and 3 mm^2. Cooling must be appropriate to the pulsing frequency. Air cooling is usually sufficient up to ~ 4 pulses/sec. At higher pulsing rates the inductor is made of silver tube, and a flow of cooling liquid (distilled water with an over pressure of 10 to 15 bar) will then ensure the required heat dissipation. The manufacture of the inductors always requires a special jig, which must be made for the work piece to be hardened. Manufacturing tolerances must be around 20 μm. At present inductors of laminated materials are being developed—e.g., silver–copper or silver–steel—to determine which make is possible to harden comparatively large surfaces.

c. *Experience with Pulse Hardening*

Pulse hardening by high frequency induction heating is the method tested most extensively, and for this reason it will be the principal method used in

4. PULSE HARDENING OF CARBON STEEL

practice for the time being. The results obtained with pulses in the millisecond range can be demonstrated by means of photomicrographs. The sudden cutoff of the high frequency energy at the end of a pulse occasionally causes undesired voltage peaks which can easily lead to a spark over between the inductor and the work piece, and to avoid this, water is used as an insulator. Running water, even tap water or salt water, has a very high breakdown voltage, up to ~ 250 kV/cm, for short pulses. Since the cooling effect of the water is clow compared to internal self-chilling, the water is not used for cooling but only for insulation. If required, salt can be added to the water when the work piece requires this treatment in preparation for later nitration or anodizing. The water must flow in surges at moderate speed through the space between inductor and workpiece. The cooling water combing out of the inductor under high pressure can also be utilized for insulation. Stationary water is unsuitable for insulation purposes because ion chains form which, at a high pulsing rate, can lead to underwater sparks.

TABLE G4c-1

CONDITIONS FOR TYPICAL PULSE HARDENING TASKS

Subject	Pulse duration (msec)	Pulses/sec rep. rate or speed
Circular saw blade 42×0.5 mm	20	1 pulse/blade
Wood cutting saw blade	3–5	10–15 pulses/sec optronically controlled by teeth
Firing pin for rifles	10–15	8–10 pulses/sec optronically controlled by pins
Blade of wood planing machine	3–5	10–15 pulses/sec
Cutting tool	3–5	1 pulse/piece
Punching tool	6–8	1 pulse/piece
Cable knife	4	1 pulse/piece
Bevel gear cutter for metal machining	5	up to 10 pulses/sec., 1 pulse/tooth
Balance shaft for watches	1	1 pulse/piece
Razor blade	1	1 pulse/knife edge

For steels with a carbide content, the duration of rectangular pulses must be selected very carefully because carbides require a time in the millisecond range to go into solution in the melting metal. Steel with $\sim 1\%$ tungsten requires a pulse duration between 20 and 50 msec at a surface temperature of $\sim 1200°C$ to dissolve the carbides. For every type of steel there is an optimum temperature and pulse duration, which has to be determined empirically. Table G4c-1 gives the pulse duration and repetition rate for typical hardening tasks.

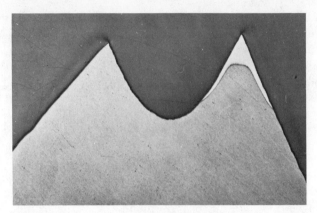

Fig. G4c-1. Pulse-hardened tooth of band saw. (With kind permission of Southern Saw, Inc., Atlanta, Georgia.)

Figure G4c-1 shows a pulse hardened tooth of a band saw. The tooth is etched with 3% alcoholic nitric acid. The pulse-hardened zone remained practically unaffected by the etching agent, indicating increased corrosion resistance.

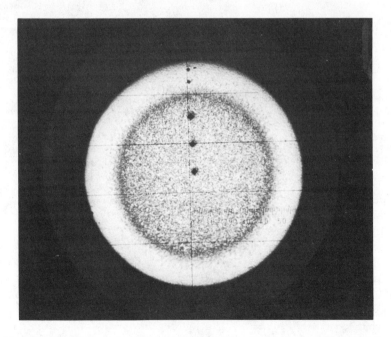

Fig. G4c-2. Micrograph of the pulse-hardened tip of the balance staff of a watch.

4. PULSE HARDENING OF CARBON STEEL

Cutting tools for wood and metal machining as well as parts of production machines subject to wear and tear are particularly suitable for pulse hardening. The form and steel of the part to be hardened determine the applicability of the method. From the wide field of application five examples have been selected where pulse hardening was carried out with good success.

Figure G4c-2 is a micrograph of what is, up to the present time, the smallest work piece which has been pulse hardened, namely, the tip of the balance staff of a watch, with a tip diameter of 0.4 mm. From the

FIG. G4c-3. Pulse-hardened counter piece for the knife of a meat grinder. (With kind permission of Southern Saw, Inc., Atlanta, Georgia.)

hardening impressions it can clearly be seen that the core of the tip has remained soft. The hardening pulse was 1 msec and the charging voltage 3.5 kV.

Figure G4c-3 shows the largest workpiece hardened so far, the counter piece for the knife of a meat grinder, with a diameter of 100 mm and a thickness of 9 mm. The holes have a diameter of 9 mm and are pulse hardened at the edges. Laboratory tests have shown that it is also possible to harden a zone 0.6 mm wide and 0.3 mm deep (~ 0.190 mm^2). The hardening pulse was 60 msec and the charging voltage 7.5 kV. Figure G4c-4 shows other examples of pulse hardening.

Example: Meat Band Saw

Material: steel (C = 0.9%)
Basic hardness: Rc 51 (VNP 0.3* 520)
Pulse hardness: Rc 69 (VPN 0.3* 1004)
Hardening time: 2 milliseconds

Example: Punch

Material: DMo 5, conventionally hardened
Basic hardness: Rc 65 (VPN 0.3* 820)
Pulse hardness: Rc 68 (VPN 0.3* 960)
Hardening time: 11 milliseconds

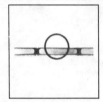

Example: Piston Ring

Material: Steel (C = 0.75%)
Basic hardness: Rc 47 (VPN 0.3* 480)
Pulse hardness: Rc 68 (VPN 0.3* 965)
Hardening time: 6 milliseconds

(* kg load)

FIG. G4c-4. Further examples of pulse hardening. a, b, c show on the left the subject, in the middle the treatment data and the hardness, and on the right the microstructure.

5. EXPLODING WIRES AND THEIR APPLICATIONS; EXPLODING WIRE SHUTTERS

a. *The Exploding Wire Process*

If a capacitor discharge is fed into a piece of wire or foil, the electrical energy first melts the metal and then vaporizes it. Metal vapor insulates like a hot gas. If ionization takes place, the metal vapor is more or less conductive. Hence, if the capacitor voltage is high enough, another breakdown occurs through the ionized metal vapor. Wire is not absolutely constant in diameter and so vaporization starts at a large number of small spots. Perhaps because only simple equipment is needed to make a wire explode, exploding wire physics is a broad field of research activity.

Oktay [1307] studied the effect of wire cross section on the first pulse of an exploding wire and determined the conditions for an optimum discharge. The circuit for the exploding wire experiment is shown in Fig. G5a-1. A 14.7 μF, 20 kV capacitor was used as the energy storage capacitor. The total inductance of the circuit was C.5 μH, and the total resistance of the circuit without the exploding wire was 10.76 mΩ. The current $i(t)$ was measured with a 2.430 mΩ inductance-free T&M current viewing resistor, and the voltage across the wire

$v(t)$ was measured with a Tektronix P-6015 high-voltage probe. Both the current and the voltage traces were recorded with a dual-beam Tektronix 545-A oscilloscope.

The characteristic current and voltage traces for an exploding wire show regions known as the first pulse, dwell, and restrike regions, respectively.

In an optimum discharge, the stored energy of the capacitor, E_0, discharges completely during the first pulse, the current and the voltage diminishing to zero. The restrike phenomenon does not occur. The experiments were conducted in air at atmospheric pressure, and Cu, Ag, and Pt wires were used. The wire cross section was varied from 4.8×10^{-5} to 3.2×10^{-3} cm^2, and the wire length was varied from 2.5 to 17.8 cm. It was determined from the results of these experiments that the initial voltage of the capacitor for the optimum discharge, V_0, is dependent only on the wire cross section and independent of the wire length.

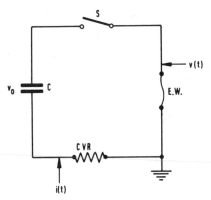

FIG. G5a-1. The exploding wire circuit.

The plots of V_0 versus the wire cross section A for Cu, Ag, and Pt wires are shown in Fig. G5a-2. The slopes of these curves indicate that V_0 is directly proportional to A.

The plots of the stored energy to the wire mass ratio, E_0/m, versus resistance of the wire at the room temperature, R_c, for the three metals are shown in Fig. G5a-3. The slopes of these curves indicate that E_0/m is inversely proportional to R_c. The energy transferred to the wire was computed for several optimum discharges by integrating the power curve obtained from the current and voltage traces. This energy was more than 90% of E_0, the energy stored in the capacitor.

The $1/R_c$ variation of E_0/m ratio is a consequence of the direct proportionality of V_0 to A. In the circuit used for these experiments, the E_0/m ratio was varied from 200 to 20,000 J/g be decreasing R_c from 200 to 2 mΩ.

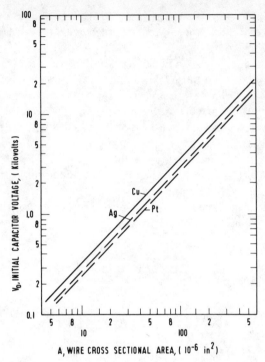

FIG. G5a-2. The initial capacitor charging voltage vs. the wire cross-section area. (The wire lengths for most of the data points were varied from 2.5 to 17.8 cm while keeping the wire cross-section area constant.)

An exact time-resolved study of the behavior of a wire during explosion is the subject of a paper by Nyberg et al. [1641].

Wires made of high boiling point materials were electrically exploded in an evacuated vessel of coaxial design. The initiation of the shunting gas discharge and the vaporization of the wire were studied by means of a single frame, Kerr cell shutter camera. It was found that the influence of the radial electric field on the shunting discharge was negligible and that the vaporization on long tungsten wires is very nonuniform. A model to explain the latter is suggested.

Jäger and Rusche [1306] made optical and electrical measurements on confined exploding wires threaded axially within tubes of dielectric material. The radiation temperature of the plasma was determined and energy and power inputs were evaluated from simultaneous current and voltage oscillograms. It was shown that with decreasing cross section of the inner diameter of the tubes, the maximum temperature of the plasma could easily be increased up to 50%. Soon after that a saturation value was reached caused by evaporation

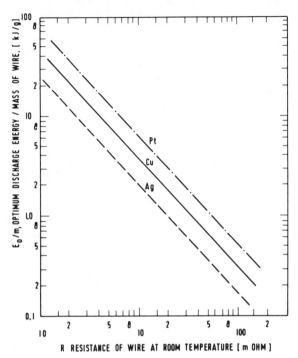

FIG. G5a-3. The stored energy for an optimum discharge to the wire mass ratio vs. the resistance of the wire at the room temperature.

of the confining material and an increase of the number of particles. According to a rough estimation pressures higher than 10^4 atm could be obtained. Their results are shown in Fig. G5-4 and in Table G5a-1. As shown in Fig. G5a-4, astrophysical temperatures are obtainable, up to 40,000°K. (15 references are given in this paper.)

A Russian paper [1530] describes an attempt, using a gigawatt for 2 msec, to generate a flash that could be used to simulate galactic processes, activate lasers and study other high intensity optical phenomena.

The inhomogenities in the wire diameter already mentioned cause the so-called "striations" during the explosion. Goronkin [1305] reported that these striations have been observed under both "slow" and "fast" explosion conditions. In the former, the temperature rise due to discharge current is slow enough to allow surface forces of the molten wire to act. Constrictions thus formed are high-resistance regions which, when severely constricted, finally sever by continued action of the surface tension. Electrical conduction through these regions takes place by breakdown of the metal or ambient atmosphere. Such "hot spots" are clearly visible by the radiation of excited metal atoms or

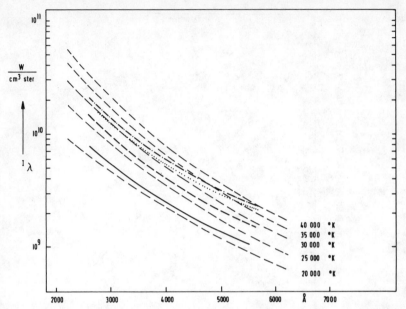

FIG. G5a-4. End-on observation: Spectral distribution of the intensity at the maximum of emission from unconfined exploding wires and wires confined in tubes with diameter d of 2–4 mm. The fine dashed curves shows black body emission at various temperatures. $-\cdot-\cdot-$ $d = 2$ mm; $\cdots\cdots$ $d = 3$ mm; $----$ $d = 4$ mm; ——— unconfined.

ionized gases at the constrictions. These striations have typical separation distances on the order of 1.0 mm. Striations observed in "slow" explosions are understood in terms of electrical breakdown in regions of high sample resistance. Fast-explosion striations may be explained in at least two ways. The primary formation process is either mechanical or electrical.

Two points should be noted concerning observed mechanical striations. First, the ionized gases will be ejected from the vicinity of the wire by rapid desorption due to heating of the wire surface and by electric diffusion due to

TABLE G5a-1

MAXIMAL TEMPERATURE AND ELECTRICAL VALUES

Diameter (mm)	I (10^4 A)	R_1 (mΩ)	R_2 (mΩ)	Conductivity (10^{-4} $\Omega \cdot$cm)	T (°K)
∞	9.7	6	6	32.4	21,500
4	8.4	34	43	8.6	27,000
3	8.2	38	41	5.4	31,500
2	8.1	44	38	2.8	34,000

the large ion concentration gradient. Secondly, the reason for the transverse nature of mechanical striations is clear. Field ionization requires a threshold field value. If the grain boundaries are randomly oriented, those boundaries perpendicular to the applied field will be the locations where the highest fields are generated. Those boundaries set at an angle other than 90° to the field will support lower fields since the effective boundary width is greater.

Other papers of interest are:

> Cassidy and Neumann [1771]; Photographic and spectroscopic studies of exploding wires in a sealed vessel. 7th International Congress on High Speed Photography, 1965, Zurich.
> Christies [1771]; Recording the surface condition of an exploding wire by external illumination, using a ruby laser. 7th International Congress on High Speed Photography, 1965, Zurich.
> Heine-Geldern [1771]; High speed photography by means of Kerr cells and exploding wires. 2nd International Congress on High Speed Photography, 1954, Paris.
> Zernow et al. [1771]; High speed photographic studies of electrically exploded metal films and wires. 5th International Congress on High Speed Photography, 1960, Washington.

b. *Exploding Bridgewire Triggering of a Flash X-Ray Discharge*

Handel and Englund-Ponterius [1042] describe their method for triggering a hard-vacuum flash x-ray (FXR) discharge, using an exploding bridgewire for discharge initiation in place of the conventional trigger electrode. The bridgewire (tungsten, diameter 0.05 mm, length 4 mm) is secured in a depression in the cathode center and extends about 2 mm beyond the cathode surface. The wire circuit (loop inductance, 230 nH) is coaxially arranged and electrically insulated from the main circuit. Examination of the x-ray output characteristics of the FXR discharge show that a more or less reproducible trigger-to-pulse time is obtained, which is not the case when an ordinary trigger electrode is used.

The FXR system which is shown in Fig. G5b-1 consisted essentially of an energy storage capacitor on which the discharge tube was coaxially arranged. The storage capacitor used was manufactured by the ASEA company, Västerås, Sweden. This capacitor, type CTU 50 kV, 0.4 μF, had an internal inductance of about 50 nH. The total inductance of the FXR system was about 70 nH. The discharge tube was provided with cylindrical plane parallel electrodes, 25 mm in diameter. A tungsten anode and an iron cathode were used. The anode–cathode spacing was 12 mm. For discharge initiation, a trigger electrode was centered in the cathode, but in the present investigation the trigger electrode was replaced by a tungsten wire. The tube was continuously evacuated by a conventional oil diffusion pump with a liquid nitrogen trap in the pumping system. The tube pressure was kept in the range $(2–5) \times 10^{-6}$ Torr.

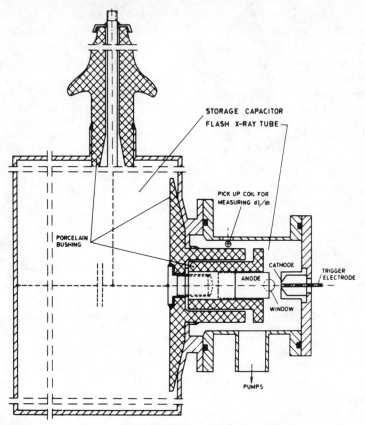

Fig. G5b-1. Cross-sectional view of low inductance flash x-ray system. Loop inductance about 70 nH. Storage capacitor 50 kV, 0.4 μF. ■ Q-ring gasket; ▲ soldered seam.

The tungsten bridgewire, length 4 mm and diameter 0.05 mm, used for discharge initiation was suspended between the edges of two brass pins. These terminals were fixed by Araldite glue in the center of a perspex bushing, which could easily be plugged in the hollow cathode. The bridgewire was arranged parallel to and about 2 mm beyond the cathode surface. The "wire plasma" produced during the wire explosion was fed through a pinhole, diameter 3 mm, in the cathode center. A stored energy of about 1 J was required to explode the tungsten wire in atmospheric pressure. At the low pressure maintained in the flash x-ray tube, however, an energy of 0.8 J was sufficient to vaporize the wire. The bridgewire circuit was also coaxially arranged and its loop inductance was calculated to be about 230 nH. The storage capacitor (10 kV, 0.1 μF) as well as the spark gap were kept in a brass can. The stored energy released was fed via a coaxial cable RG 8/U to the bridgewire.

The most important result received from this investigation is the fact that the hard x-ray pulse occurs at the same time from discharge to discharge, i.e., the trigger-to-x-ray pulse time is reproducible.

The advantage of the use of an exploding bridgewire as trigger electrode instead of a conventional trigger pin is that the first stage of the discharge is reproducible, a feature which gives a constant trigger-to-x-ray pulse time, and the "trigger delay time" (i.e., the time lag between $t = 0$ and the first significant rise of the current), is shorter.

c. Exploding Wire Light Sources

Canada et al. [1043] used a modification of the usual exploding wire light source. Shadowgraphic techniques used in the high speed photography of shock phenomena in transparent media typically require a collimated light field, which is conveniently produced by a point source and a suitable lens. The point light source is commonly provided by exploding a small wire in air. An improved variant of this method optimizes the environment of the wire (in this case gold) by immersing it in a transparent fluid which is able to retard the expansion of the plasma and thereby secure a more intense point source of longer duration. Data characterizing the fluid/wire system are presented in their paper, and measurements of intensity and spectral output are reported.

This variant of the exploding wire light source is capable of providing peak illumination equivalent to an argon bomb and exhibits greater sensitivity for resolving low pressure shock waves. Further, the improved source does not produce destructive blast waves characteristic of explosive light sources. The disadvantages of the light source are limitations on luminous duration, field size, and variation in luminous flux. However, the modified exploding wire light source provides increased duration, greater luminance, and smaller source diameter than the conventional technique. The fluid/wire technique is relatively inexpensive, simple to apply, and requires firing circuitry generally associated with high speed photographic installations.

d. Exploding Wire Shutters

Wire explosions have also been applied for the opposite purpose, i.e., as a shutter. Borucki [1280] used a Kerr cell in the well-known manner for nanosecond spectroscopy but closed the optical path after the operation by means of an exploding wire shutter of special construction. Because the sheet polarizers that are used with Kerr cell shutters all have a small residual transmission even when closed, typically with peak near 4000 Å, a capping shutter is needed if the process is accompanied by a bright and prolonged event.

Operation of the shutter may be understood by examining Fig. G5d-1. When the wire is electrically exploded, a shock wave is driven down the confining channel into which an aluminum foil flap has been placed. The flap

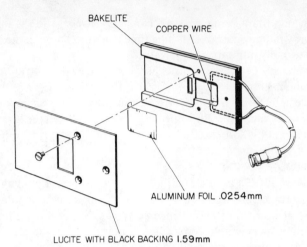

Fig. G5d-1. Sketch of the exploding wire shutter.

swings out of the path of the shock wave on its attached edge and covers an aperture in front of the spectrograph slit. The measured closing time is approximately 27 μsec with a 44-μsec delay and a 7-μsec jitter in the delay.

The German company, Impulsphysik of Hamburg, describes in an industry paper [1310] and in a related patent [1304] an exploding wire shutter of small dimensions intended for the protection of human eyes against the light flash of small nuclear explosions. This shutter uses critically dimensioned capacitor energy to explode brass or lead wire and thus, in about 30 μsec, to blacken opposed glass surfaces of an optical chamber maintained at about one tenth of normal atmospheric pressure. Just 200 J are necessary to start the procedure without generating an undesirable light flash during vaporization of the zigzagged wire.

6. High Temperature Plasma Generation by High Energy Capacitor Discharge

a. *The World of Plasma; Plasma Behavior*

An excellent general introduction to the world of plasma as the "fourth state of matter" has been written by Grad [1308]. The vaporization of an exploded wire, dealt with in Section 5, is one way of generating high plasma temperatures by capacitor discharge. It can also be done by induction using methods similar to those mentioned in Section 4. Hollister [1309] realized an electrodeless xenon plasma with the idea of obtaining a long-lived light source of very high intensity.

Short lifetimes, particularly at power levels above 1 kW, have plagued system designers employing compact-arc xenon lamps. The factor most common to each failure mode is the electrode itself. Not only do the electrode seals tend to rupture, but metal boils off the electrode and darkens the envelope. These, as well as other problems limit lamp life to a hundred or so hours at the very high power levels. Such problems are nonexistent in a new type of xenon lamp that has no electrodes. This lamp depends instead upon a high frequency electrical current conducted through its body like a single turn coil to form an arc within the body of the lamp. Having no electrodes, this lamp has inherent advantages over the conventionally constructed lamps including longer life, design flexibility, and higher power outputs. Basically the lamp consists of a loop of wire wrapped about a quartz chamber in which a gas (xenon, for example) is sealed under pressure. A high-frequency sinusoidal current passed through the wire establishes a high-frequency magnetic field which, in turn, induces a high-frequency voltage,

$$V = 2\pi f B_\perp \times \text{(area of loop)}$$

where $B_\perp$ is the component of the magnetic field that is normal to the plane containing the loop, and f is the frequency of the field.

As stated by Faraday's law, the spatial variation of an induced electric field E related to the time-rate-of-change of a magnetic flux density B is

$$\operatorname{curl} E = -\dot{B}$$

Thus the electric field associated with this voltage is azimuthal, and the first equation shows that its intensity at any point is proportional to the radius at that point. This induced field reduces to zero on the central axis, and increases radially outward to a maximum at the surface of the current-carrying wire. If the frequency of the current is held constant while the current—hence the magnetic field—is increased, the induced electric field intensity rises. Eventually, a free electron inside the chamber is accelerated to ionizing velocity in one-half cycle of the alternating field and gas breakdown occurs through the usual avalanche processes. Likewise, the current could have been fixed and the frequency increased to achieve the same result. As in the case of dc breakdown at moderate pressures (i.e., beyond the knee in the appropriate Paschen curve), a pressure increase causes an increase in the breakdown field strength requirement. At high pressures, initial electrodeless breakdown usually is accomplished by employing both high-frequency and high-voltage techniques. This lamp can be used not only for flashes but together with an oscillating feeder as a steadily burning light source.

A Russian contribution on plasma behavior during rapid variations was presented by Kiselevskiy and Minko [1641].

Extensive information on pulsed plasma dynamic processes related to formation and interaction of plasma streams with a surrounding medium and obstacles was obtained with the help of high speed photography and spectrography. The wave structure of pulsed supersonic underexpanded erosive plasma jets was studied. Some physical processes due to interactions of laser radiation with the laser-produced erosive plasma and of this plasma with a surrounding medium were investigated.

An installation with a good time and space resolution for the interferometric investigation of dense nonstationary plasma was described by Basov *et al.* [1641]. The installation consisted of a Mach–Zehnder interferometer, an electrooptical image converter camera and a ruby laser with an impulse of variable duration of 1 to 150 nsec. A classic experiment was that of Jahoda *et al.* [1219] with a 570 kJ theta-pinch plasma.

Initial measurements on the plasma produced in the Scylla IV theta-pinch of its electron density, soft x-ray emission, neutron emission, plasma shape and motion were carried out as a function of the initial deuterium pressure and bias magnetic field. Interferograms established electron densities in the range $4-7 \times 10^{16}$ cm^{-3} for the plasma at peak compression and gave azimuthally symmetric plasma shapes, which show the absence of the "rotating flute" instability as well as the absence of large trapped magnetic fields at peak compression. Streak photographs also indicated gross plasma stability. Shadow graphs in conjunction with deflection mapping of a grid pattern showed sharp plasma boundaries. Soft x-ray absorption measurements yielded plasma electron temperatures in the range 400–1200 eV, which were dependent on the magnitude of the reversed bias field and had a gross correlation with the neutron emission. Neutron emissions of 10^7 to 2×10^9 per discharge were observed and depended upon the magnitude of the reversed bias field. Neutron collimation experiments yielded a plasma length of approximately 70 cm in the 1-m mirrorless compression coil. The observed electron temperature, neutron emission, and electron density as functions of the bias magnetic field and filling pressure were consistent with the product of the plasma particle energy and the plasma density being a constant, as required for a $\beta = 1$ plasma.

The Scylla IV device was part of an extrapolation towards the longer times and higher temperatures required for a thermonuclear energy-producing plasma. The experiments reported utilized 0.8 MJ of the total 3.8-MJ energy storage of the Scylla IV system. The 1-m-long 10-cm-diam coil of the experiments was energized by three separate capacitor banks with the main 50-kV bank having an energy storage of 570 kJ. The density and configuration of the Scylla IV plasma were studied by the use of a Mach–Zehnder interferometer with ruby laser illumination at $\lambda = 6943$ Å. In addition, studies of the Scylla IV plasma with measurements of the soft x-ray emission, neutron emission, and plasma motion were carried out as a function of initial deuterium pressure and bias magnetic field. During the entire magnetic half-cycle (7.4 μsec), axial

6. HIGH TEMPERATURE PLASMA GENERATION

streak photographs indicated plasma stability and the absence of the "rotating flute" instability which is characteristic of shorter θ-pinch discharges.

Technical data are given in Table G6a-1. Some details of this apparatus have already been given in Chapter A8, especially Fig. A8a-2. See also [1308, pp. 35, 36].

TABLE Ga6-1

SCYLLA IV PARAMETERS

Parameter	Unit	P.I. bank	B_0 bank	Primary bank	Power crowbar bank
Bank voltage	kV	40	10	50	20
Bank capacity	μF	11.9	5600	454	15,000
Bank energy	kJ	10	280	570	2940
Bank inductance	nH	16.8	210	2.2	~5.5
Total inductance	nH	26.8	220	12.2	15.5
Ratio $\frac{\text{Coil inductance}}{\text{Total inductance}}$	—	0.37	0.045	0.82	0.64
Half-period	μsec	1.5	110	7.4	~50
Electric field	kV/cm	0.4	0.01	1.1	0.35
Maximum current	kA	900	1000	8600	~18,800
Maximum magnetic field	kG	11.0	12.6	93	~230

b. Neutron Production by Plasma

Important results on plasma research have been obtained with comparatively modest means in experiments in which the required current density was obtained in a small volume using correspondingly small energies. A paper by Braun et al. [1649] deals with the influence of spoke-type filaments on the formation and field structure of the focus in a 1-kJ experiment. Filaments may originate from the early stages of the breakdown which occurs across the insulator between the inner and outer electrode of the coaxial electrode arrangement.

Figures G6b-1 and G6b-2 show the two different geometries of insulation that were applied: (a) a flat insulator ring at the bottom and (b) an insulator sleeve along the inner electrode, an arrangement that is often used in focus experiments. The inner electrode is only 12 mm long; inside and outside diameters are 50 and 80 mm.

The energy level is extremely low in comparison with focus experiments which start around 10 kJ. With a low energy level, there is a better chance to observe the varying field structures within the forming focus. The paper shows pairs of 10-nsec pictures taken with two image converters simultaneously, end-on and side-on.

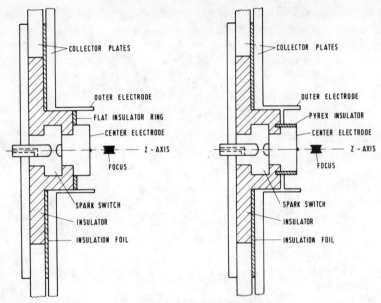

FIG. G6b-1. Geometry of flat insulation.   FIG. G6b-2. Geometry of Pyrex sleeve insulation.

A large number of pictures taken during the time of focus formation demonstrate that on the average the (a) foci appearance is more reproducible; however, intensity structures still prevail in the later phases of the focus decay. Individual (b) foci on the other hand were greatly different in structure and shape from shot to shot before formation time. The foci themselves, however, showed no internal structure after formation; they were rectangular in intensity profile, possibly because of the 5 to 10 times larger optical radiation flux densities. Foci diameters appeared comparable in the two cases.

The tests are gradually being extended up into the 10 kJ region. Filament and spike structures leading to the first appearances of neutrons will be analyzed. Present electron temperatures during the time of the focus as determined by the x-ray filter absorption method are 100–120 eV.

New aspects of the generation of neutron flashes with modest means were presented recently by Fischer *et al.* [1870]. About 4×10^7 to 2×10^8 neutrons were emitted by a plasmafocus with a small 1 kJ energy input (6.7 μF, 24 nH, 20 kV) into a 10-cm long and thin center electrode with 4 Torr deuterium. The neutron emission dropped below the threshold of observation, 5×10^6 neutrons, when the voltage was lowered to 16 kV. The silver activation neutron counter was calibrated by means of a Pu/b standard source (Radiochem. Centre, Amersham, England) assuming isotropic emission from the focus.

The results were consistent over many weeks of observation. Bostick *et al.* have also reported neutron emission at low input levels.

This high-neutron yield at such low-energy input does not appear to agree with the assumptions of a very hot, thermal plasma. Preliminary measurements by means of the x-ray absorption filter method resulted in temperatures in the several hundred electron volt range. At a temperature of 1 keV, a D-ion density of 10^{19} cm^{-3}, focus dimensions of 10×2 mm, and time of emission 100 nsec, the focus will produce 2×10^8 neutrons. However, the yield will drop to 2×10^2 with a 300-eV plasma. Selective x-ray pulses with nanosecond time structures and energies up to about 100 keV seem to appear when neutron emission is observed. The processes leading to the x-ray emission appear rather complex and need further analysis.

High resolution optical image converter photographs and also minute hot x-ray spots (3, 4) taken by a pinhole camera have backed up speculations that the neutron emission is associated with directed ion velocities, as produced in and around the focus by current filaments and hot spot field structures (3,4). The authors admit, however, that in this particular neutron experiment the optical filaments appeared to become obscured by increased background radiation when neutron emission was observed. On the other hand, the hot x-ray spots did appear at this level.

Neugebauer [1876] found previously that a hollow cylindrical beam of nitrogen clusters can be produced by expansion of nitrogen gas cut of a ring-shaped nozzle cooled by liquid nitrogen. An electrical discharge along this beam produces a plasma of corresponding shape. Since plasmas of the heavy hydrogen isotopes in configurations of this kind may be of interest for producing intense neutron flashes by nuclear fusion, it was decided to build a device for production of hollow cylindrical cluster beams of hydrogen. This paper describes the design, construction, and testing of a liquid hydrogen cooled cluster generator that produces a pulsed cluster beam of 40 mm i.d. and a pulse length of several milliseconds. In the discharge region the particle density in the beam is of the order of 10^{10} H_2-molecules/cm^3, whereas the interior contains only $\sim 1\%$ of the total amount of material concentrated in the beam. This should be an advantage for the generation of intense neutron flashes.

c. *Australian Theta-Pinch Experiments*

Five Australian papers on theta pinch should be mentioned: Bowers *et al.* [1943] describe a slow (1.6 msec to peak current) toroidal θ–Z pinch, interest being centered on "transitions" between stable and unstable states. Provided the electron temperature is above a certain critical value ($\sim 3.5 \times 10^5$ °K), the behavior is little different from that of a Tokomak operating in its so-called unstable mode. The discharge goes through repeated instability cycles (~ 400

μsec period) of oscillatory, contraction and expansion phases. The electrons are heated (to several hundred eV max) during the contraction phase and cooled during the expansion phase. Ions are accelerated (to perhaps several keV) during the expansion phase and either lost or "cooled" during the oscillatory phase.

With emphasis on the instability cycle, the basic measurements and distributions are described. An approximation for the magnetic surfaces is obtained, electron and "ion" temperatures as functions of time are determined, and particle and energy containment times τ_p and τ_e are deduced. The over-all hydromagnetic stability is shown to be consistent with the Kruskal–Shafranov limit. Both τ_p and τ_e are ~ 1 msec prior to the onset of a voltage spike. The mean value of τ_p is ~ 2 msec, but the mean value of τ_e is only ~ 100 μsec.

A schematic diagram of the experiment is given in Fig. G6c-1, and the major elements of the electrical circuit are also shown in the paper. The basic element of the experiment is a toroidal vacuum vessel of 10-cm minor radius and 40-cm major radius, fabricated from 16 gauge Inconel sheet and divided into quadrants insulated from each other by $\frac{3}{8}$ in. thick alumina spacers. A $\frac{1}{2}$ in. thick copper shell surrounds the vacuum vessel. This copper shell also consists of insulated quadrants, but is further divided into insulated "top" and "bottom" sectors. A toroidal coil encloses the copper shell, while a second coil, referred to as the primary, is wound circumferentially around the torus. A suitably insulated four-turn ion core links the primary circuit.

The primary and toroidal coils are joined in series and connected via an ignitron switch to a 92-kJ capacitor bank. A smaller bank of 4 kJ is connected, via an ignitron, to the toroidal coil only. The number and distribution of turns on both coils may be varied but nearly all the results presented in this paper were

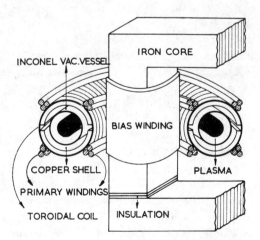

Fig. G6c-1. Schematic diagram of experiment.

obtained with a uniformly distributed toroidal coil of 172 turns, and two particular arrangements of a 4:1 primary to secondary turns ratio. An 8-turn bias coil is wound on the centered limb of the iron core.

The effective magnetic aperture is reduced by stray fields and by field perturbations. The mean maximum vertical magnetic field associated with the bias winding is only ~20 G and, although varying radially across the torus, is uniform to within 10% in the circumferential direction. There are a number of observation ports, these being 4 in. long Inconel tubes $\frac{1}{4}$ in. i.d. × $\frac{3}{8}$ in. o.d. The associated holes in the copper shell are $\frac{5}{8}$ in. diam, this being considered the maximum size which may be tolerated without introducing serious perturbations into the magnetic field configuration. The pump port is a 2-in. diam hole in the bottom of the horizontally mounted torus. During a discharge this hole is closed with a $1\frac{7}{8}$-in. diam, $\frac{1}{2}$ in. thick copper disk, subsequently referred to as the copper-plug vacuum valve. The time constant for the penetration of the axial magnetic field through the Inconel vacuum vessel is 140 μsec. This is short compared with the over-all time scale of the experiment (1.6 msec to peak current). The time constant for the penetration of stray fields through the copper shell is greater than 2.5 msec. Thus, at least up to peak current, the copper shell is essentially the "primary" for both the toroidal and the azimuthal magnetic fields. Even so calculations have shown and experimental observations confirmed that because of various perturbations the radius of the magnetic aperture is only 8.5 cm, appreciably less than the 11.25 cm of the copper shell.

Finally, over the period involved in the present series of experiments, the base pressure in the vacuum vessel varied between $\sim 5 \times 10^{-8}$ and $\sim 10^{-7}$ Torr. In operation, hydrogen gas flows continually through the system, the working pressure range being 0.4–3 mTorr. Initially the two capacitor Banks I and II are charged to preset voltages and the copper-plug vacuum-valve closed, followed by the discharge of Bank I through the toroidal coil. In addition to other effects, this enables a 200-μsec pulse of 2 MHz rf energy to produce an initial degree of ionization. Bank II is then discharged through the primary and toroidal coils in series. At this stage the switch S1 is still closed and the associated Bank I recharging current flowing through the primary coils (and hence the plasma) rapidly ionizes the gas in about 50 μsec. Under certain conditions the plasma may suddenly be extinguished, and to guard against associated voltage spikes the switch S3 is closed once ionization is complete. The elements associated with S3 are referred to as the "bypass" circuit. The final switching operation occurs when the voltage on Bank II swings negative and the switch S4 automatically clamps the main circuit.

The maximum voltage to which both banks may be charged is 10 kV. For this voltage on Bank II the toroidal magnetic field is $\sim$ 10 kG, while for a four-turn primary the maximum axial gas current i_ϕ is $\sim$ 33 kA, with $\sim$ 20% of the

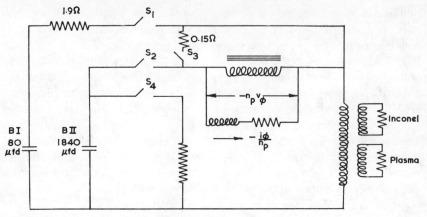

FIG. G6c-2. Basic equivalent circuit.

total current being carried by the bypass circuit. For all results presented, Bank I was operated at a fixed voltage to produce an initial toroidal magnetic field of ~ 900 G. (See Fig. G6c-2.)

Bowers et al. [1943] give seven conclusions that are interesting for plasma specialists. Also of fundamental importance is the basic work by the same author, Bowers [1875].

With the apparatus used by Bowers (Fig. G6c-1), Morton and Srinivasacharya [1945] investigated electron runaway in the slow toroidal θ-Z pinch on the basis of measurement of thick target bremsstrahlung. The method differed from those usually employed for the detection of the presence of runaway electrons in two respects. First, a target was inserted into the plasma to sample the electrons at various plasma radii and, second, the x-ray detection equipment was calibrated in terms of the energy of the electrons producing the radiation rather than in terms of the energy of the radiation itself.

The experimental results indicated that the runaway occurred mainly at a plasma radius of about 4 cm as would be expected from the radial variation in electron density and temperature. The energy distribution was obtained for electrons striking the target and estimates were made of the average runaway rate.

Irons et al. [1944] studied the spectroscopic character and energy balance of pulsed toroidal discharges. Investigations of plasma produced by a pulsed current of 40 kA peak and 20 μsec decay time in a quartz torus of major radius 32 cm and minor radius 4 cm in 30-mTorr of hydrogen are described in this paper. The electron temperature rose in 1 μsec to $\sim 4 \times 10^4$ °K and relaxed by 4 μsec to a value of $\sim 2 \times 10^4$ °K for most of the 50 μsec duration of the current pulse. Corresponding electron densities were 3×10^{15} cm^{-3}, falling to a nearby steady value of 1.5×10^{15} cm^{-3}. At later times the afterglow was followed to 8×10^{13} cm^{-3} at 180 μsec.

Dissipation of the toroidal current provided 15 to 20 times the energy required for full ionization, and temperature equilibrium resulted from atom thermal conduction loss to the wall. Although these losses could have been intrinsic, the observed asymmetries in luminosity and magnetic field showed that MHD instability was present which enhanced the wall contact.

In the proceedings of the Canberra Plasma Physics Seminar, 1972 [1874]* can be seen the yearly progress made towards higher plasma temperatures and densities (see Table G6c-1).

TABLE G6c-1

TOKAMAK PROGRESS

Date	Toroidal magnetic field (kG)	T_i (K)	T_e (K)	n (cm^{-3})	τ (msec)
1958	10	10^5	—	$\sim 10^{13}$	—
1962	20	6×10^5	—	$\sim 10^{13}$	—
1965	30	3×10^6	—	$\sim 10^{13}$	4
1968	35	10^7	—	5×10^{13}	6
1969	35	1.2×10^7	4×10^6	5×10^{13}	20
1969	British laser team confirms T_e				
1970	U.S. Tokamak (Princeton, NJ) gets similar results				
1971	50	2.4×10^7	5×10^6	4×10^{13}	17

7. Plasma Heating by Laser Beam Energy

Direct heating of plasma by capacitor discharges takes place only if the electrodes are in or on the plasma as in the exploding wire case. For magnetic heating of a plasma, the capacitor bank energy must first be transformed into a magnetic field as explained in Chapter H. Finally, the capacitor energy can be transformed with rather poor efficiency into a laser beam which interacts, with extremely high power density, with surfaces or plasmas.

A brief review of this field of research is given here, mainly because much R and D effort is being put into laser-generated plasmas attaining extremely high temperatures, with nuclear fusion as the goal. At the moment it is impossible to forecast which technique will succeed: the trapped magnetically heated plasma, the laser-heated plasma or neither.

Tonon and Rabeau [1641], at the Tenth International Congress on High

---

* This seminar report contains questions and answers and is an excellent tutorial work on plasma physics.

Speed Photography (Nice, October 1972), presented a paper on the development of a plasma created by a laser. They studied this phenomenon with the help of an interferometer joined to a high speed image converter, which assured an exposure time of 1 nsec. The results thus obtained made it possible to envisage an interferometric camera giving eight images of the plasma with a single shot. The time interval between successive images would be 10 nsec and the exposure time 1 nsec.

Hill et al. [1650] used plasmas produced by a focused TEA CO_2 laser, to study the expansion of a gas breakdown plasma. Previous studies have used laser pulses less than 100 nsec in duration. In our case, the TEA (tetraethylammonium) CO_2 laser generated appreciable amplitudes lasting several hundred nanoseconds; in this case a separate interaction process occurs during the laser pulse and a change in mechanism of expansion is observed. The laser was a helical geometric transverse excited CO_2 laser 2 m × 2.5 cm pumped with a two-stage Marx generator giving about 20 J in a 60-kV transient voltage pulse. The pulse focused by a 5-cm focal length mirror in argon at 400 Torr produced gas breakdown: a streak photograph of the expanding plasma, in time registration with the laser pulse, is shown in the paper.

A breakdown plasma appears at the lens focus 100 nsec after the pulse begins, and a nearly unidirectional expansion with an initial velocity of $\sim 10^7$ cm/sec is observed to move in a direction opposite to that of the incident laser beam. After 600 nsec, the plasma has decelerated to zero velocity which implies a much cooler plasma. At this stage radiation penetrates deep into the focal zone due to the lower absorption properties of the cooler plasma; a second plasma is generated nearer the focus which is coincident with the second peak in the laser beam intensity profile. A uniform expansion at $\sim 6 \times 10^5$ cm/sec occurs in both directions and remains constant for ~ 500 nsec.

Hugenschmidt et al. [1524] studied extensively a gas breakdown in xenon produced by a giant pulse ruby laser with a power <100 MW. Detailed information concerning the structure of the laser plasma formation and of the following expansion has been obtained by different optical methods, including Schlieren techniques (single frames and ultrahigh speed cinematography) and holography. The high quality of the holographic pictures was achieved by the use of a monomode laser. With this it was possible to visualize the plasma history and to determine the velocities of the boundary layer and of the blast wave with utmost accuracy. Investigations of the electron density and electron temperature showed that a relaxation time of about 10 nsec is necessary to establish local thermodynamic equilibrium states. After this relaxation time it is then possible to carry out thermodynamic calculations, applying the shock-wave theory, to relate the optically measured expansion velocity with the plasma parameters involved. The mean specific internal energy $\bar{\varepsilon}$, for instance, attained values in excess of 10^{12} erg/g which decayed rapidly during the first 100 nsec to $\sim 5 \cdot 10^{10}$ erg/g. By comparing the results to theoretical calculations of $\bar{\varepsilon}$ a first estimate of the temperature was obtained, taking into consideration the partial densities n_j as well as the partition functions $Z_j^{(i)}$ of the xenon atoms, the single-charged ions, and the double-charged ones. Furthermore, a two-step iteration computer program was used to give more detailed and more accurate results on the variations of the pressure, temperature, partial densities, and enthalpy as a function of time.

A special conference on "Laser interactions with matter" was held in London on January 20, 1972 [1535]. This was a joint meeting of the electronics and optical groups of the Institute of Physics. The emphasis was on what light does to matter and not what matter does to light. The first paper, presented by Heavens (University of York) was entitled "Survey of fundamentals of laser interactions." The interaction of laser radiation with matter can take place over a whole range of energy levels but in his paper Professor Heavens concentrated on high power pulses. With these, the parameters which control the phenomena occurring are radiation flux density, radiation frequency, pulse duration and repetition ratio, pressure and nature of ambient atmosphere, and the type of target material. In gaseous media at densities which are at present available, spark generation results from cascade processes and is dependent on pulse duration. Nanosecond pulses give heated plasma but shorter pulses cause less absorption. With multiple laser pulses, breakdown effects can arise from interaction between the shock wave produced by two successive pulses. In solids the predominant factors in the interaction are thermoelastic effects, localized heat absorption, quenching of fluorescence, the generation of intense lattice waves, and the dielectric breakdown of plasma formed by electron avalanche effects. In many cases the existing data are inadequate to enable a decision to be made on the dominant mechanism in any particular area.

Klewe (Central Electricity Generating Board, Southampton) read a paper on "The optical ionization of alkali metal vapours." The results of measurements made on cesium, rubidium and potassuim led to the conclusion that three-photon absorption processes are responsible for ionization.

Andrews and Pilcher (University of Essex) dealt with "Damage induced in glass by mode-locked laser pulses." Samples of optical glass were subjected to focused laser radiation at varying power levels. A Nd-glass laser which was both Q-switched and mode-locked by a dye-cell gave a train of 10 mJ energy containing 10 to 20 pulses each of 10 psec duration. Microscopic examination before and after irradiation showed the damage consisted of collinear and parallel sets of filamentary tracks with star-like fractures randomly distributed along them. These appeared at high radiation levels but for lower intensities some of these defects decreased or disappeared. Laser-induced damage in KDP was described in a short contribution by Kear of City University.

Hunt (AWRE, Aldermaston) speaking on "High power laser interactions" developed the theme that the mechanisms involving the absorption of radiation and the subsequent behavior of plasma are very much determined by the electromagnetic character of the energy source.

Higgins and Ramsden (University of Hull) gave a paper on "Development of the laser microbrobe" which is a device for microanalysis of samples of dimensions between 10 and 100 μm in diameter. To provide a quantitative analysis is the problem of establishing a systematic relationship between measured line intensities or intensity ratios with the concentration of the trace element in a matrix. A method of analysis was described by Higgins in which the plasma parameters were determined and fitted to a model which could then be used to calculate the percentage composition from measured line intensities.

The "Use of laser-produced plasma as a positive ion source" was the subject chosen by Chowdhury and Miles (University College, Swansea). In many cases applications using such a source are possible only when the various ions present can be efficiently separated. The technique used is one in which it is possible to separate the ions according to their charge-to-mass ratio and it is not energy selective; the separation is accomplished with an electrostatic deflection system on which the applied voltage varies with time. The authors suggested that compositional analysis of materials might be possible.

An interesting application of laser interaction was described by Tozer (Central Electricity Generating Board, Southampton) in his paper "Measurement of steel corrosion in a hot radioactive environment." The instrument described makes use of the fact that a beam from a ruby laser focused on a corroded surface will penetrate down to the bare metal and no further. This is because there is a large increase in reflectivity at the surface. From suitable measurements the penetration depth may be found. The instrument has to be small and run at 100°C in a radioactive encironment of ~ 10 rad/h.

"Picosecond interactions" were covered by Bradley (Queen's University, Belfast). Dr. Bradley described a means for study of resonant interactions on this time scale. The present state of picosecond generation and measurement was surveyed with discussion on ultraviolet and infrared generation and stimulated scattering.

CO_2–H_2–He gas lasers may with advantage be used as a heat source in the floating-zone-crystallization technique for oxides which melt within the range 2000–2400°C. This was the theme of Cockayne's (RRE Malvern) paper in which the problems associated with crystal growth were examined and the feasibility of using the laser for machining oxide-based items not easily machined by conventional methods was discussed and illustrated. Talking on "Some potential industrial and fusion applications of CO_2 lasers," Spalding (UKAEA Culham) initially reviewed the energies required to melt or vaporize common materials. CO_2 lasers with a power of a few watts are sufficient to cut plastics and similar materials. For metals, powers in the range 10^4–10^{15} W cm^{-2} are needed.

The final paper was given by Beach (AWRE Aldermaston). With the title "The interaction of carbon dioxide laser radiation with living tissue," this was a review of laser surgery experiments carried out by the Aldermaston Medical Physics Group which is sponsored by the Department of Health. The experiments, which were illustrated by color film, show that laser incisions in living tissue are kept to 100°C by thermostatic control. The lateral diffusion of heat was investigated using liquid cholesteric crystals painted near the incision: the diffusion is very small. The work calls for great ingenuity in the design of the laser "knife" which has the dexterity of a hand. Further work was suggested using a Nd/YAG cw laser with a frequency-doubled output.

At the VIIth International Quantum Electronics Conference 1972, Montreal,* two contributions may show the focal point of modern activities. Floux et al. [1257] reported on the heating of solid deuterium by a powerful laser pulse.

A fast-rise time neodymium laser pulse was focused on a 1-mm thick solid deuterium target. The pulse width was constant and equal to 3.5 nsec while the output energy could vary from 20 to 80 J. Evolution of the laser-produced plasma was investigated with an STL streak camera. Photographs for 8 and 17 GW incident power show clearly that the lifetime of the luminous plasma is not longer than the total pulse duration (5–6 nsec).

Interferometric measurements of the density gradient have now been undertaken as well as determination of the electron temperature by soft x-ray techniques. Hard x-rays have been detected both on scintillators and BF3 counters before the neutron pulses. This effect occurred in 70% of the shots. The neutron pulses lasted for about 100 nsec compared to the shorter lifetime of the plasma. The main pulse can be interpreted as direct neutron emission while the following tail is in agreement with the authors' measurements of elastically scattered neutrons from experimental items (30%) and the floor of the laboratory (15%). Time of flight is accounted for by nuclear DD reactions.

The neutron yields have been found to vary as E_a^2 where E_a is the absorbed energy. In order to interpret such experimental results, a two-temperature one-dimensional spherical computer calculation was used, which took into account electron conduction as well as the experimental conditions. The fast-rise time laser pulse leads to an overdense heating mechanism up to n_e values close to 10^{22} cm^{-3}. This region, heated by nonlinear heat conduction,

---

* VIIth International Quantum Electronics Conference, May 1972, Montreal, Canada. Sponsor: Joint Council on Quantum Electronics. Report edited and produced by the staff of the IEEE Editorial Department.

allows nuclear DD reactions to be produced when T_i overcomes 0.4–0.5 keV. No neutron emission can occur from the laser-heated region due to the rapid expansion of the plume. The focusing diameter has been varied from 100 to 250 μ. By comparing the experimental reflection curve with the computed ones, it is possible to follow a bidimensional effect, which develops when the incident power is increased. The computed results are in good agreement both with experimental reflected energy data as well as neutrons yield measurements.

The second paper of the same authors [1256] refers to the production of nuclear DD reactions by means of a powerful nanosecond laser pulse focused on a solid deuterium target. The heating of such a created plasma depends on the rise time of the laser pulse, and increases with the output power and the power-flux density of the target. On the other hand, maximum temperatures are bounded by the plasma expansion into vacuum.

A review of the data found in the literature is given in the paper. Both theoretical and experimental aspects are dealt with. Total neutron fluxes and absorbed laser energies as measured experimentally are compared and some discrepancy appears. This can be reduced when considering the geometries of the different experiments. The influence of the lens focal length is specially emphasized. The comparison is supported by the results obtained in Limeil, which are then presented. They are mainly concerned with reflected energy measurements, interferometric plasma density data, and total neutron fluxes measured by scintillators and BF 3 gas counters. The maximum output power was varied from 3 to 20 GW and the pulse length from 1.7 to 8 nsec. The results are conceptually extended to higher laser powers and different focusing conditions. Moreover, discussion about electron temperature determination by the well-known x-ray absorption method is tentatively set up. It is shown that photomultiplier fluctuations are strong enough to explain the larger temperature values experimentally observed.

At last, a numerical calculation based on a 1D spherical two-temperature computer code is developed. Taking into account heat-conduction mechanism and classical inverse bremsstrahlung absorption, this calculation gives an account of the reflected energies and total neutron fluxes experimentally observed.

The ion and velocity structure in a laser-produced plasma was the subject of a study by Irons *et al.* [1888]. The early expansion of the plasma produced by the laser irradiation of a polyethylene foil in vacuum was studied by making time-resolved observations at right angles to the target normal. Spectral line emission was scanned to give the spatial distribution of the ion species carbon I–VI. The lines observed were Doppler-broadened by the ion streaming motion, and this broadening was interpreted to give velocity as a function of time and of distance from the target normal. For one ion species (carbon V), a reconstruction was made of the ion trajectories. Velocities of $2-3 \times 10^7$

cm sec^{-1} were deduced from Doppler shifts observed parallel to the target normal and are in agreement with previous time of flight measurements.

A proposal for overcoming the "plasma mirror" was given by Hughes [1889] who is currently operating a 2000-J 3-nsec Nd laser.

Very high temperature plasmas can now be created from any known element by using the focused output beam of a high power laser system. In the USSR (Basov and Krokhin, 1963; Basov *et al.*, 1968) and France (Floux *et al.*, 1969) particularly, this approach is being developed with considerable success for the purpose of assessing the ability of intense laser beams to trigger thermonuclear reactors. At the somewhat lower power levels attainable at present, very high temperature laser-induced plasmas containing multicharged ions of heavy elements are being studied in attempts to establish their spectra to very short wavelengths. Such spectra will be invaluable to astrophysicists in identifying the complex spectra of stellar sources. At the same time laser-produced multicharged ions could be used to increase the effectiveness of heavy ion particle accelerators.

Unfortunately the effectiveness of existing laser systems in generating very high temperature plasmas ($>10^7$ °C) is seriously impaired by the formation of a "plasma mirror" (Basov *et al.*, 1969) at an electron density of $\sim 10^{21}$ cm^{-3}. This "mirror" reflects part of the laser pulse back into the laser system, where it is amplified in the reverse direction to a level at which optical components in the initial stages of the laser would be damaged (Basov *et al.*, 1969).

Although many techniques, including the insertion of expensive, large aperture magnetooptic and electrooptic switches between amplifier stages, have been used in other laboratories in attempts to overcome the "plasma mirror" problem, to date, the most promising approach has been the plasma optical rectifier technique shown in Fig. G7-1a, used by Kriukov and Senatsky (1970). In their system part of the main pulse is diverted into a special chamber

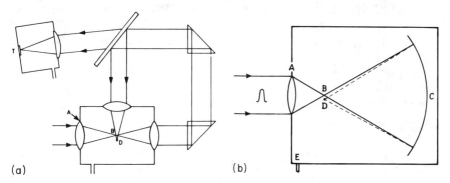

FIG. G7-1. With system (a) the optical rectifier is formed at D using a portion of the primary laser pulse, while in system (b) the optical rectifier is formed by the plasma induced by the primary laser pulse.

inserted into the amplifier chain, in which the diverted pulse creates a "plasma mirror." This in turn blocks the path of the pulse reflected from the target (T) "plasma mirror" preventing it from reentering the laser system.

A system much simpler and more efficient than this, is being incorporated into the high power laser research facility at the Australian National University (Inall and Hughes, 1970; Hughes, 1972). It is shown in Fig. G7-1b. Here the lens A brings the laser pulse to a focus at B, in the target chamber E (evacuated to avoid air breakdown). The laser pulse passes through B without interfering with the solid target D, which is placed within several millimeters of B. The high power laser reflector C brings the laser pulse to a focus inside D, ionizing the pellet target symmetrically and heating the resulting plasma to very high temperature. Any laser light reflected from the "plasma mirror" formed at D moves towards C, some of it being reflected back towards the laser via the path CBA. However, the plasma cloud formed at D, its particles traveling at up to 10^8 cm sec^{-1}, will expand to include B in a period of 10^{-9} to 10^{-8} sec. In this time the laser light reflected from the plasma will have traveled a maximum of 300 cm. Therefore, if the length of the optical delay line DCB is adjustable up to 300 cm, that is, DC is adjustable up to 150 cm, and the electron density of the plasma cloud remains above 10^{21} cm^{-3} for up to 10^{-8} sec, then the "plasma mirror" expanding around D would act as an optical shutter, effectively blocking the returning pulse form reentering the laser. To be effective for laser-induced thermonuclear research applications, the pellet D should be several millimeters in diameter, that is, it should have a particle density $>10^{21}$ cm^{-3}. To attain a thermonuclear temperature laser energies of 10^5 J would be needed, levels which are unattainable at present.

The weakest link in the optical system shown in Fig. G7-1b is the reflector C. Dielectric coated mirrors can withstand a laser power density of 5×10^8 W cm^{-2} in the nonosecond range (10^{-9} to 10^{-8} sec), resulting in an energy density of 2.5 J cm^{-2} for a pulse duration of 5×10^{-9} sec. A dielectric mirror of 30 cm diam could, therefore, withstand a total pulse energy well in excess of 1000 J, that is, more laser output energy than has been used to date for plasma excitation.

Since the volume of the target D is several cubic millimeters, the quality of the optical components is not critical, implying that large diameter optical components, possibly plastics, could effectively be used. However, the reflector C would need to be 2 m in diameter before a thermonuclear reaction could be achieved in the system shown in Fig. G7-1b, unless the loading per unit area of the reflector could be increased. A low energy (up to 50 J) but high peak power (up to 5×10^{13} W) version of the relatively simple system proposed by the author for overcoming the "plasma mirror" problem, is being incorporated into the ANU high power laser. Experiments will be carried out using this sytem to determine the properties of the "plasma mirror" particularly under the conditions proposed by Bunkin and Kazakov (1970) which demand photon number densities in excess of 10^{27} cm^{-3}.

In another paper of basic scientific character, Hughes [1890] has these conclusions:

It is to be expected that laser-induced, exothermic, thermonuclear reactions will be attained in D–T pellets during the next few years. Our aim is to study the physical processes involved in generating such superdense plasmas on the microscopic scale ($V < 10^{-4}$ cm^3) so that information will be available for further developments at Lucas Heights in an effort to meet Australia's future requirements.

In the spring of 1973, the following experiments were started at the ANU [1879]:

a. *High Power Lasers*

Amplification from the 47-mm beam aperture Neodymium glass disk amplifier powered from the HPG has been achieved, and work is currently proceeding to optimize the system to obtain extremely high peak optical powers which are required for the full implementation of experimental programs. Since then, work has progressed from preoccupation with developing the system, to its use in various experiments. The low power stages of the system are already being used in experiments to assess the reflection and absorption of laser light in high temperature, high density plasmas, multiphoton ionization and cascade processes, electron–ion separation in laser-produced plasmas, and on the spectroscopy of highly stripped ions in laser produced plasmas. Complementary studies on laser-produced breakdown in matter (using a medium power laser) have resulted in new infrared recording techniques which provide information on temperature as a function of time and position, enabling a testing of different theories on the breakdown and expansion mechanism of laser-produced plasmas to be carried out. This work is outlined in more detail in the following sections.

b. *Laser-Produced Plasmas*

Initial experiments (along the lines of those described in the first paper by Hughes referred to) involve the excitation of wire targets from the rear with respect to the incident laser beam, rather than from the front (as is standard practice in laser laboratories outside Australia). Using a specially developed target mounting and manipulating techniques for operation under vacuum conditions, it is possible to position a 100–250 μm target to an accuracy of better than 20 μm. In irradiating the targets from the rear, the plasma formed expands into the path of the incoming laser beam, satisfying the conditions required to overcome the difficult problem of "back-reflected" laser light from the high density plasma which has so far beset the application of lasers to plasma heating. In operation, the complete laser pulse passes close to the target on its forward journey, and is then reflected back accurately onto the target, the full pulse being trapped between target and rear reflector. (While these tests were being carried out, it was realized that the same basic system with only slight modification, might be used to provide a "tailored" laser pulse of the type required for the optimum compression of deuterium targets in the quest for "break-even" in self-sustained thermonuclear reactions.)

Initial experiments carried out with available optical components have confirmed the concept in practice. To date, the optical power which has been used to irradiate tungsten targets is estimated as $\sim 10^{14}$ W/cm^2 (0.5 J for 10^{-9} sec per 2.5×10^{-6} cm^2). Surprisingly, relatively little "back-reflected" light from the rear of the target could be detected with the apparatus used, and this very favorable situation is being investigated to determine the distribution of the light scattered from the high-density plasma.

c. *Multiphoton Ionization and Cascade Processes*

Experiments designed to examine in detail multiphoton ionization of "isolated" atoms by intense laser pulses and to examine cascade processes when the atoms are not "isolated" are being developed. A laser beam of nano to picosecond pulses is focused onto an atomic beam (thermal), the products of the interaction being examined after ion–electron separation and ion acceleration. Combined with measurement of the laser photon flux and atomic beam density, cross sections for multiphoton processes can be measured. An ion analysis system allows the number of ions of each charge state to be measured. As the atom density in the beam is increased, cascade ionization processes will begin and the transition can be studied.

d. *Electron Ion Separation in Laser-Produced Plasmas*

An experiment to determine whether significant charge separation will occur in the early stages of a laser-produced plasma when a very high dc electric field is applied radially on a plasma is being attempted. Apart from its relevance to ion source technology, this may produce other interaction changes in the plasma expansion. A high field ($>10^7$ V/cm) is produced at a sharp pointed target surface whose end radius is ~ 5 μm, standing at a potential of $V_0 = 10^4$ V. On this target is focused an intense laser beam pulse which produces expanding plasma and hopefully, electron–ion separation in the early stages of the plasma expansion. An ion analysis and detection system similar to that used in the multiphoton ionization work is employed, but energy as well as charge state analysis is performed, and some applied voltages, V_0, need to be pulsed.

e. *The Spectroscopy of Highly Stripped Ions in a Laser-Produced Plasma*

The plasma formed when a high power laser pulse is focused onto a solid target *in vacuo* offers certain almost unique opportunities as a light source for spectroscopy. For example, any element can be used as a target material (that is, the plasma is not restricted to those elements which are available in gaseous form), and with the high electron density ($> 10^{19}$ cm^{-3}) and high electron temperature (> 100 eV), it is possible to strip atoms of many electrons, down to and including those in the most tightly bound, innermost electronic states.

The program involves studying the spectra of atoms stripped of all but one or two electrons. Initial study is with laser pulses of up to 5×10^9 W available in a 5×10^{-9} sec pulse from the final rod amplifier of the high power laser system on the second sequence elements, lithium to fluorine. As higher powers from the disk amplifier become available, the study will be extended. Topics of interest include:

(1) Spectral line identification, leading to a more precise determination of atomic energy levels.

(2) Stark broadening, particularly of lines from high quantum levels with unit charge of principal quantum number, leading to a better understanding of the processes of stark broadening in these particular transitions which are of interest in connection with the n_a lines of hydrogen observed in the radio spectrum of certain nebulae.

(3) The contribution of recombination to the density decay, leading to a better appreciation of the potential of a laser-produced plasma as an ion source, and to a comparison with theory.

f. *Laser-Produced Breakdown in Matter*

Project objectives include the development of experimental techniques and the extension of theoretical understanding of nonlinear optical processes such as breakdown in matter, in preparation for experiments planned with the high power laser system. Because of operation of this system in the infrared region, it is also necessary to develop suitable recording techniques.

Following completion of the experimental setup, the heating and expansion of laser-produced microplasmas in gases has been investigated in detail, the results being analyzed with the assistance of a series of computer programs. Important parameters in such plasmas are temperature and particle density. Theoretical investigations on the mechanism of laser light absorption in these plasmas has led to the concept of using a probe beam to measure the absorption, so allowing calculation of plasma temperature and an estimate of its density, if the plasma emission at the particular probe beam wavelength is monitored. Only a very general assumption (local kinetic equilibrium) has to be made for the temperature measurement. To obtain the density, however, the absorption and emission mechanism needs to be specified.

Focusing the probe beam and spatially filtering the plasma emission allows the probing of the microplasma at different positions. From the photographically recorded emission and absorption versus time traces and the known position, temperature as a function of time and position can be calculated. This has been achieved with a series of computer programs and novel picture processing techniques developed by the Information Science Group, which allow information to be fed into the computer directly from the photographs, and resulting in computer-plotted pseudo-three-dimensional diagrams as output.

It is planned to use these results for testing different theories on the breakdown and expansion mechanism of laser-produced plasmas.

g. *Magnetic Confinement of Plasmas*

Over-all objective of this project is to produce a valid description of a magnetically confined high-temperature plasma in a toroidal system. Emphasis is placed on investigations of plasma properties and behavior which have relevance to C.T.R. The work is in complementary association with the laser-produced plasma studies, which is preceded historically; collaboration between the corresponding groups is expected to be mutually advantageous.

The current experimental programs are based on the LT-3 version of the Liley Torus, a Tokamak-like device which came into operation as LT-1 in 1965. The aims of the main experiments in this program are:

(a) the investigation of magnetic surfaces associated with plasma confinement in Tokamaks,

(b) the measurement of the toroidal magnetic field within the plasma, and

(c) the estimation of ion energies or collective motions of ions in the plasma.

There is a continuing program of spectroscopy employing spectrographs, monochromators, and multichannel devices referred to as filterscopes (which are essentially systems of dielectric interference filters and photomultipliers used to measure radiation from oxygen ion impurities present in the plasma). Spectroscopic observations are used to monitor the general condition of the plasma, to obtain reasonably direct information about its geometry, and to obtain estimates of the electron temperature.

H. Conversion of capacitor energy into magnetic fields

1. Pulsed Magnetic Fields, Concentration

Since the publication of "High Speed Pulse Technology" (1965), not many new ideas have been realized for the generation of high magnetic fields.

As a commercially available item, Advanced Kinetics, Inc. [1646] developed a high field pulsed electromagnet for applications in plasma and nuclear physics, optics and solid-state experiments, magnetic metal forming, magnetochemistry and biophysics. The system consists of multiturn solenoids which are energized from capacitor banks. Pulse times can be selected from the microsecond to millisecond region. The available bore diameter is 0.6–7.5 cm with a coil length of 2–60 cm. Electrical insulation is provided for operation up to 10 kV (up to 20 kV in long coils).

This series of pulsed coils was developed to provide transient magnetic fields up to the region of 500,000 Oe. The transient pressures produced require a careful selection of coil material and housing. The HFM coils are constructed of the most appropriate type of beryllium–copper subjected to special heat treatment to optimize tensile strength and electrical conductivity. The retaining shells are of massive stainless steel, and use vacuum-cast, glass-impregnated epoxy for insulation. Flexible current cables allow mounting of the coils in any position and within any geometry of associated equipment. Two basic coil types are offered: The open-bore (HFM-O) for optical experiments or other uses requiring unobstructed line of sight, and the single-end open-bore (HFM-C) which is accessible for insertion of samples only from one end. Typical specifications are shown in Table H1-1.

Another kind of a pulsed magnet from the same manufacturer generates strong fields over plane surfaces. The main applications of these high field pulsed spiral magnets [1646] are plasma physics, magnetoacoustics, magnetic hammering, magnetic impacting, and magnetic acceleration. The system consists of a multiturn massive spiral in the form of an integral retaining shell. Current coupling is provided via coaxial cables. Available frontal diameters are from 10 to 100 cm (40 in.). The coil is insulated for operation at 10 kV (20 kV in large coils).

This series of pulsed coils was developed to provide transient magnetic fields up to the region of 250,000 Oe. The field is available in a circular plane at the face of the coil. The magnetic field lines are directed radially from the coil center to the outer circumference. Enclosed flux return path is provided. Typical rise times are from 10 to 500 μsec depending on inductance of configuration and capacitor bank size. The transient pressures produced require a careful selection of coil material and housing. The HFSM coils are constructed of the most appropriate type of beryllium–copper subjected to special heat treatment to optimize tensile strength and electrical conductivity. The shells are of massive stainless steel and vacuum cast glass-impregnated epoxy for insulation. Specifications of the HFSM coils are shown in Table H1-2.

TABLE H1-1

DETAILED SPECIFICATIONS OF PULSED ELECTROMAGNETS[a]

Magnetic field intensity	300,000 Oe maximum
Repetition rate	1 pulse/10 min convection-cooled coil
	1 pulse/3 min liquid-cooled coil
Maximum operating voltage	
coils up to 10 cm bore length	10,000 V
coils 10–60 cm bore length	20,000 V
Bore diameter	0.6–7.5 cm ($\frac{1}{4}$–3 in.)
Bore length	2–60 cm (1–24 in.)
Inductance	0.5–100 μH
Flux return path	Internal within retaining shell
Bore type	Open-bore type HFM-O
	Single-ended open-bore type HFM-C
	Bore with nylon insulator cover
Cooling	Convection- and water-cooled types
Cable connectors (coaxial type)	4–20, according to bore size
Linear turn density	1.5–5 turns/cm
Mechanical pressure (radial compression)	Up to 150,000 psi

[a] Adkin [1646].

Still another method is available, mainly for high field experiments using variable bore size. The instruments are named "high field flux concentrator magnets" [1646]. The system consists of a massive pulsed flux generator with retaining frame. It has replaceable bore inserts of diameters from 0.6 to 10 cm. The insert bore length is 0.5–20 cm. Air or water-cooled models are provided.

This series of coils was developed to provide transient magnetic fields up to 250,000 Oe in bore diameters of various sizes. Simple replacement of an appropriately machined insert of a specially heat-treated beryllium–copper alloy allows experimentation with bore sizes ranging from the diameter of the unloaded flux generator coil (10–20 cm) to the smallest insert bore of 0.6 cm. The field intensities available by changing bore size can be evaluated approximately from flux conservation. Typical field rise times are from 10 to 100 μsec depending on coil inductance and capacitor bank size. The inserts are available with constant diameter cross section and must be contoured to achieve required magnetic profiles. The flux generator is heavily compressed in a retaining frame which can be used in any position and at any separation from the energy storage source. Flux generator coils are connected to capacitor banks to produce the energizing current pulse (see Table H1-3).

TABLE H1-2

DETAILED SPECIFICATIONS OF PULSED SPIRAL MAGNETS (ADKIN)

Magnetic field intensity	250,000 Oe maximum
Repetition rate	1 pulse/5 min convection-cooled
	1 pulse/min water-cooled
Maximum operating voltage	
coils up to 6 in. diam	10,000 V
coils 6–24 in. diam	20,000 V
Frontal diameter	10–100 cm (4–40 in.)
Inductance	1–100 μH unloaded
	0.1–10 μH loaded with conducting plate
Typical magnetic fields with 12,000 J capacitor bank	
frontal diameter 15 cm (6 in.)	40,000 Oe unloaded
	130,000 Oe loaded with conducting plate
frontal diameter 30 cm (12 in.)	14,000 Oe unloaded
	65,000 Oe loaded with conducting plate
Flux return path	Internal only
Cooling	Convection- and water-cooled types
Cable connectors (coaxial type)	1–14
Radial turn density	1–8 turns/cm
Mechanical pressure (normal to working face)	Up to 50,000 psi

A very high field magnet taking several megajoules has been constructed at the ANU (Australian National University, Canberra). In view of the available energy from the ANU homopolar generator (up to 418 MJ [1891 p. 2 ff.]), the design was based on an input voltage up to 800 V only. The features of the high field magnet laboratory at the ANU have been described in detail by Carden [1891].

Very high magnetic fields are generated exclusively by water-cooled "air-cored" electromagnets. Iron-cored electromagnets are limited by the magnetic saturation property of ferrous materials to ~40 kG. Superconducting electromagnets, devices which make use of the fact that certain exotic materials at very low temperatures have zero electrical resistance, are limited to just over 100 kG. Above this figure the most economical means of generation

TABLE H1-3

DETAILED SPECIFICATIONS OF HIGH FIELD FLUX CONCENTRATOR MAGNETS (ADKIN)

Magnetic field intensity	250,000 Oe maximum
Repetition rate	1 pulse/5 min convection-cooled coil
	1 pulse/30 sec water-cooled coil
Maximum operating voltage	10,000–20,000 V, according to coil size
Inductance without insert	1–20 μH, according to coil size
Inductance with insert	0.5–5 μH, according to coil size
Typical magnetic field with 18,000 J capacitor bank	
Insert dimensions:	
i.d. 2.54 cm (1 in.)	125,000 Oe
length 2.54 cm (1 in.)	
i.d. 1.27 cm ($\frac{1}{2}$ in.)	210,000 Oe
length 2.54 cm (1 in.)	
Flux generator coil bore diameter	10–20 cm (4–8 in.)
Axial length	7.5–20 cm (3–8 in.)
Ratio of flux generator	3:1 short coils
Bore length to insert bore length	1.5:1 long coils
(optimized for efficiency)	
Insert bore diameter	0.6–10 cm ($\frac{1}{4}$–4 in.)

appears to be with copper or copper alloy conducting material formed into a solenoid configuration. For other than very short duration fields, water cooling is essential. Electric current densities of up to 40 kA/cm^2 and power densities of up to 3 kW/cm^3 are common. Field intensities, dimensions, and powers of the largest available electromagnets are shown in Table H1-4.

TABLE H1-4

DATA FOR THE LARGEST HIGH FIELD MAGNETS[a]

Country	Number of high field magnet laboratories	Highest available field (kG)	Highest available power (MW)	
			continuous	pulsed
Australia	1	165		300 for 1 sec, 5 for 1 min
Europe (excluding USSR)	2	90	4	
Japan	1	120	4	
UK	3	130	3.5	
USA	7	205	8	32 for 4 sec
USSR	?	?	?	?

[a] See Ref. [1967].

A particularly successful design of a magnet has been developed by an American, Francis Bitter. The Bitter design is essentially of a single layer solenoid. In its simplest form, each turn consists of a set of thin copper disks, each with a central hole and a radial slit. The disks are interleaved with each other and with similar disks of insulation to form a continuous helix as illustrated in Fig. H1-1, the overlap between adjacent disks being up to 90°. Specially shaped end plates are used to make contact with the end disks, thus facilitating connection to the power source and providing a means of axially clamping the stack of disks together. On energizing such a stack, an accessible magnetic field occurs in the central hole, largely in the axial direction. More recently, Carden [1892] published information on the testing of the ANU 30 T high field magnet at Canberra.

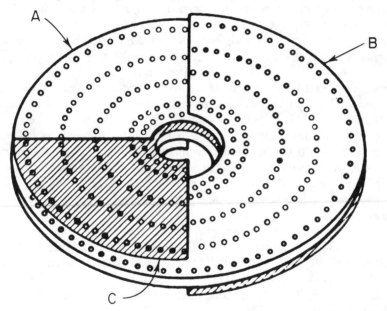

FIG. H1-1. One turn of a typical Bitter disk comprising two copper disks A and B and one insulating disk C, all split and interleaved. Holes are for water cooling.

Interest has been shown in applying the inner coil design to the inner coils of "hybrid" combinations with superconducting coils. It appears that there are limitations to Bitter disk type constructions in this situation where the requirement is for compactness and high strength.

Since the magnet was first energized in September 1970, a total of 370 cycles have been produced, 72 of which were above 20 T and 15 above 27.5 T. One quarter of these have been used in experimental programs involving the study of magnetic phase transitions in Cr, V_2O_3, and V_2O_3 doped with Cr.

Early this year a department of solid-state physics was established largely to extend the use made of the high field facilities. It is hoped that the new magnet will enable members of this department, and interested physicists elsewhere, to investigate new and rewarding areas of research not previously accessible.

The instrumentation of the ANU 300 kG magnet was described by Carden and Whelan in 1969 [1893]. The main problems were concerned with displacement caused by mechanical forces between the conducting disks. In addition, the required flow and heat exchange of the high-pressure cooling water had to be determined by exact measurements.

In its latest version, the ANU 30 T electromagnet has an inner and an outer solenoid, each having special problems in design and construction as described by Carden in three papers [1894–1896].

One paper [1896] gives an account of the major considerations in design especially with reference to the subcoils. Studies were made of the electromagnetic forces associated with every part of the current path, the heating and temperature of each part, the flow of water through each passage as well as the practicality of manufacturing each part and the means of assembling the parts. Most of these studies were concerned with steady-state energization but considerable attention was also paid to transient energization. In many complex situations mockups were constructed and studied under stimulated conditions. The conclusion that might be reached now following successful testing is that the design is fundamentally sound. Given this conclusion, Carden believes that the same methods may be applied to produce more intense magnetic fields in conventional electromagnets and to produce compact and efficient solenoids for operating within superconducting solenoids. The methods may be applicable to superconducting solenoids themselves now that the magnetic fileds being achieved by these devices are sufficiently intense to cause designers to give close consideration to the mechanical stresses within them. Copper wires containing filamentary superconducting material seem to be ideally suited for forming into subcoils while the enhanced access for helium between subcoils might alleviate problems with stability and rapid field changes. The subcoils might of course have to be energized in series necessitating layers of insulation between them, but this requirement need not cause great difficulty.

The ANU report for 1972 [1879] mentions that current control using feedback methods was used for the first time in the laboratory to achieve fields higher than 20 T stable within 1% for periods up to 3 sec. Using these fields, Dr. E. R. Vance of the Department of Solid-State Physics photographically measured the longitudinal Zeeman effect on the strong absorption lines due to U^{4+} in zircon. Dr. R. Dingle of the same Department commenced experiments to measure the reflective characteristics of PbI_2, GaP, and GaAs, at liquid helium temperatures in fields in excess of 15 T.

2. Electromagnetic Plasma Propulsion

Like any gas, a plasma can be described by conventional macroscopic parameters such as pressure, temperature, density, and so on. But, being significantly ionized—above 1% is a rough rule of thumb—a plasma conducts electricity far better than any ordinary gas. It therefore behaves in a much more anisotropic way when in the presence of a magnetic field. Theoretical treatment of this anisotropic behavior predicts the occurrence within a plasma of many intriguing phenomena unknown in other states of matter. Some of these are of theoretical interest in astrophysics, geophysics, and communications. Others offer practical possibilities, among them MHD power, microwave generation at gigacycle frequencies, and controlled thermonuclear fusion Gottlieb [1326] shows how a plasma is different and how these differences determine applications (see also the tutorial papers by Spitzer, Jr. [1329] and Frieman [1331]).

Boreham [1897] gives an explanation of a repetitive impulse device which employs a traveling conduction wave to accelerate the gas. This device is aimed at producing high gas velocities with reduced average heat transfer. (In comparison it has been found in electric arc heater experiments that rotating the arc over cold electrodes considerably reduces the local average heat flux to these electrodes.) Also, relatively high mass flow rates would be obtainable. The main problem, however, is that although high peak values of velocity are obtained, the intermediate values are low, and therefore the average effect must be determined. In order to do this a simple system has been analyzed, and expressions obtained for the specific impulse, thrust, and power.

A schematic diagram of the device plus annular arc injector is shown in

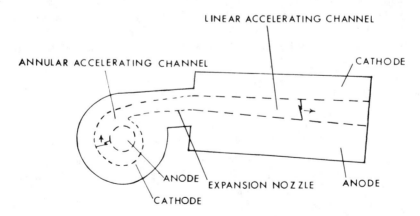

FIG. H2-1. Traveling conduction wave accelerator schematic.

Fig. H2-1. The main accelerating section is a linear channel consisting of a pair of electrodes separated by insulating side walls. A constant voltage is applied across the electrodes and perpendicularly to this, a constant magnetic field is maintained, as in a conventional $j \times B$ configuration. (The actual values of the magnetic field which would be required, however, are expected to be much lower than those used in a conventional $j \times B$ device.) The traveling conduction wave is presumed to have a square wave shape which is produced by introducing to the channel a series of ionized gas slugs with a known constant repetition frequency. These slugs are separated by regions of cooler gas with negligible electrical conductivity, thus completing the square wave shape. As each ionized slug enters the channel an electric arc is struck and subsequently accelerated by the applied Lorentz force. A constant current is maintained through the arc by means of a series inductance. It is envisaged that a rotating arc accelerator would be used as the source of both the ionized gas slugs and the gas to be accelerated.

The performance of the device, including the influence of the linear channel length with plasma speeds of up to 10^5 m/sec, can be seen in Fig. H2-2. It may be useful, while reading Ref. [1897], to consider the state of the art 10 yr ago as presented in the paper of Smy [1048], which describes the design and performance of a low attenuation electromagnetic shock tube. At that time 10^4 m/sec at 20 kV and 50 μF was the highest available speed with argon, and the paper reached the conclusion that the shock driving mechanism was predominantly electromagnetic in nature but that at low downstream gas pressures ($\leqslant 1$ Torr g), this mechanism was augmented by the wave action which accompanied energy heating of the driver gas.

Schmidt and Kaeppeler [1328] described a gas dynamic shock tube for studying plasma–magnetic shocks propagating into a not fully ionized plasma of relatively high density. Two consecutive shock waves were generated in argon at 5 Torr g initial pressure. The first shockwave with a Mach number of $M \approx 10$ was used for ionizing the argon gas in the test section where a transverse magnetic field of 15 kG was switched on. The second shock wave entered this test section shortly behind the first. The jump of the magnetic field was to be measured by magnetic probes. With the aid of a Mach–Zehnder interferometer the electron density and neutral gas density were determined. Schmidt and Kaeppeler discuss a photoelectric method for registering the fringe displacement, permitting a high-time resolution.

In a continuation of these experiments, the transverse magnetic field in the test section was to be applied. The section of the shock tube in the range of the magnetic field was constructed of an appropriate plastic. For the magnets, two Bitter coils were used in a Helmholtz arrangement with 40 windings each made of 0.5-mm copper sheet with 0.2-mm insulation between. The interior diameter of the coils was 80 mm, the exterior diameter 200 mm. The coils

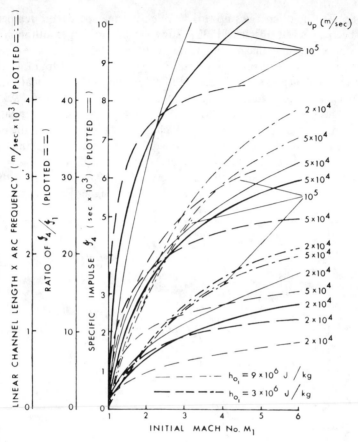

FIG. H2-2. Variation of specific impulse, ratio of final specific impulse to initial specific impulse, and linear channel length with initial Mach number.

were switched to a battery of 20 kV and 12.5 kJ. It was contemplated that the strength of the magnetic field in the test section would be varied in order to ascertain at what magnitude of the magnetic field plasma magnetic shocks vanish completely.

Moving or rotating plasmas also lose energy by electromagnetic radiation. The radiation of a rotating, charged particle in a plasmon field with an external magnetic field was the subject of a paper by Räuchle [1327]. The spectral intensity distribution of the radiation which is emitted by a rotating electron in a magnetic field under the influence of statistical fluctuations of the electric field has been calculated. It is shown that it depends on the Fourier transform of the fluctuating plasma potential. One obtains maxima of the spectral distribution function at harmonics of the cyclotron frequency.

Even with small energies up to 1 kJ, high plasma speeds and temperatures have been achieved. Fischer [1733] showed this in his paper although he was more concerned with plasma filament structures.

High resolution (0.1 mm), 5–10 nsec image-converter photographs demonstrate the existence of strongly directed fields extending into a small energy of 1 kJ plasma focus, resulting from filaments which can be traced back to the early stages of the predischarge. Neutron emission appears to originate from areas of minute size, hot x-ray spots which can be observed within the filament structures. Such microscopic details would be obscured by background radiation with larger levels of applied energy within 10–1000 kJ.

Finally, plasma propulsion has an increasing interest in the field of space vehicles. As speeds of plasma particles are much higher than those resulting

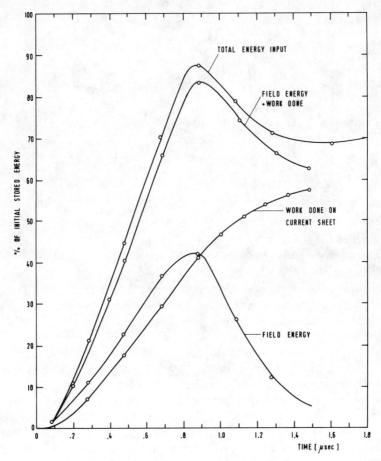

FIG. H2-3. Energy inventory for the pulse line.

from chemical processes, space vehicles making long trip can save fuel weight by using ion source materials. As early as 1965, Gooding *et al.* [1325] evaluated the pulsed coaxial plasma gun as a propeller for space propulsion applications. The pursuit of efficiency had led to the development of a low loss, light weight energy storage capacitor which exhibited pulse line behavior. An energy inventory shows that the plasma resistive losses are small and that the accelerator efficiency is not limited by poor energy transfer from the energy source to the plasma. The thermal efficiency (exhaust energy/stored energy) has been increased to 45% at an average exhaust velocity of about 7 cm/μsec. A theoretical discussion shows that a significantly higher efficiency cannot be expected with the present mode of operation, but that two other modes are available which in principle have no such limitation. Figure H2-3 shows the energy inventory for the pulse line. It can be seen that in theory extremely high efficiencies are obtainable.

3. Charging of Permanent Magnets

At first sight, there would appear to be no difficulty in charging a magnet. In practice, the component to be magnetized has to be brought into the center of a magnetizing coil. A unipolar current pulse would result in a perfect magnetic charge if the hard magnetic material e.g., a sintered magnet had no electrical conductivity. But, in a conductive magnet, a current is induced by the rise and decay of current in the magnetizing coil, and this results in an incomplete charge. The only way to prevent this difficulty with pulsed magnet chargers is to cause the current to decay exponentially, so that all time derivatives of this current also decay exponentially. The decay must be exponential (i.e., rate of decay proportional to e^{-t}) since critical damping is realized. Schmied [1898] considered this problem theoretically and designed a relatively small magnetizer known as the "Magneto–Impulsa Q" [1333] in which a type of a Q-switch is used by adding the following half-wave of a beginning oscillation to the first pulse shape of the magnetizing current in such a way that an almost exactly exponential decay is obtained. Without this exponential decay, he found on the poles of advanced types of magnet inhomogenities caused by induced secondary currents in the magnet. With his circuit, precision measurements show an entirely homogenous field strength. He found that the pulse duration of the current must be longer than

$$T = 0.8 D^2 (B/H_s) \cdot \rho$$

where D is the maximum diameter of the magnet, H_s the saturating field strength, B the induced field in the magnet, and ρ the electric conductivity of the material. To obtain the highest magnetic remanence, it is necessary to ensure that

$$dH/dt < 1 - H_s(dH/dt)_K$$

where H_s is the minimum required magnetic field strength and the expression $(dH/dt)_K$ the critical value of variation of field strength. If this value is exceeded, parasitic currents cause unacceptable demagnetization.

With Oerstit 450 (made by Deutsche Edelstahlwerke, Dortmund) and Koerzit, the permissible value of $(dH/dt)_K$ is 5×10^6 A/cm·sec. The circuit provides that a capacitor discharge feeds a special pulse transformer in such a way that all the energy is stored in the transformer, a silicon diode being used to prevent the energy flowing back into the capacitor. The discharge of the transformer takes place as an exponential decay. The transformer must be dimensioned to suit the largest magnet to be charged. The result is that even very anisotropic hard magnetic materials of Al–Ni–Co base can now all be charged without loss due to induced current. (For earlier and additional literature see Refs. [1946–1950].)

4. Faraday (Magnetooptical) Effect and Its Use

The Faraday rotation of the plane of polarized light caused by a magnetic field can be used to measure the current in conductors. A Faraday cell is placed between crossed polarizers in the path of a light beam (using a narrow band filter if the light source is not monochromatic), and the amount of light transmitted is a measure of the current in the adjacent conductor.

Jaecklin [1334] of the Brown Boveri Co., Baden, Switzerland, applied this technique to the measurement of the current in high-voltage power lines. He used a power-stabilized HeNe laser as light source, initially with Schott Flint glass SF 59 as Faraday effect material. Because of the low Verdet constant of this glass, he used multiple transmission through the same glass block in the way shown in Fig. H4-1. Fiber optics light guides made it possible to have the laser and photodiodes at a safe distance from the power line. Higher sensitivity was obtained by using a YIG crystal (yttrium–iron–garnet), which has a very high Verdet constant but is transparent only in the infrared (see [1951, 1952]). Wild [1544] developed after the same principle a phasemeter for photoelectric measurement of magnetic fields. The fact that Faraday rotation depends on wavelength can be used to separate one spectral line from another. Sansalone [1285] describes a compact Faraday rotation isolator using this principle.

In a number of laser applications, as well as in systems involving precise measurements, it is necessary to isolate the laser from any light reflected from components along the path of the beam. Methods of achieving this isolation include the use of attenuation padding or of a quarter-wave-polarizer combination and Faraday rotation. The first technique is undesirable because of the loss of signal; the second requires an application where circularly polarized light can be used. The Faraday rotation isolator, on the other hand, can produce a large backward to forward loss ratio for linearly polarized light and is desirable where practical. Because of the small Verdet constants associated with most transparent materials, achieving the required 45° rotation of the direction of polarization requires high magnetic fields or long path lengths in the Faraday material, or both. Generally, isolators constructed with a material such as lead glass use water-cooled solenoids to produce the required magnetic field. These isolators are bulky and consume a large amount of power.

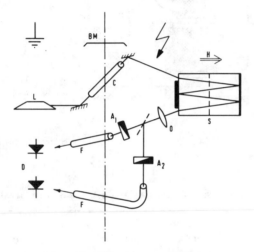

Fig. H4-1. Current measurement using Faraday effect. H is the magnetic field caused by the current being measured. The sensor S is a block of Schott glass SF59 which the laser beam traverses several times before passing through the lens O and, via a beam splitter, into the two analyzers A, and A_2 and being measured by the photodiodes at dc. F and F are glass fiber light guides.

Sansalone's isolator uses terbium aluminum garnet (TAG) as the Faraday rotation material and small high field permanent magnets made of copper–rare earth alloys. The present version of the isolator measures 3.49 cm in diameter by 3.81 cm in over-all length as shown in Fig. H4-2. The isolator provides effective isolation from 4880 to 5145 Å. The best isolation obtained to date is 26 dB with a 0.4-dB forward loss. Table H4-1 shows the superior behavior of the TAG material as compared to lead glass.

Reference is made to the use of the Faraday effect for shutters in "High Speed Pulse Technology," Vol. I, pp. 444–450. Another application of the

Faraday effect is to eye protection shutters for use against the light of nuclear explosions. Faraday cells have also been used as high speed photographic shutters as described by Edgerton and Germeshausen [1771, p. 42], and Pfund [1771, p. 71].

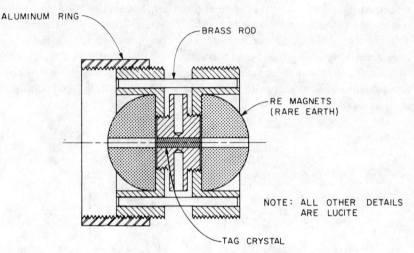

FIG. H4-2. Cutaway view of the optical isolator.

TABLE H4-1

FIELD VERSUS WAVELENGTH FOR TAG AND LEAD GLASS

λ (Å)	$B \times l$ (TAG)(Oe-cm)	$B \times l$ (lead glass)(Oe-cm)
4880	2128	14,595
5145	2520	16,364
5320	2660	18,000
6328	4310	27,000

5. Capping and Uncapping Shutters

The basis of operation of the so-called collapsing foil shutters is the electromagnetic force between a coil surrounding a cylinder of copper or aluminum foil due to the induced secondary current (see "High Speed Pulse Technology," Vol. I, pp. 450–453).

The collapsing foil system of Adkin [1646] consists of an energy storage and discharge module coupled to a high-field solenoid magnet of 2.5–12.5 cm (1–5 in.) bore diameter. The transient magnetic field collapses a cylindrical

5. CAPPING AND UNCAPPING SHUTTERS

foil towards the bore center at velocities in excess of 10^4 cm/sec. The system can be used in a number of applications: the transient compression of magnetic fields, the closure of large aperture optical systems (the bore is initially unobstructed and can be made opaque to soft x rays after collapse), the generation of high speed flows. Collapse time depends on the bore diameter used and can be computed from the collapse speed. Typically, a 10-cm- (4-in.) diam system can be closed in 150 μsec.

The energy storage module (3000 J) is discharged via a low-inductance line into a high field coil of special profile. The trigger unit can be activated by a pushbutton switch or by external trigger pulse; a synchronizing output pulse is available for triggering related events. Initiation of the sequence occurs with microsecond accuracy. For improved timing flexibility, a delay generator is provided as an option. The shutter is connected to the control module by a 10-ft-long flexible line and can be located at an appropriate position in the optical path. The aluminum foil used is easily replaced after each firing, thus allowing rapid sequencing of operation. Larger bore systems are available. (See Table H5-1 for data given).

TABLE H5-1

ENERGY STORAGE SUPPLY AND CONTROL MODULE OF THE ADKIN COLLAPSING FOIL SHUTTER[a]

Maximum energy storage	3000 J
Maximum working voltage	20 kV
Maximum output current	100 kA
Total system inductance	150 nH
Output cable length	10 ft
Maximum repetition rate	1 shot/3 min
	Delay: ~5 μsec from sync-in pulse

[a] See Ref. [1646].

Other applications of collapsing foil systems are the generation of acoustic shocks and the collapse of fast moving metallic diaphragms. Photographic applications of collapsing foil shutters can be seen in papers by Allan [1771, p. 81], and by Früngel and Lohse [1771, p. 171].

The same principle can also be used another way for an uncapping shutter. Webster and Thomas [1771, p. 84] realized an electromagnetic highspeed uncapping shutter, shown diagrammatically in Fig. H5-1, that is capable of uncapping a 2.7 in. diam aperture in 200, 180, or 125 μsec.

When heavy current is passed through a metal foil, each current filament in the foil, having its own magnetic field, reacts with other current filaments in that foil. Each filament carries current in the same direction, hence the resultant attracted forces cause the foil to be compressed into a rod. A foil of dimensions 0.001 in. thick by 1.5 in. wide can be compressed into a rod of ~ 0.1 in. diam. If two separate foils are mounted side by side in the same plane, insulated from each other, and connected to carry current in opposite directions, a field of mutual repulsion will be set up causing the foils to move sharply apart and collapse under a second impulse of current. This principle formed the basis of our original experiment. However, a shutter based on this simple principle will open only a 1.5-in. aperture taking some 200 μsec to complete this action. High speed camera photographs indicate the outside edges of the foils tend to move inwards in a pincushion action.

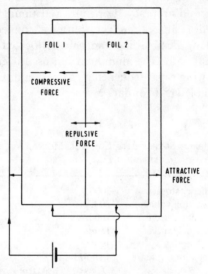

Fig. H5-1. Collapsing Al foil uncapping shutter. Current in all filaments of one foil is in the same direction and the resulting attracted forces collapse foil. Opposite currents in two foils cause mutual repulsion, which is reinforced by attraction to conductors forming tides of aperture.

It was a logical development of this experiment to place the assembly of two foils between rigid conductors (Fig. H5-1), which in turn carry current in the same direction as the adjacent foils, thus setting up three magnetic fields: (1) compressive forces within each foil; (2) attraction of each foil to its adjacent conductor; and (3) mutual repulsion between the foils. Aluminum was selected as the foil material on the basis of its low resistivity, density, and stiffness. A foil 0.001 in. thick has very little strength and is easily damaged.

Finally Impulsphysik GmbH, Hamburg, has developed a small-sized collapsing foil shutter as described in Ref. [1335] for the eyepiece of telescopes to protect the eye of the observer against the light emission of nuclear explosions. For this limited diameter of less than 20 mm, 200 J of energy is sufficient to give a foil closure speed of ~1000 m/sec. The coil and its discharge capacitor oscillate, first at high power amplitude followed by slow damping during shutter operation. Figure H5-2 shows such a shutter after closure taken at 110,000 frames/sec.

FIG. H5-2. Collapsing foil shutter in the closed condition. The foil was originally a cylinder concentric with the optical axis. The plane of the coil is at the end point of the axis.

6. Magnetically Driven Hypervelocity Macroparticle Accelerators

Hypervelocity macroparticle research may, very generally, be divided into two separate areas of interest: the production of hypervelocity macroparticles (that is, the theory, techniques, and "hardware" of macroparticle acceleration), and the uses to which hypervelocity macroparticles might be put (such as hypervelocity impact and the associated physics of penetration, crater formation, and material behavior). Such a division is of course arbitrary and the two cannot be entirely divorced. Accelerators must produce macroparticles with properties that are of interest in experiments which might be conducted with them.

In 1972, Barber [1899] published a book on this subject. His investigation was primarily concerned with the production of hypervelocity macroparticles and he refers to their uses only to place the study in perspective. After a brief specification of hypervelocity macroparticles, the paper provides background by describing the homopolar generator (HPG) operated by the Department of Engineering Physics at the Australian National University, and then discusses the aims and objectives of the project.

H. CONVERSION OF CAPACITOR ENERGY

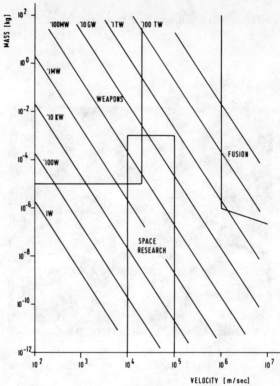

FIG. H6-1. The peak acceleration power required to constantly acclerate a particle over 1 meter as a function of particle mass and velocity for a range of masses and velocities of interest.

Figures H6-1 and H6-2 give relevant power and energy relationships of hypervelocity macroparticles and the particular ranges of interest are shown for their application on weapons, space research, and fusion technology.

In Fig. H6-1, the acceleration distance x is chosen to be 1 m. Equivalent powers for other distances may be found by simply dividing by the distance in meters. For a particle of 1 gm mass and a velocity of 10^5 m/sec, the peak power is in excess of 1 MW. Such large amounts of energy and power place heavy demands on energy stores. Multimegajoule capacitor banks are both rare and expensive. One of the most readily available energy stores that meets these requirements is chemical explosives which can have stored energies of several megajoules per kilogram. This probably accounts for the widespread use and success of chemical explosives for macroparticle acceleration.

In view of the information shown in Figs. H6-1 and H6-2, it is interesting to consider the power and energy requirements of some of the possible applications of hypervelocity macroparticles. The advent of the space age has stimulated interest in the region shown as "space research" for such purposes as hypervelocity flight simulation and meteroid flight/spacecraft damage, and the possibility of reaching fusion temperatures with hypervelocity impact has been studied extensively. The approximate range of velocities and masses predicted to be necessary to attain nuclear fusion is also identified in Figs. H6-1 and H6-2. Present accelerators are capable of reaching only the lower velocity range of space research interest (up to $\sim 2 \times 10^4$ m/sec), with fusion still far beyond present capabilities.

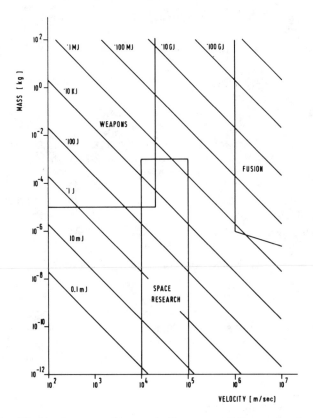

FIG. H6-2. The kinetic energy of hypervelocity macroparticles as a function of particle mass and velocity for a range of masses and velocities of interest.

Electrical accelerators have, in general, produced low masses, low velocities or both. The homopolar generator (HPG) operated by the Department of Engineering Physics at The Australian National University (ANU) is capable, in principle at least, of providing an opportunity to investigate high-mass, high-velocity electromagnetic acceleration.

The principal design features of the components of the system described by Barber [1899] are:

(1) Clamp switch—to close in 10 msec and carry a peak current of 800,000 A;
(2) Inductor—to carry a peak current of 500,000 A and store 2.5 MJ of energy;
(3) Rail switch—to switch the inductor current into the rail gun in the order of 20 μsec;
(4) Rail gun—to conform to the design principles derived from fundamental considerations;
(5) Dump resistor—to absorb the energy remaining in the system after gun operation;
(6) Ballistic range—to provide a target area and facilities for in-flight observation of the projectile;
(7) Instrumentation—to provide for the measurement of basic system operating parameters such as currents, voltages, temperatures, and deflections and to provide facilities for observation of gun performance (e.g., high speed photography of the projectile, projectile position detection, and gun voltage measurements).

A discussion of the results obtained for the various devices in an experimental program follows.

The clamp switch successfully closed on up to 460,000 A and carried peak currents of up to 650,000 A. Several hundred operations at various currents produced no noticeable deterioration in performance and required no maintenance. This demonstrates the basic soundness of the switch and appears to indicate the importance of multiple, flexible contacts in the design of high current dc switches.

Successful operation of the inductor with currents of up to 460,000 A, despite the increased heat load from the lengthened pulse used, indicates the adequacy of inductor design. The balanced termination used did appear to reduce the unwinding torques and there were no apparent signs of mechanical overstressing or failure.

One of the most important results of the rail switch tests was the demonstration of general rail gun behavior. That is, the rail switch performed very much as predicted from simulations. Instability, electrical contacts, and friction were apparently not serious problems (at these accelerations at least) which was encouraging evidence with regard to gun operation.

The switch was also able to switch high currents very rapidly and the measured rate of 4×10^9 A/sec was very close to the predicted rate of 5×10^9 A sec, the difference being attributable to the uncertainty in estimating the switching inductance. The ability of the switch to perform its functions with acceptable minimum maintenance was also demonstrated. The cotton wool deceleration method proved to be extremely efficient and highly satisfactory, although no complete explanation of its action has evolved.

Experiments with the rail gun itself were basically of an exploratory nature and demonstrated clearly that detailed understanding and optimization of performance would require a very considerable expenditure of time and resources to arrive at any degree of finality; however, the following important aspects of high-performance rail gun operation have been identified:

(1) The electrical contact between projectile and rails appears to be one of the most important factors. Two methods of obtaining good electrical contact were briefly investigated.

(a) By employing multiple individual contacts, similar to the method derived for the rail switch slug. Experiments indicate that this form of construction provides better contact than the solid projectiles; however, the consequent mechanical weakness of the projectile may result in mechanical failure. Conversely, solid projectiles appear to be strong enough to withstand the acceleration forces, but usually have such poor contacts that they may be destroyed by arcing.

(b) By the use of low melting point or liquid contacts (such as solder). A notable success (shot 9) was achieved by using a combination of solder and the laminated projectiles described above; however, the role of the solder is still not clearly understood. Solder did appear to improve the performance of a solid projectile, however, so the possibilities of liquid contacts still remain of interest.

(2) The strength of the projectile as mentioned in (1) appears to be of importance. Although the forces exerted on the projectile are, theoretically, body forces or nearly so, large stresses could still arise in the projectile, especially during current buildup. Thus, it is suggested that, at least at the beginning of acceleration, the projectile must be relatively strong. However, it is also apparent that strength cannot be achieved at the expense of good contact: strong, flexible, multiple, contacts appear to be required.

(3) Instabilities have not, as yet, been conclusively demonstrated. The vertical instability appears to be not important, although the effects it would produce (such as destruction of the Perspex location blocks) are often masked by "explosive" damage. Instrumentation at present is inadequate to directly detect Rayleigh–Taylor instabilities. However, rail damage evidence indicates that "tumbling" is rare, but "blow-through" on one side may occur. The latter is more likely to be due to the projectile sticking on one rail with subsequent mechanical deformation, allowing "blow-through" to occur. Thus it is difficult, at present, to differentiate between strength problems and instabilities.

(4) The presence of a good vacuum appears to be an important factor in the gun for two reasons; rapid expansion in high-temperature arcs causes extensive structural damage to the gun and higher air pressures (up to ~ 5 Torr) probably result in lower breakdown voltages between the rails. The experiments conducted to date indicate that even lower pressures than those presently obtained (100 mTorr) may be required to ensure the breakdown voltages required for maximum operation (>10 kV).

(5) Other aspects of interest that have been observed in the operation of the gun are:

(a) A surprising lack of extensive rail damage, even on apparently unsuccessful shots. On shots in which even short accelerations have taken place, rail damage is even less, thus indicating that if operation can be further improved, rail damage may become negligible and rail replacement may not be a serious problem. The lack of rail damage is attributed to the very high velocity of arcs in the gun, resulting in very little heat being transferred to the rails.

(b) The operation of the installation is safe and reliable. In this respect the rail gun facility has proved to be very acceptable.

(6) The maximum performance achieved by the present installation in terms of solid projectiles being accelerated to high velocities was for shot 9 in which a 3.3-g projectile was accelerated to about 1800 m/sec. Other shots may have achieved similar performance but verification was less conclusive. Had shot 9 operated on the long rails rather than the short ones, a velocity of nearly 6000 m/sec would have been reached, which is well into the hypervelocity region. It is confidently felt that, when some of the problems identified have been investigated, this order of performance will be achieved. With further effort, much improved performance (20,000 m/sec) does not seem improbable.

7. Magnetically Driven Gas Accelerators

A study is under way in the Department of Engineering Physics at the Australian National University to determine the feasibility of using the HPG to power a hypersonic wind tunnel. Theoretical analysis has shown that this can be done using a traveling conduction wave gas accelerator and that this method may be able to provide a better performance than existing steady-state (thermal arc jet, crossed field, hall current) accelerators. The theory has been completed and a rotary gas accelerator built for use as an injector for the linear gas accelerator, testing of which has been delayed by difficutlies encountered in making water-tight cooling passages in the electrodes of the rotary accelerator, and in obtaining good quality boron nitride insulators.

8. Megagauss Fields

The generation of magnetic fields above 500 kG, even of short duration and in small volumes, presents two main technical difficulties: the storage of the large energy involved and the pressure exerted on the coil system by the high magnetic fields. In addition to this, the very high current densities in the current-carrying skin layer of the conductors will result in melting the metal; hence the coil system remains damaged even if the pressure problem is solved. These difficulties are well known and have been described by, among others, Furth, Levine, and Waniek, and, more recently, by Gordienko and Shneerson [1954, 1955].

Gordienko and Shneerson [1955] describe one possible method to overcome these difficulties and to generate larger fields in bigger volumes, namely, by magnetic flux compression combined with the use of a chemical explosive as the energy source. It is based on the following principle.

A closed electrical system in which a current I_0 threads a magnetic flux

$$\phi = L_0 I_0$$

is transformed by external forces in such a way as to diminish the initial inductance L_0. If there are no flux losses during compression (which only holds for perfect conductors), the current increases with decreasing inductance as

$$I/I_0 = L_0/L = \gamma_L$$

where γ_L is the compression factor. The total energy E of the system increases by the same factor, since

$$E/E_0 = LI^2/L_0 I_0^2 = L_0/L = \gamma_L$$

This means that the work done by the external (mechanical) forces is transformed into electromagnetic energy. The magnetic field linked with the current is multiplied by a factor γ_H, which, in general, is space-dependent. For the simple compression systems considered further on, however, its mean value is very near to γ_L.

To limit the diffusion of the magnetic fields through the conductor and to avoid the ejection of the molten metallic surface, it is necessary to limit the implosion time of the megagauss compression experiments to the order of some tens of microseconds. These time scale and considerations of pressure and total energy suggest the use of a chemical explosive as the energy source for magnetic compression experiments.

Fowler *et al.* [1956, 1957] have shown that explosive flux compression is feasible and have given results and some details of their cylindrical compression experiment. These appear to be the first and only experimental contributions to this problem which have been published. The results of the cylindrical implosion experiments of Gordienko and Shneerson agree well with those given by Caird et al. Despite considerable efforts, however, it was never possible to reproduce the very high fields (14.6 MG) reported by Fowler et al. and the same thing happened to the French group.

Some recent theoretical studies on the basic principles of flux compression indicate that there are effects which can set a limit to the maximum fields attainable by this method. Quantitative indications, however, can be given only by further experimental evidence.

Unlike other groups working at present in this field, Gordienko and Shneerson were not involved in any military application and were therefore able to describe in detail all essential parts of their experiments. Their experiments were carried out in the open air a few meters from a small bunker, which housed the experimental equipment. The experimental assembly, consisting of explosive charge, liner, coils, probes, mirror, light source, photomultiplier, etc., was mounted on a simple wooden stand and placed on the firing table. The table served to protect the end connectors of the electrical wiring and of the high-voltage line. Only the last 1–2 m of every cable, viz., the part from these connectors to the charge, was destroyed by the explosive action. Rugged electronic equipment could also be placed below the firing table without being damaged. This arrangement has been used successfully with up to 4 kg of explosives. More details about the experimental facilities are given in a report published by Herlach et al. [1958].

The initial field is generated by the discharge of a 540-μF condenser bank (22 kJ at 9 kV) through a suitable coil system. Recently, the bank has been enlarged to 1340 μF, giving 55 kJ at 9 kV. The bank and its three electrode spark gap are housed in a weather- and blast-proof steel case placed behind a protecting wall outside the bunker. Twenty-seven coaxial cables RG-8/U in parallel, 8 m long, connect the bank with the high-voltage terminal under the firing table. The total inductance of the circuit at this point is 150 nH.

The coils are a critical element of the experiment. They ought to be simple and inexpensive (as all other parts destroy each time) but should not break down electrically or mechanically in the first quarter-period of the discharge, when the magnetic field is built up. The design of the coils is discussed together with the experiments, since the coils vary in form and dimensions for each type of charge. However, they are generally made with Formvar-coated copper wire of rectangular 2×10 or 3×5 mm section, in close turn-by-turn winding. An annular metallic sheet of ~ 2 turns, insulated with glass fiber and Araldite, takes up the radial strain during discharge. Araldite or concrete fills the interspace between the coil and the metal cylinder.

The experiments use essentially two types of explosive: precision-cast composition B (40% TNT, 60% RDX), and plastic explosive sheets of 2, 4, and 8 mm thickness, now commercially available. For the simultaneous ignition, special detonators, which are also commercially available, were developed with a time jitter lower than 0.1 μsec when properly fired. For the simultaneous ignition of up to 24 detonators, a firing unit with a condenser of 3 μF charged to 5 kV is used.

The magnetic fields are measured by small pickup coils (Fig. H8-1). The probes are connected to the oscilloscopes by about 12 m of coaxial cable. The frequency response of the probe–cable system has been measured; it is essentially flat up to 10 MHz. To display the magnetic field signal on the scope, the probe signal is integrated by a simple RC circuit with a calibrated time constant of 500 μsec placed at the oscilloscope input. The integrated signals have been checked to agree with graphically integrated direct probe signals.

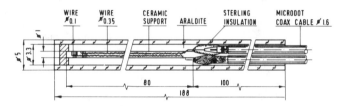

FIG. H8-1. Up to four magnetic probes can be used with the field measuring arrangement shown here. The number of turns range generally from 5 to 20 for each pick-up coil. All dimensions are in millimeters.

The probes are calibrated by comparison with a set of standard probes with precisely known geometrical dimensions in a stabilized high-frequency field of 400 kHz. Depending on the type of probe, the accuracy of this calibration is within 1 to 5%. In a few experiments the magnetic fields have been measured also by using the Faraday effect, confirming the probe measurement within the experimental error of ∼7%. Optical diagnostics (framing camera and Kerr cells) and electronic measurements have always been applied simultaneously to all experiments, to avoid, as much as possible, misinterpretation of results. For the detonation problem different solutions have been adopted, as can be seen, together with other details, from the following description of the experiments reported in Table H8-1. The implosion dynamics of the liner have been studied theoretically, assuming the liner material to be both compressible and incompressible. Compression device is of 25°C.

In this charge, as shown in Fig. H8-2, one single detonator starts the detonation process, which is then transformed by the explosive system to a nearly cylindrical, converging detonation wave. When free access to both sides of the compression volume is required, the charge is ignited by 24 detonators regularly spaced along a circle at the position AA′ (type 25 B).

TABLE H8-1

RESULTS OF CYLINDRICAL COMPRESSION EXPERIMENTS

Device	Shot No.	B_0 (kG)	B_{max}[c] (MG)	$\gamma = B_{max}/B_0$
25 B and C[a]	347	32	3.7 (3.5)	120
	376	33	3.0 (3.0)	90
	381	33	3.6 (2.7)	110
	472	44	3.4 (3.1)	77
25 M[b]	525	52	3.0 (2.7)	58
	526	53	3.6 (2.6)	68
	527	53	2.8 (2.5)	53

[a] Charge: composition B; o.d. 180 mm, i.d. 76 mm; length 70 mm. Liner: brass, slotted; thickness 3 mm, length 160 mm.

[b] Charge: composition B; o.d. 180 mm, i.d. 76 mm; length 25 mm. Liner: stainless steel AISI 304; thickness 2 mm, length 160 mm.

[c] The figures in parentheses correspond to magnetic probes which are 5–10 mm out of the center plane.

Starting from $B_0 = 30$–40 kG, final fields of up to 3.7 ± 0.3 MG in a volume of about 1.5 cm^3 (useful diameter 0.6 cm, length 5 cm) have been measured. Results obtained with this charge are, in general, in the range 3.4–3.6 MG.

For the power P which is transferred from the liner to the field, we can write (per centimeter length of the charge)

$$P = 2\pi r p |dr/dt|$$

where r is the radius of the liner and $p = B^2/8\pi$ the pressure of the magnetic field B. With the following set of experimental figures, at the point of probe destruction, $r = 0.3$ cm, $p = 5.4 \times 10^{11}$ dyn cm^{-2} ($B = 3.7$ MG), $v = 0.39$ cm μsec^{-1}, the power amounts to

$$P = 4 \times 10^{10} \text{ W cm}^{-1}$$

or, for the whole compression region (5 cm),

$$P_{tot} = 0.2 \times 10^{12} \text{ W}$$

Since the relevant physical quantity in these systems is the current rather than the magnetic field, they represent explosive-driven current generators. Within the possibilities and limitations given by relation (1), they deliver a current pulse out of a coaxial line.

Figure H8-3 shows the coaxial compression device. Due to the relatively high inductance of the condenser bank used, initial currents of only 380 kA have been induced in the system. The two compression experiments made

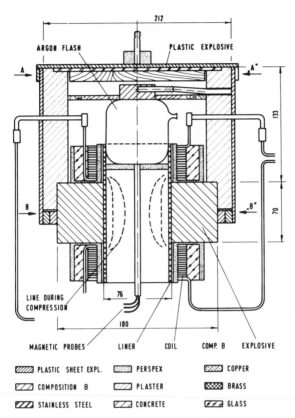

FIG. H8-2. Compression device 25 C. The detonation process, resulting in a quasi-cylindrical implosion, is started by one detonator. The inner flash is made up from an argon-filled rubber balloon shocked by a small explosive charge (composition B). All dimensions are in millimeters and all parts of the drawing are scaled to correspond with the original device; this applies to all following figures together with the symbols of materials.

with this device have given final currents of 1.8 and 2.4 MA. This corresponds to a final current line density at the surface of the inner conductor (0/6 mm) of 0.95 and 1.27 MA cm^{-1} and a magnetic field of 1.2 and 1.6 MG, which is similar to what has been produced with the "one-turn solenoid" device. The compression factor for shot No. 566 is 5.8, slightly less than half of what could be expected from the geometry of the charge. Evaluations show that nearly half the magnetic flux has been lost during compression and that maximum current is achieved before the liner closes completely on the measuring chamber (corresponding to the liner position drawn in Fig. H8-3 translated 8 mm toward the axis). Further experiments on this compression device and a fast condenser bank for the initial current will certainly result in appreciable progress.

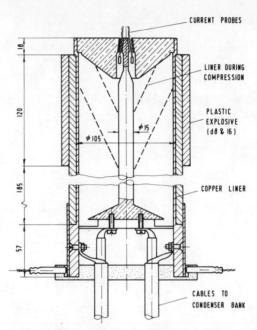

Fig. H8-3. Section through a coaxial compression device, consisting of a compression volume and a measuring chamber. The initial current is fed in through four coaxial cables, 12 simultaneously fired detonators start the detonation process, which first closes the system electrically and then compresses the trapped flux.

The flux compression method seems to be at present the only way to produce fields above 1 MG within useful volumes, i.e., of some cubic centimeter. The maximum measured fields reported here do not exceed 4 MG. From results obtained, it is still difficult to deduce whether this limit could be overcome by increasing the technical effort (e.g., augmenting the initial magnetic field), or whether it is due to physical reasons (e.g., anomalous field diffusion). But even in the optimistic case of no physical limit in this range, it would be difficult to generate magnetic fields in excess of 8 MG.

The electromagnetic energy of the system increases the same compression factor as the field. This means that by the flux compression method, it is possible to concentrate large electromagnetic energies into small inductive loads (this last point being both a limitation and a merit of the method). In the experiments, the largest measured compression factor was 120. A low inductive condenser bank can feed up to 80–90% of its energy into a suitable compression device. If, for instance, such a bank stores 200 kJ, more than 15 MJ electrical energy could be concentrated in the one-turn solenoid. This is at least one order of magnitude more than that which can be achieved by other electromagnetic energy sources.

A last remark can be made as to the efficiency of energy transformation from chemical to magnetic energy. In the most favorable case, i.e., at turnaround, efficiency reached 10%. This efficiency has, so far, no practical importance; the cost of the explosive, in fact, is small compared to the rest of the experiment, which is blown up.

Bitter [1046] recently published a summary about ultrastrong magnetic fields. He carries the discussion forward and describes methods that have already created, for brief periods, fields in the neighborhood of 1×10^7 G and appear potentially capable of achieving fields exceeding 1×10^8 G. These methods have been pioneered at the Los Alamos Scientific Laboratory by Fowler, Caird, Garn, and Thomson. It is believed that such fields can provide new insights into the structure of matter. Ultrastrong magnetic fields should distort matter in various characteristic ways, and one would like to see if the distortions predicted by theory are actually observed under experimental conditions. Beyond that one may well encounter new and previously unsuspected phenomena.

However, there is an opportunity to do more than simply explore the response of matter to magnetic fields of a new order of intensity. Magnetic fields have a little-appreciated property that can be exploited along with the production of ultrastrong fields: They transmit forces with the velocity of light. The only method currently available for creating ultrastrong magnetic fields employs explosive charges arranged in a ring so as to produce an implosion and thereby compress a preexisting magnetic field. Ultrastrong fields proiduced in this way can probably exert or withstand pressures of many millions of atmospheres. Transmitted with the speed of light, they can momentarily provide a shielded experimental chamber in the midst of an explosion, and they can also provide a window through which the events taking place in the chamber can be observed with precision and in detail. In other words, the fields should make it possible to study matter under far more extreme conditions than can be provided by magnetic fields alone, conditions that normally exist only in the interor of planets and stars. This seems the most exciting prospect presented by the effort to create ultrastrong magnetic fields.

"Magnetic atoms" (Fig. H8-4) could be produced by exposing electrons to an ultrastrong magnetic field. The term signifies that an electron, which is obliged to travel in a circular orbit in a magnetic field, would have an orbit of atomic dimensions if placed in a very high field. When traveling in orbits, electrons characteristically emit radiation, known as "cyclotron" radiation. The smaller the orbit, the shorter the wavelength. As shown in Fig. H8-5, the wavelength of cyclotron radiation is about 2,000 times longer for protons than for electrons in the same magnetic field. In the region between 125–250 million gauss, the cyclotron radiation of electrons lies in the visible part of the spectrum.

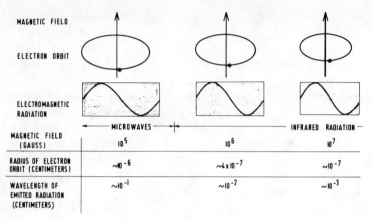

Fig. H8-4. Magnetic atoms.

The compression of a hydrogen atom by an ultrastrong magnetic field, as shown in Fig. H8-6, depends on the atom's state of excitation, expressed by a quantum number n, which represents angular momentum. In the absence of a magnetic field the radius of the atom increases as the square of the value of n, as shown by the electron orbits depeicted in black. If a magnetic field of 100 million gauss were applied, the orbits would have approximately the dimensions shown in dotted lines. The scale of the drawing is arbitrary below a radius of 10^{-9} cm.

Magnetic compression of hydrogen atom is shown in Fig. H8-7 for various energy states and for magnetic fields that range in strength from 10^6 to 10^9 G. For hydrogen in its lowest energy state (when $n = 1$), compression is slight even at 10^9 G.

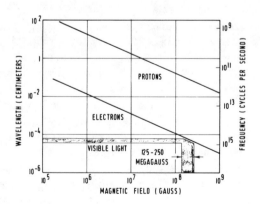

Fig. H8-5. Wavelength of cyclotron radiation.

8. MEGAGAUSS FIELDS

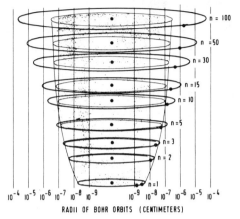

FIG. H8-6. Compression of hydrogen atom.

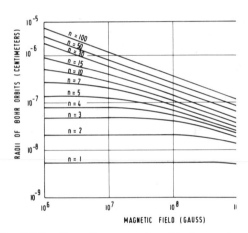

FIG. H8-7. Magnetic compression of hydrogen atom.

I. Conversion of capacitive stored energy into acoustic pulses

1. Acoustic Pulses

a. *Remote Sensing of Wind and Temperature by Acoustic Echo Techniques*

At the better equipped airports, the pilot of an aircraft about to take off or land is given reports of surface wind and visibility, cloud height, etc. In addition, information on wind shear and turbulence in the lowest layers especially between the surface and 200 ft. is sometimes called for. Wind shear in the form of an increasing headwind during the descent along the glide path, for example, tends to make the aircraft overshoot the runway, while the same conditions reduce the rate of climb of an aircraft after takeoff. Turbulence may increase the instantaneous wind shear and at the same time make the aircraft more difficult to control. However, so far, no practical methods for measuring wind shear or turbulence at airports, with the required continuity, have yet been developed and it is necessary to rely on occasional subjective reports by other pilots. Since anemometers mounted on towers close to the climb and descent paths would constitute unacceptable obstructions, efforts have been concentrated on remote sensing systems.

Acoustic echo sounding shows some promise as a tool for these measurements, since the necessary scatterers (temperature and wind fluctuation) practically always appear to be available, although perhaps not in "dead calm" conditions. Several observation points are needed, or conical scanning from one point since special problems are encountered. Some of these problems are: Range is likely to be not much more than 1 km. Absorption can be minimized by using low frequencies, but even so, a range of 10 km is likely to be possible only in exceptional circumstances. Ambient noise (natural and man made) interferes seriously. Very sharply directional antennas, designed also to be insensitive to the noise of rain and hail, could help, however, the beam can be bent by wind and temperature gradients, thus making it difficult to make the necessary corrections.

Sound waves transmitted into the atmosphere and scattered back to microphone receivers are more sensitive to changes in wind and temperature than high-powered radar signals or optical signals (see Table IIa-1 and compare with Table IIa-3) [1669]. An acoustic radar, or echo sounder, operates with considerably higher power return efficiency than a conventional

1. ACOUSTIC PULSES

TABLE IIa-1

EFFECT OF CHANGES IN METEOROLOGICAL ELEMENTS ON ACOUSTIC, RADIO, AND OPTICAL REFRACTIVE INDEX

Magnitude of parameter deviation	Change in refractive index (N units)		
	Acoustic	Radio	Optical
1°C fluctuation in temperature	1700	1	1
1-m/sec variation in wind speed	3000	Negligible	Negligible
1-mB change in water vapor pressure	140	4	0.04

radar because sound waves are scattered by the atmosphere about a million times more strongly than radio waves. This is in part a result of the difference in their modes of transmission: Sound waves depend on compression and rarefaction of air molecules to propagate, whereas electromagnetic waves require no propagating medium at all and in fact travel best in a vacuum. The strong absorption of sound waves by the atmosphere limits the range of acoustic sounding to distances of a few kilometers, but the technique is ideal for studies of the boundary layer. Echo sounders can be constructed for a fraction of the cost of modern meteorological radars.

One of the first to attempt atmospheric sounding with acoustic radar was McAllister [1666] of the Australian Defense Science Service. In experiments conducted near Adelaide in early 1968, McAllister used an array of 196 8-in. (20-cm) commercial loudspeakers to transmit pulses of 900-Hz sound vertically and to receive the scattered echo returns. Pulses lasting 50 msec were transmitted every 10 sec at a peak pulse power of 500 W. The returned echoes were amplified and recorded on a facsimile recorder; the recording speed was adjusted to give an effective height range of 1.675 m. Table IIa-2 shows the parameters of McAllister's experiment.

TABLE IIa-2

PARAMETERS OF ACOUSTIC SOUNDER

Carrier frequency	900 Hz
Pulse length	50 msec
Pulse repetition rate	0.1 pulse/sec
Transmitted power	500 W
Transmit/receiver transducer	Square array of 196, 8 in. loudspeakers
Receiver	Tuner audio amplifier, 100 Hz bandwidth
Recorder	Facsimile receiver
1. Paper speed	3 in./h
2. Paper width	8 in.

From the soundings McAllister could identify high pressure ridges, temperature inversions, turbulent regions within inversion layers, the structure and boundaries between maritime and continental air layers, vertical oscillations of boundary surfaces, and various other atmospheric features. He concluded that it is feasible to obtain a continuous record of the height, movement, and spatial distribution of inhomogeneities in the temperature structure of the lower troposphere by means of acoustic sounding. In subsequent experiments, in April 1969 McAllister *et al.* [1665] obtained results in the form of calibrated acoustic echoes against temperature and wind data.

In 1970 Wescott *et al.* [1658] of ESSA (Environmental Science Services Administration), Boulder Colorado, reported on similar work, including acoustic Doppler radar.

During the last part of September and the first part of October 1969, the Acoustic Group of the Wave Propagation Laboratory (WPL) conducted experimental observations at Haswell, Colorado, of the scatter of acoustic waves by thermal fluctuations and turbulent velocity motions in the boundary layer of the lower atmosphere. This report includes a brief discussion of the theory of atmospheric scatter of sound waves, a description of the experimental apparatus, and a résumé of results obtained so far.

The tone burst and timing generator serves as the master control center of the system. The tone burst part of this block is a fast-acting electronic switch that, when closed, passes the signal output of the audio oscillator to the input of the power amplifier. The duration of the switch closure or tone burst is adjustable. Typical settings range from 20 to 200 msec and are accurate to a small fraction of a microsecond. The timing generator contains controls for setting the repetition rate of the tone bursts. Typical repetition rates in field experiments range from 0.1 to 0.5 Hz. During each burst, the power amplifier delivers 40 W through the back-to-back diodes of the transmit–receive circuit to the acoustic transducer or driver. The driver is coupled acoustically to a reflector-horn antenna, which in turn radiates a tone burst of acoustic energy into the atmosphere. As soon as the tone burst is completed, the subthreshold high impedance of the back-to-back diodes comes into play to prevent low-level hum and noise of the power amplifier from reaching the acoustic transducer. During the tone burst, another set of back-to-back diodes in the transmit–receive circuit shorts out the input of the receiver variable-gain preamplifier to prevent overload and damage to this extremely sensitive component. At the completion of the tone burst, the subthreshold impedance of this set of diodes comes into play and they act as an open circuit (no load) across the very small echo signal about to be received (Fig. Ia-1).

During reception of echoes, the acoustic transducer acts as a microphone, transformer-coupled to the variable-gain preamplifier. The transformer has a high step-up ratio and increases the microphone Johnson noise voltage to a

1. ACOUSTIC PULSES

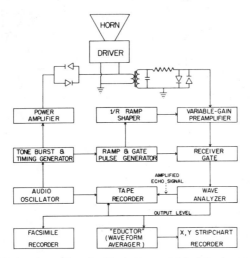

FIG. IIa-1. Acoustic radar block diagram. Mark I echo sounder.

level that exceeds the input self-noise of the preamplifier. This insures that the receiving system does not degrade the ultimate signal-to-noise ratio at the microphone terminals.

The gain of the preamplifier is electronically controlled by the $1/R$ ramp shaper, which modifies the shape of a linear ramp furnished by the ramp and gate pulse generator, so that the preamplifier gain is in turn varied in a manner to compensate for the spherical divergence or spreading loss of acoustic echoes as they are received from progressively longer ranges. The voltage gain is varied inversely with range $(1/R)$ rather than as the inverse square of range $(1/R)^2$ because the voltage signal developed by the receiving transducer is proportional to acoustic pressure rather than acoustic power.

The output of the variable-gain preamplifier is applied to a receiver gate. This gate, or electronic switch, is opened and closed by rectangular pulses from the ramp and gate pulse generator. Pulse timing is controlled by the tone burst and timing generator. The purpose of the receiver gate is to suppress the small amount of signal that leaks through the transmit–receive circuit during the transmitted tone burst and would otherwise reach the main part of the receiver (wave analyzer). Since the driver "rings" for some 100 msec after termination of the high power tone burst, the opening of the receiver gate is delayed for a controllable time (typically ~ 180 msec) until the ringing has stopped and the driver has fully recovered its sensitivity as a microphone. This also serves to prevent overloading the receiver with any close-in echoes from nearby structures, vans, etc. As soon as the receiver gate has opened, amplified echo signals from the preamplifier are able to reach the main receiver unit, or wave analyzer.

The wave analyzer operates as a receiver tuned to the carrier frequency of the acoustic tone burst radiated into the atmosphere. Its bandwidth is set to accomodate the radiated bandwidth and is therefore determined by the duration of the tone burst. A 100-Hz bandwidth generally has been used to accommodate the major sideband frequencies generated by tone bursts lasting 20 msec. The wave analyzer has two types of output signal, both of them tape recorded. One of the signals is an amplified version of the received carrier and sidebands. This is the type of signal that will be used for determining Doppler shifts of frequency between a transmitted signal and its received echo. The other signal is the detected level of the received echo and is displayed as functions of altitude and time of day by means of the facsimile recorder.

An alternate display of receiver output level is indicated by the blocks labeled "eductor" and X–Y stripchart recorder. The "eductor" is an electronic instrument that obtains the waveshape of a repetitive signal obscured by noise by sampling and accumulating the signal level at corresponding points along consecutive sweeps of a noisy waveform. The noise, being random, sometimes adds, sometimes subtracts from the accumulating signal; i.e., the true signal accumulates while the noise averages out to nearly zero. Selecting an accumulation time not appreciably longer than the time in which the repetitive waveshape of the original signal changes significantly yields an optimum combination of improved signal-to-noise ratio without substantial loss of time resolution.

The improved echo signal from the "eductor" is applied to the X–Y stripchart recorder that plots signal level (proportional to echo strength) along the Y-axis and time (proportional to height above local terrain) along the X-axis. The incrementally advancing stripchart of the recorder permits nesting of consecutive traces at regular intervals of real time. The advantage of this display over facsimile recording is that it is quantitative rather than qualitative, allowing direct and accurate reading of echo signal strength. The "eductor"–recorder combination was tested in the field with borrowed equipment. The results shown in Figs. I1a-2 and 3 indicate that the combination is a worthwhile addition to the echo sounder system.

Some months later Little [1657] found that typical height resolutions attainable would be on the order of one-half of the pulse length, i.e., about 10 m if a 60-msec pulse is used. Spatial wavenumbers could be explored over the range $4\pi/\lambda_{min}$ (~ 400 m^{-1}) to $\sim 10^{-2}$ m^{-1} in the case of mechanical turbulence, and from 400 to 40 m^{-1} for the temperature inhomogeneities. The difference is due to the ability of the Doppler radar to measure the radial velocities at different points along a beam simultaneously. Time resolutions for successive independent measurements to heights of 1500 m would be ~ 10 sec.

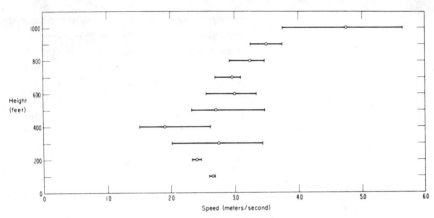

FIG. IIa-2. Profile of wind speed.

Limitations to the acoustic technique include its limited range (to perhaps 10 km) and the probability of serious loss of sensitivity due to increased noise level during periods of strong wind, hail, or rain. Other difficulties include the low information rate of the order 1 pulse $(10 \text{ sec})^{-1}$ and the strong refraction effects due to wind and temperature gradients.

Despite the somewhat unknown magnitude of these limitations, it is concluded that the acoustic echo sounding technique could be developed to measure (to heights of up to at least 1500 m): (1) the vertical profile of wind speed and direction (by utilizing a Doppler system); (2) the vertical profile of humidity (by means of a multiwavelength system); (3) the location and

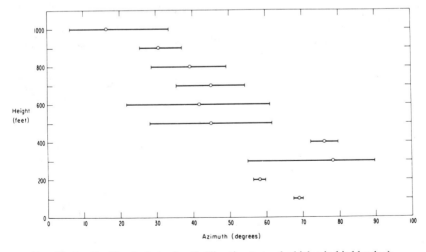

FIG. IIa-3. Profile of wind azimuth (direction toward which wind is blowing).

intensity of temperature inversions (by using a monostatic system to study the echo power and frequency spectrum, as a function of elevation angle, range, and wavelength); (4) the three-dimensional spectrum of temperature inhomogeneity (by using a monostatic system to study the echo power as a function of wavelength, azimuth, and elevation); and (5) the three-dimensional spectrum of mechanical turbulence (by using both monostatic and bistatic systems to study the echo power as a function of scatter angle, wavelength, and direction). These potentialities suggest that the acoustic sounding technique will prove a uniquely valuable remote-sensing concept for atmospheric boundary layer investigations.

Derr and Little [1336] presented a comparison of ultrasensitive microwave radar, lidar, and acoustic echo sounding as techniques for atmospheric studies. From this analysis the conclusion emerges that a remote-sensing facility consisting of these instruments probing the same volume of the atmosphere can measure many of the meteorologically significant parameters necessary to increase our understanding of the structure and dynamics of the clear lower atmosphere, and provide the spatial and temporal density of measurements necessary for weather forecasting.

The sensitivity of the three techniques to deviations of different atmospheric parameters is shown in Table IIa-3 (compare with Table IIa-1). The authors recommend the use of the three methods in large scale meteorological experiments and plan to develop laser Doppler lidars at some later date.

At the 7th International Symposium on Remote Sensing of the Environment, 1971, Beran and Willmarth [1663] presented a paper indicating that acoustic echo sounding had considerable promise as a remote sensor of various low-level atmospheric phenomena. The feasibility of one application, the derivation of vertical winds from the Doppler shift on a monostatic acoustic sounder, had been demonstrated in an earlier experiment by the authors. This later paper describes a more complex arrangement involving the use of

TABLE IIa-3

RELATIVE VALUES OF DETECTABLE DEVIATION C_n FOR ACOUSTIC, RADIO, AND OPTICAL WAVES[a]

Atmospheric condition	Acoustic $C_n \times 10^6$	Radio $C_n \times 10^6$	Optical $C_n \times 10^6$
Temperature $C_T = 1\ °K$	1720	1.26	0.93
Water vapor pressure $C_e = 1\ mB$	138	4.50	0.04
Wind velocity $C_V = 1\ msec^{-1}$	0–4000	$(0-2) \times 10^{-6}$	$(0-2) \times 10^{-6}$

[a] The very small values of C_n indicated for the velocity fluctuations in the electromagnetic cases are those due to Fizeau drag.

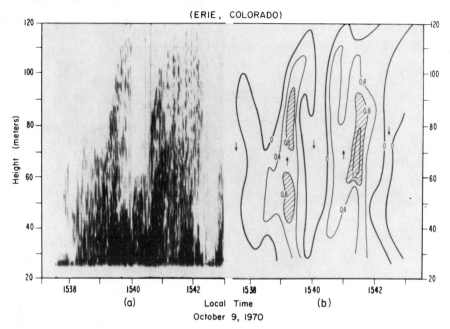

FIG. IIa-4. Acoustic echoes (left) and Doppler-derived vertical velocity field during thermal plume activity. Contour interval is 0.4 msec^{-1}; regions of vertical velocity >0.8 msec^{-1} are shaded; arrows indicate direction of vertical component.

bistatic acoustic sounding to measure a component of the horizontal wind. The feasibility of this approach is demonstrated (see Fig. IIa-4) and found to be superior to a monostatic system. The results also suggest that refraction due to temperature or wind gradients along the tilted beam of a bistatic sounder can result in an incorrect estimate of the real wind (see Fig. IIa-5). Their parameters can be seen in Table IIa-4.

The close spacing of the antennas (80 m) during this exercise made it

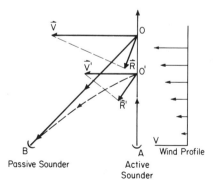

FIG. IIa-5. Geometry demonstrating the effect of refraction on a sloping acoustic beam. For the case shown the Doppler derived horizontal wind would be an overestimate.

TABLE IIa-4

ACOUSTIC ECHO SOUNDER PARAMETERS USED FOR BISTATIC DOPPLER EXPERIMENT

Peak power	7.5 W (acoustic)
Pulse width	100 msec
Pulse repetition frequency	2 sec^{-1}
Maximum range	340 m
Carrier frequencies	2750 and 3750 Hz
Antenna diameters	122 cm
Antenna beam width	$\pm 2°$ to 3 dB points
Beam directions	Vertical, 30°, 45°, and 60°
Receiver bandwidth	100 Hz (vertical antenna)
	1000 Hz (tilted antenna)
System noise level in 100 Hz bandwidth	$\sim 10^{-18}$ W
Scattering volume at 100 m altitude	Intersection of two cylinders ~ 7 m diam

impossible to determine the maximum altitude at which the horizontal wind might be measured. The 60° tilt angle does, however, appear to be approaching the maximum usable limit. The trade off between tilt angle, the angle of the measured component to the horizontal, and the varying intensity of returns at different scatter angles in order to achieve best results at a given range of altitudes must be based on the requirements for each application and an optimum arrangement for all cases cannot be predicted. The signal strength at a tilt of 60° (an altitude of 139 m) during this experiment was more than ample for determining the frequency spectra, and, for the equipment used, the winds at altitudes of at least 500 m should be within reach.

In still another paper the same authors together with Little [1660] report again on acoustic Doppler measurements of vertical velocities in the atmosphere during thermal plume activity and a nocturnal inversion. The existence of breaking wave formations within the inversion was clearly established. Table IIa-5 gives the parameters of their modified system. It may be noted that the frequency and pulse width had reached 4000 Hz and 100 msec, compared with 900 Hz and 50 msec in McAllister's work [1666] mentioned earlier.

About the same time Simmons et al. delivered the first expanded NOAA report [1659] (National Oceanic and Atmospheric Administration) on acoustic echo sounding as related to air pollution in urban environments, taking into account 30 references. A significant reduction of both radiated and received sidelobes was achieved by constructing a semianechoic, semi-opaque shield around their Mark II antenna. Comparative measurements first were made of the 90° and 80° sidelobes radiated from (a) the unshielded Mark II horn-reflector antenna, and (b) the same antenna shielded by an admittedly crude anechoic structure of hay bales. The structure approximated

TABLE IIa-5

ACOUSTIC ECHO SOUNDER PARAMETERS USED DURING THE
DOPPLER EXPERIMENT

Peak power	6.4 W (acoustic)
Pulse width	100 msec
Pulse repetition frequency	1 sec^{-1}
Maximum range	170 m
Carrier frequency	4,000 Hz
Antenna diameter	122 cm
Antenna beam width	$\pm 2°$ to 3 dB points
Beam direction	Vertical
Receiver bandwidth	100 Hz
System noise level in 100 Hz bandwidth	$\sim 10^{-18}$ W
Scattering volume at 100 m altitude	Cylindrical, ~ 7 m diam by 17 m deep

an open-top cylinder with a diameter three times that of the 4-ft antenna aperture, and with a height extending to 2 aperture diameters above the antenna. The data obtained from these two sets of measurements show that the anechoic shield provides a minimum improvement of 22 dB in suppression of the 90° sidelobe, and a minimum of 16 dB increased suppression of the 80° sidelobe at frequencies from 1 to 5 kHz.

Measurements were also made of the effectiveness of the anechoic shield in reducing the level of urban ambient acoustic noise received by the antenna. The results show an improvement of about 17 dB in noise rejection at frequencies from 1 to 3 kHz. The results also show only 6 to 7 dB of improvement at frequencies above 3 kHz. This is because the ambient noise at these frequencies was so much reduced by the shielding that it approached the irreducible thermal noise level of the antenna transducer.

The results of the "haybale" experiment indicate that it will be safe to operate echo sounders in urban and suburban areas without creating a noise pollution problem. It is therefore planned to develop a mobile, steerable acoustic antenna with built-in anechoic shielding for use in populated areas. The well-developed state of this equipment is evident from Fig. IIa-6.

The following is an extract from the conclusions of this report:

The operation of acoustic echo sounders, to monitor atmospheric structure in the boundary layer, is entirely feasible in a variety of urban and industrial environments, if proper attention is given to acoustic antenna design. The side lobe patterns of the antenna must be limited 97 dB below the main lobe gain if performance of the sounder is to be limited only by nominal Johnson noise in the acoustic transducer, in the most noisy industrial environments. Anechoic absorbing structures surrounding horn-reflector antennas approach

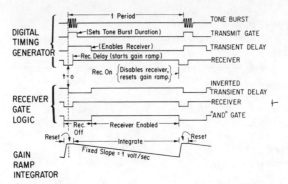

Fig. IIa-6. Mark II echo sounder control signals.

this performance, and further improvement is possible by tapering the antenna feed. But even with available and proven antenna designs, satisfactory sounder performance in urban areas can be obtained if some care is used in locating the equipment as remotely as possible from sources of noise.

The improvement in performance mentioned in the NOAA report was mainly the result of the directional gain of the shielded acoustic antenna. The principle used is described by Strand [1667]. The experiment by Simmons and Wescott [1659] was conducted with an antenna surrounded by absorbing material and the sound signal at 80° and 90° from the axis of the antenna was compared with that which would be received in the absence of any shielding. A theoretical numerical model for this experiment is derived in this article, and numerical results arising from this model are presented.

The geometrical configuration being considered is shown in Fig. IIa-7a. A planar antenna with circular symmetry is assumed. The dimensions studied for the configuration of Fig. IIa-7a are as follows (all dimensions are in feet):

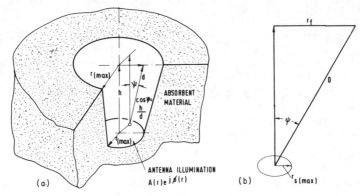

Fig. IIa-7. (a) Assumed geometry of shielded antenna. (b) Farfield approximation $h \gg r_s(\max)$.

$2 \leqslant h \leqslant 12$, $r_{(\max)} = 2$, $r_{s(\max)} = 4$ and 6, and wavelength $\lambda = 0.5$, corresponding to a frequency of ~ 2 kHz.

The procedure used to calculate farfield gain pattern may be summarized as follows (see Fig. I1a-7b): Kirchoff's integral, which may be used for acoustic waves as well as electromagnetic waves, is calculated to derive an "equivalent illumination" across the horizontal aperture at height h above the shielded antenna. This equivalent illumination is then truncated at $r_s = r_{s(\max)}$. A two-dimensional Fourier transform (derived from Kirchhoff's integral) is then used to obtain the farfield resulting from the equivalent illumination. Finally, the absolute value of the farfield signal is converted to decibels below the center value to obtain the gain pattern.

This theoretical study can be helpful in many future experiments with a highly directional antenna, Hooke et al. [1664] presented and compared the acoustic sounder and microbarograph records of atmospheric waves propagating in the planetary boundary layer over Table Mountain, Colorado. The two observing techniques are complementary in that the array provides wave amplitude, horizontal phase speed, direction, and wavelength, while the sounder provides a detailed picture of temporal changes in the structure of the lowermost kilometer or so of the earth's atmosphere.

With such information, and with the use of Doppler acoustic sounding [1663], it should then be possible to relate the height fluctuations of these strata to the wave amplitude, to determine the directions and magnitudes of the associated wave energies and momentum fluxes, and determine the energy exchanges between the wave, the turbulence, and the background flow. Then for the first time, these energy exchanges might be used as an additional input into atmospheric computations. To revert to the monitoring of the environment around airports previously mentioned, Beran [1661] suggested that on the basis of confirmed measurements with acoustic techniques and an extension of these ideas, one could envision systems similar to that shown in Fig. I1a-8. Here a fixed, vertically pointing antenna, in conjunction with two orthogonally positioned scanning antennas could provide a continual real time record of the inversion height, the turbulent intensity, an indication of the presence of wing tip vortices, and the vertical profile of the total wind vector.

In spite of these developments it appears that acoustic sounding techniques have a serious competitor in the FM–cw radar sounder. Richter et al. [1773] report on their second generation FM–cw radar sounder, which has scanning capability and is fully mobile. The sounder is now used as a remote sensor with unprecedented range resolution for fine scale structural details in the lower troposphere. Design considerations and performance characteristics of this new sounder are presented in their paper and examples of observations of clear air atmospheric scattering layers and structure of rain are given.

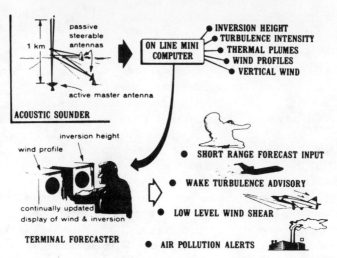

Fig. IIa-8. Schematic of monitoring system which could provide real time information about meteorological conditions near an airport.

b. *Acoustic Air Pollution Monitoring*

Beran *et al.* [1662] also applied acoustic echo techniques to air pollution monitoring. The conclusions of their paper are that, while full-scale testing in an urban environment remains to be completed, the results to date clearly indicate the potential of an acoustic sounder as a monitor of important meteorological parameters that contribute to undesirable air pollution situations. At present it appears that the worst effect of a high noise environment on the operation of a sounder will be periodic losses of record, a factor which can be significantly reduced by improved anechoic shielding and proper selection of operating frequencies. In the unlikely event that as much as half of the record is lost, the amount of usable information would still be orders of magnitude greater than is available from conventional measuring techniques.

The importance of knowing the low-level winds associated with an inversion also suggests that the Doppler capability should be incorporated into any operational system. The alternative would be to produce a backlog of acoustic records taken under various types of boundary layer structure and then use a pattern recognition technique to relate the observed records to conditions which might produce hazardous air pollution events. The Doppler capability would require slightly more complex equipment, but the additional information would seem to be worth the extra investment. Analog techniques, now under development, will allow rapid extraction of the Doppler shift at predetermined intervals above the ground. This information can then be converted into a real time display of the wind profile, updated every two or three minutes.

The number and location of monitoring stations in an urban environment must be a function of such factors as, the local terrain, the pattern of residential versus commercial zoning, and the mesoscale climatology of the city, each case being analyzed individually. If, for example, convergence into the urban heat island is a factor, stations could be placed at strategic locations around the perimeter of a city and the rate of inflow monitored continuously. In the final analysis, especially when compared with other types of remote sensors, the inexpensive acoustic sounder, combining its wind and temperature measuring capability, seems well suited to the task of monitoring the mesoscale urban environment. Hall [1656] has presented a paper of similar content.

c. *Acoustic Pulses in Physiology*

While very sophisticated antennas and amplifiers are needed in human techniques for remote acquisition of information, nature often obtains her results with small but tricky and often ingenious systems. Physiological research on acoustic communication between animals indicates that the sounds used are mainly of a nonsinusoidal flip–flop character. Nocke [1707] shows the ability of the cricket ear to code different sound stimulus parameters compared with the song parameters of the natural cricket songs. The following is a summary of the results:

(1) The threshold curve of the whole tympanal nerve shows an optimum near 4 and 14 kHz for male and female crickets.

(2) Near 2 kHz sound as well as vibration stimuli are answered by the same type of unit. In this range a strict division between the function of the sound and vibration reception organs is not possible.

(3) Sound stimuli near 4 kHz are answered by numerous spikes of mostly small amplitudes. On the other hand, 14-kHz stimuli are answered most frequently by only one unit, though a maximum of three units has been recorded. The coding features of this 14-kHz (HF) unit have been analyzed.

(4) The anatomical facts, the simultaneous sound stimulus experiments and the ablations of the tympanal membranes indicate that the cricket ear is able to discriminate pitched sounds near 4 and 14 kHz.

(5) Both the spectrograms of the calling and rivalry songs have a main peak near 4 kHz and a secondary peak near 14 kHz. The dominant components of the courtship song show a maximum only near 14 kHz and ultrasonic components up to 100 kHz.

(6) The absolute sound level has been measured for the three cricket songs. From this, the theoretical range of the cricket songs has been calculated.

(7) The 4-kHz optimum of the threshold curve matches the main peak of the sound frequency spectrogram for the calling and rivalry songs. The

14-kHz optimum of the threshold curve matches the main peak of the courtship song spectrogram as well as the secondary peak of the calling and rivalry songs.

After more detailed study of the physics of reception and transmission of this directional communication (the same author was at the time engaged on locust research at the ANU, Canberra), it might be feasible to make acoustic remote-sensing equipment more compact, or perhaps to make an entirely different approach.

2. Acoustic Pulses and Electroceramics

a. *Capacitance Microphones*

Microphones should cover either a broad frequency band with low noise level and without significant resonance or should use their own resonance to obtain peak sensitivity at a certain frequency. For 30 MHz operation and small bandwidth, Meeks *et al.* [1341] describe two capacitance microphones for the measurement of ultrasonic properties of solids. Two capacitive transducer designs have been used in their laboratory for amplitude measurements of longitudinal ultrasonic waves in solids. The designs are simple, can be fabricated easily, are rugged, and should prove to be a useful ultrasonic laboratory item. Each of the microphones can be conveniently altered to be used with a variety of specimen sizes, and displacement amplitudes as small as 10^{-14} m have been measured for 30 MHz frequency longitudinal waves. One of them is shown in Fig. I2a-1. As an example application of the transducer, measurement of phase shift upon reflection from an x-cut quartz transducer bonded to an $SrTiO_3$ crystal is given.

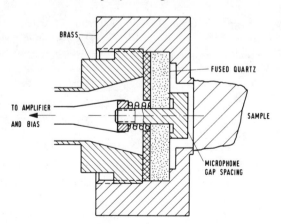

Fig. I2a-1. An assembled capacitive microphone.

b. *Acoustic Pulses in Technology*

In the field of science electroceramic substances are used for such purposes as for example, the ultraprecise movement of interferometer mirrors as shown in Fig. I2b-1. Here the effect is used to lock a piezoelectrically scanned interferometer to one mode of a multimode laser to obtain a single mode source.

The technique of ultrasonic inspection is at an almost fully developed stage, as can be seen by the modest steps represented by new patents in this field (see e.g. [1337]). Modern equipment changes the testing frequency automatically. The following is a typical claim of such a patent [1337, claim 12].

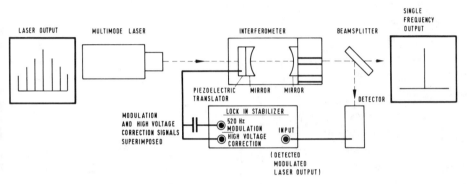

FIG. I2b-1. Piezoelectrically scanned interferometer [1051].

Apparatus for ultrasonic testing which comprises rate generator means arranged to generate periodic electrical timing signals; pulser means connected to receive said timing signals and generate a multiple frequency electrical pulse in response to each timing signal; ultrasonic transducer means connected to receive each of said electrical pulses and generate an ultrasonic multiple frequency mechanical wave corresponding thereto, said transducer means being responsive to mechanical echo waves received during the intervals between said pulses to generate electrical echo signals corresponding thereto; a plurality of bandpass receiver means, each connected to receive said electrical echo signals and amplify only those frequencies within a preselected different frequency range; a plurality of amplifier gain compensator means, each operable from said rate generator means and connected to sequentially activate a different one of said bandpass receiver means; and indicator means connected to receive the passed frequencies from said bandpass receiver means and provide a physical indication thereof.

c. *Acoustic Pulse Technology of Hydrophones*

The development of better electroceramic materials enables the industry to manufacture improved deep sea and general purpose hydrophones.

TABLE I2c-1

SPECIFICATIONS OF FOUR NUS CORP. HYDROPHONES

	Model 1120	Model 1140	Model 1100	Model 1110
Element sensitivity, dB re 1 V/μbar	−88	−92	−78	−90
Element material	Lead metaniobate		Lead zirconate-titanate	
Element electrostatic shield	No	Yes	Yes	Yes
Preamplifier voltage gain, dB[a]	20	20	20	20
Transmission transformer (2:1 stepdown)[b]	Yes	No	No	No
Operating sensitivity, dB re 1 V/μbar	−74	−72	−58	−70
Self-noise		See noise curves		
Frequency response,				
Low end ±3 dB (Hz)	5	5	5	5
High end ±3 dB (kHz)	10	50	50	100
Directivity[c]				
Omnidirectional in horizontal plane ±2 dB (kHz)	10	50	100	150
Frequency at which −3 dB points occur in vertical plane at angle of ±135° about hydrophone axis (kHz)	5	30	40	50

2. ACOUSTIC PULSES AND ELECTROCERAMICS 397

Maximum acoustic level, dB re 1 μbar	+74	+78	+64	+76
Operating temperature range	−2° C to +30° C		−2° C to +50° C	
Maximum sensitivity change over temperature range of −2° C to +30° C	½ dB	½ dB	—	—
Maximum sensitivity change over pressure range of 0–10,000 psi	½ dB	½ dB	—	—
Calibration provisions	10 Ω, 1% resistor between element and low side of preamplifier input		35 Ω in series with 100 μF	
Output impedance	50 Ω			
Maximum output voltage, volts RMS (open circuit)	1.0	2.0	2.0	2.0
Maximum output current, milliamperes RMS (short circuit)	3	1.5	1.5	1.5
Maximum operating depth, feet	36,000	10,000	3000	3000
Power requirements		12 ± 1 V at 4.5 mA		
Housing material		Type 316 stainless steel		
Size (excluding connector)	4 o.d. × 12¾ in. long		2 o.f. × 10½ in. long	
Weight (lb)	23	5	5	5

[a] Preamplifiers are available with voltage gains of 10 dB and 14 dB[c]
[b] Transmission transformer provides balanced output.
[c] Hydrophone axis vertical.

398 I. CONVERSION OF CAPACITIVE STORED ENERGY

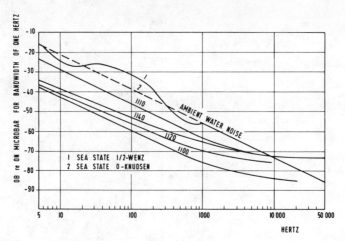

FIG. I2c-1. Self-noise of four NUS Corp. microphones at a range of frequencies, as compared with ambient water noise.

Figure I2c-1 shows hydrophone self-noise versus frequency for the hydrophones of NUS Corp., Underwater Systems Division [1346], and for comparison the ambient water noise.

All models are equipped with low-noise FET preamplifiers, power supply noise isolation filters and electrical calibration circuitry. The model 1120 is also equipped with a wideband transmission transformer for coupling to a balanced line. Transformers can be added to other models, if desired, with a nominal increase in price. The housings and all external hardware are fabricated from type 316 stainless steel for maximum resistance to corrosion. NUS Hydrophones can be used to measure ambient sea noise levels, to observe fish and animal sounds, to receive signals in acoustic propagation studies and for geophysical applications such as seismic sounding.

Table I2c-1 shows the specifications of the four hydrophones to which Fig. I2c-1 applies. As can be seen, Lead metaniobate, lead zirconate, and lead titanate are used here as the most advanced electroceramics.

d. *Conversion of Acoustic Pulse Energy into Electric Pulses*

The use of the piezoelectric effect to generate electric sparks by shock wave excitation was first reported by Vollrath and Schall [1959].

If a pressure wave or shock wave strikes piezoelectric substances such as crystalline quartz or electroceramic materials, it produces measureable electrostatic charges at a voltage that depends on the maximum pressure. A number of sintered mixtures, especially of metal oxides, have been developed, having different characteristics of brittleness, elasticity, and Coulomb number. If such electrostatic pressure transducers are used in practice, the elasticity of the material must be adapted to the particular shock wave. Typical watt-second numbers per cubic centimeter are available for each substance.

The principle of the method used by Vollrath and Schall is as follows:

An explosive charge produces the shock wave and the impact of this on the electroceramic material produces a correspondingly high-voltage pulse or an electric spark. The efficiency of this energy transformation can be very high because of the low loss by deformation in the electroceramic material. If a number of electroceramic blocks are connected in series so that the shock wave passes through all the member in predetermined microsecond intervals, a number of electric sparks will be produced in sequence. If each spark gap is used as a light source for high speed photography, such generators can be used for producing the limited series of light flashes required in Cranz–Schardin high speed photographic equipment. The high-voltage spark can also be used for igniting another explosive, an arrangement which could have applications in the weapons field. For example the shock wave could be produced when the nose of the projectile struck another object. The very high pressure, with steep rise, would cause the electroceramic material to produce an electrical ignition pulse, which would fire the explosive charge of the projectile.

Another, more practical, application for everyday use is in electroceramic transducers feeding spark plugs of car engines. Here the synchronized pressure pulses are supplied from the camshaft.

Figure I2d-1 shows a simple arrangement to produce a delayed electric spark by detonation of a small lead azide detonator. The detonator is on the left, normally having a mass of 20 mg; it is housed in a steel box. A 60-mm-long water buffer transports the generated shock wave to the right-hand end. The electroceramic generator is backed by a copper plate. The shock wave needs ~40 μsec to pass through the water so that the electric spark at the spark gap F is delayed by this time.

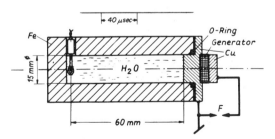

Fig. I2d-1. Shock wave activated electroceramic spark generator. The spark is delayed by the time taken for the shock wave to pass through the H$_2$O buffer.

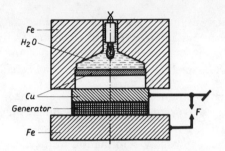

FIG. I2d-2. High intensity spark generator. The shock wave from the detonator in the head (20–40 mg lead azide) passes the H_2O buffer and accelerates the upper copper plate which strikes the lower copper plate, and causes the electroceramic generator to produce the high-voltage pulse which feeds spark gap F.

For one-time use the arrangement in Fig. I2d-2 is preferable. Here the lead azide detonator is shown in the upper part of the figure inside a steel block. The shock wave passes through an acoustically formed water buffer and it reaches all parts of a copper plate almost simultaneously. The entire copper plate then moves against another copper plate and in the electroceramic disk a very steep high power electric pulse is generated, igniting the spark gap F. The electroceramic material is backed by a steel block.

Figure I2d-3 shows a multiple spark arrangement of the kind described

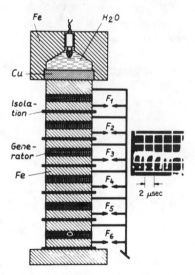

FIG. I2d-3. Electroceramic multiple-spark generator. The shock wave produced by the detonator in the head passes the six electroceramic disks which are sandwiched between steel plates. A 2-μsec delay between sparks has been realized, corresponding to 500,000 frames/sec high speed photography.

earlier. The detonator inside the water buffer at the top of the figure produces the shock wave which passes the copper plate Cu and continues through six electroceramic disks, each sandwiched between steel disks. The light intensity of up to six sparks is good enough for high speed photography; the delay between the sparks depends on the shock wave velocity and can be as short as a few microseconds. Normally the best electroceramic materials are the so-called ferroelectrics because they are at the same time soft ferromagnetic materials with a very small hysteresis loop. Often barium titanate or lead zirconate are used.

e. *Generation of Acoustic Waves Using Electrodynamic Transducers*

The method of generating shock waves by electrodynamic transducers is still the same as described in "High Speed Pulse Technology," Volume I, pp. 460 ff.

An ultrasonic atomizer using piezoceramics and able to spray 100 liters/h has been described by Martner [1338]. To obtain highest efficiency, it operates in resonance. The spraying rate then requires only a 100-W generator. For pulsing movements the efficiency would drop. Here again barium titanate, lead niobate, and metaniobate are the electroceramic materials used. They are artificially polarized by applying 16 kV/cm potential between the electrodes in the presence of a slowly decreasing temperature gradient.

3. Advanced Sonar Techniques

a. *Acoustic Data Acquisition Systems*

The development of electroceramic ultrasonic transmitters and hydrophones together with sophisticated IC-circuitry has boosted the sonar technique. For this fairly broad field some typical examples may be reported. Brown [1177, 1178] gives a survey of the activities of the Navy in the biological sciences.

The U.S. Navy Underwater Sound Laboratory, in New London, Connecticut, has been engaged for some years in the study of the acoustics of deep scattering layers. Surface-borne techniques have limited the scope of some experiments, but with the development of Deep Research Vehicles (DRV), definite advantages over the surface investigations have been found (Assard and Hassell, 1966 [1960]; Barham, 1963, 1966 [1961, 1962]; Fisch and Dullea, 1966 [1963]). A program was developed at the Laboratory to utilize a DRV to obtain, concurrently, acoustic measurements and samples of organisms of the deep scattering layers.

Acoustic ranging pulses and other research tools are used to detect biologic phenomena and previously undiscovered worlds of life. At each depth a

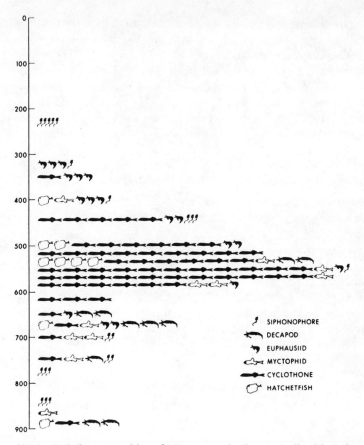

FIG. I3a-1. Relative composition of scatterers vs depth, ascent, dive 26, station 6.

different family of organisms is found to cause the acoustic scatter. Figure I3a-1 [1178] is a typical example of results obtained with the help of strobe light illumination and controlled nets.

The dominant components of volume reverberation are generally attributed to resonant scattering of sound by the gas-filled swim bladders of certain fishes, of the marine fauna, composing the Deep Scattering Layers (DSL) in the ocean. Maintaining layered distribution in an effort to remain at fixed light levels, most of the organisms in the DSL have habits of diurnal vertical migration during which their bladders experience changes in pressure. In the Western North Atlantic, resonant scattering has been observed from ~ 3 to 20 kHz. Since DSL occur in much of the world's ocean area, volume reverberation due to the DSL can be a factor limiting the performance of active sonar systems.

The special acoustic instrumentation developed for use on the small submersible is shown in Fig. I3a-2 [1179]. It consists of a "flextensional" ceramic transducer mounted at the focal point of a paraboloidal reflector system, which is constructed of compliant, hollow aluminum tubes filled with air and pressure-equalized to permit operation under varying hydrostatic pressures with changing depth. The assembly has an effective aperture 1.5 m in diameter and produces a conically symmetric beam 15–20° wide at the half-power points over a frequency band of 1500 Hz centered at 4 kHz. It serves as both a source and a receiver in these measurements, with a source index level of 109 dB/1 μbar. Its effective weight in water is only 70 lb, including a four-tank scuba air supply.

FIG. I3a-2. Deepstar 4000 carrying paraboloidal array.

During a measurement, at least 20 horizontal transmissions were made at each of three frequencies (3.25, 4.00, and 4.75 kHz) and two pulse lengths (40 and 120 msec). As indicated in the block diagram in Fig. I3a-3, a keying pulse generated by the 5-kHz clock triggered the transmitter. The reverberation was tape-recorded for 10 sec following each pulse. In the data reduction process, the pulse from the source-level-monitor hydrophone was used to start the input sampling program of the Univac 1230 computer. After being filtered and envelope-detected, the reverberation signal was converted to digital form by the multiverter at a rate of 500 samples/sec. The output of the multiverter was then transferred to the computer for averaging and computation of scattering strength. The scattering strength data were

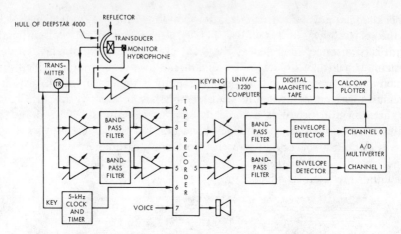

FIG. I3a-3. Block diagram of data acquisition and reduction system.

tape-recorded in digital form for possible future reference and displayed graphically by a CALCOMP plotter for the purpose of analysis.

Bennett [1655] describes some of the more important current developments in the electronics field which are aimed at producing new or improved equipment for use in deep-sea fisheries. In the main, the article concentrates on developments in the British fishing industry, in which the author has been working for the last ten years. However, reference is made to developments in other countries where, for certain applications, the development of new electronic aids has been even more striking than in the British Isles. Although the scope of the article has been confined to deep-sea fishing, many of the developments described apply equally well to the inshore or near-water fleets of Britain and other countries. In fact, the private ownership of the small vessels making up these inshore fleets is one of the more important reasons why they have been at least as quick to appreciate and adopt new technical developments as the operators of deep-sea vessels.

Quite apart from the improved detection capabilities of modern sounders, it is important to observe that their displays have been made much easier to read and interpret. The increasing complexity of modern fishing operations is imposing an extremely high workload on the trawler captain, and this whole question of simplified presentation, leading through to automatic data processing, is currently receiving attention in Britain and elsewhere. Although, as mentioned above, the modern sounders are able to give a more quantitative indication of the likely catch rate than the earlier versions, the relationship between echo-sounder presentation and the catchable quantities of fish present underneath the vessel is still crude and highly variable.

The difficulties involved in detecting fish close to the bottom with anything other than a vertical echo sounder are mentioned by Barnett [1654] because of the very considerable amount of work which has been carried out in the British Isles in recent years on very advanced sector-scanning sonar for horizontal detection. The problem can be seen by reference to Fig. I3a-4, which shows that, unlike vertical sounding in which there is a small range distance allowing the target echo to be separated from the sea bottom, the fish are now at the same range as part of the bottom itself. This immediately raises the problem of discrimination between the fish and bottom echoes.

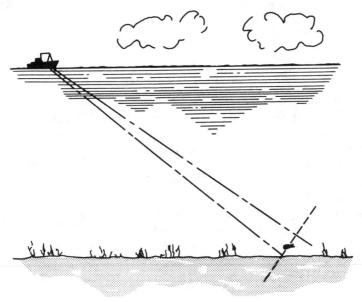

Fig. I3a-4. The acoustic problem in horizontal searching for fish close to the sea bottom (not to scale).

The need for narrow beams has made it necessary to use high frequencies, often greater than 200 kHz, to maintain a reasonable size of transducer, and this has introduced severe limitations on range. Furthermore, to ensure adequate coverage, it has been necessary to sweep the narrow beams through a considerable angle, and fairly complex within-pulse scanning systems have had to be adopted. Also, it has been necessary to introduce accurate stabilisation against ship motion, since the beams are so narrow (see Chapter I3b). The result has been the development of "sector-scanning" sonar of very considerable cost and complexity, which at the present time is suitable only for research and military purposes. However, it is possible that these research programs will result in viable commercial equipment in the foreseeable future.

TABLE 13a-1

CHARACTERISTICS OF AN OMNIDIRECTIONAL SCANNING SONAR, BY C-TECH LTD.

Operating frequency	30 kHz	Detection range	Dependent on sonar conditions such as thermal layers, sea state, etc. Under good sonar conditions ranges in excess of 2000 m can be achieved
Pulse length	3–30 msec		
Transmitter power	6 kW		
Beam width (−3 dB)			
Vertical	All modes 9°		
Horizontal	OMNI transmit 360°		
	Receiving 10°		
	All DT modes 10°		
Receiver sensitivity	1 μV	Scanning rate	555 Hz
Source level	OMNI, 117 dB/1 μbar	Beam rotation	360°
	All DT modes, 127 dB/1 μbar at 1 yd	Speaker tone	800 Hz (adjustable)

With stationary components C-Tech Ltd. [1651] has realized an omnidirectional scanning sonar. It provides effective search, tracking, and classification by means of two display channels. The omnidirectional scanning channel produces a continuous display of acoustic reception in all directions on a cathode ray tube indicator. The searchlight channel produces an audio response from any desired single direction. The characteristics of the system are shown in Table 13a-1.

The receiver transmitter unit encloses the preamplifiers, the electronic scanner, the transmitters, and associated power supplies. The cabinet is rectangular, fabricated steel with hinged front door. It is designed for bulkhead mounting with four bolts through the rear wall. The connector panel is located in the bottom of the cabinet for plug in of transducer cables and cables which interconnect this unit with the Control Indicator.

The transducer is a right circular cylinder consisting of 36 active transducer staves spaced equally around a cadmium-plated steel frame and enclosed by a ¼-in. thick rubber boot. The active material is lead zirconate, lead titanate resulting in an over-all transducer efficiency of better than 50%. The transducer is mounted by bolting its upper flange to a mating flange fixed to the hull or to the shaft of a hoist. Interconnection is provided by two 25-ft-long cables which plug in to the receiver/transmitter unit. The half-width angle of the rotating beam can be varied during operation and the electrostrictive transmitters and receivers are electronically gated to provide the required range and rotational scanning behavior.

The Omnidirectional Scanning (OMNI) mode of operation is used during medium and short range search and tracking operations. At the start of each echo ranging cycle a multielement cylindrical transducer transmits acoustic signals of high power and short duration. The intensity of this acoustical pulse is the same at all bearings (360°) and within a vertical angle of about

10°. During the reception period, which follows immediately the end of the transmission of the acoustical pulse, the area is scanned rapidly by the receiver and scanner network. An all-around picture of the underwater area is displayed on a cathode ray tube or plan position indicator (PPI). Thus, the size, shape, and exact position of all targets, at all bearings, are easily determined with minimum interpretation.

The Rotational Directional Transmission (RDT) mode of operation is used during long range search and tracking maneuvers. Instead of transmission in all directions at the same instant, acoustic signals are sent out simultaneously in three 10° beams, 120° apart. The three beams are then electronically rotated 120° effectively achieving 360° coverage. After the completion of the transmission cycle, the receiver rapidly scans the whole 360° area and the information is again displayed on the PPI. With increased effective acoustical pressure, longer range operation is possible. Because transmission is not simultaneous at all bearings, range errors of up to 20 m may occur. This fact, coupled with the extrahigh intensity acoustical pressure, makes this mode of operation not advisable at short ranges.

During the Steered Directional Transmission (SDT) mode of operation, the model LSS-30 performs like a searchlight sonar detecting underwater objects at both short and long ranges within a narrow trainable bearing sector. Three narrow acoustic beams 120° apart are transmitted. The operator controls the bearing of the beams. With the SDT mode of operation, both video and audio displays are available. During video display mode after each transmission, the receiver scans the full 360° at a frequency of 555 Hz and the signals are displayed on the PPI. During audio display, the scanner is stopped and the receiver is locked in position with only one of the narrow transmitted beams. The echoes are indicated audibly with the speaker and on a sound range recorder (optional). Both the transmitted and received beams are electronically steerable by the operator through 360°. SDT is an investigative high directivity mode of operation used primarily to investigate and concentrate on targets within a narrow sector initially located with the OMNI or RDT modes of operation.

Stepped-Steered Directional Transmission (SSDT) is used to automatically scan an 80° sector of the forward 180°. The same three narrow acoustic transmitted and received beams are employed as in audio SDT but after each transmission–receive cycle, the bearing is automatically stepped 5° over an 80° sector. The center bearing of the 80° sector is controlled by the operator. This mode of operation is used for search operations where only an 80° sector is of interest and where it is desired to have the information displayed audibly and on the recorder (optional).

In all modes of operation the range scale is selected by a five-position switch and may be set from 250 to 4000 m. In the model LSS-30 (PT) only, the vertical position of both the transmitting and receiving beams may be tilted by a manual control.

A smaller type of equipment is the acoustic ray trace indicator of Plessey Co. Ltd. [1050]. Developed originally for the Royal Navy (as Sonar Type 2014), it is an analog device which computes sonar ray paths from bathythermograph or sound velocity meter charts and displays predicted paths on a cathode ray tube. Features of the MS 100 include: immediate presentation of sonar ray paths; temperature/depth or sound velocity/depth data can be set-in in less than one minute; applicable to all types of sonar; accurate and reliable in any water conditions; equipped with special Polaroid recording camera.

The "diver-held sonar" by Burnett Electronics Laboratory [1347] is a self-contained portable sonar equipment, completely transistorized and utilizing printed circuitry. This gives the equipment compactness, dependability, and economy of operation. Easily installed or removed, the standard flashlight batteries are available internationally and provide power for ~ 20 h of continuous use, or 50–60 h of intermittent operation.

The "diver-held sonar" is hemispherical in shape and is ~ 12 in. in diameter. It can be operated by one diver, but has provision for an extra set of earphones for a "buddy" operation.

b. *Stabilization of Shipborne Sonar*

A system for stabilizing a shipborne sonar transducer was presented by Barnett [1654].

The development in recent years of high-power and high-resolution sonar transducers has led to a more critical requirement for the mechanical stabilization of such transducers. A new stabilization system, using gyroscopes as a reference, was recently developed by S. G. Brown Ltd., under a contract from the UK Ministry of Agriculture, Fisheries & Food. The Ministry, which holds the design rights of the system, now has it installed on the British Fisheries research vessel "Clione." The main design features of the stabilization mechanism are described in this article and some indication given as to how the various problems encountered were overcome.

c. *Acoustic Pingers*

Another useful acoustic aid to maritime research is the long-life pinger beacon of Burnett Electronics Laboratory [1347]. It is an underwater sonic transmitter especially designed to aid in the immediate and precise relocation of a position or preplaced object after extended time periods. Using a battery selected for reliability and long life, the unit provides a highpower, omnidirectional signal. It is buoyant and when anchored, will not become buried in bottom sediment. Its specifications are shown in Table I3c-1. Its one-year

TABLE 13c-1

SPECIFICATIONS OF UNDERWATER PINGER BEACON

Ping frequency	38 kHz nominal
Repetition rate	~12 msec on; 1.3 sec off
Operating range	~1 mi
	Range depends on sensitivity of receiving device as well as power output of pinger. The 1 mi figure is typical of the range when using the Burnett model 512 underwater acoustic receiving system
Operating depth	300 ft standard (optional gases for greater depths)
Battery life	~1 yr
	Actual current drain is low enough so that unit should operate 2 yr on one set of batteries. However, reliable shelf life is ~1 yr.
Dimensions	3 ft over-all length; 4 in. diam
Weight in air	10.75 lb
Weight in water	3.5 lb buoyant

operation with a detection range of 1 mi makes it a most useful accessory for picking up oceanographic equipment.

A dual-frequency pinger model 238 is also available. Another useful device is the underwater acoustic switch which generates pulse bursts of 320 msec every 2.5 sec and can be operated by pushbutton at a range of 1 mi. The acoustic switch is trimmed to respond only if actuated by the associated interrogator. The half-power angle of the interrogator is $\sim \pm 30°$. Small distress pingers with 17 different frequency channels from 31 to 44 kHz are available for emergency identification signaling, e.g., by trapped swimmers. Other items produced by Burnett Electronics Laboratory are: sonar transducer, broad band/narrow band sonar receiver; adjustable pulse tone pinger; environment indicators; deep submergence, pulse-type ,hydrobeacon; portable, directional, deckside scanning sonar set; missile pinger, multipurpose pinger, mine pinger, instrumentation pinger.

A fishing net monitoring system has been developed by EMI [1049] to enable appropriate parameters to be monitored while a trawl is in progress by passing signals from sensing devices on the trawl via an acoustic link to a visual display on the bridge of the trawler.

In Ref. [1344], a brief account of Soviet oceanographic activities may be found. Technical details are not presented.

4. Underwater Shock Waves from Sparks

a. *Underwater Spark Gaps with Metal Electrodes*

This section deals with further developments of the techniques described in "High Speed Pulse Technology," Volume I, pp. 470–503 and p. 537.

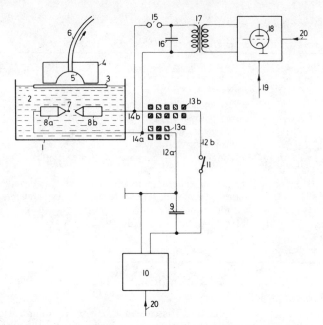

FIG. I4a-1. Circuit arrangement for liquid pressure forming of sheet metal.

Underwater spark gaps have a very high breakdown voltage. The electrolytic conductivity is nearly independent of that voltage. Normal salt sea water needs ~ 100–200 kV/cm to cause an underwater spark. The ignition pulse must be long enough to permit the development of an avalanche but short enough in rise time to inhibit unnecessary loss in the conductive phase before the breakdown. One of the main difficulties in economical operation of hydrospark metal-forming processes has been the lack of suitable circuits. Früngel [1359] proposed the solution shown in Fig. I4a-1.

In the figure, a vessel or tank 1 contains a liquid 2 such as, for example, water. A metal sheet or other suitable formable material 3 is positioned in the water 2 just beneath the surface. A mold 4 has an inner surface configuration 5. A vacuum is maintained between the surfaces 3 and 5 by means of a vacuum pump 6a, connected by a tube 6. A pair of spark electrodes 8a and 8b are positioned in the water at a distance 7 from each other. When a spark is produced across the spark gap 7, it produces a pressure wave or shock wave in the water. Electrical energy is supplied for the spark by a capacitor 9, charged by a power supply 10. The switch 11 keeps the discharge circuit open until the capacitor is ready to discharge.

The leads 12a and 12b are electrical conductors of large diameter and as free from inductance as possible in order to enable them to carry the discharge current of up to 10,000 A. A number of spaced ferrite rings 13a and 13b, positioned around the leads 12a and 12b provide current-dependent inductance. The ferrite rings introduce very high inductance into the main discharge circuit in the initial condition when the main discharge current is not flowing. The very high inductance is substantially eliminated upon magnetic saturation of the ferrite rings when the discharge current flows.

The main spark between the electrodes 8a and 8b is produced by a very high intensity pulse of very short duration which is supplied by a trigger capacitor 16 via an auxiliary spark gap 15. In operation, the main capacitor 9 is charged to a charge voltage of, for example, 1000 V, the switch 11 being open. If this switch were closed during this phase the current passing through the electrolyte would not be enough to cause magnetic saturation of the ferrite rings but it would reduce the charging rate of the main capacitor. The pulse generator 18 is then energized, for example by firing a thyratron, and charges the trigger capacitor via the pulse transformer 17. The trigger capacitor is charged to its full voltage of 60 to 80 kV in 10^{-5} sec or less. Just before maximum voltage is reached, the auxiliary gap 15 sparks over. The very high peak voltage produced by the discharging of the trigger capacitor produces a spark across the main gap 7. This spark has a relatively low energy and produces only a weak pressure wave that will form nothing more substantial than, for example, tin foil. Upon the production of a spark across the auxiliary gap, the switch 11 is closed either manually or by, for example, a relay or an electronic switch such as an ignitron, energized by the discharge current of the trigger capacitor. When the switch 11 is closed, the main capacitor 9 discharges through its discharge circuit which is completed by the spark across the main spark gap 7. This spark is usually maintained for about 10^{-4} sec, because of the ionization of the liquid. The pressure wave thus produced is exceptionally effective in forming the metal sheet 3.

The circuit arrangement permits the selection of the charge voltage of the main capacitor as desired, without variation of the spacing between the spark electrodes, so that any magnitude of charge voltage, low, medium, or high may be utilized. This permits great variation of the energy, ensuring optimum deformation of the metal sheet. (Section 15 deals further with metal forming using shock waves.) The considerably greater energy provided by this circuit arrangement results in great economy of operation as compared with earlier methods, which used a charge voltage of 20 kV instead of the modest 1 or 2 kV of this new system. There is thus a considerable saving of electrical input energy or power.

Spark gaps must be designed differently depending on whether the spark is to be produced in a liquid or in a gas. If a system designed to produce a spark in a gas is used instead to produce a spark in water, the operation of the system is unpredictable and unreliable. This is due to the high electrolytic conductivity of the water between the spark electrodes. Even if a liquid other than water is utilized, the spark is produced after a few discharges due to diffusion of the electrolyte in the liquid. If a discharge current of ∼10 A is produced between the spark electrodes by a charge voltage of 1 kV, the liquid has a resistance of ∼100 Ω and at a charge voltage of 60 kV, the discharge current is ∼6000 A. This would require a pulse generator which can provide a peak power of 360 MW. The pulse produced needs a duration of 1 μsec, so that the power requirement is 360 W-sec.

The impedance of the leads between the pulse transformer 17 or trigger capacitor 16 and the main spark electrodes should be equal to, or less than, the impedance of the spark produced across the main spark gap. If the leads are not equal, the discharge current would be dissipated in inductance and there would be no effective spark produced. The peak power mentioned above is easily attained when the trigger capacitor comprises a bank of capacitors having low inductance and the spark gap circuit has a low impedance. The construction of suitable spark gaps is described in "High Speed Pulse Technology," Volume I, Chapter I3.

b. *Underwater Spark Gaps with Water Electrodes*

An electrodeless spark sound source has been reported on by Wright [1349], who was apparently unaware of similar work referred to already in "High Speed Pulse Technology," Volume I, Chapter I3, pp. 476–503. His electrodeless spark exhibits, in principle, the same behavior as described by Früngel and Bailitis in a paper mentioned in that book. Figure I4b-1 shows the construction of the primary electrodeless spark device and Fig. I4b-2,

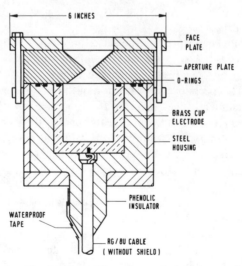

FIG. I4b-1. Construction of the primary electrodeless spark device.

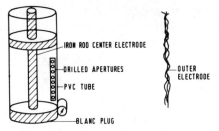

FIG. I4b-2. Construction of the secondary multiaperture tubular device.

the multiaperture tubular device (see Figs. I3c-15 and 21 of Volume I). The characteristics of these devices are given in Table I4b-1. The results presented by Wright are very precise.

The discharge energy of the electrodeless spark is provided by a capacitor. A bank of capacitors which represent an 18,000-J total energy storage capability is used. Capacitances from 2 to 40 μF are available in 2-μF increments at voltages up to 30 kV and capacitances from 1 to 10 μF in 1-μF increments are available at voltages up to 60 kV. Other capacitances are, of course, possible by means of various series–parallel combinations of the ten, double-pole, center-ground capacitors. The bank of capacitors is charged by a 60-kV 10-mA dc power supply. Charges of either polarity are possible. The total bank capacity of 18,000 J can be charged in 2 min.

The stored electrical energy may be switched either manually with a simple laboratory-fabricated manual switch, or electrically, using an E.G.&G. model GP-12A triggered spark gap. Virtually all of the triggering to date

TABLE I4b-1

CHARACTERISTICS OF PRIMARY AND MULTIAPERTURE DEVICES[a]

	Primary electrodeless device	Eight-aperture secondary device
Polarity of charge to inner electrode	Positive	Positive
Applied voltage	15 kV	6 kV
Capacitance	10 μF	10 μF
Water salinity	50 parts/thousand[b]	50 parts/thousand[b]
Aperture length	0.116 in.	0.035 in.
Aperture diameter	0.188 in.	0.060 in.
Aperture-to-electrode spacing	0.78 in.	0.5 in.
Aperture-to-aperture spacing	—	0.5 in.

[a] Used by Wright [1349].
[b] By weight.

has been accomplished manually since manual switching is less complicated. The energy stored on the capacitors is transmitted to the electrodeless device through approximately 50 ft of either one or two parallel G 8/U cables. The inductance of the storage bank and transmission wire has been measured at 4 μH.

The laboratory procedure is as follows:

(1) The electrodeless device is assembled with an appropriate aperture plate and mounted either near the viewing port, near the midpoint of the wood-stave tank, or in the pressure cell.

(2) The capacitor bank is charged to some voltage.

(3) The charged capacitors are switched manually onto the load.

(4) The initial rise of voltage on the electrodeless device triggers the oscilloscopes which record the outputs from the various transducers.

Successive variations of electrical and geometric parameters from these standard conditions resulted in empirical relationships between the acoustic output and the parameter being considered. Surveys were conducted for both electrical and geometric parameters, using the primary electrodeless device and, for geometric parameters, also using the tubular multihole device. The results are described below.

1. Electrical Parameters

a. *Polarity of charge to the inner electrode.* The pressure at one yard was measured as a function of time for both positive and negative polarities. The pressure–time waveform was similar in both cases; however, the amplitude of the pressures for a positive charge on the inner electrode was considerably greater than for a negative charge and represented a gain of ~ 3 dB. The implication that a somewhat greater pressure is developed on the electrically negative side of the aperture was substantiated by constructing a thin diaphragm-like aperture plate and discharging an energy sufficient to fracture it. The diaphragm was seen to be forced, almost intact, into the positively charged cup electrode. Additional evidence of the asymmetry of the spark is provided by high speed photographs showing that the negatively charged side of the spark is considerably more dendritic in character than the relatively cloud-like spherical surface of the positive side.

b. *Applied voltage.* The second electrical parameter examined was voltage. Standard conditions were maintained for each of the parameters, except the voltage, which was varied from the threshold of sparking to the point of mechanical fracture of the aperture plate. The peak pressures observed are plotted in Fig. I4b-3 as a function of applied voltage. The peak pressure increased to a degree corresponding to ~ 40 dB for an applied voltage increase from 6 to 12 kV and another 20 dB for a further voltage increase to 24 kV.

4. UNDERWATER SHOCK WAVES FROM SPARKS

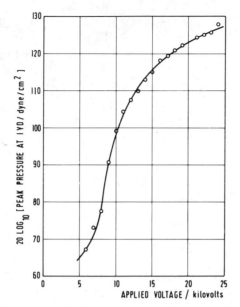

FIG. I4b-3. Dependence of peak acoustic pressure upon voltage. $C = 10$ μF; standard single aperture; salinity, 50 parts/thousand.

The increase of peak pressure resulting from an increase of applied voltage is fairly linear above 12 kV and is of a magnitude corresponding to ~ 1.7 dB/kV. The upper limit of 24 kV corresponded to a point of aperture plate failure.

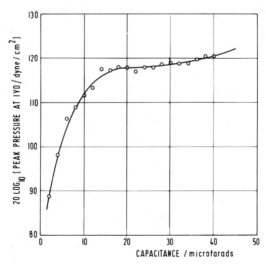

FIG. I4b-4. Dependence of peak acoustic pressure upon capacitance for the primary device. $V = 15$ kV; standard single aperture; salinity, 50 parts/thousand.

c. *Capacitance.* A pressure–time record at one yard was recorded at a fixed 15 kV, for a range of capacitances from 2 to 40 μF. The peak pressures observed have been plotted in Fig. I4b-4. The rate of increase of peak pressure with respect to capacitance was the greatest for capacitances up to 14 μF. The approximately threefold increase in applied electrical energy represented by the step from 14 to 40 μF resulted in only a 3-dB increase in peak pressure.

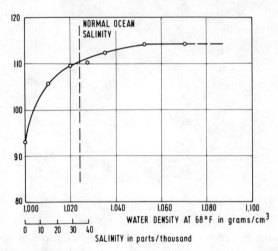

FIG. I4b-5. Dependence of peak acoustic pressure upon salinity for the primary device. $V = 15$ kV; $C = 10$ μF; standard single aperture.

d. *Salinity (conductivity).* The electrolyte used in the electrodeless spark has been a simulated ocean water composition. The pressure has been recorded for a standard discharge in electrolytes ranging in salinity from tap water to a solution of density 1.070 g/cm^3 at 68°F. The variation of observed peak pressures as a function of salinity is shown in Fig. I4b-5.

2. Geometric Parameters

a. *Aperture length.* The pressure at one yard has been recorded for a series of apertures under "standard" electrical conditions. A circular right cylinder aperture geometry was maintained whenever possible. A survey was accomplished for both the primary device and the secondary eight-aperture tubular device. The peak pressures recorded have been plotted in Fig. I4b-6 for each of these cases. It can be seen that the shorter the aperture length the greater the peak pressure generated. The increases realized with short aperture lengths have been the result of decreasing the electrolytic resistance in series with the growing spark. This correlation between aperture length and peak pressure has been found to be generally true; however, goemetries will be

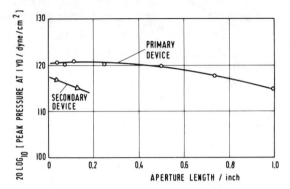

Fig. I4b-6. Dependence of peak acoustic pressure upon aperture length. All parameters other than aperture length maintained at standard conditions.

described later which result in improved electroacoustic efficiency, but which have a modest aperture length.

b. *Aperture diameter.* Pressure–time waveforms at one yard have been recorded for discharges using apertures of a variety of diameters. Hole diameters were varied from 0.0135 to 0.50 in. in the case of the primary device, and from 0.0135 to 0.25 in. in the case of the secondary device. The peak acoustic pressures observed are plotted in Fig. I4b-7. In both cases, the smaller the aperture diameter the greater the acoustic pressures. An optimum diameter for operation at 15 kV and 10 μF was found to be a surprisingly small 0.020 in.

c. *Proximity: aperture-to-electrode.* Pressure–time waveforms were recorded for discharges with varied aperture–electrode spacing. The inner cup electrode was artificially built up with brass washers to vary the aperture–electrode proximity. The outer face plate–aperture distance remained at the normal 0.78 in. An eight-aperture 2-in. diam tubular device was also tested.

d. *Proximity: aperture-to-aperture.* The evident increase in peak pressure with increased spacing implies a mutual interaction of the spark discharges. Subsequent experiments show that devices of this tubular construction (in which the inner electrolyte is common to all apertures) interact seriously, unless the aperture-to-aperture spacing is large compared to the aperture-to-electrode spacing. Electrical interaction between apertures can be eliminated by separating the inner electrolyte of the units one from another, i.e. utilizing discrete units.

The author summarizes his findings as follows: Electrodeless spark research has included a series of parametric surveys and, more recently, analytical and experimental studies of the physics of the discharge. The parametric surveys showed that in the range of operating conditions considered, the most

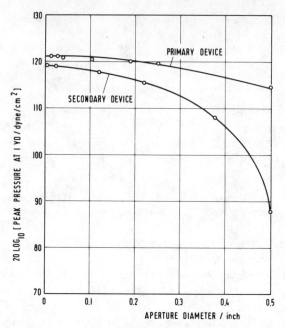

Fig. I4b-7. Dependence of peak acoustic pressure upon aperture diameter. All parameters other than aperture diameter maintained at standard conditions.

efficient discharges are produced by: capacitances generally in the region 6–12 μF, the greatest voltage available, the more conductive electrolytes, and geometric design which minimizes the electrolyte resistance. There is evidence that an impressed hydrostatic pressure decreases the peak pressure of high energy discharges to a degree which corresponds to $\sim$1 dB/1000 psi applied.

The electrodeless spark is a relatively omnidirectional source. Structural masking by the device produces, at high intensities, a broad band, 6 dB down, total beam width of 120°, and a 15 dB front-to-back ratio. Low energy discharges are less directional. Measurements of the spark indicate transient temperatures approaching 30,000°K and pressures in the spark of 20,000 psi.

The dynamic equations of motion of a spherical gas bubble are discussed and those aspects developed which are pertinent to the electrodeless spark dynamics. It is concluded that the electrodeless spark enjoys the unique potential to suppress the intensity of the secondary collapse, or bubble pulse. Experimental data are presented which support the conclusions of the analysis of bubble pulse suppression. Data are submitted, as well, from electrically parallel, arrayed discharges, which verify the absence of mutual loading of array elements.

The electrical and operating criteria necessary to ensure efficient and reproducible discharges are discussed. Under these condtiions the electrodeless spark has been found to:

(1) Be extremely reproducible;
(2) Routinely generate peak pressures up to 134 dB/μbar at 1 yard, and occasionally generate peak pressures corresponding to 141 dB/μbar;
(3) Have a pulse length of 50 μsec typically, 5–500 μsec observed;
(4) Vary in time of ignition by less than 1 μsec;
(5) Have an electroacoustic efficiency of up to 3%;
(6) Be suitable for arrayed operation, i.e., have no observed mutual interaction of simultaneous discharges;
(7) Apparently have the potential to suppress, and presumably to essentially eliminate, the secondary collapse, or bubble pulse.

Two new, relatively unexplored modes of operation are described which involve (1) operation of the device with a partially gas-filled inner electrode, and (2) control of the point of spark ignition by the use of a bubble or bubble screen.

c. *Steep Rise Underwater Sparks*

Schmied [1964] produced an ignited steep rise underwater spark at small energies. The paper (in German) describes an underwater impulse sound source with a stored capacitor energy of ~25 W. Since the time for the discharge is ~2 μsec the converted electric power amounts to ~12 MW. The special construction of the spark gap and the choice of "Vulkullan" as insulating material for the sliding discharge made it possible to reach a life expectancy of 200,000 pulses for this arrangement with a pulse rate of 3/sec. Figure I4c-1 shows the principle involved. Schmied used 0.5 μF at 10 kV and found a spark resistance of 60 mΩ at 1-mm electrode distance. Due to the 10 kV inductance of only 50 nH, he realized 12.5 MW electric peak power and 2 μsec pulse duration. The annular slit between the inner disk and the outer

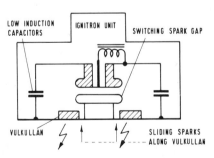

FIG. I4c-1. Underwater spark source of Schmied [1964].

420 I. CONVERSION OF CAPACITIVE STORED ENERGY

ring provided a very long path for the sliding spark. Several million sparks can be generated without readjustment.

d. *Physical Behavior during the Rise Phase of Underwater Sparks*

Kuzhekin [1652] published a dissertation on the behavior of underwater spark gaps as a function of pulse voltage and pulse current load. His paper gives a fundamental theory of the breakdown mechanism of underwater spark avalanches (in German). Twenty-five references can be found in this paper, most of them in the Russian literature. The author gives practical formulas

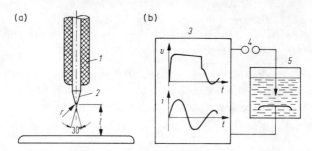

FIG. I4d-1. Principle of Kuzhekin experiments on underwater sparks [1652]. (a) Electrode configuration; (b) basic circuitry of the test gap: 1, insulation of the electrode; 2, zone of contact between metal and water; 3, voltage–current generator; 4, switching gap; 5, test vessel; r, radius of the metal tip; l, gap width.

for the calculation of the gap width, depending on pulse voltage, the rise time of the power pulse avalanche and the maximal value of the current pulse. He calculates, also the energy needed to bring the water in the channel up to boiling point, which also has a great influence on the gap width. The principle of his experimental arrangement is shown in Fig. I4d-1, for the benefit of those who have similar problems.

5. Shock Waves in High Pressure Physics and Metal Forming

a. *Material Behavior at Very High Shock Pressures*

A prerequisite for understanding metal forming and cracking processes is a knowledge of the shock compression curves of metals. The excellent Russian paper of Al'tshuler [1339], translated into English about the use of shock waves in high-pressure physics gives all data available in the year 1965 from shock adiabats up to the dynamic strength of materials and may be strongly recommended for reading. 169 references are given. Figures I5a-1 and -2 show the behavior of 32 metals.

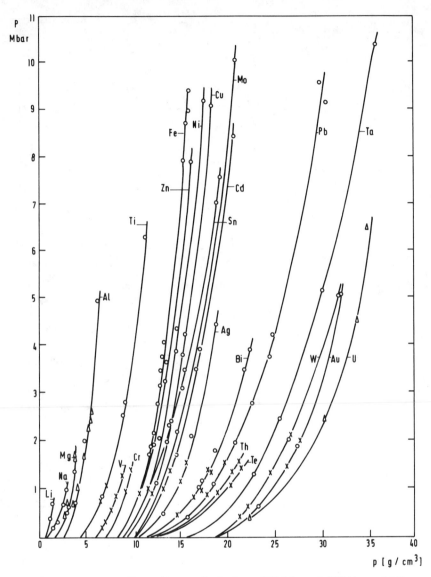

FIG. 15a-1. Shock adiabats of 23 metals. ⊙, data of Walsh *et al.* (USA); ×, McQueen and Marsh (USA); ○, Soviet investigators; △, experimental points of Skidmore and Morris, England [1339].

By way of an example, Table 15a-1 gives in detail data obtained with samples of copper and cadmium, at different iron striker speeds. The densities of the samples were more than doubled by the high pressures. The increase in the

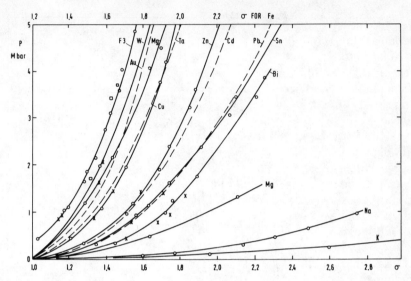

FIG. I5a-2. Shock compression curves of 32 metals as functions of their relative density $\sigma = p/p_0 = v_0/V$. ×, data of McQueen and Marsh; ○, experimental points obtained in the Soviet Union; ■, data of Skidmore and Morris [1339].

wave velocities is quantitatively characterized by the sharp increase in the average modulus of shock compression $\rho_0 D^2 = \Gamma v_0/(v_0-v)$: at the largest impact speeds, the per-unit internal energies of the investigated material are very large. For cadmium and copper they exceed by 6–7 times the energy of explosion of trotyl. Expansion from these states is accompanied by explosion and atomization of the metals. Even greater concentrations of thermal energy

TABLE I5a-1

PARAMETERS OF SHOCK COMPRESSION OF COPPER AND CADMIUM BY STRONG SHOCK WAVES[a]

	Copper					Cadmium				
W_y, km/sec	D, km/sec	U, km/sec	σ	P, mbar	$E \cdot 10^{-3}$, J/g	D, km/sec	U, km/sec	σ	P, mbar	$E \cdot 10^{-3}$, J/g
5.60	8.00	2.71	1.506	1.95	3.67	7.16	2.88	1.604	1.79	4.15
9.10	10.58	4.43	1.720	4.19	9.80	9.55	4.67	1.958	3.85	10.90
14.68	14.20	7.15	2.014	9.07	25.60	12.99	7.40	2.362	8.41	28.08

[a] W_y, wave velocity; D, velocity of the perturbation boundary; U, mass velocity of the substance; σ, relative density; P, shock pressure; E, stored compression energy.

were attained by shock compression of metals with lower density, with initial per unit volume $v_{00} = mv_0$.

At high pressure many metals show a two-wave configuration. For example, between 130 and 220 kbar and in a time of 0.2 μsec a phase transformation occurs in the structure of iron. The measurements of the required very high pressures have been done with sophisticated explosives. The ultrahigh speed machining reported by Arndt [1965] takes advantage of the behavior of metals as described in the Figs. I5b-1 and -2.

A rigorous analysis of ultrahigh speed machining (cutting speeds $> \sim 10,000$ m/min) must account for interrelationships between metal-cutting phenomena, dynamics, material behavior at high strain rates, dynamic plasticity and fracture phenomena, dislocation theory, stress waves, and metallurgical effects. Simplifying assumptions are clearly necessary before an analysis becomes possible. Previous studies of UHSM have been mainly of an empirical nature. They were stimulated by the obvious advantages of UHSM concerning faster production, and the possible improvement in the cutting mechanism, leading to various beneficial effects, such as, for example, lower cutting forces. An investigation into both the theoretical and practical effects of UHSM has been started at Monash University. Preliminary experimental results are reported elsewhere [1966, 1967]. The present paper deals with some of the theoretical aspects of UHSM.

A new theory for determining the cutting forces at ultrahigh speeds is presented. It accounts for an exponential decrease in shear volume (caused by the onset of adiabatic shear, melting or microcracks), a strain-rate-dependent increase in yield stress, and inertia effects. This theory predicts that for low melting point, low-density materials $(V_i > V_a)$* the resultant cutting force may exhibit an "adiabatic trough," in which for a range of cutting speeds, situated near 20,000 m/min, the resultant cutting force falls below that found at conventional speeds. As the situation $V_i < V_a$ is approached, this trough becomes more and more insignificant, and the resultant cutting force is always high. However, its material-strength-dependent component decreases with increasing speed, so that it would still be possible to cut the material, provided that inertia effects could be counteracted.

b. *Electrohydraulic Metal Forming*

The foregoing is a good introduction to electrohydraulic forming, in which the spark energy is nothing more than an electrically produced energy shock with the same behavior as an explosive. Its main advantage is that with small workpieces, no special precautions as with explosives, are necessary. Duncan

---

* V_i is the isothermal cutting speed under normal slow conditions, V_a is the specific speed at which adiabatic shear will commence.

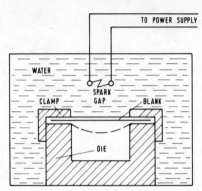

Fig. I5b-1. Basic electrohydraulic forming setup.

and Johnson devoted a section to electrohydraulic forming in their paper on high energy rate working of metals [1348].

Although the same basic equipment is required for this process as for magnetic forming, the transfer of energy to the workpiece is achieved in a very different way. Two electrodes, acting as a spark gap, are located in the vicinity of the workpiece (Fig. I5b-1). When the capacitor bank is discharged, an arc is established across the gap and the extremely rapid local heating in the vicinity of the arc creates a shock wave. The surrounding medium, which is generally water, transmits the energy to the workpiece. The process is therefore very similar to explosive forming in that kinetic energy is imparted to the workpiece; but the energy level from the electrical method is generally very much less than that available from explosives.

The operation can be carried out either with or without an initiating wire joining the two electrodes; both have advantages under certain circumstances. An initiating wire has to be replaced after every forming operation, and a system without this disadvantage obviously lends itself to more rapid operation and high production rates. On the other hand, there is some evidence to show that more consistent results can be achieved when a wire is used, and there are applications where large spark gaps are necessary and a wire is essential in order to guide the spark and to ensure that the pressure is directed as required onto the workpiece.

A certain amount of fundamental work has been carried out on both systems and as with explosive forming, the free forming of a circular blank securely clamped at the edges has been used as the test operation. Figure I5b-2 shows some results obtained by Kirk [1973] which indicate that without an initiating wire there is an optimum gap which varies with voltage for maximum deformation, and under these conditions the discharge is critically damped. This work also showed that, for comparatively low energy levels, the maximum deformation at optimum gap widths was obtained with a voltage of 30 kV.

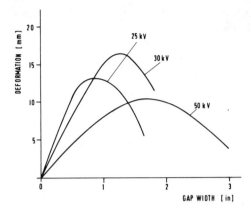

FIG. I5b-2. Optimum gap width results. Kirk [1973].

Work with an initiating wire has been reported by Kegg [1970] and Kapitza [1975]. Figure I5b-3 shows that an increase in forming efficiency, i.e., depth of deformation, was still evident with gaps considerably greater than those observed by Kirk [1973] to be the optimum without a wire. The diameter of the wire, rather than the material from which the wire is made, has been shown to be a critical factor, as the wire diameter decreases, forming efficiency increases. This has been explained by the fact that the energy required to vaporize the wire will increase with diameter and with larger wire diameters, less energy will be available for forming.

As would be expected, the deformation of a circular blank increases as energy of discharge increases. This is shown in Fig. I5b-4 together with an average of the results of similar tests using the magnetic forming process. It is interesting to note that at low energy levels, that is up to 3000 J, the

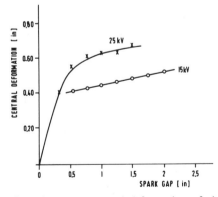

FIG. I5b-3. Effect of spark gap on central deformation of clamped circular blank. Capacitance, 10.5 μF; 1.5 in. standard off; steel wire, 0.010 in. diam.

electrohydraulic process appears to be more efficient than the magnetic method. At higher energies, however, the reverse may be true. It is hoped that this point will be investigated up to energies of 20,000 J when suitable equipment is in operation.

So far as equipment for carrying out electrohydraulic forming is concerned, there is already a 10,000-J unit commercially available and it is understood that a 60,000-J unit will be on the market shortly. From a direct comparison of available energy, a typical high explosive liberates ~5000 J/g; therefore a 10-kJ capacitor bank is equivalent in terms of energy released to only 2 g of explosive. The size of explosive charge which is necessary to form a 6-in-diam. pressure vessel end from 0.125 in. mild steel material is ~28 g which, assuming an equally efficient operation in both cases, would require a capacitor bank capable of delivering 140,000 J. Furthermore it has been estimated that a megajoule bank—equivalent to 12 1/2 oz of explosive—would require about 2500 sq ft of floor space.

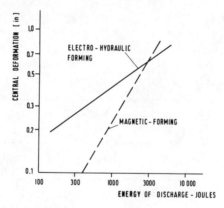

FIG. I5b-4. Effect of discharge energy on central deformation of clamped circular blank.

From this it would seem that the applications for electrohydraulic forming, as with magnetic forming, will be limited to small components. Operations such as the bulging of tubes and the forming of flat blanks could be suitable especially where the numbers are such that cheap onepiece tooling is a consideration. Compared with magnetic forming, it has the disadvantage that there must be water between the workpiece and the electrodes, but the advantages that the efficiency of forming is unaffected by the conductivity of the work material and no complicated coils are required.

A survey of electrohydraulic forming was presented by Duncan and Johnson [1348]. Metal forming with the aid of the pressure field produced by a high-energy spark appears to have been demonstrated first about ten years ago [1968, 1969]. Since that time, the process has been studied extensively in a number of countries. Among those who have investigated the process experimentally are the Cincinnati Milling Machine Co. [1970] and Republic Aviation Corp. [1971] in the United States and N.E.L. [1972], R.A.R.D.E. [1973], and the universities of Manchester [1974] and Belfast in Great Britain. The effects of the basic parameters of the process are now generally understood and machines are available for employing this process on a production basis. One machine, manufactured by the Cincinnati Shaper Co. has a maximum storage energy capacity of 121 kJ and a smaller machine of 25 kJ. A still smaller machine (10 kJ) is manufactured by James Scott (Electronic Engineering) Ltd. These machines are used chiefly for forming components from thin sheet and thin-wall tube. A simple diagram of the process is given in Fig. 15b-5. A bank of high-energy capacitors is charged by a high-tension transformer–rectifier unit.

At a predetermined charge level a switch is closed allowing the capacitors to discharge through an underwater spark gap. The pressure waves produced by the spark travel through the water and rapidly accelerate the blank which moves with a considerable velocity to fill the die cavity. The space between

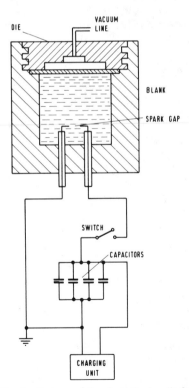

FIG. 15b-5. Simple diagram of the electrohydraulic forming system.

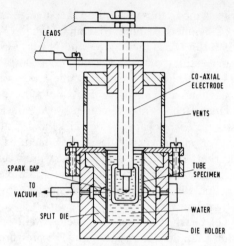

FIG. I5b-6. Arrangement of experimental forming of tubular components.

the blank and the die is evacuated. In some cases, an open tank of water is used instead of the closed chamber shown and also a fine wire may be placed across the spark gap. For forming tubular components, the spark gap is generally placed centrally inside the tube as shown in Fig. I5b-6.

A number of methods are available for storing energy to obtain very high current pulses. Electrochemical systems cannot deliver stored energy fast enough, but specially constructed generators have been used in the past, notably by Kapitza [1975]. For very high energy systems, superconducting inductions coils might provide an attractive storage method (Wiederhold [1976]), but it appears that for systems storing up to 100 kJ, the best method is that of capacitor banks.

The discharge circuit used in electrohydraulic forming equipment is generally constructed from heavy copper strip and arranged so that the inductance of the circuit is as low as possible. In some instances multiple coaxial cables are used. The discharge switch must be capable of carrying the very high current without introducing serious losses and it must also permit very short current rise times. A review of switching devices employed is given in Ref. [1972].

The electrodes and die equipment must be of robust construction to withstand the shock loading and the electrodes are insulated to prevent breakdown to the container walls. The equipment used by the authors to free form sheet metal blanks experimentally is shown in Fig. I5b-7. In production equipment, cycle times of 10 sec to 1/2 min are usual and equipment for rapidly opening and closing the die is necessary.

The physical nature of an underwater spark and the pressure field associated

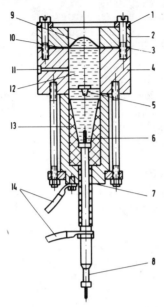

Fig. I5b-7. Die for free forming sheet specimens. 1. Clamping ring; 2. die ring; 3. bleed; 4. cylinder; 5. fixed electrode; 6. adjustable electrode; 6. adjustable electrode; 7. electrode holder; 8. micrometer screw; 9. specimen; 10. locating ring; 11. filling hole; 12. water; 13. insulation; 14. leads.

with it has been studied extensively [1977–1981]. When a high voltage is applied across an underwater spark gap, a bubble of ionized vapor is formed between the electrodes. The rate at which the energy is discharged also has some effect on the amount of energy in the pressure wave phase [1982], but in the range of inductance encountered in typical discharge units, this is not so significant.

An experimental investigation of the free forming of sheet metal blanks using the die shown in Fig. I5b-7 was carried out using the 1250 J storage bank. The results are given in the following paragraphs.

The actual work done in forming cannot be calculated without at least a knowledge of the stress–strain relation for the material at the strain rates involved. An estimate was made, however, using the static stress–strain relation and the strain distribution of deformed blanks. The process efficiency, i.e.,

$$\frac{\text{plastic work of deformation}}{\text{electrical energy discharged}} \times 100\%$$

was found to be quite low in this work. The maximum observed was ~7%.

The experimental investigation indicates the similarity between electrohydraulic and chemical explosive processes as metal working techniques. The advantages to be obtained by using explosive forming methods should apply equally well to the electrohydraulic technique. In addition, electrohydraulic forming can be carried out in a normal production area and repetitive working is possible. The scale of forming is limited, however, by the difficulty of storing large quantities of electrical energy. The maximum storage energy used at present is $\sim 120,000$ J and this is equivalent to only $\frac{3}{4}$ oz (20 g) of high explosive. There are, however, a large number of sheet metal components which can be formed with energies below this limit.

The deformation process of sheet metal by underwater spark discharges was also investigated by Bodenseher and Schmied [1984]. It was possible to find out the time dependence of the delivery of energy to the underwater

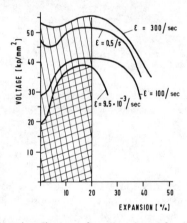

FIG. I5b-8. Pressure-elongation diagram of unalloyed steel for different elongation speeds.

spark gap on the deformation of the sheet metal, and an attempt was made to estimate the resistance of the underwater spark. Besides it was possible to formulate the relation between the energy discharged into the spark gap and the maximum velocity of deformation of the sheet metal. Finally, the mechanical efficiency of the underwater discharge applied to sheet metal deformation was estimated. In 1972, Balzerowiak and Prümmer [1360] published a very sound paper (in German) on the practice of electrohydraulic metal forming.

Figure I5b-8 gives typical data on the effect of speed of elongation on the intensity of the pressure wave, and Fig. I5b-9 shows the basic circuitry. Figures I5b-10 and I5b-11 show the application of a reflector to obtain an increase in efficiency. The authors give some examples and show failures. Hydrospark forming has been introduced with success mainly in the aircraft industry and

5. HIGH PRESSURE PHYSICS AND METAL FORMING

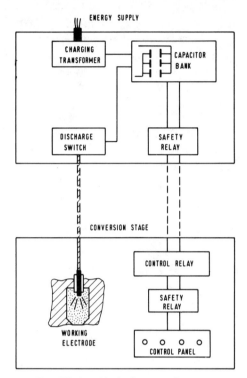

FIG. I5b-9. Block diagram of system.

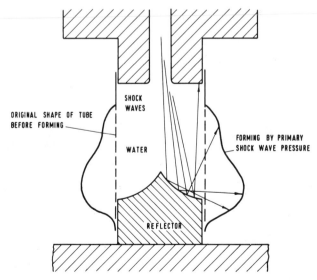

FIG. I5b-10. Importance of the reflector in electrohydraulic widening of a tube.

especially for stainless steel and other high tensile strength metal sheets when small radius curvature is required. Seventeen references are given, of which Refs. [1985] and [1986] may be mentioned here.

A Japanese report on a study of the utilization of generated energies in hydrospark and explosive forming is presented by Nishiyama and Inoue [1362]. The authors explain that in hydrospark and explosive forming, it is necessary to transform effectively any generated energy into energy for the plastic working of a blank. This paper describes the mechanism of energy transmission involved in plastic deformation and the utilization of generated energies. A high speed camera was frequently employed to observe conditions under water.

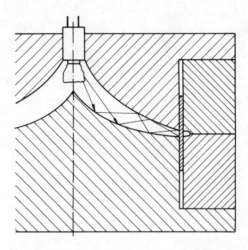

FIG. I5b-11. Arrangement of a reflector to generate ring constrictions in tubes.

The main points in this paper are as follows:

(1) In the free forming of a circular blank, the polar deflection varies with the stand-off distance and the distance between the water surface and the charge. A result of observation is that the behavior of the gas bubble greatly influences the deflection of the blank.

(2) By adopting the forming method that uses an air bag over a blank, the efficiency of energy utilization for deformation is improved remarkably. In this case, the primary shock wave has little effect on the deformation, and the kinetic energy of the water which is taken up from the gas bubble energy is responsible for the deformation.

(3) When both the pressure pulse and water hammer act together, the deformed shape of a blank becomes round. The character of the deforming process in this case is such as to establish a round profile at an early stage.

(4) The high pressure, whose duration is of the order of a millisecond, is created by the water hammer, which is generated by the water jet and a rapid gas expansion.

(5) Generally, the primary shock wave is not essential for deformation but the pressure pulse as a shock wave can form a circular blank into a conical shape.

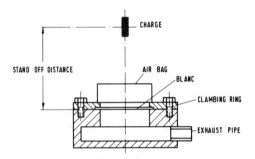

FIG. I5b-12. Schematic view when using an air bag.

They recommend the use of an air bag as shown in Fig. I5b-12 to improve efficiency. Figure I5b-13 indicates the increased deformation obtainable with this arrangement. Figure I5b-14 shows graphically the progress of the deformation with time, when an air bag is used.

In another paper [1361] the same authors studied the effects of hydrostatic head and an air bag over a specimen using high speed photography. Figure I5b-15 shows the equipment for high speed photography. The camera employed is the Simazu high speed camera. The water tank has a window

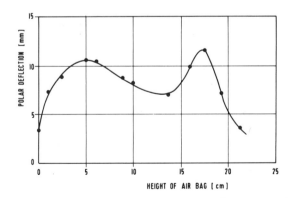

FIG. I5b-13. Polar deflection against height of air bag for a copper blank (0.7 mm thickness) using a No. 6 electric detonator at a stand-off distance of 22 cm and a charge depth of 33 cm.

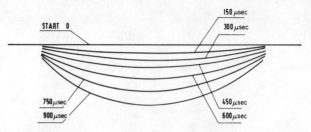

Fig. I5b-14. Deforming process on using an air-bag whose height is 1 cm in explosive forming with a No. 6 electric detonator at a stand-off distance of 22 cm and a charge depth of 33 cm. (Blank: copper, 0.7 mm thickness; i.d. of die: 50 mm.)

enabling photographs to be taken. Illumination is provided by four lamps rated at 1 kW. The film used was Eastman 4- ×, type 7224, 16-mm black and white negative film. The die used for free forming is made of a transparent acrylic acid resin, its internal diameter being 50 mm. The interval between the primary pressure pulse and the secondary pressure pulse is ~25 msec, which is equal to the first period of oscillation of the gas bubble. With change of the charge depth, this first period also changes (see Fig. I5b-16).

The behavior of the gas bubble generated by exploding wire is shown in the paper, where the electric energy used is 1–5 kJ. First the gas energy, E_0, is transformed into internal energy of the air, and then it is released rapidly. At this moment, both the pressure pulse and the water hammer act against the blank. With the air bag, the ratio of the first deformed volume to the second deformed volume varies with the height of the air bag. The air bag is constructed in polyethylene sheet of thickness 0.02 mm on a cardboard frame and its shape is cylindrical. (see Figs. I5b-17–20.)

The authors conclude that in free forming, the first deformation bears no relation to the hydrostatic head, while the second deformation is more influenced by the hydrostatic head. In the forming method with an air bag

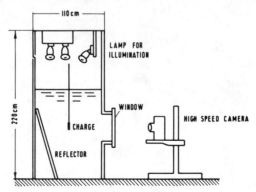

Fig. I5b-15. Equipment for high speed photography.

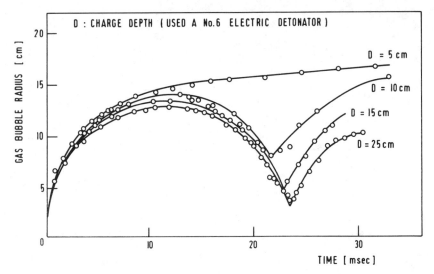

FIG. 15b-16. Radius–time curves for charge depth.

over the specimen, the kinetic energy of water which is taken from the energy of the gas, namely, E_0, is made use of. When both the pressure pulse and the water hammer act together, the effect on deformation is exceedingly improved. There are essential common points on the mechanism of the energy transformation between explosive forming and electrohydraulic forming.

Two other Japanese authors, Oyane and Masaki [1343] also made a fundamental study of electrospark forming. In this paper the analyses of deformation of the cylindrical shell and the circular diaphragm in electrospark forming were considered from the Newtonian equation of motion, the maximum shear–stress theory, and a few assumptions. The applicability

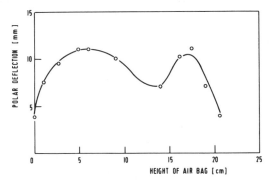

FIG. 15b-17. Polar deflection against the height of air bag for copper blanks (0.7 mm thickness) using a No. 6 electric detonator at stand-off 22 cm, charge depth 33 cm.

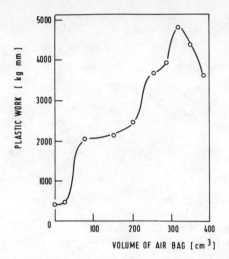

FIG. I5b-18. Plastic work against the volume of air bag for copper blanks (0.4 mm thickness) using a No. 6 electric detonator at stand-off distance 25 cm, charge depth 33 cm, height of air bag 4–5 cm constant.

of the analyses was confirmed from the experiments with specimens of aluminum, copper, and steel. Finally, the laws of similarity were obtained on the basis of the results of analyses and experiments. The formulas derived herein can be used to estimate the effects of changing major variables of the

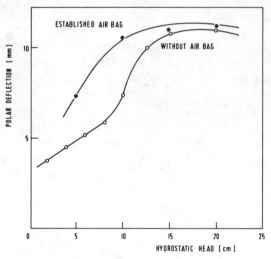

FIG. I5b-19. Polar deflection against hydrostatic head under the air bag (height, 1 cm) in the electrohydraulic forming. Stand-off distance, 9 cm. Capacitance, 30 μF; 10-kV wire used. Copper, 0.2 mm diam. 20 mm length. Blank copper, 0.4 mm thickness.

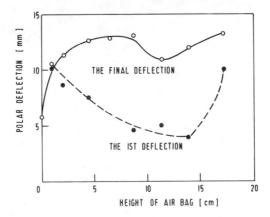

FIG. 15b-20. Polar deflection against height of air bag using a No. 6 electric detonator at 22 cm stand off distance, and 33 cm charge depth. Blank used: copper 0.7 mm thickness (from the result of observation by high speed photography).

process on the amount of final deformation, and to understand the contribution of each effect in determination of pressure magnitude. It should be borne in mind that these formulas were obtained for charged energy up to 4 kJ and have to be verified for higher charged energy. Tests for charged energy more than 4 kJ have been carried out and more general formulas have been developed, including the effects of conditions in the forming chamber on the maximum deflection. These results will be reported in a later paper.

Other useful papers refer to explosives and show how to handle reflectors. An excellent paper of R. Schall, president of the German–French Research Institute at St. Louis, can be found in the famous book by Vollrath and Thomer [1703 1] in Chapter 7, pp. 890–907, along with 116 other references.

Still another useful publication is available (in German) by Held [1052]. Held shows how, by using electrical capacitor-ignited detonators, to obtain desired shock waves of special types. His results are based on 10 μsec multi-exposures with an image-converter camera. The many excellent pictures in this paper can contribute greatly to a better understanding of high-pressure physics. A related technology is the observation of metal cracking processes. References [1987–2004] deal with the motion analysis of these events with the help of high speed photography.

Bibliography[1]

1007.* Tobe Deutschmann Lab. Canton, Massachusetts. Low inductance energy storage capacitors. Bull. EB365-20.
1008.* Fansteel Metallurgical Corp. Tantalum capacitors. Industrial paper.
1009.* Edgerton, Germeshausen, and Grier, Inc., Boston, Massachusetts, Krytrons, Thyratrons. Industrial papers.
1010.* English Electric Valve Comp. Limited. Lincoln, England. Electronic Valve and Component data.
1011.* Bacchi, H., and Pauwels, J. C. Study of a very low jitter spark gap: Application to the realization of a high speed camera using a biplanar shutter tube with exposure time of 0.3 nsec. *Proc. Int. Congr. High-Speed Photography, 9th, Denver,* pp. 488-492. Soc. Motion Picture Television Eng., New York (1970).
1012.* Pulse Generators. Tektronix. 25 years of excellence 1946-1971.
1013. Isolex, high voltage instruments. Industrial paper. Impulsphysik GmbH, Hamburg.
1014.* Marine radar pulse-coded transponder. Radar recording systems. Industrial paper, Emi Electronics Limited. Wells, Somerset.
1015. Früngel, F., and Thorwart, W. Spark tracing method progress in the analysis of gaseous flows. *Proc. Int. Congr. High-Speed Photography, 9th, Denver* pp. 166-170. Soc. Motion Picture Television Eog., New York (1970).
1016. Guinard, N. W. Naval Res. Lab., Washington. Radar detection of oil spills. *Joint Conf. Sensing Environm. Pollutants, Palo Alto* 71-1072, 1-8 (1971).
1017. Isopuls 10, pulse generator. Industrial paper. Impulsphysik GmbH, Hamburg.
1018.* Rudenko, N. S. A new method of high voltage supply to streamer chambers. *Sov. Phys. JETP* **22**, No. 5, 959-962 (1966).
1019.* Clement, G., Eschard, G., Hazan, J. P., and Polaert, R. Construction of a shutter tube operating with an exposure time below 300 psec. *Proc. Int. Congr. High-Speed Photography, 9th, Denver* pp. 498-501. Soc. Motion Picture Television Eng., New York (1970).
1020.* Marillaeu, J., Saint-Mleux, M., Maindron, R., and Garcin, G. Very-high speed three frames electronic camera with storage tubes, slow scanning and magnetic recording. *Proc. Int. Congr. High-Speed Photography, Denver* pp.216-223. Soc. Motion Picture Television Eng., New York (1970).

[1] *Key to symbols;* * These references have been abstracted in the text. † These references have been paraphrased in the text. ‡ These references have been quoted directly or very slightly paraphrased.

1021.* Blanchet, M.   High-tension generator and attenuator of pulses in the nanosecond and subnanosecond range. *Proc. Int. Congr. High-Speed Photography, 9th, Denver* pp. 112–119. Soc. Motion Picture Television Eng., New York (1970).
1022.* Preonas, D. D., and Swift, H. F.   A high-intensity point light source. *Proc. Int. Congr. High-Speed Photography, 9th, Denver* pp. 148–152. Soc. Motion Picture Television Eng., New York (1970).
1023.* Rowlands, R. E., and Wentz, J. L.   A low-voltage pockels cell having high repetition rates and short exposure durations. *Proc. Int. Congr. High-Speed Photography, 9th, Denver* pp. 50–57. Soc. Motion Picture Television Eng., New York (1970).
1024. Marks, A. M.   Light control by dipole glass window. Rep., The glass industry (1965).
1025.* Kerr cells, minikerr, standardkerr, HV-kerr.   Industrial paper. Impulsphysik GmbH, Hamburg.
1026.† The Variosens.   Oceanology international 72. Impulsphysilk GmbH, Hamburg.
1027.* Duguay, M. A., and Hansen, J. W.   Bell Telephone Lab. Inc., Murray Hill, New Jersey. Ultrahigh speed photography of picosecond light pulses.
1028.* Martin, J. C.   Multichannel gaps. SSWA/JCM/703/27 (1970).
1029.* Martin, J. C.   Nanosecond pulse techniques. SSWA/JCM/704/49 (1970).
1030.* Ares Facility.   Sandia Base, Albuquerque, New Mexico.
1031.‡ Mattson, A.   Ten nanosecond radiography at 1200 kV. *Proc. Int. Congr. High-Speed Photography, 9th, Denver* pp. 272–276. Soc. Motion Picture Television Eng., (1970).
1032.* Bergon, J. C., and Constant, J.   Application of electrooptical recording equipment to short-pulse-length x-ray flash investigations of explosive processes. *Proc. Int. Congr. High-Speed Photography, 9th, Denver* pp. 277–282. Soc. Motion Picture Television Eng., New York (1970).
1033.* Schaaffs, W., and Krehl, P.   Flash radiography of shock wave processes generated by double discharges from series-and parallel-connected sparks gaps. *Proc. Int. Congr. High-Speed-Photography, 9th, Denver* pp. 244–250. Soc. Motion Pictures Television Eng., New York (1970).
1034. Lankford, J. L.   Application of laser and flash x-ray techniques in hypervelocity ablation-erosion investigations in a hyperballistics range. *Proc. Int. Congr. High-Speed Photography, 9th, Denver* pp. 97–103. Soc. Motion Picture Television Eng., New York (1970).
1035.* Baykov, A. P. *et al.*   Pulse x-ray system with television method of image visualization. *Proc. Int. Congr. High-Speed Photography, 9th, Denver* pp. 283–286. Soc. Motion Picture Television Eng. (1970).
1036. Früngel, F.   Spark photography, Part 1: Operating principles. Part 2: Equipment and results. *Rep. Visual* **8**, No. 1/2 (1970).
1037. Ion Physics Corp., Burlington, Massachusetts.   Addendum to paper on neutron radiography with a 2.5 MeV Van de Graaff accelerator (June 1970).
1038.‡ Graybill, S. E., and Uglum, J. R.   Ion Physics Corp. Burlington, Massachussetts. Observation of energetic ions from a beam-generated plasma. *Appl. Phys.* **41**, No. 1, 236–240 (1970).
1039.‡ Viguier, Ph., and Bourdarot, G.   High-speed cineradiography at high energy. *Proc. Int. Congr. High-Speed Photography, 9th, Denver* pp.315–320. Soc. Motion Picture Television Eng., New York (1970).
1040. Früngel, F.   Introduction to the principle of stored energy resistance welding. Industrial paper. Impulsphysik GmbH, Hamburg.

1041. Impulse welding machines amd pneumatic impulse welding equipments. Industrial papers. Impulsphysik GmbH, Hamburg.
1042.* Handel, S. K., and Englund-Ponterius, A. Exploding bridgewire triggering of a flash x-ray discharge. *Proc. Int. Congr. High-Speed Photography*, *9th*, *Denver* pp.268–271. Soc. Motion Picture Television Eng., New York (1970).
1043.* Canada, C. E., Warren, T. W., and Rhoton, N. O. Modification of the exploding wire light source. *Proc. Int. Congr. High-Speed Photography*, *9th*, *Denver* pp.251–257. Soc. Motion Picture Television Eng., New York (1970).
1044. Leighton, R. L. Gas velocity measurement employing high-speed schlieren observation of laser-induced breakdown phenomena. *Proc. Int. Congr. High-Speed Photography*, *9th*, *Denver* pp. 93–96. Soc. Motion Picture Television Eng., New York (1970).
1045. Magnet Charger. Industrial paper, Impulsphysik GmbH, Hamburg.
1046.* Bitter, Fr. Ultrastrong magnetic fields. Inst. Technol. Nat. Magn. Lab., Massachusetts. Industrial paper.
1047. Herlach, F., and Knoepfel, H. Megagauss fields generated in explosive-driven flux compression devices. *Rev. Sci. Instrum.* **36**, No. 8, 1088–1095 (1965).
1048.* Smy, P. R. Design and performance of a low attenuation electromagnetic shock tube. *Rev. Sci. Instrum.* **36**, No. 9, 1334–1339 (1965).
1049. Fish net monitoring. Industrial paper, Emi Electronics Limited. Camberley.
1050.* Plessey at oceanology 72. Southampton. News from Plessey, E 1230 (March 1972).
1051.* Lansing Res. Corp. Ithaca, New York. Catalog E (1972).
1052.† Held, M. Simultane Initiierung. Kolloquim elektr. Detonatoren und Zünder. ISL (1967).
1053. de Benedictis, L. C. Diffuse optical pumping system for lasers. Assigned to Union Carbide Corp. (July 1968).
1054. Edgerton, Germeshausen, and Grier, Inc., Boston, Massachusetts. Custom equipment division.
1055. Gurevich, I. M. Some characteristics of thermal plasma in flash tubes. *Proc. Int. Congr. High-Speed Photography*, *9th*, *Denver* pp. 294–298. Soc. Motion Picture Television Eng., New York (1970).
1056. Edgerton, H., Macroberts, V. E., and Crossen, K. R. Small area flash lamps. *Proc. Int. Congr. High-Speed photography*, *9th*, *Denver* pp. 237–243. Soc. Motion Picture Television Eng., New York (1970).
1057. Confocal lens eliminates spark flash parallax. Industrial paper, Impulsphysik GmbH, Hamburg.
1058. "Nanolite" Nanosecond spark light sources. Industrial paper Impulsphysik GmbH, Hamburg.
1059. "Chronolite 8" multiple spark camera. Industrial paper, Impulsphysik GmbH, Hamburg.
1060. Thorwart, W., Suarez, F., and Patzke, H.-G. Chronolite 8-a versatile slow-motion spark camera. Explosivstoffe, 3. Erwin Barth Verlag KG, Mannheim (1969).
1061. Hartmann, G. Production of trains of luminous impulses of less than one nanosecond duration. *Proc. Int. Congr. High-Speed Photography*, *9th*, *Denver* pp. 258–262. Soc. Motion Picture Television Eng., New York (1970).
1062. Früngel, F., Thorwart, W., and Patzke, H.-G. High-speed cinematography exposed by nanosecond flashes. *Proc. Int. Congr. High-Speed Photography*, *9th*, *Denver* pp. 534–35. Soc. Motion Picture Television Eng., New Year (1970).
1063. Michel, L., and Fischer, H. Vacuum ultraviolet emission from density spark discharges in argon and helium. *Appl. Opt. Vol.* **11**, No. 4, 899–906 (1972).

1064.* Hollow electrode speed arc-lamp recharging. Industrial paper, Impulsphysik GmbH, Hamburg.
1065. Internal gas pressure controls photolamp intensity. Industrial paper, Impulsphysik GmbH, Hamburg.
1066. Charnaya, F. A. Flash discharge radiation in vacuum ultraviolet. *Proc. Int. Congr. High-Speed Photography*, *9th, Denver* pp. 287–293. Soc. Motion Picture Television Eng., New York (1970).
1067. Wittwer, H.-J. Use of the afterglow of pulsating light sources to increase the information content of high-speed films. *Proc. Int. Congr. High-Speed Photography*, *9th, Denver* pp. 263–167. Soc. Motion Picture Television Eng., New York (1970).
1068. Strobokin applications at photoelastic models. High speed physics information service, Impulsphysik GmbH, Hamburg.
1069. Unknown phenomena? Strobokin precision analysis, International Impulsephysics Assoc., Hamburg.
1070. Sound, light or voltage triggers strobe control unit. Industrial paper, Impulsphysik GmbH, Hamburg.
1071. Experiments with Strobokin control unit with increased output energy. Rep. Impulsphysik, GmbH Hamburg 56.
1072. Bubble flow up to the critical pressure. High-speed physics information service, Impulsphysik, GmbH, Hamburg.
1073. Nano Twin, Monoflash. Strobokin flash lamp, Strobodrum drum camera, Strobokin power pack, Strobokin control unit, Industrial paper, Impulsphysik GmbH, Hamburg.
1074. High Speed photography. High speed physics information service, Impulsphysik GmbH, Hamburg.
1075. Früngel, F., and Thorwart, W. Spark tracing method progress in the analysis of gaseous flows. *Proc. Int. Congr. High-Speed Photography*, *9th, Denver* pp. 165–174. Soc. Motion Picture Television Eng., New York (1970).
1076. Vanjukov, M. P., Evdokimov, S. V., and Nilov, E. V. *Proc. Int. Congr. High-Speed Photography*, *9th, Denver* pp. 75–78. Soc. Motion Picture Television Eng., New York (1970).
1077. Chabannes, F., and Milot, E. Application of phase-locked solid lasers to high-speed cinematography. *Proc. Int. Congr. High-Speed Photography*, *9th, Denver* pp. 68–74. Soc. Motion Picture Television Eng., New York (1970).
1078. Modular Laser System. Industrial paper, Impulsphysik GmbH, Hamburg.
1079. Früngel, F., and Schultz, P. Holography for 3-D vector analysis of rapid events. *Opt. Technol.* **2**, No. 1, 32–35 (1970).
1080. Hildum. J. S., and Cooper, J. A bright continuum source for the vacuum ultraviolet. *Rev. Sci. Instrum.* **43**, No. 4, 699–701 (1972).
1081. Optical Antiflash device. U. S. At. Energy Comm. Germantown, Brit. Patent 1, 066, 022 (1967).
1082. Barstow, F. E., and Lilliott, C. Flashblindness protective apparatus. U.S. Patent 3, 152, 215 (1964).
1083. Abegg, M. T., and Leslie, W. B. Antiflash device. U.S. Patent 3, 342, 540 (1967).
1084. Pike, C. T., and Teaneck, N. J. Exploding wire high-speed shutter for protecting a human viewer. U.S. Patent 3, 360, 328 (1967).
1085.* Marks, A. M., and Marks, M. M. Electrooptic responsive flashblindness controlling device. U.S. Patent 3, 245, 315 (1966).
1086. Marks Polarized Corp., Whitestone. "Varad."
1087. Cast, J. C. Wax jet shutter. U.S. Patent 3, 133, 458 (1964).

1088. Campbell, A. W.   Camera shutter. U.S. Patent 2, 470, 139 (1949).
1089. Elf Eye-protective devices.   Bermite Powder Comp.
1090. Brewster, J. L., Barbour, J. P., Charbonnier, D. M., and Grundhauser, F. J.   A new technique for ultrabright nanosecond flash light generation. *Proc. Int. Congr. High-Speed Photography*, 9th, Denver pp. 303–309. Soc. Motion Picture Television Eng., New York (1970).
1091. Früngel, F., and Oddie, G. J. W.   Military Aspects of Visibility. Presented at the NATO Panel XII "Visibility," Paper No. 48 (1969).
1092. Früngel, F.   Lidar Technik und Sichtweitenmessung. Impulsphysik GmbH, Hamburg.
1093. AEG   Appareil de mesure de visibilité. Enregistreur de dispersion AEG/FFM.
1094. Vogt, H.   Visibility measurement using backscattered light. *J. Atmos. Sci.* **25**, No. 5, 912–918 (1968).
1095. Früngel, F.   The "Videograph" backscatter fog detector and visibility meter. *Bull. de l'A.I.S.M.* No. 40 (1969).
1096. Oddie, G. J. W.   The Transmissometer. Impulsphysik GmbH, Hamburg.
1097. Oddie, G. J. W.   Runway Visual Range. ICAO Bull. (1970).
1098. Impulsphysik GmbH, Hamburg.   Fog detector Fumosens. Industrial paper.
1099. Impulsphysik GmbH, Hamburg.   Videograph, Skopograph, Skopolog, Ceilograph, Ceilolux, Ceiloskop. Industrial papers.
1100. Impulsphysik GmbH, Hamburg.   RVR, the Impulsphysics Runway Visual Range System. Industrial paper.
1101. International Impulsephysics Assoc., Hamburg.   Videograph Visibility Meter. Industrial paper.
1102. International Impulsephysics Assoc., Hamburg.   Visibility? Airfield Safety Instruments. Industrial paper.
1103. International Impulsephysics Assoc., Hamburg.   The Videograph and its Application. Industrial paper.
1104. International Impulsephysics Assoc., Hamburg.   Notes on Experience with the Use of Remote Skopograph Transmissometers as an Aid to Forecasting at Copenhagen/Kastrup.
1105. International Impulsephysics Assoc., Hamburg.   Report on the Installation and Checking of Transmissometer RVR Systems at Zurich-Kloten Airport.
1106. Impulsphysik GmbH, Hamburg.   The Impulsphysik Meteorological Instruments. Industrial paper.
1107. International Impulsephysics Assoc., Hamburg.   Remote-control signs and fog detection device provide warning system for motorists. Better Roads (1970).
1108. Gates, J. W. C., Hall, G. R. N., and Ross, I. N.   High-speed holographic recording of transilluminated events. *Proc. Int. Congr. High-Speed Photography*, 9th, Denver pp. 4–10. Soc. Motion Picture Television Eng., New York (1970).
1109. Holtz, E.   Laser having a telescopic system related to the output reflector to keep the direction of the output beam constant. Asst. to Carl Zeiss-Stiftung (1970).
1110. Earnshaw, K. B., and Hernandez, E. N.   Two-laser optical distance-measuring instrument that corrects for the atmospheric index of refraction. *Appl. Opt.* **11**, No. 4, 749–754 (1972).
1111. Hirth, A., and Vollrath, K.   Interferometry by means of holography in two wavelengths. *Proc. Int. Congr. High-Speed Photography*, 9th, Denver pp. 11–15. Soc. Motion Picture Television Eng., New York (1970).
1112. Lapp, M., Goldmann, L. M., and Penney, C. M.   Raman scattering from flames. *Science* **175**, 1112–1115 (1972).

1113. Penney, C. M., Goldman, L. M., and Lapp, M.   Raman scattering cross sections. Tech. Informat. Ser. General Electric, No. 72CRD042 (1972).
1114. Kildal, H., and Byer, R. L.   Comparison of laser methods for the remote detection of atmosphere pollutants. *Proc. IEEE* **59**, 1644–1663 (1971).
1115. Ley, J. M. *et al.*   Solid-state subnanosecond light switch. *Proc. Inst. Elect. Eng.* **117**, No. 6 (1970).
1116. Früngel, F., Knütel, W., and Suarez, J. F.   A light pulse fluorometer for measurements of Rhodamine-B concentration in situ. BMBW Forschungsbericht M 71–01 (1971).
1117. Früngel, F, Knütel, W., and Suarez, J. F.   Pulse-light fluorometer in oceanological measuring techniques. *S. Meerestechnik* **6**, (1971).
1118. Früngel, F. Koch, and Suarez, J. F.   Measurement Technique of Very High Dilution of Rhodamine B for Marking Currents and Locating Pollution. International Impulsephysics Assoc., Hamburg.
1119. Impulsphysik GmbH, Hamburg.   "Variosens" In situ fluorometer turbidity meter.
1120. International Impulsephysics Assoc., Hamburg.   Can the distribution of industrial waste water in lakes, rivers or oceans be forecast or controlled?
1121. International Impulsephysics Assoc., Hamburg.   "Variosens" optoelectronic instrument for in-situ measurement of turbidity, extinction, and fluorescence.
1122. Früngel, F., Knütel, W., Suarez, J. F. and Rudolph, J.   Development and application of the Variosens—in situ instrumentation for the fluorescent tracer technology and sand-caused turbidity. *Symp. No. 9, Phys. Processes responsible Dispersal Pollutants Sea with Special Reference Nearshore Zone* (1972).
1123. Impulsphysik GmbH, Hamburg.   Variosens. Industrial paper.
1124. Measures, R. M., and Bristow, M.   The development of a laser fluorosensor for remote environmental probing. *Joint Conf. Sensing Environ. Pollutants, Palo Alto, Calafornia.* (1971).
1125. Ballinger, D. G.   Laboratory methods for the measurements of pollutants in water and waste effluents. *Joint Conf. Sensing Environ. Pollutants, Palo Alto, California* (1971).
1126. Chandler, Ph. B.   Oil pollution surveillance. *Joint. Conf Sensing Environ. Pollutants, Palo Alto, California* (1971).
1127. Arvesen, J. C., Weaver, E. C., and Millard, J. P.   Rapid assessment of water pollution by airborne measurement of chlorophyll. *Joint Conf. Sensing Environ. Pollutants, Palo Alto, California* (1971).
1128. Multiple scattering of laser light from turbid water. *Joint Conf. Sensing Environ. Pollutants , Palo Alto,* California (1971).
1129. Digital multimeter.   Tektronix, 25 years of excellence 1946–1971, p. 61.
1130. Reference information.   Tektronix, 25 years of excellence 1946–1971, pp. 20–33.
1131. Isolex high voltage instruments. Industrial paper, Impulsphysik GmbH, Hamburg.
1132. Capdevielle, R. Ch., Jamey, P. O., and Pelte, Ch, P.   Recent developments on the chronometry of multiple intervals of time. *Proc. Int. Congr. High-Speed Photography, 9th, Denver* pp. 505–509. Soc. Motion Picture Television Eng. (1970).
1133. Reference Information.   Tektronix, 25 years of excellence 1946–1971, p. 201.
1134. Dual-Beam Oscilloscope. Tektronix, 25 years of excellence 1946–1971, pp. 137–139.
1135. Mishin, G. I., and Borisof, V. N.   Problems of precision measurements of time intervals between exposure moments. *Proc. Int. Congr. High-Speed Photography, 9th, Denver* pp. 374–377. Soc. Motion Picture Television Eng. (1970).

1136. Pellinen, D. G., and Spence, P. W.   A nanosecond rise time megampere current monitor. *Rev. Sci. Instrum.* **42**, No. 11, 1669–1701 (1971).
1137. Van de Heem.   The Hague, Transistorized counters. Industrial paper.
1138. Berkeley Nucleonics Cop.   Pulse Generator. Industrial paper.
1139. Edgerton, Germeshausen, and Grier, Boston.   Traveling wave oscilloscope. Industrial paper.
1140. Electrooptical Instrum., Inc. Monrovia.   Trigger delay generator. Industrial paper.
1141. Philips. Equipments scientificques pour industrie.   Industrial paper.
1142. Impulsphysik GmbH, Hamburg.   Kerr cells. Industrial paper.
1143. Impulsphysik GmbH, Hamburg.   Sound, light or voltage triggers strobe control unit. Industrial paper.
1144. Khakimov, Kh. Sh., and Sukhanov, Y. A.   Transistorized pulse amplifier with automatic noise-level stabilization. *Opt. Technol.* **38**, No. 6, 346–348 (1971).
1145. Schechtmann, A.   A logarithmic amplifier. Philips, Eindhoven.
1146. Durig, R. F. *et al.*   System for the detection of high intensity light flashes. U.S. Patent 3, 321, 630 (1967).
1147. Shardanand,   Compact photomultiplier housing with controlled cooling. *Rev. Sci. Instrum.* **43**, No. 4, 641–643 (1972).
1148. Impulsphysik GmbH, Hamburg.   Hydrophone. Industrial paper.
1149. Johnson, R. T.   Method and apparatus for measuring fast neutron fluences with cadmium sulfide or cadmium selenide. U.S. Patent 3, 588, 505 (1971).
1150. Mark, J. C.   The detection of nuclear explosions. Los Alamos Scientific Laboratory.
1151. Cottermann, R. W.   Elektromagnetisches Erkennen von Kernwaffendetonationen ITT Kellogg Commun. Syst., Chicago, Illinois.
1152. Dickinson, H., and Tamarkin, P.   Systems for the detection and identification of nuclear explosions in the atmosphere and in space. *Proc. IEEE.* **53**, No. 12, 1921–1933 (1965).
1153. Burrill, E. A.   Neutron Production and Protection. High Voltage Eng. Corp., Burlington.
1154. Caccia, F., Guekos, G. L., Günthard, H. H., Strutt, M. J. O., and Wild, U.   Multiframe high-speed image converter camera. Sci. Publ. Dep. of Advan. Elect. Eng. Swiss Fed. Inst. Technol., Zürich (1971).
1155. Strutt, M. J. O.   Current noise and photon noise of semiconductor diode injection lasers. Sci. Publ. Dep. Advan. Elec. Eng. Swiss Fed. Inst. Technol. Zurich (1971).
1156.* Schwickardi, G.   Dissertation of the Swiss Fed. Inst. Technol., Zurich. No. 4800 E. T. H.: Long term life test of power–(pnpn)–thyristors for the determination of their real-time behavior (1971).
1157. Abdel-Latif, M., and Strutt, M. J. O.   Comments on "a new integrated gate eliminating line reflections in high-speed digital systems. Sci. Publ. Dep. Advan. Elec. Eng. Swiss Fed. Inst. Technol., Zurich (1971).
1158. Bunkenburg, J.   Fast coaxial flashlamp pumped liquid dye laser for photolysis. *Rev. Sci. Instrum.* **43**, No. 3, 497–499 (1972).
1159. Uetz, H., and Nounou, M. R.   Gleitriebungsuntersuchungen über Reibmartensitbildung in Zusammenhang mit Grenzschichttemperatur und Verschleiß bei Weicheisen und Stahl C 45. *J. Mater.Technol.* **3 2**, 64–68 (1972).
1160. Schwab, A. J.   Low-resistance shunts for impulse currents. Rep. from *IEEE Trans. Power Apparatus Syst.* **PAS-90**, No. 5, 2251–2257 (1971).

1161. Derr, V. E., and Little, C. G. A comparison of remote sensing of the clear atmosphere by optical, radio, and acoustic radar techniques. *Appl. Opt.* **9**, No. 9 (1970).
1162. Dugger, P. H., Enis, C. P., and Hill, J. W. Laser high-speed photography for accurate measurements of the contours of models in hypervelocity flight within an aeroballistic range. Rep. from Arom, Inc., Arnold Air Force Station, Tennessee, pp. 326–335.
1163. Rompe, R., and Steenbeck, M. "Ergebnisse der Plasmaphysik und der Gaselektronik." Adademie-Verlag, Berlin (1971).
  a.† pp. 255–258. Einführung, Impulslichtquellen.
  b. pp. 266–310. Die Entwicklung der Entladung. Der elektrische Durchschlag des Gases.
  c. pp. 335–352 Strahlungseigenschaften rohrförmiger Impulslampen.
  d. pp. 353–371 Strahlungseigenschaften kugelförmiger Impulslampen.
  e. pp. 372–391. Die Selbstabsorption im Entladungsplasma.
  f. pp. 412–422. Frequenzverhalten.
  g. pp. 423–428. Der Einfluß der Elektroden auf Wirkungsgrad und Lebensdauer.
1164. Newton, A. M., and Tokarski, J. M. J. Optical Effects of atmospheric turbulence. *Proc. Int. Conf. Electro-Opt., Brighton, England* pp.329–336 (1972).
1165. Gates, J. W. C., Hall, R. G. N., and Ross, I.N. Pulsed lasers for optical measurement. *Proc. Int. Conf. Electro-Opt. Brighton, England* pp.1–8 (1972).
1166. Killick, D. E., and Bateman, D. A. Transmissometry. *Proc. Int. Conf. Electro-Opt., Brighton, England*, pp. 319–328 (1972).
1167. Paton, J. O. G. Gunfire simulation by gas laser. *Proc. Int. Conf. Electro-Opt., Brighton, England*, pp. 8–13 (1972).
1168. Holzman, M. A. Application of atmospheric light scattering for contrast analysis in electro-optical detection systems. *Proc. Int. Conf. Electro-Opt., Brighton, England* 341 (1972).
1169.† Ahmed, M. S., and Ley, J. M. Design and development of an electrically tunable birefringent filter using dihydrogen phosphate type crystals. *Proc. Int. Conf. Electro-Opt., Brighton, England*, pp.237–246 (1971).
1170.* Kennedy, W., Waddell, A. J., and Waddell, P. Stroboscopic holography using an electro-optic modulator. *Proc. Int. Conf. Electro-Opt., Brighton, England*, pp. 201–206 (1971).
1171. Hepner, G. Digital light deflector with prisms and polarization switch based on the pockels effect with transverse field. *Proc. Int. Conf. Electro-Opt., Brighton, England*, pp. 223–228 (1971).
1172. Rolls, W. H., Eddington, R. J., and Simkins, R. S. Lead–tin telluride photovoltaic detectors and arrays for the 8–14 micron window. *Proc. Int. Conf. Electro-Opt., Brighton, England*, pp. 87–92 (1971).
1173. McCormick, M. P. Simultaneous multiple wavelength laser radar measurements of the lower atmosphere. *Proc. Int. Conf. Electro-Opt., Brighton, England*, pp.495–512 (1971).
1174.‡ Lee, M. D. Solid state driving of electro-optic light deflectors at high frequencies *Proc. Int. Conf. Electro-Opt., Brighton, England*, pp. 207–218 (1971).
1175.† Hill, B. Information exchange for large channel numbers by spatially and time modulated laser beams. *Proc. Int. Conf. Electro-Opt, Brighton, England*, pp. 219–222 (1971).
1176.* Cuchy, Z., and Landovsky, J. Czechoslovak electro-optical modulators. *Proc. Int. Conf. Electro-Opt, Brighton, England*, pp. 229–236 (1971).

1177.† Brown, Ch. L. Jr.   The use of multiple sampling plankton nets. *Rep. Bio-Sci.* **18**, No. 10, 962 (1968).
1178.* Brown, Ch. L., Jr.   Biological oceanographic investigations from a deep submarine. Dept. of the Navy Underwater Syst. Center. NUSC Rep. No. 4069 (July 1971).
1179.* Dullea, R. K., and Fisch, N. P.   Volume reverberation studies with deep research vehicle Deepstar 4000. Dept. of the Navy Underwater Syst. Center. NUSC Rep. No. 4037 (April 1971).
1180.† Robra, J.   Ein Programm zur Berechnung der Elemente des Stoßkreises für beliebige Stoß- und Schaltspannungen. *Bull. ASE* **63**, 6, 274–277 (1972).
1181.† English Electric Valve Co. Ltd.   Catalog Vacuum Capacitors 1971 and Industrial paper (1972).
1182. Edgerton, Germeshausen, and Grier, Inc., Boston, Massachusetts.   Xenon flashtubes, dye laser flashtubes. Industrial paper.
1183. Edgerton, H. E. *et al.*, U.S. Patent 3, 148, 307 (Sept. 1964).
1184. Germeshausen, K. J.   Electric system embodying cold-cathode gaseous discharge device. U.S. Patent 2, 492, 142 (Dec. 1949).
1185. Edgerton, Germeshausen, and Grier, Inc., Boston, Massachusetts.   Flash device and method. U.S. Patent 2, 939, 984 (June 1960).
1186. Germeshausen, K. J.   Gaseous discharge device having a trigger electrode and a light-producing spark gap to facilitate breakdown between the trigger electrode and one of the principal electrodes. U.S. Patent 3, 350, 602 (Oct. 1967).
1187. Germeshausen, K. J.   Two-electrode spark gap with interposed insulator. U.S. Patent 3, 356, 888 (Dec. 1967).
1188. EG&G, Inc., Bedford.   Gas Mixture for electric flashtubes. U.S. Patent 3, 399, 147 (Aug. 1968).
1189. Germeshausen, K. J.   Gaseous-discharge device. U.S. Patent 2, 812, 465 (Nov. 1957).
1190. Germehausen, K. J.   Gaseous-discharge device and method of making the same. U.S. Patent 2, 756, 361 (July 1956).
1191. Edgerton, H. E.   Flash tube and apparatus. U.S. Patent 2, 919, 369 (Dec. 1959).
1192. Edgerton, Germeshausen, and Grier, Inc., Boston, Massachusetts.   Flash tube and system. U.S. Patent 3, 075, 121 (Jan. 1963).
1193. Edgerton, Germeshausen, and Grier, Inc., Boston, Massachusetts.   Gaseous-discharge device and system. U.S. Patent 2, 977, 508 (Mar. 1961).
1194. Edgerton, Germeshausen, and Grier, Inc., Boston, Massachusetts.   Electric-discharge device. U.S. Patent 2, 943, 222 (June 1960).
1195. Germeshausen, K. J.   Blitzlichtentladungslampe. Deutsches Patent 1, 065, 092, März (1960).
1196.* Campbell, J. D., and Kasper, J. V. V.   A microsecond pulse generator employing series coupled SCR's. *Rev. Sci. Instrum.* **43**, No. 4, 619–621 (1972).
1197.† Walz, L.   Geregelte Impulsgeneratoren für Halbleiter-Zeilen-Endstufen. *Funkschau*, **8**, 246–248 (1972).
1198.* Crouch, J. H., and Risk, W. S.   A compact high speed low impedance blumlein line for high voltage pulse shaping. *Rev. Sci. Instrum.* **43**, No. 4, 632–637 (1972).
1199.† Kubis, R., and Strassburg, J.   Triggerbarer Laborgenerator für positive und negative Rechteckimpulse. *Wiss. T. TH Ilmenau* **2**, 119–121 (1971).
1200. Kroschel, K.   Zeitdiskrete Detektoren zur Laufzeitmessung gestörter Signale. *Inst. f. Nachrichtensyst. Univ. Karlsruhe* **5**, 259–262 (1972).
1201. Maass, H. F.   Diagram for short-line fault tests. A practical aid to the determination of actual test conditions for e.h.v. circuit-breakers. *Elec. Rev.* **28**, 582–583 (1972).

1202. Howton, J. S. Maintenance of airport lighting. Light and Lighting, pp. 130–131 (April 1972).
1203. Stephens, H. P., Donaldson, A. B., and Heckman. R. C. A passive calorimeter for pulsed electron beam energy measurement. *Rev. Sci. Instrum.* **43**, No. 4, 614–619 (1972).
1204. Hönscheidt, W. Optisch-elektronisches Temperaturmeßgerät *Elektron.-Anzeiger* **4**, Jg. No. 5, 103–104 (1972).
1205. Oddie, G. J. W. Videograph calibration. Information Service.
1206. Bespalov, V. I. *et al.* Single-crystal, electro optic shutter for Q-switch lasers emitting unpolarized radiation. *Opt. Technol.* **38**, No. 12, 739–741 (1971).
1207. Asnis, L. N. *et al.* Heterodyne Detection with a Ge-Hg photoresistor at high frequencies. *Opt. Technol.* **38**, No. 12, 729–730 (1971).
1208. Apollo Lasers, Inc., Arizona. Description and general performance specifications. Industrial paper.
1209. English Electric Valve Co. Ltd. The international broadcasting convention. New Range of flash tubes. Industrial paper.
1210. Hook, W. R. Xenon flashlamp triggering for laser applications. *IEEE. Trans. Electron. Devices* **ED-19**, No. 3, 308–314 (1972).
1211. Fill, E. E., and v. Finckenstein, Graf K. A comparison of the performance of different laser amplifier media. *IEEE J. Quantum Electron.*, 24–27 (1972).
1212. Monsanto Electronic Special Products, Cupertino. Green and yellow solid state lamps. Industrial papers.
1213. Tektronix, Inc. Oscilloscopes, Cameras, Electro-Optic Products. Industrial papers (July 1972).
1214. Bradley, D. J. Recent development in dye lasers and their applications. *Proc. Int. Conf. Electro-Opt.*, *Brighton, England* pp.1–8 (1971).
1215. Glicksman, R. Recent progress in injection lasers. *Proc. Int. Conf. Electro-Opt.*, *Brighton, England* pp. 9–20 (1971).
1216. Weihrauch, G., Maurer, R., and Späth, E. Rußverschluß für die 24-Funken-Kamera nach Cranz-Schardin. Rapport-Technique, ISL, RT 15/71
1217. Schwertl, M., and Stenzel, A. Eine flächenförmige Funkenentladung hoher Lichtstärke und kurzer Dauer. ISL, Rapport Technique RT 12/71.
1218. Schwertl, M., and Stenzel, A. Eine Vielkanalentladung hoher Lichtstärke und kurzer Dauer.—A multichannel discharge of high light intensity and short duration. ISL, Deutsch-Französisches Forschungsinstitute- Saint-Louis.
1219. Johoda, F. C. *et al.* Plasma experiments with a 570-kJ theka-pinch. *J. Appl. Phys.* **35**, No. 8, 2351–2354 (1964).
1220. Bacchi, H., and Pauwels, J. C. Study of a very low jitter spark-gap: Application to the realization of a high-speed camera using a biplanar shutter tube with exposure time of 300 ps. *Proc. Int. Congr. High-Speed Photography.*, *9th, Denver* pp. 488–492 (1970).
1221.† Fischer, H. *et al.* Time resolved spectral opacity profiles and absorption coefficients in nanosecond spark channels. European Office of Aerospace Res., U.S. Air Force Contract F61052-68-C-0039, Progr. Rep. No. 9 (Febr. 1971).
1222.* Lafferty, J. M. Triggered vacuum gaps. *Proc. IEEE* **54**, No. 1, 23–32 (1966).
1223.† Bayle, P., and Schmied, H. Analyse des mecanismes de formation du streamer. Centre de physique atomique, Toulouse, France (May 1972).
1224.* Milde, H. ION Phys. Corp., Burlington. Industrial paper, Flash X-ray and EMP Simulators. Proposal No. 720715 (July 1972).
1225.† Schmied, H., Rohrbach, F., and Piuz, F. Hydrogen streamer chamber development: First Results. Eur. Organization for Nucl. Res., EMSA/TC-L/Int. 72-7 (May 1972).

1226.* The "Les Renardiéres" Group. Breakdown phenomena of 5 m and 10 m rod–plane gaps in air with positive switching impulses. International Conference on large high tension Electric Systems, No. 33–15 (Aug-Sept. 1972).
1227.* Leroy, G., Gallet, G., and Simon, M. F.   "Les Renardiéres" UHV Lab., Original aspects in its design and operation. ETZ-A. Bd. 93, H. 7, pp. 410–414 (1972).
1228. Maier, H. H.   Gepulste Photomultiplier. Physikalische Verhandlungen Frühjahrstagung März, Fachausschuß Kurzzeitphysik, S. 152 (1972).
1229.* Forgacs, R. I.   A fast thermal pulser for metallic phase transformation studies. Rev. Sci. Instrum. **43**, No. 2, 302–306 (1972).
1230. Ortec, an EG & G Co.   Biomedical research instruments. Bull. WS–201.
1231.* Velonex, Santa Clara, California.   High power pulse generators Industrial paper (Aug. 1972).
1232.* RCA Solis State Div.   Microwave power transistors. Industrial paper (Jan. 1972).
1233.† Bonin, E. L.   Drivers for optical diodes. Electronics **77**–82 (1964).
1234. Oldenberg, O.   Mechanism of the short-duration nitrogen afterglow. J. Opt. Soc. Amer. **61**, No. 8, 1092–1098 (1971).
1235. v. d. Piepen, H., and Schroeder, W.   Radial resolution of an argon jet guided spark. Z. Angew. Phys. **31**, H. 4, 189–193 (1971).
1236. Hoppe, W., and Menzel, M.   Hochleistungsröntgenröhre. Z. Angew. Phys. **31**, H. 5–6, 343–345 (1971).
1237. Wolter, H.   Teleobjektive für Röntgenstrahlenteleskope und Neutronenkameras, Z. Angew. Phys. **31**, H. 3, 152–155 (1971).
1238. Pai-Lien-Lu.   An inexpensive bright short duration point light source. Picatinny Arsenal, Dover, 07801, 1526 (June 1971).
1239. Miyachi, I., and JayaRam. K.   A stabilized metal vapor arc device. Rev. Sci. Instrum. **42**, No. 7, 1002–1005 (1971).
1240. Marlier, S. F., and McIntosh, R. E.   Experimental apparatus to investigate transient nonlinear effects in a plasma. Rev. Sci. Instrum. **42**, No. 7, 1038–1042 (1971).
1241. Grasberger, W. H. et. al   X-ray measurements of laser-produced plasma. Int. Quantum Electron. Conf., 9th, Montreal p. 34 (1972).
1242. Yamanaka, C., and Nagao, Y.   Improvement of damage threshold of glass laser. Int. Quantum Electron. Conf. 7th, Montreal pp.17–18 (1972).
1243. Hordvik, A., and Schlossberg, H.   Luminescence from $LiNbO_3$. Int. Quantum Electron. Conf., 7th, Montreal pp.75–76 (1972).
1244. Kielich, S., Lalanne, J. R., and Martin, F. B.   Cooperative second harmonic laser light scattering in liquid benzene and in liquid carbon disulphide. Int. Quantum Electron. Conf. 7th, Montreal pp. 1–2 (1972).
1245. Taran, J. P. E., and Thomas, J. M. R.   The influence of phase mismatch on pulse shapes in second-harmonic generation. Int. Quantum Electron. Conf. 7th, Montreal p. 5 (1972).
1246. Drexhage, K.-H.   Design of laser dyes. Int. Quantum Electron. Conf. 7th, Montreal p. 8 (1972).
1247. Alfano, R. R., and Shapiro, S. L.   Dynamics of ultrafast decay processes of the erythrosin molecule in different solvents. Int. Quantum Electron. Conf. 7th, Montreal p. 10 (1972).
1248. Baardsen, E. I., and Terhune, R. W.   Detection of OH in the atmosphere using a dye laser. Int. Quantum Electron. Conf., 7th, Montreal p. 10 (1972).
1249. Kressel, H.   Recent trends in room temperature laser diodes. Int. Quantum Electron. Conf., 7th, Montreal p.14 (1972).

1250. de Witte, O. The Rhodamine 6G dye laser experimental studies and theoretical analysis. *Int. Quantum Electron. Conf.*, *7th, Montreal* p.12 (1972).
1251. Arthurs, E. G., Bradley, D. J., New, C., and Roddie, A. G. Picosecond dye-laser pulses frequency tunable between 580 and 700 nm. *Int. Quantum Electron. Conf.*, *7th, Montreal* p. 12 (1972).
1252. Sasaki, T., and Nagoa, Y. Improvement of damage threshold of glass laser. *Int. Quantum Electron. Conf.*, *7th, Montreal* pp. 17–18 (1972).
1253. Hodgson, R. T., and Dreyfus, R. W. Vacuum UV laser action at 1161–1240 A. *Int. Quantum Electron. Conf.*, *7th Montreal* p. 20 (1972).
1254.* Lax, B. Prescription for soft x-ray laser. *Int. Quantum Electron. Conf.*, *7th, Montreal* pp. 21–22 (1972).
1255.* Zarowin, C. B. Possible vacuum ultraviolet/soft x-ray stimulated emission from optical transitions of multi-ionized atoms in very high current density discharges. *Int. Quantum Electron. Conf.*, *7th, Montreal* p. 22 (1972).
1256.* Floux, F. *et al.* Heating of solid deuterium by a powerful laser pulse production of nuclear DD reactions. *Int. Quantum Electron. Conf.*, *7th, Montreal* p. 33 (1972).
1257.* Floux, F. *et al.* Heating of solid deuterium by a powerful laser pulse. *Int. Quantum Electron. Conf.*, *7th, Montreal* p.36 (1972).
1258. Kobayasi, T. *et al.* TEA UV nitrogen laser and its application to high-sensitive remote pulsed-Raman spectroscopy of atmospheric pollutants. *Int. Quantum Electron. Conf.*, *7th, Montreal* pp. 61–62 (1972).
1259. McNice, G. T., Derr, V. E., and Schwiesow, R. L. The study of Raman, resonance Raman, and fluorescence spectra by tunable dye lasers. *Int. Quantum Electron. Conf.*, *7th, Montreal* p. 62 (1972).
1260. Rheault, F. *et al.* 2–GW peak-power generation from a TEA CO_2 Oscillator-amplifier laser. *Int. Quantum Electron. Conf.*, *7th, Montreal* p.76 (1972).
1261. Hidson, D. J., and Makios, V. A high-voltage, high-pressure TEA CO_2 laser. *Int. Quantum Electron. Conf.*, *7th, Montreal* pp. 76–77 (1972).
1262. Winterberg, F. Magnetically insulated transformer for attaining ultrahigh voltages. *Rev. Sci. Instrum.* **41**, No. 12, 1756–1763 (1970).
1263.* Alexander, T. A. *et al.* High dc power, solid-state switch for pulsing an arc lamp. *Rev. Sci. Instrum.* **36**, No. 12, 1707–1709 (1965).
1264.* Landecker, K., Skattebol, L. V., and Gowdie, D. R. R. Single-spark ring transmitter. *Proc. IEEE* **59**, No. 7, pp. 1082–1090 (1971).
1265.† Landecker, K., and Imrie, K. S. A novel type of high pulse transmitter. *Aust. J. Phys.* **12**, 638–654 (1960).
1266. Landecker, K. Means for the generation of very large pulses of radio frequency waves. Australian Patent 233. 302, British Patent 834, 337, and U.S. Patent 3, 011, 051.
1267. Imrie, K. S. Investigation into operation and design parameters of a novel type of spark transmitter and on some aspects of the propagation characteristics of the produced pulses. Ph. D. dissertation, Univ. of New England, Armidale, N. S. W. Australia (1961).
1268. Dolphin, L. T., and Wickersham, A. F. The generation of megawatt peak power by modern spark transmitter techniques. Office of Naval Res. and Advan. Res. Proj. Agency, Washington, D. C. No. 4178(00) ARPA Order 463 (1966).
1269. Fraser. M. P. Project mapole (magnetic dipole spark transmitter) M. I. T. Lincoln Lab., Lexington, Massachusetts. Note 1967-24 (1967).
1270. Landecker, K. Australian Patent 268, 894, British Patent 1, 050, 626, and U.S. Patent 3, 317, 839.

1271. Skattebol, L. V. High power spark transmitters and their application to radar. Ph.D. dissertation, Univ. of New England, Armidale, N.S.W. Australia (1966).
1272. Craggs, J. D., and Meek, J. M. High voltage lab. "Techniques." Butterworths, London and Washington, D.C. (1954).
1273. Ramo, S., and Whinnery, J. R. "Fields and Waves in Modern Radio," 2nd ed., p. 395. Wiley, New York (1952).
1274. Glagelewa-Arkadiewa, A. Eine neue Strahlungsquelle der kurzen elektromagnetischen Wellen von ultraherzscher Frequenz.—A new source of radiation of short electromagnetic waves of ultra-Hertzian frequency. *Z. Phys.* **24**, 153–165 (1924).
1275. Moulin, E. B. The radiation from large circular loops. *J. Inst. Eng.* **93**, 345–351 (1946).
1276. Imrie, K. S. Simple peak voltmeter for nanosecond pulses. *J. Sci. Instrum.* **39**, 172 (1962).
1277.* Hoffman, G. W. A nanosecond temperature-jump apparatus. *Rev. Sci. Instrum.* **42**, No. 11, 1643–1647 (1971).
1278.* Binns, D. S. *et al.* Breakdown in sulphur hexaflouride and nitrogen under direct and impulse voltages. *Proc. IEEE* **116**, No. 11, 1962–1968 (1969).
1279.* Uman, M. A., Orville, R. E., and Sletten, A. M. Four-meter sparks in air. *J. Appl. Phys.* **39**, No. 11, 5162–5168 (1968).
1280.* Borucki, W. J. Kerr-cell-shuttered $f/1.5$ stigmatic spectrograph for nanosecond exposures. *Appl. Opt.* **9**, No. 2, 259–264 (1970).
1281. OSA Spring Meeting 1967:
    Session 8–2 Wideband laser communication systems.
    Session 8–3 Multiplexing and demultiplexing techniques for an optical pulse code modulation (PCM) transmission system.
    Session 8–4 $3.39\text{-}\mu$ infrared optical heterodyne communication system.
    Session 8–5 The folded-back light communication system utilizing polarization characteristics.
    Session 8–6 Fading and Polarization noise of a PCM/PL system.
    Session 9–1 Attenuation by percipitation of laser beams at 0.63, 3.5, and 10.6μ.
    Session 9–6 A dual wavelength optical distance measuring instrument which measures air density.
    Session 9–7 Comparison of scattering of laser and monochromatic incoherent light in fog.
    Session 9–8 Laser beam propagation in the atmosphere.
    Session 10–1 Electroopic effects in strontium barium niobate.
1282.† Andre, Mi., and Haas, Ph. Etude dóbturateurs a cellule de Kerr. Rapport CEA-R. 3927, Centre d'Etudes de Bruyères-le-Chatel (1970)
1283.† Cassidy, E. Ch. Pulsed laser Kerr system polarimeter for electrooptical fringe pattern measurement of transient electrical parameters. *Rev. Sci. Instrum.* **43**, No. 6 866 (1972).
1284.* Chen, F. S. *et al.* Light modulation and beam deflection with potassium tantalate-niobate crystals. *J. Appl. Phys.* **37**, No. 1, 388–398 (1966).
1285.* Sansalone, F. J. Compact optical isolator. *Appl. Opt.* **10**, No. 10, 2329–2331 (1971).
1286. Ortec, an EG&G Comp. A new stimulator, a 2-Kanal stimulator. Industrial paper, No. 4797.
1287.‡ Reifsnider, K. Decent development in the use of high-gain optical image intensifiers for high-speed photography of dynamic x-ray diffraction patterns. *Soc. Motion Picture Television Eng.*, **80**, 18–21 (1971).

1288.† Moody, N. F. *et al.* A survey of medical gamma-ray cameras. *Proc. IEEE* **58**, No. 2, 217–242 (1970).
1289. A Gordon Publication. Laboratory Equipment. X-ray fluorescence analyzer for quantitative analysis of a wide range of elements. Vol. 5, No.2 (1968).
1290. Fexitron, A new, portable flash x-ray system. Industrial paper.
1291.* Cohen, L. *et al.* Study of the x-rays produced by a vacuum spark. *J. Opt. Soc. Amer.* **58**, No. 6, 843–846 (1968).
1292. Eichhorn, Fr., and Dilthey, U. Röntgen-Hochgeschwindigkeitsphotographie von Lichtbogenbewegung und Werkstoffübergang beim Ulterpulverschweißen. *VDI-Z* **113**, No. 1 33–38 (1971).
1293. Field Emission Corp. McMinnville. Industrial papers, Electron pulses.
1294.* ION Phys. Corp. Burlington, Massachusetts. Industrial papers:
   a.* Flash x-ray/pulsed electron systems
   b. Accessory specification FX-25-A-100 external electron beam handling system
   c. Model FX-75 flash x-ray system. Spec. No. 167-75
   d. Model FX-45 flash x-ray system. Spec. No. 167-45
   e. Model FX-25 flash x-ray system. Spec. No. 167-25
   f.* IPC flash x-ray service facility now offers low jitter electronic pulse trigger
   g.* Flash x-ray systems for radiation research, medicine, and industry.
1295. High Voltage Engineering, Burlington, Massachusetts. 2.5 MeV radiographic accelerator. Industrial paper.
1296. High Voltage Engineering, Burlington, Massachusetts. Van de Graaff Supervoltage radiography. Industrial paper.
1297. Weldmatic, Monrovia, California. Miniature welding head. Industrial paper.
1298. Kollmann, E. Electrode holder arrangement for welding machines. U.S. Patent 3, 238, 352 (1966).
1299. Continental Carbon, Inc. Cleveland, Ohio. Resistors. Industrial paper.
1300.† South London Electrical Equipment Co. Ltd. Welding machines. Industrial paper.
1301.* Baldwin-Spembly, Dartford, England. Welding apparatus. Industrial paper.
1302.† Hughes Aircraft Comp., Oceanside, California. Electronic Welding Equipment. Industrial papers.
1303. Impulsphysik GmbH, Hamburg. Pulse Hardening Generator. Impulsa H. Industrial paper.
1304. Impulsphysik GmbH, Hamburg. Improvements in means for rapidly inhibiting the passage of light. Brit. Patent 1, 133, 025 (1967).
1305.* Goronkin, H. Striations in fast wire explosions. Phys. Dept., Temple Univ., Philadelphia, Pennsylvania (1968).
1306.* Jäger, H., and Rusche, D. Optische und elektrische Messungen an verdämmten Drahtexplosionen. *Z. Angew. Phys.* Bd. **26**, H.3, 231–238 (1969).
1307.* Oktay, E. Effect of wire cross section on the first pulse of an exploding wire. *Rev. Sci. Instrum.* **36**, No. 9, 1327–1328 (1965).
1308.† Grad, H. Plasma. *Physics Today*, Dec., 34–44 (1969).
1309.* Hollister, D. D. A xenon lamp with two less electrodes. *Electro-Opt. Sys. Design*, pp. 26–30 (1971).
1310. Impulsphysik GmbH, Hamburg. Exploding wire shutters camera. Industrial paper.
1311.† Schwab, A. "Hochspannungsmeßtechnik." Springer, Berlin (1969).
1312.† Schwab, A. "High Voltage Measurement Techniques." Cambridge Univ. Press, London and New York, and MIT-Press, Boston, Massachusetts (1971).

1313.† Lochte-Holtgreven, W. "Plasma-Diagnostics." Wiley, New York (1968).
1314.† Hudlestone, R. H. "Plasma Diagnostic Techniques." Academic Press, New York (1965).
1315.† Zaengl, W. Der Stoßspannungsteiler mit Zuleitung. *Bull. SEV* **61**, 1003–1017 (1970).
1316.† Pedersen, Aa. Dynamic Properties of Impulse Measuring Systems. *IEEE Trans. PAS* 1424–1432 (1970).
1317.† Lührmann, H. Fremdfeldbeeinflussung kapazitiver Spannungsteiler. *ETZ A* **91**, 332–335 (1970).
1318.† Thomas, R. T. High-impulse current and voltage measurement. *IEEE Trans. IM* pp. 102–117 (1970).
1319.† Zaengl, W. Zur Ermittlung der vollständigen Übertragungseigenschaften eines Stoßspannungsmeßkreises. *ETZ A* **90**, 457–62 (1969).
1320.† Hylten, Cavallius, N. R. Calibration and checking methods of rapid high-voltage impulse measuring circuits. *IEEE Trans. PAS*, 1393–1403 (1970).
1321.† Thomas, R. J. High-voltage pulse reflection type attenuators with subnanosecond response. *IEEE Trans. IM*, 146–154 (1967).
1322.† Thomas, R. F. Response of capacitive voltage dividers. *Microwaves* **6**, 50–53 (1967).
1323.† Newi, G. A high impedance nanosecond rise time probe for measuring high-voltage pulses. *IEEE Trans. PAS* 1780–1786 (1969).
1324.† Lord, H. W. High frequency transient voltage measuring techniques. *Elect. Eng.* **82**, 121–124 (1963).
1325.* Gooding, T., Hayworth, B. R., Larson, A. V., and Ashby, D. E. T. F. Development of a coaxial plasma gun for space propulsion. General Dynamics: Astronautics. Contract NAS 3-2594 (1965).
1326.* Gottlieb, M. B. Plasma- the fourth state. *Int. Sci. Technol.* Aug. 44–50 (1965).
1327.* Räuchle, E. The radiation of a rotating, charged particle in a plasmon field with an external magnetic field. *Inst. Hochtemperaturforsch., TH Stuttgart* pp. 500–503.
1328.* Schmidt, H. and Kaeppeler, H. J. Investigation of plasma-magnetic shock waves with a Mach-Zehnder-Interferometer. *Inst. Hochtemperaturforsch., TH Stuttgart*, 777–783.
1329.† Spitzer, L. Jr. 1. Experimental Plasmas. *Physics Today*, Dec., 33–38 (1965).
1330. TRW Instruments, El Segundo, California. A multipurpose uv continuum radiation source system. *Physics Today*, Dec. 39 (1965).
1331.† Frieman, E. A. 2. Plasma Theory. *Physics. Today*, Dec., 40 (1965).
1332. Hewlett Packard Sanborn Div. Instrumentation specs in 250 KC tape recording. Industrial paper.
1333. Magnetic Charger. Industrial paper, Impulsphysik GmbH, Hamburg.
1334.† Jaecklin, A. A. Magnetooptische Strommessung in Hochspannungsleitungen. *Laser angew. Strahlentech.*, No. 4, 11–13 (1971).
1335. Impulsphysik GmbH, Hamburg. Improvements in or relating to means for effecting rapid closure of light paths in optical apparatus. Brit. Patent 1, 147, 283 (1969).
1336.* Derr, V. E., and Little, C. G. A comparison of remote sensing of the clear atmosphere by optical, radio, and acoustic radar techniques. *Appl. Opt.* **9**, No. 9, 1976–1992 (1970).
1337.* Automation Industries, Inc. Improvements in or relating to a method and apparatus for ultrasonic inspection. Brit. Patent 1. 034. 724 (1963).

1338.* Martner, J. G. Stanford Res. Inst., Menlo Park, California. An ultrasonic atomizer capable of high rates. API Conference Paper CP 66-5 (1966).
1339.* Al'tshuler, L. V. Use of shock waves in high-pressure physics. *Sov. Phys. Usp.* **8**, No. 1, 52-91 (1965).
1340. Trevelyan, B. The electro-optic effect in proustite. *Opto-Electron.* **1**, 9-12 (1969).
1341. Meeks, E. L., Peters, R. D., and Arnold, T. R. Capacitance microphones for measurement of ultrasonic properties of solids. *Rev. Sci. Instrum.* **42**, No. 10, 1446-1449 (1971).
1342. Schönback, K. Explosive erosion in high current sparks. *Z. Angew. Phys.* **32**, H4, 253-257 (1971).
1343. Oyane, M., and Masaki, S. Fundamental study of electrospark forming. *Bull. ISME* 621, 776, 537, 528, pp. 366-372
1344.† Some questions of Soviet oceanological reseach. The national report of the Soviet delegation to Oceanology International 72 at Brighton (March 1972).
1345. Kuzhekin, I. Verhalten von Funkenstrecken unter Wasser bei Impulsspannung und Impulsstrombeanspruchung. *ETZ-A* **93**, H. 7, 404-409 (1972).
1346.* Underwater Systems Division, Paramus. Deep sea and general purpose hydrophones. Industrial paper.
1347.* Burnett Electronics Lab., Inc., San Diego. Underwater sound/search systems. Industrial paper.
1348.* High energy rate working of metals. A NATO Advanced Study Institute, Oslo. Vol. 1 (Sept. 1964).
1349.* Underwater electrodeless spark sound source. Technical operating report 21-R-138. 1 AVCO Space Systems Division, Lowell, Massachusetts (July 1966).
1350. Sondericker, J. H. The bubble chamber technique for photographing interactions of high-energy particles. *J. SMPTE* **79**, 222-225 (1970).
1351. Lincoln, K. A. Thermal radiation characteristics of xenon flashtubes. *Appl. Opt.* **3**, No. 3, 405-412 (1964).
1352. Austin, R. W. Assessing underwater visibility. *Opt. Spectra*, May, 33-39 (1970).
1353. Früngel, F. Gas filled envelope for a spark gap or the like. U.S. Patent 3, 377, 496 (Apr. 1968).
1354. Lidholt, L. R., and Wladimiroff, W. W. Scintillator and dye lasers in the range 350 to 600 nm pumped by a 200 kW nitrogen laser. *Opto-Electron.* **2**, 21-28 (1970).
1355. LODIF ? Long distance infrared flash - a new instrument for infrared night photography and observation. Impulsphysik GmbH, Hamburg. Industrial paper.
1356. Nanosecond light source. Pek Lab., Palo Alto. Industrial paper.
1357. Rose, D. Thousand million candlepower flash lamps. *Opt. Spectra*, Feb., 43-47 (1970).
1358. Harrington, Fr. D. Summaries of papers on several light sources and a framing drum spectrograph. *J. SMPTE* **75**, 355-356 (1966).
1359. Früngel, F. Circuit arrangement for liquid pressure forming sheet metal. U.S. Patent 3, 520, 162 (July 1970).
1360.* Balzerowiak, H. -P., and Prümmer, R. Das elektrohydraulicshe Umformverfahren in der Praxis. *Z. Wirts. Fertigung* **67**, H. 5, 246-250 (1972).
1361.* Nishiyama, U., and Inoue, T. Research on explosive and electrohydraulic forming using high speed photography. *C.I.R.P.* **XVI** 123-130 (1968).
1362.* Nishiyama, U., and Inoue, T. A study of the utilization of generated energies in hydro-spark forming and explosive forming. *Int. J. Mech. Sci.* **10**, 740-486 (1968).
1363. Strobodrum drum camera. Impulsphysik GmbH, Hamburg. Industrial paper.

1364. Wiebelt, J. A., and Williamson, R. D.   A simple multiple spark system. *Rev. Sci. Instrum.* **43**, No. 6, 899–901 (1972).
1365. Waddell, J. H.   What is high speed photography? *Res. Develop.*, 70–71 (1968).
1366. Zavoisky, E. K., and Fanchenko, S. D.   Image converter high speed photography with 10^{-9}–10^{-14} sec time resolution. *Appl. Opt.* **4**, No. 9, 1155–1167 (1965).
1367. Dearing, L. M.   Thin-probe pulsed-light photometer for measurement and calibration of timing and other pulsed-light sources. SMPTE Preprint No. 104-45 (Nov. 1968).
1368. Andreyev, S. I. *et al.*   Surface spark discharge as a course of intensive light flashes. Phys. Inst. of K. Marx Univ., Leipzig (1964).
1369. Continuum radiation source.   TRW Instruments. Bull. CRS-1. Industrial paper.
1370. Ericsson, K. G., and Lidholt, L. R.   Ultraviolet source with repetitive subnanosecond kilowatt pulses. *Appl. Opt.* **7**, No. 1, 211 (1968).
1371. Goncz, J. H., and Newell, P. B.   Spectra of pulsed and continuous xenon discharges. *J. Opt. Soc. Amer.* **56**, No. 1, 87–92 (1966).
1372. Capillary high pressure mercury arc lamps.   Pek Labs., Palo Alto, California. Industrial paper.
1373. Bracco, D. J., and Weisberger, S.   Determination of the spectral emission of commercial flashbulbs by emission spectrographic techniques. *Appl. Opt.* **5**, No. 8, 1275–1279 (1966).
1374. Barnes, Fr. S.   Physical characteristics of xenon flashtubes. *J. SMPTE* **73**, 569–573 (1964).
1375. Birdsall, D. H., and Ping, D. E.   Bakeable, pulsed gas value for plasma physics experiments. *Rev. Sci. Instrum.* **36**, No. 12, 1777–1778 (1965).
1376. Hagen, W. F.   Linear lamps for lunar laser light. *Opt. Spectra*, 22–25 (1971).
1377. Clay, W. G. *et al.*   Application of nanosecond light pulses to ballistic range measurements. *AIAA J.* 1966 copyright by the American Inst. of Aeronautics. Technical Notes, pp. 346–65 (Feb. 1967).
1378. Yguerabide, J.   Generation and detection of subnanosecond light pulses: Application to luminescence studies. *Rev. Sci. Instrum.* **36**, No. 12, 1734–1742 (1965).
1379. Lieber, A. J., and Sutphin, H. D.   Nanosecond high resolution framing camera. *Rev. Sci. Instrum.* **42**, No. 11, 1663–1669 (1971).
1380. Pollack, S. A.   Short-duration light pulse during electrical breakdown in gases. *J. Appl. Phys.* **36**, No. 11, 3459–3465 (1965).
1381. Nucletron GmbH, Eimac Div. San Carlos, California.   Xenon lamps. Industrial papers.
1382. Hodgson, B. W., and Keene, J. P.   Some characteristics of a pulsed xenon lamp for use as a light source in kinetic spectrophotometry. *Rev. Sci. Instrum.* **43**, No. 3, 493–496 (1972).
1383. Impulsphysik GmbH, Hamburg.   Modular laser system. Industrial papers.
1384. Früngel, F.   Discharge lamp arrangement for lighting purposes. U.S. Patent 3, 529, 208 (Sept. 1970).
1385. Westley, R., and Woolley, J. H.   Shock cell noise-mechanisms, the near field sound pressures associated with a spinning screech mode. Conference on Current developments in Sonic Fatigue, Southampton (July 1970).
1386. Fischer, H., and Gallagher, C. C.   Electrode processes in a high current 20 nanosecond arc *Rep. Annu. Conf. Phys. Electron.*, *26th*, pp. 385–394 (March 1966).
1387. Tschinke, M.   An inexpensive ultrahigh-speed photographic technique. *Strain*, pp. 10–13 (Oct. 1968).
1388. Merzkirch, W. F.   Making flows visible. *Int. Sci. Technol.* 46–56 (1966).

1389. Farello, G. E.　　Droplets sprayed and deposited on heating walls. *Pres. Int. Heat Transfer Conf.*, *10th, Philadelphia* (Aug. 1968).
1390. Kachel, V.　　Methoden zur Analyse und Korrektur apparativ bedingter Meßfehler beim elektronischen Verfahren zur Teilchengrößenbestimmung nach Coulter. Dissertation D 83, Techn. Univ. Berlin, Fachbereich für Verfahrenstechnik (1972).
1391. Kachel, V., Metzger, H., and Ruhenstroth-Bauer, G.　　Der Einfluß der Partikeldurchtrittsbahn auf die Volumenverteilungskurven nach dem Coulter-Verfahren. *Z. Ges. Exp. Med.* **153**, 331–347 (1970).
1392. Unilux, Inc., New York.　　Videostrobe 800 System. Industrial papers.
1393. Mansberg, Hy. Schlesinger, Jules.　　Electronic flying spot particle analyzer. *The Pulse of Long Island* (Oct. 1963).
1394. ASEA Waser cloud ceilometer. Industrial paper, ASEA, Västeras.
1395. Ralston, J. M.　　An apparatus for measuring quantum efficiency in electroluminescent devices over a wide current range. *Rev. Sci. Instrum.* **43**, No. 6, 876–878 (1972).
1396. Chance, B., Graham, N., and Mayer, D.　　A time sharing fluorometer for the readout of intracellular oxidation-reduction states of NADH and flavoprotein. *Rev. Sci. Instrum.* **42**, No. 7, 951–957 (1971).
1397. Glicksman, R.　　Technology and design of GaAs laser and noncoherent IR-emitting diodes. RCA/Solid State Div., Somerville (Oct. 1970).
1398. Kerr, M. A.　　Electroluminescent diodes as timing signal recorders in high framerate cameras. *SMPTE Technical Conf. 104th*, pp. 104–144 (1968).
1399. Ninomiya, Y., and Motoki, T.　　$LiNbO_3$ light modulator. *Rev. Sci. Instrum.* **43**, No. 3, 519–524 (1972).
1400. Kogelschatz, U., and Newman, E. H.　　Laserstrahl als Temperaturfühler. Laser and angew. *Strahlentechnik* No. 4, 9–11 (1971).
1401. Fine, S., and Hansen, W. P.　　Optical second harmonic generation in biological systems. *Appl. Opt.* **10**, No. 10, 2350–2353 (1971).
1402. Kornstein, E., and Wetzstein, H.　　Blue-green high-powered light extends underwater visibility. *Electronics*, 140–150 (Oct. 1968).
1403. Berger, S. B. *et al.*　　Faraday rotation of rare-earth (III) phosphate glasses. *Phys. Rev.* **133**, No. 3A, A723–727 (1964).
1404. Hessel, K. R.　　Use of photochromic glasses and films in optical processing applications. Sandia Lab., Sc-DR-71 0829 (1971).
1405. Electro-Optical Instruments, Inc., Pasadena, California.　　Industrial papers. Cappings hutters and energy dump systems, and Basic intergrated Kerr cell modulator.
1406. Impulsphysik GmbH, Hamburg.　　Improvements in and relating to optical shutters. Brit. Patent 1, 260, 692 (1972).
1407. Melhart, L. J.　　Pulsed-eddy current motivated shutter. U.S. Patent 3, 382, 785 (1968).
1408. Sansalone, F. J.　　Compact optical isolator. *Appl. Opt.* **10**, No. 10, 2329–2331 (1971)
1409. Roberts, J. R.　　An image intensifying shutter for observing transient plasmas. *Appl. Opt.* **4**, No. 9, 1179–1183 (1965).
1410. McMahan, W. H.　　High-power, visible-output gas-dynamic lasers. *Opt. Spectra*, 30–34 (Dec. 1971).
1411. RCA Radio Corp. of America.　　Camera tubes. Industrial papers.
1412. Mullard Ltd., and Courtney-Pratt, J. S., Esq.　　Sequential picture image converter. Winston Electron. Limited.
1413. Hofmeister, F.　　Multiple frame image converter unit with up to seven frames per trigger using solid state design. *Rev. Sci. Instrum.* **42**, No. 7, 1043–1048 (1971).

1414. ITT Laboratories, Fort Wayne, Indiana. Industrial paper. Image converter tube (and image intensifier tube).
1415. Miyashiro, S. Image electron multiplication by silicon target. *IEEE*, 2080–2081 (Nov. 1969).
1416. Kaufman, I. Self-scanned optical sensor using elastic surface waves. *IEEE*, 2081–2082 (Nov. 1969).
1417. EMI Electronics Ltd., Hayes Middlesex, England. Special valves and tubes, and 4-stage image intensifiers. Industrial papers.
1418. Yu, F. T. S. Statistical brightness gain of a channeltron image intensifier. *Appl. Opt.* **7**, No. 8, 1601–1607 (1968).
1419. Duguay, M. A., and Mattick, A. T. Ultrahigh speed photography of picosecond light pulses and echoes. *Appl. Opt.* **10**, No. 9, 2162–2170 (1971).
1420. Mainster, M. A. *et al.* Retinal-temperature increases produced by intense light sources. *J. Opt. Soc. Amer.* **60**, No. 2, 264–270 (1970).
1421. Gates, J. W. C., Hall, R. G. N., and Ross, I. N. Holographic recording using frequency-doubled radiation at 530 nm. Div. Opt. Metrol., Nat. Phys. Lab. Teddington, pp. 89–94 (Oct. 1969).
1422. Yin, P. K. L., and Long, R. K. Atmospheric absorption at the line center of P(20) CO_2 laser radiation. *Appl. Opt.* **7**, No. 8, 1551–1554 (1968).
1423. Curcio, J. A., and Knestrick, G. L. Transmission of ruby laser light through water. U.S. Naval Res. Lab., Washington, D.C., NRL Report 5941 (1963).
1424. Früngel, F. Arrangement for controlling marine warning lights as a function of fog density. U.S. Patent 3, 576, 557 (1971).
1425. Früngel, F. Light receiver housing having inclined mirror mounted on tiltable shaft. U.S. Patent 3, 544, 797 (1970).
1426. Früngel, F. Incipient fog detecting system and method. U.S. Patent 3, 415, 984 (1968).
1427. RCA Radio Corp. of America. Industrial papers RCA Photomultiplier and image tubes, ICE-269 series.
1428. EG & G. Industrial papers, Photodiode application notes.
1429. EMI Electronics Ltd., Hayes Middlesex, England. Industrial papers, Photomultiplier tubes.
1430. Freeman, Dr., and Hall, F. Jr. Laser measurements of turbidity in the atmosphere. *Opt. Spectral*, 67–70 (July/Aug. 1970).
1431. Früngel, F. Method and arrangement for measuring the density of natural fog in the free atmosphere using light source which is also a flashing warning beacon. U.S. Patent 3, 672, 775 (1972).
1432. Impulsphysics Association, Hamburg. Laser fog bank detector. Meteorological information service, (1972).
1433. Dodge, J. Electronic fog signal system on Scotland sea buoy. *Bull. l'A.I.S.M.* **45** (1970).
1434. Laser Diode Laboratories, Inc., Metuchen, New Jersey. GaAs crystal, injection lasers and arrays, laser pulse generators, led's, gated viewing systems and custom laser systems. Industrial papers.
1435. Brandenburg, W. M., and Neu, J. T. Undirectional reflectance on imperfectly diffuse surfaces. *J. Opt. Soc. Amer.* **56**, No. 1, 97–103 (1966).
1436. Reisman, E., Cumming, G., and Bartky, Ch. Comparison of fog scattered laser and monochromatic incoherent light. *Appl. Opt.* **6**, No. 11, 1969–1972 (1969).
1437. Whitman, A. M., and Beran, M. J. Beam spread of laser propagating in a random medium. *J. Opt. Soc. Amer.* **60**, No. 12, 1595–1602 (1970).

1438. Hall, F. F., and Ageno, H. Y.   Absolute calibration of a laser system for atmospheric probing. *Appl. Opt.* **9**, No. 8, 1820–1824 (1970).
1439. Westendorf, W.   Method and arrangement for testing of visibility measuring arrangements. U.S. Patent 3, 668, 674 (1972).
1440. Früngel, F.   Optical calibrating device including light diffusers arranged to simulate a path of back-scattered light. U.S. Patent 3, 598, 492 (1971).
1441. Früngel, F.   Geodetic ranging system. U.S. Patent 3, 443, 095 (1969).
1442. Hagard, A.   Slant visibility measurement with Lidar. Försvarets Forskningsanstalt, Stockholm. F O A 2 Rapport A 2554-E1 (1972).
1443. Witt, G., and Lundin, A.   Laser beam backscatter sounding of the upper atmosphere. Inst. in Stockholm, Rep. AP-2, UDC 535, 81 (1971).
1444. Ericsson Militärelektronikdivision.   Laser systems for cloud altitude measurements. S. -41487, Stockholm (1965); Ericsson Review No. 2, Rep. 1499 (1965).
1445. Buck, A. L.   Effects of the atmosphere on laser beam propagation. *Appl. Opt.* **6**, No. 4, 703–713 (1967).
1446. Blau, H. H. *et al.*   A prototype cloud physics laser nephelometer. *Appl. Opt.* **9**, No. 8, 1798–1803 (1970).
1447. Plass, G. N., and Kattawar, G. W.   Reflection of light pulses from clouds. *Appl. Opt.* **10**, No. 10, 2304–2310 (1971).
1448. Impulsphysik GmbH, Hamburg.   Lichtimpulssender für Wolkenhöhenmesser. DP 1 648 240 (1970).
1449. Früngel, F.   Optical radiation pulse control receiver. U.S. Patent 3, 516, 751 (1970).
1450. United Detector Technology, Santa Monica, California.   Industrial papers, Schottky barrier standard and long line series.
1451. Yura, H. T.   Small-angle scattering of light by ocean water. *Appl. Opt.* **10**, No. 1, 114–118 (1971).
1452. Tyler, J. E. *et al.*   Predicted optical properties for clear natural water. *J. Opt. Soc. Amer.* **62**, No. 1, 83–91 (1972).
1453. Tyler, J. E., and Smith, R.   Submersible spectroradiometer. *J. Opt. Soc. Amer.* **56**, No. 10, 1390–1396 (1966).
1454. Pritchard, D. W., and Carpenter, J. H.   Measurements of turbulent diffusion in estuarine and inshore waters. *Proc. Symp. Tidal Rivers, Helsinki, July, 1960.* Int. Assoc. Sci. Hydrol. No. 20 (1960).
1455. Pollio, J.   Photogrammetric applications to underseas tasks. SMPTE No. 104–27 (1968).
1456. Neuymin, H. G.   Inhomogeneities of optical properties in deep ocean waters. *J. Opt. Soc. Amer.* **60**, No. 5, 690–693 (1970).
1457. Kullenberg, G.   Scattering of light by Sargasso sea water. *Deep-sea Res.* **15**, 423–432 (1968).
1458. Stevens, N. B., Horman, M. H., and Dodd, E. E.   The determination of atmospheric transmissivity by backscatter from a pulsed light system. Motorola Riverside Res. Lab., ASTIA Doc. No. 133602 (1957).
1459. Früngel, F.   Visual range measuring system including a plurality of spaced lamps. U.S. Patent 3, 393, 321 (1968).
1460. Doeg, C.   Photography underwater. mt 3, No. 4 (Aug. 1972).
1461. Robben, F.   Noise in the measurement of light with photomultipliers. *Appl. Opt.* **10**, 776–796 (1971).
1462. Rees, J. D., and Givens, M. P.   Variation of time-of-flight of electrons through a photomultiplier. *J. Opt. Soc. Amer.* **56**, No. 1, 93–95 (1966).

1463. Palmer, R. E.   An improved method for measuring photoemission electron energy distribution curves. *Rev. Sci. Instrum.* **42**, No. 10, 1450–1452 (1971).
1464. McIntyre, R. J., and Webb, P. P., and Springings, H. C.   Solid-state detectors for laser applications. Corporale Engineering Services, 32–39 (1969).
1465. Philco, Microelectronics Div. Spring City.   Industrial paper, Photodetectors? – We cover the spectrum.
1466. RCA Radio Corp. of America.   Industrial paper, RCA Photomultipliers would have fascinated thomson. *Appl. Opt.* **10**, No. 9, p. A15 (0000).
1467. Philips.   Industrial papers: A 4-decade linear density scale photometer equipped with the XP1110 photomultiplier tube. Application Information No. 318, and A logarithmic amplifier. Application Information No. 86.
1468. Mayskaya, K. A., and Privalova, V. Ye.   Effect of light on efficient photocathodes with current drain. Physics Inst. K. Marx Univ., Leipzig (Feb. 1964).
1469. Land, P. L.   A discussion of the region of linear operation of photomultipliers. *Rev. Sci. Instrum.* **42**, No. 4, 420–425 (1971).
1470. Judson, Res. and Mfg. Co. Conshohocken, Pennsylvania.   Photovoltaic indium antimonide, infrared detectors. Industrial papers.
1471. Felmy, R. A.   An expanding role for the GaAs laser. *Opt. Spectra*, 26–29 (Dec. 1971).
1472. Cooney, J.   Measurements separating the gaseous and aerosol components of laser atmospheric backscatter. *Nature (London)* **224**, 1098–1099 (1969).
1473. Shaw, S. A., Grant, G. R., and Gunter, W. D., Jr.   Optical enhancement of photomultipliers at ultraviolet wavelengths. *Appl. Opt.* **10**, No. 11, 2559 (1971).
1474. Kildal, H., and Byer, R. L.   Comparison of laser methods for the remote detection of atmospheric pollutants. *IEEE* **59**, No. 12, 1644–1663 (1971).
1475. Smith, J. B.   An amplifier for electron multiplier pulse counting applications. *Rev. Sci. Instrum.* **43**, No. 3, 488–492 (1972).
1476. Früngel, F.   Arrangement for measuring the concentration of fluorescent materials in air and water. U.S. Patent 3, 666, 945 (1972).
1477. Früngel, F.   Method of and apparatus for determining the presence and/or quantity of a fluorescent substance in a volume in water. Brit. Patent 1, 282, 748 (1972).
1478. Conner, W. D. *et al.*   Optical properties and visual effects of smoke-stack plumes. Public Health Service, Cincinnati, Ohio, PB 174, 705 (1967).
1479. Hinkley, E. D.   Tunable infrared lasers and their applications to air pollution measurements. MIT Lincoln Lab., Lexington, Massachusetts (0000).
1480. Hering, W. S., Muench, H. St., and Brown, H. A.   Field test of a forward scatter visibility meter. Air Force Systems Command, AFCRL-71-0317 (1971).
1481. Schwab, A.   Messung schnell veränderlicher Spannungen und Ströme. Hochspannungsinstitut TH Karlsruhe.
1482. Schwab, A.   Precision capacitive voltage divider for impulse voltage measurement. *IEEE* **60**, Jan./Febr. (1972).
1483. Zaengl, W.   Zur Ermittlung der vollständigen Übertragungseigenschaften eines Stoßspannungskreises. *ETZ A* **90**, 457–462 (1969).
1484. Zaengl, W.   Der Stoßspannungsteiler mit Zuleitung. *Bull. SEV* **61**, 1003–1017 (1970).
1485. Creed, F. C., and Collins, M. M. C.   The step response of measuring systems for high-impulse voltages. *IEEE Trans.* **PAS 86**, 1408–1420 (1967).
1486. Hylten-Cavallius, N. R., and Vaugham, L.   Calibration and checking methods of rapid high-voltage impulse measuring circuits. *IEEE Trans.* **PAS 89**, 1303–1403 (1970).

1487. Pedersen, A., and Lausen, P.   Dynamic properties of impulse measuring systems. *IEEE Trans.* **PAS 90**, 1424–1432 (1971).
1488. Schwab. A,   "Hochspannungsmeßtechnik." Springer, Berlin (1969).
1489. Sscwab, A.   "High-voltage Measurement Techniques." MIT Press, Cambridge, Massuchusetts (1971).
1490. Zaengl, W.   Das Messen hoher, rasch veränderlicher Stoßspannungen Dissertation thesis, Univ. Munich.
1491. Zaengl, W.   Ein neuer Teiler für steile Stoßspannungen. *Bull. SEV* **56**, 232–240 (1965).
1492. Feser, K.   Ein neuer Spannungsteiler für die Messung hoher Stoßspannungen und hoher Wechselspannungen. *Bull. SEV* **16** (1971).
1493. Cassidy, E. C. et al.   Development and Evaluation of electrooptical high-voltage pulse measurement techniques. *IEEE Trans.* **IM 19**, 395–402 (1970).
1494. Greif, G.   Pruefling und Normal zugleich "Hartmann und Braun Meßwerte" **11**. pp. 30–33 (1970).
1495. Keller, A.   Konstanz der Kapazitaet von Pressgaskondensatoren. *ETZ A* **80**, 757–761 (1959).
1496. Kusters, N. L., and Petersons, O.   The voltage coefficients of precision capacitors. *IEEE Trans.* **CE 69**, 601–611 (1963).
1497. Betzler, K., Weller, T., and Conradt, R.   Improvement of photon counting by means of a pulse height analyzer. *Rev. Sci. Instrum.* **42**, No. 11, 1594–1596 (1971).
1498. Micro Instrument Company, Hawthorne, California.   Industrial papers. Peak voltage memory paks and A-D converters, high voltage surge-transient monitors systems; real surge-current three phase meter.
1499. Dearing, L. M. et al.   Thin-probe pulsed-light photometer for measurement and calibration of timing and other pulsed-light sources. *J. SMPTE* **78**, 718–721 (1969).
1500. DiDomenico, M. et al.   High speed photodetection in germainum and silicon cartridge type point-contract photodiodes. *Appl. Opt.* **4**, No. 6, 677–685 (1965).
1501. Linford, R. M. F.   The application of an ultraviolet sensitive fire detector in a manned space vehicle. NASA Contract NAS9-6555 (1971).
1502. ITT Laboratories, Fort Wayne, Indiana.   Industrial papers: Multiplier phototube, coaxial phototube, high-current phototube.
1503. Bendix Electro-Optics Div., Ann Arbor, Michigan.   Industrial papers, Continuous dynode electron multiplier.
1504. Continental Electric Co., Chicago, Illinois.   Industrial paper, Phototubes, lead sulfide.
1505. United Detector Technology, Santa Monica, California.   Industrial paper, Photometer/radiometer, laser power meter, 30-A linear displacement display.
1506. Centronic Works, Croydon, England.   Industrial papers, Photodiodes, photomultiplier tubes.
1507. CBS Laboratories, Stamford, Connecticut.   Industrial papers: Photomultiplier tubes, CL-1010 Photomultiplier, CL-1011, etc. Special purpose tubes.
1508. Furcinitti, P. et al.   Measurement of the light-field amplitude correlation function through joint-photon-count distributions. *J. Opt. Soc. Amer.* **62**, No. 6, 792–796 (1972).
1509. Vorob'yeva, O. B., Mostovskiy, A. A., and Stuchniskiy, G. B.   The secondary electron emission of multi-alkali photocathodes. Phys. Inst. Karl Marx Univ., Leipzig, pp. 414–418 (Febr. 1964).
1510. Birth, G. S., and DeWitt, D. P.   Further comments on the areal sensitivity of end-on photomultipliers. *Appl. Opt.* **10**, No. 3 687–691 (1971).

1511. Korneff, T.   Optical pyrometer with microsecond resolution time. *Rev. Sci. Instrum.* **42**, No. 11, 1561–1565 (1971).
1512. Taylor, H. R. *et al.*   Recording flash x-ray burst times with scintillation crystals. *Rev. Sci. Instrum.* **42**, No. 11, 1627–1629 (1971).
1513. Wood, D. S.   A portable calorimeter for x-ray measurement of fluence and front surface dose simultaneously. *Rev. Sci. Instrum.* **43**, No. 6, 908–910 (1972).
1514. Willis, R. G.   A simplified orders-of-scattering technique for calculating the absorption of x-rays. *J. Appl. Phys.* **39**, No. 11, 5116–5121 (1968).
1515. Fixatron Field Emission Corp.   New research, design tool instant x-rays for workbench or laboratory. Industrial paper.
1516. KSP Industries, Inc., Alexandria, Virginia.   Pressure transducer. Industrial paper.
1517. Früngel, F., Ettenreich, L., and Andre, K. -H.   Impulshärten von Stahl. VDI-Z. 114. Nr. 14 u. 16 (1972).
1518. Hermet, P.   Design of a rangefinder for military purposes. *Appl. Opt.* **11**, No. 2, 273–276 (1972).
1519. Cooke, C. R.   Automatic laser tracking and ranging system. *Appl. Opt.* **11**, No. 2, 277–284 (1972).
1520. Staron, M.   Control optimization of a laser automatic tracking system: Influence of the space-time returns of the echoes. *Appl. Opt.* **11**, No. 2, 285–290 (1972).
1521. Flom, T.   Spaceborne laser radar. *Appl. Opt.* **11**, No. 2, 291–299 (1972).
1522. Lehr, C. G. *et al.*   Transportable lunar-ranging system. *Appl. Opt.* **11**, No. 2, 300–304 (1972).
1523. Chrzanowski, A., Ahmed, F., and Kurz, B.   New laser applications in geodetic and engineering surveys. *Appl. Opt.* **11**, No. 2, 319–330 (1972).
1524. Hugenschmidt, M., Vollrath, K., and Hirth, A.   Schlieren-cinemato-graphic and holographic diagnostic of a laser-produced plasma in xenon. *Appl. Opt.* **11**, No. 2, 339–344 (1972).
1525. Mead, S. W. *et al.*   Preliminary measurements of x-ray and neutron emission from laser-produced plasmas. *Appl. Opt.* **11**, No. 2, 345–352 (1972).
1526. Eichler, H. *et al.*   Power requirements and resolution of real-time holograms in saturable absorbers and absorbing liquids. *Appl. Opt.* **11**, No. 2, 372–275 (1972).
1527. Hall, R. T., and Rawcliff, R. D.   Model for the visible-light scattering properties of clouds. *Appl. Opt.* **11**, No. 2, 468–469 (1972).
1528. Allen, D. H. *et al.*   Logarithmic detector for pulsed lasers. *Appl. Opt.* **11**, No. 2, 476–477 (1972).
1529. Edgerton, H. E.   "Electronic Flash Strobe." McGraw-Hill, New York (1970).
    a.   pp. 6–15. Theory of the electronic flash lamp.
    b.   pp. 20–23 Theory of the electronic flash lamp.
    c.   pp. 26–29. Theory of the electronic flash lamp.
    d.   pp. 30–47. Theory of the electronic flash lamp.
    e.   pp. 48–60. Spectral output of flash lamps.
    f.*   pp. 62–95. Circuits for electronic flash equipment.
    g.   pp. 96–102. Electronic flash lighting requirements for photography.
    h.   p. 120. Electronic flash equipment (single flash).
    i.   pp. 123–127. Short duration xenon flash lamps.
    j.*   pp. 136–139. Electronic flash equipment of short exposure time.
    k.   pp. 143–151. Electronic flash equipment for nature photography.
    l.   pp. 152–158. Electronic flash equipment for nature photography.
    m.   pp. 165–167. The Stroboscope.

n. pp. 174–177. The Stroboscope.
o.† pp. 177–181 .The Stroboscope, Hydrogen thyratron modulator.
p. pp. 182–185. The Stroboscope.
q. pp. 198–201. The Stroboscope, Multiflash photography.
r. p. 206. The Stroboscope.
s. pp. 219–233. Exposure calculations and special photography.
t. pp. 235–249. Techniques of light measurement.
u. pp. 286–288. Specialized applications.
v. pp. 300–311. Specialized applications. Laser stimulators.
w. pp. 325–331. Specialized applications, Photochromic goggle system.

1530.† Optics and laser technology. *Opt. Laser Tech.* **4**, No. 2, 62–65 (1972): Optical voltage sensor. 1 GW flash. Laser in estuary sand-wave study.

1531. Gates, J. W. C., Hall, R. G. N., and Ross, I. N. Holographic interferometry of impact-loaded objects using a double-pulse laser. *Opt. Laser Technol.* **4**, No. 2, 72–75 (1972).

1532. Tanner, L. H. A holographic interferometer and fringe analyzer, and their use for the study of supersonic flow. *Opt. Laser Technol.* **4**, No. 2, 66–71 (1972).

1533. Sandia Lab., Albuquerque, New Mexico. Ferro-electric ceramic stores high resolution images. *Opt. Laser Technol.* **4**, No. 2, 91–92 (1972).

1534. New products. Ceramic crystals; neodymium laser used to monitor smoke density; Piezo-optical polarizer; High-gain laser amplifiers; High-speed vacuum photodiodes: High power lasers for cutting and welding. *Opt. Laser Technol.* **4**, No. 2, 93–97 (1972).

1535.* Laser interactions with matter. Conference at the Inst. of Electrical Engineers, Savoy Place, London., Jan. *Opt. Laser Technol.* **4**, No. 2, 98–99 (1972).

1536. Schwart, H. J., and Hora, H. Laser interaction and related plasma phenomena. *Opt. Laser Technol.* **4**, No. 2, 101 (1972).

1537. Curran, R. J. Ocean color determination through a scattering atmosphere. *Appl. Opt.* **11**, No. 8, 1857–1866 (1972).

1538. Granatstein, V. L. et. al. Depolarization of laser light scattered from turbid water. *Appl. Opt.* **11**, No. 8, 1870–1871 (1972).

1539. Amoss, J., and Davidson, Fr. Detection of weak optical images with photon counting techniques. *Appl. Opt.* **11**, No. 8, 1793–1800 (1972).

1540. Bridge, N. K., and Porter, G. Primary photoprocesses in quinones and dyes. I. Spectroscopic detection of intermediates. *Proc. Roy. Soc. Bd.* **244A**, 259–275 (1958).

1541. Bridge, N. K., and Porter, G. Primary photoprocesses in quinones and dyes. II. Kinetic studies. *Proc. Roy. Soc. Bd.* **244A**, 276–288 (1958).

1542. Porter, G. Description of a standard flash photolysis equipment. *In* "Techniques of organic chemistry," VIII, Part II, pp. 1062–1072. Interscience, New York (1963).

1543. Rabinowitch, E. et al. Transfert d'energie et photosynthese. *J. Chim. Phys.* **55**, 927–933 (1958).

1544. Wild, P. J. A phasemeter for photoelectric measurement of magnetic fields. *Rev. Sci. Instrum.* **41**, No. 8, 1163–1167 (1970).

1545. Claesson, S., and Linqvist, L. A flash photolysis apparatus for photochemical studies at very high light intensities and energies. *Arkiv Kemi* **11**, No. 60, 535–559 (1957).

1546. Claesson, S., and Linqvist, L. A fast photolysis flash lamp for very high light intensities. *Arkiv Kemi* **12**, No. 1, 1–8 (1957).

1547. Linqvist, L. A flash photolysis study of fluorescein. *Arkiv Kemi* **16**, No. 8, 79–138 (1960).

1548.‡ Claesson, S., Lindqvist, L., and Strong, R. L.   A fast 50 kV, 8 kJ flash photolysis apparatus. *Arkiv Kemi*, **22**, No. 21, 245–251 (1964).
1549. Sharp, L. E., and Wetherell, A. T.   High power pulsed HNC laser. *Appl. Opt.* **11**, No. 8, 1737–1741 (1972).
1550. Everett, P. N., and Cantor, A. J.   Long-range holography. *Appl. Opt.* **11**, No. 2, 1697–1699 (1972).
1551. Johnson, R. H., and Holshouser, D. F.   Application of a mode-locked laser to holography. *Appl. Opt.* **11**, No. 8, 1708–1715 (1972).
1552. Harris, F. S.   ACS Symp. Occasion Centennial Rayleigh Scattering Theory, Washington, Sept. (1971). *Appl. Opt.* **11**, No. 8, 1887–1889 (1972).
1553. Cook, Ch. S., Bethke, G. W., and Conner, W. D.   Remote measurement of smoke plume transmittance using lidar. *Appl. Opt.* **11**, No. 8, 1742–1748 (1972).
1554. Wood, D. S.   A depth dose calorimeter for high intensity, low voltage flash x-ray machines. *Rev. Sci. Instrum.* **43**, No. 8, 1094–1096 (1972).
1555. Schuch, R. L., and Kelly, J. G.   A compact faraday cup array for measurement of current distribution from pulsed electron beams. *Rev. Sci. Instrum.* **43**, No. 8, 1097–1099 (1972).
1556. Hager, N. E. Jr.   High speed thermal analysis with thin-foil calorimeter. *Rev. Sci. Instrum.* **43**, No. 8, 1116–1122 (1972).
1557. Conway, J. C.   An improved Cranz-Schardin high speed camera for two-dimensional photomechanics. *Rev. Sci. Instrum.* **43**, No. 8, 1172–1174 (1972).
1558. Pellinen, D. G.   Small combination x-ray calorimeters. *Rev. Sci. Instrum.* **43**, No. 8, 1181–1184 (1972).
1559.* Hviid, Th., and Nielsen, S. O.   35 volt, 180 ampere pulse generator with droop control for pulsing xenon arcs. *Rev. Sci. Instrum.* **43**, No. 8, 1198–1199 (1972).
1560.* Singer, I. L., and Ems, S.   A fast high current pulser suitable for low temperatures. *Rev. Sci. Instrum.* **43**, No. 8, 1204–1205 (1972).
1561. Small, J. G., and Ashari, R.   A simple pulsed nitrogen 3371 Å laser with a modified blumlein excitation method. *Rev. Sci. Instrum.* **43**, No. 8, 1205–1206 (1972).
1562. Chu, T. S.   Bell Telephone Lab., Holmdel. Attenuation by percipitation of laser beams at 0, 63 μ, 3, 5 μ and 10. μ. pp. 30–35 (June 1967).
1563.† Böcker, H.   The electromagnetic wave of the high-voltage impulse. *Int. Symp. Hochspannungstech.*, *TU München, März* pp.9–16 (1972).
1564.† Moeller, J., Steinbigler, H., and Weiß, P.   Potential gradients on the surface of shielding electrodes for ultra high voltages. *Int. Symp. Hochspannungstech. TU München, März* pp.36–46 (1972).
1565.* Weiss, P.   Field strength effects in two-dielectric arrangements. *Int. Symp. Hochspannungstech.*, *TU München, März* pp. 73–80 (1972).
1566. Aubry, J., Claverie, P., and Cristescu, D.   Contribution to electric measurement technique by means of probes. *Int. Symp. Hochspannungstech.*, *TU München, März* pp. 81–85 (1972).
1567. Esposti, G. D. *et al.*   Considerations on the maximum allowable surface gradients on the electrodes of the high voltage apparatus. *Int. Symp. Hochspannungstech.*, *TU München, März* pp. 91–97 (1972).
1568. Früngel, F., and Ebeling, D.   A new method for investigation of the corona—onset on aerials or HV-conductors. *Int. Symp. Hochspannungstech.*, *TU München, März* pp. 98–103 (1972).
1569.* Anderson, R. A., Harrison, J., and Stine, R. D.   A low inductance, compact, modularized Marx generator. *Int. Symp. Hochspannungstech.*, *TU München, März* pp. 111–115 (1972).

1570.* Feser, K., and Rodewald, A.   A triggered multiple chopping gap for high lighting and high switching surges. *Int. Symp. Hochspannungstech., TU München, März* pp. 124–131 (1972).
1571.* Finsterwalder, F., Zacke, P., Fischer, A., and Böcker, H.   Rotating symmetrical spark gaps for triggered chopping of high impulse voltages. *Int. Symp. Hochspannungstech., TU München, März* pp. 132–138 (1972).
1572. Früngel, F., and Ebeling, D.   Technology of pulse generation with differential pulse transformers. *Int. Symp. Hochspannungstech., TU München, März* pp. 139–146 (1972).
1573.* Guraraj, B. I.   Choice of circuit constants for long-duration current generator for surge diverter testing. *Int. Symp. Hochspannungstech., TU München, März* pp. 147–152 (1972).
1574.† Kolb, A., Graham, F., and Wooten, R.   High voltage output switches for fast Marx generator pulse discharge systems. *Int. Symp. Hochspannungstech., TU München März* pp. 161–167 (1972).
1575.† Müller, W.   The numerical computation of the spatial voltage distribution in the transformer windings during voltage surges with respect to time. *Int. Symp. Hochspannungstech., TU München, März* pp. 176–183 (1972).
1576.† Pflanz, H. M.   Generation of high voltages on energizing capacitive circuits. *Int. Symp. Hochspannungstech., TU München, März* pp. 184–190 (1972).
1577.* Rodewald, A.   A new triggered multiple gap system for all kinds of voltages. *Int. Symp. Hochspannungstech., TU München, März* pp. 199–203 (1972).
1578. Arrighi, R., and Deaumont, B.   Free-potential measurement at Fontenay high voltage laboratory. *Int. Symp. Hochspannungstech., TU München, März* pp. 204–210 (1972).
1579.† Blasius, P. *et al.*   The influence of extended test circuits on the waveshape of lightning impulse. *Int. Symp. Hochspannungstech., TU München, März* pp. 211–218 (1972).
1580. Feser, K., and Rodewald, A.   The transfer characteristics of damped capacitive voltage dividers above 1 MV. *Int. Symp. Hochspannungstech., TU München, März* pp. 219–224 (1972).
1581.* Freitag, J., and Schiweck, L.   A compressed-gas capacitor with intermediate electrode for the measurement. *Int. Symp. Hochspannungstech., TU München, März* pp. 225–231 (1972).
1582.† Heymann, F. G.   Surge propagation on a 400 kV transmission line. *Int. Symp. Hochspannungstech., TU München, März* pp.232–238 (1972).
1583. Krawczyński, R.   Calibration of high voltage impulse measuring systems in complete test circuits. *Int. Symp. Hochspannungstech., TU München, März* pp. 239–244 (1972).
1584. Richardson, A. V., and Ryan, H. M.   Computer-aided analysis of an impulse voltage measuring system. *Int. Symp. Hochspannungstech., TU München, März* pp. 245–251 (1972).
1585. Schwab, A. J., and Pagel, J. H. W.   Precision capacitive voltage divider. *Int. Symp. Hochspannungstech., TU München, März* pp. 252–259 (1972).
1586.* Aked, A., and McAllister, I. W.   The effect of impulse voltage wavetail duration on the prebreakdown phenomena. *Int. Symp. Hochspannungstech., TU München, März* pp. 260–262 (1972).
1587.* Böcker, H., and Fischer, A.   Streamer and leader in long spark gaps with impulse voltage. *Int. Symp. Hochspannungstech., TU München, März* pp. 265–272 (1972).

1588.* Cookson, A. H., and Farish, O.   Breakdown of wire–plane, disk–plane and rod–plane gaps in air. *Int. Symp. Hochspannungstech.*, *TU München*, *März* pp.273–278 (1972).
1589.* Gallimberti, I., and Stassinopoulos, C. A.   Development of positive discharges in atmospheric air. *Int. Symp. Hochspannungstech.*, *TU München*, *März* pp. 279–285 (1972)
1590.* Hepworth, J. K., Klewe, R. C., and Tozer, B. A.   A model of impulse breakdown in divergent field geometries. *Int. Symp. Hochspannungstech.*, *TU München*, *März* pp. 286–291 (1972).
1591.* Jones, B., and Whittington, H. W.   Influence of the shape of the testing waveform on breakdown of long air gaps. *Int. Symp. Hochspannungstech.*, *TU München*, *März* pp. 292–297 (1972).
1592.* Krasser, G.   Flashovers in air on outdoor bushings and control of same by means of potential grading. *Int. Symp. Hochspannungstech.*, *TU München*, *März* pp. 298–305 (1972).
1593.* Lennertz, H.   Analysis of serious connection of inhomogeneous electrode systems. *Int. Symp. Hochspannungstech.*, *TU München*, *März* pp. 306–313 (1972).
1594.* Levitov, V. I., Bazelian, E. M., and Volkova, O. V.   The peculiarities of discharge developed in long gaps with negative voltage polarity. *Int. Symp. Hochspannungstech.*, *TU München*, *März* pp. 314–318 (1972).
1595.* Mosch, W., Lemke, E., and Fahd, I.   Breakdown tests on inhomogeneous air gaps with switching surges superimposed on dc voltage. *Int. Symp. Hochspannungstech.*, *TU München*, *März* pp. 319–326 (1972).
1596.* Newi, G.   Variation of breakdown voltage of sphere gaps in air at negative impulse voltages. *Int. Symp. Hochspannungstach.*, *TU München*, *März* pp.327–338 (1972).
1597.* Banford, H. M., and Tedford, D. J.   Prebreakdown phenomena in nitrogen and air at high values of "pd." *Int. Symp. Hochspannungstech.*, *TU München*, *März* pp. 339–342 (1972).
1598.* Cookson, A. H., and Farish, O.   Particle-initiated breakdown between coaxial electrodes in compressed SF_6. *Int. Symp. Hochspannungstech.*, *TU München*, *März* pp. 343–349 (1972).
1599.* Flieux, R. *et al*.   Influence of air conductivity on corona and breakdown voltage of a point-plane gap. *Int. Symp. Hochspannungstech.*, *TU München*, *März* pp. 350–355 (1972).
1600.* Ganger, B.   Electric strength of air with high alternating and impulse voltages and pressures up to 100 kp/cm_2. *Int. Symp. Hochspannungstech.*, *TU München*, *März* pp. 356–362 (1972).
1601.* Hasse, P.   The development of the electric breakdown in uniform-field in sulphur hexafluoride. *Int. Symp. Hochspannungstech.*, *TU München*, *März* pp. 363–370 (1972).
1602.* Mosch, W., and Hauschild, W.   A condition for SF_6-breakdown in slightly nonuniform fields. *Int. Symp. Hochspannungstech.*, *TU München*, *März* pp. 371–377 (1972).
1603.* Oppermann, G.   The Paschen's curve and the limitations of the Paschen's law of sulphur hexafluoride. *Int. Symp. Hochspannungstech.*, *TU München*, *März* pp. 378–385 (1972).
1604.* Takuma, T., Watanabe, T., Kita, K., and Aoshima, Y.   Discharge development of long gaps in SF_6 gas. *Int. Symp. Hochspannungstech.*, *TU München*, *März* pp. 386–390 (1972).

1605.* Teich, T. H., and Sangi, B.   Discharge parameters for some electronegative gases and emission of radiation from electron avalanches. *Int. Symp. Hochspannungstech., TU München, März* pp. 391–395 (1972).
1606.* Badran, I. M. S. *et al.*   Internal discharges in paper insulation under direct-voltage conditions. *Int. Symp. Hochspannungstech., TU München, März* pp. 396–402 (1972).
1607.* Hossam-Eldin, A. A., Pearmain, A. -J., and Salvage, B.   Internal discharges in impregnated paper: phenomena at low temperatures. *Int. Symp. Hochspannungstech., TU München, März* pp.403–407 (1972).
1608.* Kindij, E.   Electric breakdown anisotropy. *Int. Symp. Hochspannungstech., TU München, März* pp. 408–413 (1972).
1609.* Ieda, M., Sawa, G., and Miyairi, K.   Dielectric breakdown of polyethylene films at cryogenic temperature. *Int. Symp. Hochspannungstech., TU München, März* pp. 414–420 (1972).
1610.* Mosch, W., Pilling, J., Tschacher, B., and Eckholz, K.   Treeing inception time (channel inception time) and breakdown time as a criterion for long-time behavior of solid insulations. *Int. Symp. Hochspannungstech., TU München, März* pp. 421–427 (1972).
1611.* Nawata, M., Kawamura, H., and Ieda, M.   Voltage and temperature dependence of treeing breakdown in plastic insulators. *Int. Symp. Hochspannungstech., TU München, März* pp. 428–434 (1972).
1612.* Narayana Rao, Y.   Breakdown mechanisms in plastic insulation under the influence of corona discharges. *Int. Symp. Hochspannungstech., TU München, März* pp. 435–442 (1972).
1613.* Ryder, D. M., Wood, J. W., and Hogg, W. K.   Partial discharge experiments on small samples of epoxy resin bonded mica turbine generator stator bar insulation. *Int. Symp. Hochspannungstech., TU München, März* pp. 450–456 (1972).
1614.* Schirr, J.   Influence of mechanical stress on the growth of predischarge channels in epoxy resin. *Int. Symp. Hochspannungstech., TU München, März* pp. 457–464 (1972).
1615.* Balakrishna, R., and Raju G. R.   A study of lichtenberg figures in transformer oil with impulse potentials. *Int. Symp. Hochspannungstech., TU München, März* pp.471–475 (1972).
1616.* Beyer, M., and Bitsch, R.   The influence of different load of gases and moisture on the electrical strength of insulating oils in the inhomogenous field under alternating current. *Int. Symp. Hochspannungstech., TU München, März* pp. 476–483 (1972)
1617.* Németh, E.   Polarization of large time constants in insulating materials. Determination of the time constant distribution by measuring the absorption current. *Int. Symp. Hochspannungstech., TU Munchen, März* pp. 484–490 (1972).
1618.* Penneck, R. J., and Swinmurn, J. C.   Polymeric materials for use in polluted high voltage environments. *Int. Symp. Hochspannungstech., TU München, März* pp. 491–497 (1972).
1619. Rumeli, A.   Discharge propagation over polluted insulating surfaces. *Int. Symp. Hochspannungstech., TU München, März* pp. 498–503 (1972).
1620.* Rumeli, A.   Preventing discharge growth over polluted insulating surfaces. *Int. Symp. Hochspannungstech., TU München, März* pp. 504–509 (1972).
1621. Veverka, A.   Measurement of partial discharges in high-voltage transformers at induced voltage. *Int. Symp. Hochspannungstech., TU München, März* pp. 514–520 (1972).
1622.* Vlastós, A. E.   Prebreakdown discharges on insulator surfaces. *Int. Symp. Hochspannungstech., TU München, März* pp. 521–526 (1972).

1623.* Widmann, W. The impulse strength of creepage discharge arrangements in transformer oil. *Int. Symp. Hochspannungstech.*, *TU München, März* pp. 527–534 (1972).
1624.* Pillsticker, M. Laser triggered high-voltage spark gaps. *Int. Symp. Hochspannungstech.*, *TU München, März* pp. 619–626 (1972).
1625. v. d. Piepen, H., and Schroeder, W. W. A time- and space-resolved spectroscopic study of an argon jet guided spark discharge. *Int. Symp. Hochspannungstech.*, *TU München, März* pp. 612–618 (1972).
1626. Gasparini, F., and Macchiaroli, B. Glow discharge probes to evidence the bubbles in the fluidized bed reactors. *Int. Symp. Hochspannungstech.*, *TU München, März* pp. 607–611 (1972).
1627.* Burden, R. A., and James, T. E. Premature breakdown and electrode erosion studies in a high current spark gap switch. *Int. Symp. Hochspannungstech.*, *TU München, März* pp. 587–590 (1972).
1628.* Dokopoulos, P., and Lochter, M. Solid dielectric switches in plasma-physics. *Int. Symp. Hochspannungstech.*, *TU München, März* pp. 591–598 (1972).
1629.* Dokopoulos, P., and Steudle, W. Application of water for insulating high voltage pulse systems. *Int. Symp. Hochspannungstech.*, *TU München, März* pp. 599–606 (1972).
1630. Hartley Measurements Limited. Hartley Wintney, Rapid discharge capacitors. Industrial papers.
1631. Impulsphysik GmbH, Hamburg. Can the distribution of industrial waste water in lakes, rivers or oceans be forecast or controlled? Industrial papers.
1632. Impulsphysik GmbH, Hamburg. High energy-flash lamps duration PL 1 and PL 2; Technical data of x-ray flash instrumentation; Chronokerr.
1633.* General Electric, Inc. Flash tubes. Industrial papers.
1634. Andreyev, S. I., Vanyukov, M. P., and Daniel, E. V. Surface spark discharge as a source of intensive light flashes. Phys. Inst. of Karl Marx Univ., Leipzig (1964).
1635. Séguin, J. N., and Carswell, A. I. Fluorescence and gain measurements in a CO_2 TEA amplifier. *Int Quantum Electron. Conf. 7th*, Montreal pp. 77–78 (1972).
1636.* Kasuya, K., and Murasaki, T. A new method for current "crowbar," utilizing reflected shock waves. *Rev. Sci. Instrum.* **43**, No. 10, 1481 (1972).
1637.* Milam, D., Gallagher, C. C., Bradbury, R. A., and Bliss, E. S. Switching jitter in spark gap triggered by TEM_{00}-mode mode-locked ruby laser. *Rev. Sci. Instrum.* **43**, No. 10, 1482–1484 (1972).
1638. Omni Ray A. G. Zürich, Multipliers. Modulators and Detectors. Industrial paper.
1639. Kuper, G. Lidar-measurement by means of laser with a view towards Ramanspectroscopy. *Eur. Electro-Opt. Markets Technol. Conf. Exhibition, 1st*, Geneva 13. -15. Sept. (1972).
1640.* Official exhibition catalog. *Eur. Electro-Opt. Markets Technol. Conf. Exhibition*, *1st Geneva*, 13–15 Sept. (1972).
1641.* 10ème Congres International de Cinematographie Ultra-rapide. Resumes. 25.–30. Sept. Nice-France (1972).
1642. Ginzburg, V. M. et al. Holographic methods in the research of Diesel fuel supply. *Congr. Int. Cinematogra. Ultra-Rapide, 10th, 26—30 Sept., Nice-France* (1972).
1643. Patzke, H.-G., and Thorwart, W. A simple method for isochronous representation of isobaric and isoclinic zones in dichronic fluids (liquid crystals) by means of microsecond flashes (Monoflash). *Congr. Int. Cinematogra. Ultra-Rapide. 26–30 Sept., Nice-France* (1972).

1644.* Berkeley Nucleonics Corp. Berkeley, California. Generators. Industrial paper.
1645.* Matson, G. B. A precision current pulse generator for NMR self-diffusion measurements by the pulsed gradient technique. *Rev. Sci. Instrum.* **43**, No. 10, 1504–1508 (1972).
1646.* Advanced Kinetics, Inc., Costa Mesa, California. High energy, high current pulse transformer, high field pulsed spiral magnets, high field pulsed electromagnets, high field flux concentrator magnets, fast pulsed gas valve system, collapsing foil system, magnetic flux pick-up probes, high speed infrared detector system. Industrial papers.
1647.† Marugin, A. M., and Ovchinnikov, V. M. The effect of electrode arrangement and z-cut KDP crystal configuration on the control voltage for switching an electro-optic shutter. *Opt. Technol.* **39**, No. 1, 4–5 (1972).
1648. Abbot, Ch. E., and Cannon, Th. W. A droplet generator with electronic control of size, production rate, and charge. *Rev. Sci. Instrum.* **43**, No. 9, 1313–1317 (1972).
1649.* Braun, K., Fischer, H., and Michel, L. Filaments in a 1 kJ plasma focus experiment. *High-Beta Conf., Garching, Germany* (July 1972).
1650.* Hill, G. A., James, D. J., and Ramsden, S. A. Studies of plasmas produced by a focussed TEA CO_2 laser. Technical note. *J. SMPTE* **81**, 618 (1972).
1651.* C-Tech Ltd. Montreal Rd. Cornwall, Ontario. Installation planning data C-Tech Ltd. Model LSS-30 (P) and (PT) omnisonar. Rep. No. 102 U.
1652.† Kuzhekin, I. Verhalten von Funkenstrecken unter Wasser bei Impulsspannung und Impulsstrombeanspruchung. *ETZ-A* **93**, H. 7, 404–409 (1972).
1653. ESB Inc. Philadelphia, Pennsylvania. Sea space power systems. Industrial papers.
1654.* Barnett, D. A system for stabilizing a shipborne sonar transducer. mt 3 No. 5, pp. 202–205 (1972).
1655.* Bennett, R. Electronic aids for deep-sea fishing. mt 3. No. 5, pp. 206–211 (1972).
1656.† Hall, Fr. F. Acoustic remote sensing of temperature and velocity structure in the atmosphere. *Statist. Methods Instrument. Geophys.*, Oslo, pp. 167–179 (1971).
1657.* Little, C. G. Acoustic sounding of the lower atmosphere. *Meteorolog. Monogr.* **11**, No. 33, pp. 397–404 (1970).
1658.* Wescott, J. W., Simmons, W. R., and Little, C. G. Acoustic echo-sounding measurements of temperature and wind fluctuations. ESSA technical memorandum, ESSA ERLTM-WPL-5 (Jan. 1970).
1659.* Simmons, W. R., Wescott, J. W., and Hall, F. F. Jr. Acoustic echo sounding as related to air pollution in urban environments. NOAA Tech. Rep. ERL 216-WPL 17 (May 1971).
1660.* Beran, D. W., Little, C. G., and Willmarth, B. C. Acoustic doppler measurements of vertical velocities in the atmosphere. *Nature (London)* **230**, 160–162 (1971).
1661.* Beran, D. W. Acoustic: A new approach for monitoring the environment near airports. *J. Aircr.* **8**, No. 11, 934–936 (1971).
1662.* Beran, D. W., Hall, F. F. Jr., Wescott, J. W., and Neff, W. D. Application of an acoustic sounder to air pollution monitoring. *Air Pollut. Turbulence Diffusion Symp.*, New Mexico. (Dec. 1971).
1663.* Beran, D. W., and Willmarth, B. C. Doppler winds from a bistatic acoustic sounder. *Proc. Int. Symp. Remote Sensing Environm., 7th*, No. 10259-1-X. pp. 1699–1713 (May 1971).
1664.* Hooke, W. H., Young, J. M., and Beran, D. W. Atmospheric waves observed in the planetary boundary layer using an acoustic sounder and a microbarograph array. *Boundary-Layer Meteorol.* **2**, 371–380 (1972).

1665.* McAllister, L. G. et al. Acoustic sounding—a new approach to the study of atmospheric structure. *IEEE* **57**, No. 4, 579–587 (1969).
1666.* McAllister, L. G. Acoustic sounding of the lower troposphere. *J. Atmos. Terr. Phys.* **30**, 1439–1440 (1968).
1667.* Strand, O. N. Numerical study of the gain pattern of a shielded acoustic antenna. *J. Acoust. Soc. Amer.* **49**, No. 6, Part I, 1698–1703 (1971).
1668. Hall, F. F., Wescott, J. W., and Simmons, W. R. Acoustic echo sounding of atmospheric thermal and wind structure. *Pro. Int. Symp. Remote Sensing Environm.*, 7th, No. 12059-1-X, pp. 1715–1731 (May 1971).
1669.* Beran, D. W. Remote sensing with sound waves. NOAA Study., pp. 1–11.
1670. Optics Applications: A simplified form of interferometry, and proposed laser measurement system for determining surface contour: a concept. *Opt. Spectra*, 33–34 (1972).
1671. Gyori, R. P., and Scobey, Fr. J. Some design considerations for electrolytic silver recovery from photographic fixing baths. *J. SMPTE* **81**, 603–605 (1972).
1672. Edgerton, Germeshausen and Grier, Boston, Massachusetts. Blitzlichtgasentladungslampe mit rohrförmigem Entladungsgefäß. DBP 1065 091 (1962).
1673. Halbedl, G. Generatorgesteuerte und stimmgesteuerte Laryngo-Stroboskopie. *Medizintechnik* **12**, A, H 3. 85–89 (1972).
1674. Fischer, H., and Schwanzer, W. Spark lightpulses of extremely short duration. *Int. High Speed Photo.-Congr.* 10th, 1972, Nice. TH Darmstadt, Contract AF F-61062-68-C-0039 (1972).
1675. Electro-Optics Application: A new high speed photographic technique. *Opt. Spectra*, p. 32 (Aug. 1972).
1676. Temkin, S., and Reichman, J. M. A new technique to photograph small particles in motion. *Rev. Sci. Instrum.* **43**, No. 10, 1456–1459 (1972).
1677. Smoot, G. F., Buffington, A., and Smith, L. H. Spatial spark jitter measurements of highly charged nuclei for optical spark chambers. *Rev. Sci. Instrum.* **43**, No. 9, 1285–1286 (1972).
1678. Vanyukov, M. P. et al. High-power giant-pulse laser with unstable cavity. *Opt. Technol.* **39**, No. 1, 51–52 (1972).
1679. Strohwald, H. Ein elektronenoptischer Verschluß für Öffnungszeiten von 5.10^{-11} s. Inst. f. Plasmaforschung, Univ. Stuttgart.
1680. Close, D. H. Holographic imagery. Industrial Research, pp. 34–37 (Aug. 1972).
1681. Gates, J. W. C. Holography, industry and the rebirth of optics. National Physics Lab. Teddington, Midx, pp. 173–191 (1971).
1682. Gates, J. W. C., Hall, R. G. N., and Ross, I. N. Pulsed lasers for optical measurement. *Proc. Electro-Opt. Syst. Design Conf., Brighton, England* (1972).
1683. Gates, J. W. C. Holographic elements for projection. *Proc. Electro-Opt. Syst. Design Conf., Brighton, England* (1972).
1684. Nakajima, K. An experimental investigation of the air swirl motion and combustion in the swirl chamber of Diesel engines. *Congr. Int. FISITA*, *12th*, No. 1–02, Kyoto Univ., Japan (1964).
1685. Früngel, F. Messung der dreidimensionalen Luftverwirbelung nach dem sog. sprak tracing Verfahren. VDI-Bericht Nr. 146, pp. 181–186 (1970).
1686. Impulsphysik GmbH, Hamburg. The Strobokin spark tracing method. Industrial paper.
1687. Bomelburg, H. J., Herzog, H. J., and Weske, J. R. The electric spark method for quantitative measurements in flowing gases. *Z. Flugwissensch.* **7**, 322–329 (1959).

1688. Fister, W. Photographische Erfassung der Strömungsverhältnisse in Turbomaschinen. *Umschau Wissensch. Tech.* No. 4/67, 130 (1967).
1689. Fister, W. Versuche zur Erfassung der Strömungsverhältnisse an Radiallaufrädern. Triebwerks-Aerodynamik der Turbomaschinen. Teil II. No. 63–01, pp. 106–129 (1963).
1690. Fister, W. Sichtbarmachung der Strömungen in Radialverdichterstufen, besonders der Relativströmung in rotierenden Laufrädern, durch Funkenblitz. *S. Brennstoff-Kraft-Wärme.* 18, No. 9, 425–429 (1966).
1691. Gutenhoffnungshütte Sterkrade, Die Möglichkeit und die Ausrüstung der neuen GHH-Prüffelder des Maschinenbaus. Industrial paper. Untersuchungen zur Weiterentwicklung von radialen Turboverdichtern.
1692. König, G., and Ullmann, W. Gesichtspunkte bei der Sichtbarmachung von Strömungen in kleinen Bauelementen mit der Funkenblitzmethode. *Conf. Pneumatische Strahlelemente*, 2nd, Dresden No. 3/3 (1968).
1693. Stanford Res. Inst. Menlo Park, California. Visibility measurement for aircraft landing operations. SRI Project 8301, AFCRL-70-0598 (Nov. 1970).
1694. Uzgiris, E. E. Sensitive optical heterodyne method for light scattering studies. *Rev. Sci. Instrum.* 43, No. 9, 1383–1385 (1972).
1695. Laboratores de Marcoussis, France. TCV 29 laser rangefinder for tank. Industrial paper.
1696. Burak, I. et al. Q-switched CO_2 lasers with variable pulse delay. *Rev. Sci. Instrum.* 43, No. 9, 1390–1392 (1972).
1697. Shipman, J. D., Jr. Traveling wave excitation of high power gas lasers. *Appl. Phys. Lett.* 10, No. 1, 3–5 (1967).
1698. Leonard, D. A. Saturation of the molecular nitrogen second positive laser transition. *Appl. Phys. Lett.* 7, No. 1, 4–7 (1965).
1699. Hodgson, R. T. Vacuum-ultraviolet laser action observed in the lyman bands of molecular hydrogen. *Phys. Rev. Lett.* 25, No. 8, 494–497 (1970).
1700. Smith, S. D. Tunable infra-red light sources. Physics Dept., Herriot-Watt Univ., Edinburgh (Oct. 1972).
1701.* Ion Physics Corp., Burlington, Massachusetts. Report on development of five EMP generators. Contract No. F29601-69-C-0138 (1970).
1702.* Palva, V. Research on long air gap discharges at les renardiéres. Chairman of study committee No. 33, Electra, pp. 53–157.
1703.* Vollrath, K., and Thomer, G. "Kurzzeitphysik." Deutsch-Französisches Forschungsinstitut Saint-Louis, Springer-Verlag, New York (1967).
   a. Stenzel, A. Messung kurzer Zeiten, pp. 10–25.
   b. Stenzel, A. Abzählverfahren, elektronisch zählende Zeitmesse, pp. 26–45.
   c. Baldinger, E., and Spycher, U. Oszillographen und Verstärker, pp. 46–75.
   d. Vollrath, K. Funkenlichtquellen und Hochfrequenz-Funkenkinematographie, pp. 108–165.
   e. Unger, H. -G. Steuerung und Modulation, pp. 198–206.
   f.* Müller, W. Elektro-optische Verschlüsse, pp. 207–300.
   g.† Thomer, G. Röntgenblitztechnik, pp. 328–366.
   h. Rössler, F. Temperaturmessungen, pp. 436–441.
   i. Golsmith, W. Dynamic Photoelasticity, pp. 593–658.
   j. Eichelberger, R., and Kineke J. H., Jr. Hypervelocity Impact, pp. 659–692.
   k. Oertel, H. Messungen im Hyperschallstoßrohr, pp. 815–816.
   l.† Schall, R. Detonationsphysik, pp. 890–907.
   m.† Andelfinger, C. Kurzzeitmeßmethoden in der Plasmaphysik pp. 985–1035.

1704.* ITT Intermetall GmbH, Freiburg.   Halbleiterbauelemente. Industrial paper. (1972/73).
1705.* Advanced Kinetics, Inc., Costa Mesa, California.   Electrolytic capacitor storage bank. Industrial paper.
1706.* Advanced Kinetics, Inc., Costa Mesa, California   High voltage capacitor banks. Industrial paper.
1707.* Nocke, H.   Physiological aspect of sound communication in crickets (Gryllus campestris L.) *J. Comp. Physiol.* **80**, 141–162 (1972).
1708.† Raether, H.   "Electron Avalanches and Breakdown in Gases." Butterworths, London and Washington, D. C. (1964).
1709.† Klewe, K. C., and Tozer, B. A.   Impulse breakdown of a point-plane gap in SF_6. *IEE Conf. Gas Discharges, London* pp. 36–41 (1970).
1710. Ion Physics Corp., Burlington, Massachusetts.   Proposal for 5 kJ accelerator for the Weizmann Institute, Rehovoth, Israel. Proposal No. 721020 (Oct. 1972).
1711.† Cookson, A. H., and Farish, O.   Motion of spherical metal particles and arc breakdown in compressed SF_6. *Annu. Rep., Conf. Elect. Insulat. Dielect. Phenomena* (1971).
1712.* Small, J. G., and Ashari, R.   A simple pulsed nitrogen 3371 A laser with a modified Blumlein excitation method. *Rev. Sci. Instrum.* **43**, No. 8, p. 1205 (1972).
1713.* Früngel, F., and Hartung, R.   Ultrapulse Welding.—Impulsschweißen. Draht-Fachzeitschrift 1972/9, p. 552 (1972).
1714. Milde, H., Cloud, R. W., and Philip, S. F.   Barium absorption pumps. *Trans. vacuum symp.*, *8th*, p. 352 (1961).
1715. M. S. Thesis   Production and Control of a High Energy Neutral Particle Beam. Massachusetts Inst. Technol. (1962).
1716. Ph.D. Thesis   Measurement of current density distribution and emittance of a positively charged ion beam. Tech. Inst. Gras (1966).
1717. Milde, H., and Gallant, H.   Electrodeless conductivity measurement within the reaction zone of a combustion chamber. *MHD Power Generat. Conf.* (1966).
1718. Milde, H., Mulcahy, M. J., and Bell, W. R.   Insulation breakdown and switching in high pressure gases, *Elect. Insulat. Conf.*, *8th*, *Los Angeles, California* (Dec. 1968).
1719. Milde, H., and Moriarty, J. J.   Switching of fast high voltage pulse generators. High Voltage Technology Seminar, Boston, Massachusetts (Sept. 1969).
1720. Milde, H., Moriarty, J. J., Bettis, J. R., and Guenther, A. H.   Precise laser initiated closure of multimegavolt spark gaps. *Rev. Sci. Instrum.* **42**, No. 12, 1767 (1971).
1721. Milde, H., and Rose, P. H.   Transient effects in UHV tandem accelerators. *Particle Accelerator Conf.*, *Chicago* (1971).
1722. Methods for Evaluating visual performance aspects of lighting (CIE)   *J. Opt. Soc. Amer.* **62**, No. 11, 1333 (1972). (A survey paper of the society).
1723.* Ion Physics Corp., Burlington, Massachusetts.   Specifications for a high voltage pulse generator to drive a large electron gun. Industrial paper.
1724.† Schaaffs, W.   Röntgenblitzröhre zur Untersuchung von schnell ablaufenden Prozessen. Meßtechnik 9/1972, pp. 247–251 (1972).
1725.† Krehl, P.   Erzeugung von Röntgenblitz-Interferenzen. Meßtechnik 9/1972, pp. 252–256 (1972).
1726.* Schaaffs, W., and Krehl, P.   Röntgenblitzuntersuchungen über die Entstehung einer Stoßwelle aus einer durch Stoßwellenprozesse zur Erstarrung gebrachten Flüssigkeit. Acistica, Hirzel-Verlag, Stuttgart. **23**, H.2, pp. 99–107 (1970).
1727.† Schaaffs, W., and Krehl, P.   Röntgenographische Untersuchungen über das Verhalten kondensierter Materie bei sehr hohen Verdichtungen und Drucken. *Z. Naturforsch.* **27a**, H.5, 804–808 (1972).

1728. Schaaffs, W., and Krehl, P. Untersuchungen von Problemen der nichtlinearen Akustik in Flüssigkeiten mit Hilfe von Röntgenblitzen. *Int. Congr. Acoust.*, *7th*, *Budapest* 20 M 1, pp. 21–24 (1971).
1729. Lauterborn, W. High speed photography of laser-induced breakdown in liquids. *Appl. Phys. Lett.* **21**, No. 1, 27–29 (1972).
1730.* Schaaffs, W. Eine Röntgenblitzanlage zur Untersuchung der Kavitationsbildung und Lichtbogenlöschung in Hochspannubgsleistungsschaltern. Meßtechnik 5/70, pp. 85–92.
1731. Lauterborn, W., Hinsch, K., and Bader, F. Holography of bubbles in water as a method to study cavitation bubble dynamics. *Acustica* **26**, H. 3, 170–171 (1972).
1732.† Johnson, K. I. The Welding Institute. Union Carbide Welding Products.
1733.* Fischer, H., and Bostick, W. Are filaments instrumental in neutron. AFCRL Air Force Cambridge Res. Lab. (Sept. 1972).
1734. Orthotron, Longjumeau. Industrial papers.
1735. RCA Electronic Co., Harrison, New Jersey. Infrared-emitting diodes, injection lasers, Silicon photodetectors. Industrial paper.
1736. Ruppersberg, G. H. Principles and methods for the automatic measurement of visibility. *Bull. A.I.S.M.* No. 31 (Jan. 1967).
1737. Gsänger, M. Lithium formate monohydrate. Industrial paper.
1738. Lawrence, T. R., Wilson, D. J., and Craven, C. E. A laser velocimeter for remote wind sensing. *Rev. Sci. Instrum.* **43**, No. 3, 512–518 (1972).
1739. Früngel, F. Gerät zum Messem der Trübung industrieller Abwasser. *Z. Ind. Fertig.* **62**, No. 8, 471 (1972).
1740. Vaughan, R. W. et al. A simple, low power, multiple pulse NMR spectrometer. *Rev. Sci. Instrum.* **43**, No. 9, 1356–1364 (1972).
1741. Knütel & Co., Hamburg 56. Prüfen, Einstellen, Überwachen von Widerstandsschweißmaschinen. Industrial paper.
1742. Cudney, R. A., Phelps, C. T., and Barreto, E. An undergrounded electronic field meter. *Rev. Sci. Instrum.* **43**, No. 9, 1372–1373 (1972).
1743. Sutphin, H. D. Subnanosecond high voltage attenuator. *Rev. Sci. Instrum.* **43**, No. 10, 1535–1536 (1972).
1744. *Optics Laser Technol.* Optical voltage sensor, p. 62; Photomultiplier sensitivity enhancement, p. 4 (Apr. 1972).
Photomultiplier sensitivity enhancement.
1745. Nishida, K., and Nakajima, M. Temperature dependence and stabilization of avalanche photodiodes. *Rev. Sci. Instrum.* **43**, No. 9, 1345–1350 (1972).
1746. Rossetto, M., and Mauzerall, D. A simple nanosecond gate for side window photomultipliers and echoes in such photomultipliers. *Rev. Sci. Instrum.* **43**, No. 9, 1244–1246 (1972).
1747. Scientific Service Co., Somerset. New Jersey. Vacuum photodiodes, sapphire window seals. Industrial papers.
1748. Rohde & Schwarz, Hamburg. Digital-photometer/Radiometer. Industrial papers Tektronix.
1749. Judson Res. and Mfg. Co., Conshohocken, Pennsylvania. Laser detector. Industrial paper.
1750. Oriel Corp. Stamford. Radiometric and photometric instruments. Industrial papers.
1751. Gunn, St. R. A tubular calorimeter for high power laser pulses. *Rev. Sci. Instrum.* **43**, No. 10, 1523–1526 (1972).
1752. Wacker GmbH, Frankfurt. Laser detector. Industrial paper.

1753. Rofin Ltd., Windhill, Bishops Stortford, Herts.   Pyroelectric infrared detector. Industrial papers.
1754. Merkelo, H., and Hartman, S. R.   Mode-locked lasers: measurements of very fast radiative decay in fluorescent systems. *Science* **164**, 301–302 (1969).
1755. Wood, K. E., and Cassis, R.   Factors in the design of a power transfer system for inductively coupled oceanographic sensors. Oceanology International, Brighton, England, pp. 66–76 (1972).
1756. Graefe, V.   A high speed digitizer for the measurement of microstructures in the ocean. Oceanology International 72, Brighton, England, pp. 31–33 (1972).
1757. Joslin, C. W.   Self-powered neutron detectors. *Nuc. Eng. Int.*, pp. 399–401 (May 1972).
1758. Bader, H. *et al.*   Theory of coincidence counts and simple practical methods of coincidence count correction for optical and resistive pulse particle counters. *Rev. Sci. Instrum.* **43**, No. 10, 1407–1412 (1972).
1759. v. d. Does, de Bye, *et al.*   Multichannel photon counter for luminescence decay measurements. *Rev. Sci. Instrum.* **43**, No. 10, 1468–1474 (1972).
1760.† Malmberg, J. H.   *Rev. Sci. Instrum.* **28**, 1027 (1957).
1761.† Kerns, Q. A., Kisten, F. A., and Cox, G. C.   *Rev. Sci. Instrum.* **30**, 31 (1959).
1762.† D'Alessio, J. T., Ludwig, P. K., and Burton, M.   Ultraviolet lamp for the generation of intense, constant-shape pulses in the subnanosecond region. *Rev. Sci. Instrum.* **35**, 1015 (1964).
1763. English Electric Valve Co. Chelmsford, England.   Ignitrons. Industrial papers.
1764.* Leutron GmbH, Stuttgart.   Discharge tubes for the protection against lighting and induction from power lines. Industrial papers.
1765. Rossetto, M., and Mauzerall, D.   A simple nanosecond gate for side window photomultipliers and echoes in such photomultipliers. *Rev. Sci. Instrum.* **43**, No. 9, 1244–1246 (1972).
1766.* Fischer, H., and Bostick, W.   Are filaments instrumental in neutron production in the small energy plasma focus. AFCRL No. 8,1–3162 (Sept. 1972).
1767.* Passner, A.   Laser-driven high power pulsed x-ray tube. *Rev. Sci. Instrum.* **43**, No. 11, pp. 1640–1643 (1972).
1768.* Friedman M., and Ury, M.   Microsecond duration intense relativistic electron beams. *Rev. Sci. Instrum.* **43**, No. 11, 1659–1661 (1972).
1769.* Thermo Magnetic, Inc., Woburn, Massachusetts.   Electromagnetic solid state joining. Industrial paper (1971).
1770. Dangel, J., and Jostan, J. L.   Ein 90-Nanosekunden Versuchsspeicher mit galvanisch hergestellten, dünnen NiFeSe-Speicherschichten. Wissenschaftliche Berichte AEG-Telefunken (1972).
1771.* Index to the *Proc. Int. Congr. High-Speed Photography, 9th, 1952–1970.*   Royal Aircraft Establishment, Farnborough, England.
1772. Bunkenburg, J.   An 11 MW 6.8 J flash lamp pumped coaxial liquid dye laser. *Rev. Sci. Instrum.* **43**, No. 11, 1611–1612 (1972).
1773.* Richter, J. H., Jensen, D. R., and Phares, M. L.   Scanning FM-cw radar sounder. *Rev. Sci. Instrum.* **43**, No. 11, 1623–1625 (1972).
1774. Pellinen, D.   A segmented faraday cup to measure kA/cm^2 electron beam distributions. *Rev. Sci. Instrum.* **43**, No. 11, 1654–1658 (1972).
1775. Yamashita, M.   Light output control of a pulsed light source for use in stabilizing scintillation detectors. *Rev. Sci. Instrum.* **43**, No. 11, 1718–1721.(1972).
1776. Becker, H., Dietz, E., and Gerhardt, U.   Preparation and characteristics of a channel electron multiplier. *Rev. Sci. Instrum.* **43**, No. 11, 1587–1589 (1972).

1777. Massey, G. A., and Yarborough, J. M. High average power operation and nonlinear optical generation with the Nd:YAlO$_3$ laser. *Appl. Phys. Lett.* **18**, No. 12, 576–579 (1971).
1778. Kogelschatz, U., and Schneider, W. R. Quantitative Schlieren techniques applied to high current arc investigations. *Appl. Opt.* **11**, No. 8, 1827–1830 (1972).
1779. Monsanto, Cupertino, California. The photometry of LED's, a primer in photometry. Industrial paper AN 601.
1780. Pratapa Reddy, V. C. V. *et al.* Decade rate multiplier. *Proc. Lett.* Dec. 759 (1971).
1781. Alcock, A. J. Experiments on self-focusing in laser-produced plasmas. 2nd workshop on laser interaction and related plasma phenomena. Rensselaer Polytechnic Inst., Hartford, Aug.-Sept. (1971).
1782. Andreyev, S. I. *et al.* A gigawatt pulsed xenon lamp. *Opt. Tech.* **39**, No. 5, 265–266 (1972).
1783. Magnus, D. E. *et al.* Environmental limitations of optical sensors for guideway surveillance. *Opt. Eng.* **11**, No. 2, 54–60 (1972).
1784. Blaise M. P. The effective intensity of short flashes. *Bull. A.I.S.M.* No. 52, (1972).
1785. Harrison, H., Herbert, J., and Waggoner, A. P. Mie-theory computations of lidar and nephelometric scattering parameters for power law aerosols. *Appl. Opt.* **11**, No. 12, 2880–2885 (1972).
1786. Waggoner, A. P., Ahlquist, N. C., and Charlson, R. J. Measurement of the aerosol total scatter-backscatter ratio. *Appl. Opt.* **11**, No. 12, 2886–2889 (1972).
1787. Plass, G. N., and Kattawar, G. W. Degree and direction of polarization of multiple scattered light 2: Earth's atmosphere with aerosols. *Appl. Opt.* **11**, No. 12, 2866–2879 (1972).
1788. Kattawar, G. W., and Plass, G. N. Degree and direction of polarization of multiple scattered light 2: Homogeneous cloud layers. *Appl. Opt.* **11**, No. 12, 2851–2865 (1972).
1789. Spänkuch, D. Information content of extinction and scattered-light measurements for the determination of the size distribution of scattering particles. *Appl. Opt.* **11**, No. 12, 2844–2850 (1972).
1790. Hutcheson, L. D., and Hughes, R. S. Rapid transverse acoustooptic tuning. *Appl. Opt.* **11**, No. 12, 2981–2983.
1791. United Detector Technology, Inc. Santa Monica, California. Photodetectors. Industrial papers.
1792. Gee, T. H. An introduction to the laser. AGARD LS–49, p. 1–1 (1971).
1793. Gee, T. H. Principles of holography. AGARD LS–49, p. 2–1 (1971).
1794. Gee, T. H. Mathematical methods in coherent optical systems analysis. AGARD LS–49, p. 3–1 (1971).
1795. Tanner, L. H. Effects of coherence on flow visualization methods. AGARD LS–49, p. 4–1 (1971).
1796. Trolinger, J. D. Aerodynamic holography. AGARD LS–49, p. 5–1 (1971).
1797. Upatnieks, J. Experimental holography. AGARD LS–49, p. 6–1 (1971).
1798. Upatnieks, J., and Leonard, C. D. Characteristics of dielectric holograms. AGARD LS-49, p. 7–1 (1971).
1799. Wuerker, R. F. Applications of pulsed laser holography. AGARD, LS–49, p. 8–1 (1972).
1800. Wuerker, R. F., and Heflinger, L. O. Pulsed laser holography. AGARD LS–49, p. 8A–1 (1971).
1801. Wuerker, R. F., and Heflinger, L. O. Ruby laser holography. AGARD LS–49, p. 8B–1 (1971).

1802. Koch, B. Laser beam probing for aerodynamic flow field analysis. AGARD LS-49, p. 9–1 (1971).
1803. Büchl, K. Laser—a light source for high speed photography. AGARD LS-49, p. 10–1 (1971).
1804. Lennert, A. E. *et al.* Laser metrology. AGARD LS-49, p. 11–1 (1971).
1805. Lennert, A. E. *et al.* Application of dual scatter laser doppler velocimeters for wind tunnel measurements. AGARD LS-49, p. 12–1 (1971).
1806.† Farrall, G. A. Low voltage firing characteristics of a triggered vacuum gap. Submitted for publication.
1807.† Aked, A., and McAllister, I. W. International conference on gas discharges. *IEEE Conf. Publ.* 70, 279–283 (1970).
1808.† Pellinen, D. G., and Smith I. A reliable multimegavolt voltage divider. *Rev. Sci. Instrum.* 43, No. 2, 299–301 (1972).
1809.† James, T. E. A high current 60 kV multiple arc spark gap switch of 1.7 nH inductance. CLM-P. 212 Culham Lab. (1969).
1810.† Deutsch, F. *J. Phys.* **D1**, 1711 (1968).
1811.† Karlyn, D. C. *Trans. Amer. Soc. Metals* **62**, 288 (1969).
1812.† Hodgson, B. W., and Keene, J. P. *Rev. Sci. Instrum.* **43**, 493 (1972).
1813.* Otis, G. High trigger current structure for double-discharge TEA lasers. *Rev. Sci. Instrum.* **43**, No. 11, 1621–1623 (1972).
1814.* v. d. Veeke, A. A. High voltage pulse amplifier drives capacitive loads with short rise times. *Rev. Sci. Instrum.* **43**, No. 11, 1702–1703 (1972).
1815.† Alcock, A. J. *et al.* The application of laser-triggered spark gaps to electrooptical image converter cameras. *Int. High Speed Congr.*, *9th, Denver*, p. 191 (1970).
1816.† Capdevielle, R. C. *et al.* Recent developments on the chronometry of multiple intervals of time. *Int. High Speed Congr.*, *9th, Denver*, p. 505 (1970).
1817.† Laviron, E., and Delmare, C. Realization of an image converter with a 300-psec exposure time. *Int. High Speed Congr.*, *9th, Denver*, p. 198 (1970).
1818.† Eigen, M., and De Maeyer, L. *In* "Technique of Organic Chemistry" (A. Weissberger, ed). Wiley (Interscience), New York (1963).
1819.† Czerlinski, G., and Eigen, M. *Z. Electrochem.* **63**, 652 (1959).
1820.† Rabl, C. R. In preparation.
1821.† Hoffmann, H., Yeager, E., and Stuehr, J. *Rev. Sci. Instrum.* **39**, 649 (1968).
1822.† Crooks, J. E., Robinson, B. H., and Tregloan, P. A. *Int. Conf. Mech. Reactions Solution, Canterbury,* July, p. T3 (1970).
1823.† Rigler, R., Jost, A., and De Maeyer, L. *Exp. Cell. Res* **62**, 197 (1970).
1824.† Eigen, M., and De Maeyer, L. *Exp. Cell Res.* **62**, 994 (1970).
1825.† Bacchi, W., and Blanchet, M. Générateurs dímpulsions haute tension de 1 ns à mihauteur. *L'onde électrique* **48**, No. 494, (1968)
1826.† Persson, A. A multiple Kerr cell camera. *Int. High Speed Congr.*, *7th, Zürich* (1965).
1827.† Lyot, B. Un monochromateur a grand champ utilisant les interferences en lumiere polarisée. *C. R. Acad. Sci. Paris* **197**, 1593–1595 (1933).
1828.* Zarem, A. M. *et al.* Millimicrosecond Kerr cell camera shutter. *Rev. Sci. Instrum.* **29**, 1041–1044 (1958).
1829.* Hull, A. J. Thesis, Univ. of Mexico (1956).
1830. Funke, D. Ein digitales Verfahren zur Messung der elektrischen Energie von impulsförmigen Vorgängen. Berichte des Schering-Instituts für Hochspannungstechnik und Hochspannunsanlagen, Tech. Univ., Hannover (1972).
1831. Painter, E., and Harting, E. Development of a lyman capillary discharge. *J. Opt. Soc. Amer.* **62**, No. 12, 1455–1458 (1972).

1832. Orthotron, Longjumeau, Strobohertz. Industrial paper.
1833. Früngel, F., Knütel, W. et al. Development and application of the Variosens in situ instrumentation for the fluorescent tracer technology and sand-caused turbidity. *Symp. Phys. Proc. Responsible Dispersal Pollutants Sea Spec. Ref. Nearshore Zone* No. 9 (1972).
1834. Inall, E. K., and Hughes, J. L. Trigger pulses for a powerful laser system. Laser group, Aust. Nat. Univ., Canberra, ACT. *Search* 2, No. 3 (1971).
1835. Pfeiffer, W. Berechnung und Aufbau eines Impulsverstärkers sehr kurzer Anstiegszeit. *Int. Elektron. Rundsch.* No. 3, 57–60 (1972).
1836. Robra, J. Ein ultraschneller Analog-Digital-Wandler für Abtastsysteme. Institut für Hochspannungs- und Meßtechnik, Techn. Universität Darmstadt., *Meßtechnik* 8, 235–238 (1972).
1837. Pfeiffer, W. Aufbau und Anwendung kapazitiver Spannungsteiler extrem kurzer Anstiegszeit für gasisolierte Koaxialsysteme. *ETZ- A* Bd. 94, 2 (1973).
1838. Pfeiffer, W. Messung von Impulsströmen extrem kurzer Anstiegszeit in koaxialen Rohrsystemen. *ETZ* (to be published).
1839. Pfeiffer, W. Nanosekundenimpulsverstärker mit hoher Ausgangsspannung. *Int. Elektron. Rundschau* No. 2, 34–38 (1972).
1840. Pfeiffer, W. Ein einfacher Impulsgenerator für Reflexionsfaktorund Sprungübertragungsmessungen. *Int. Elektron. Rundschau* No. 11, 268–272 (1971).
1841. Newstead, G., and Cleary, J. Preliminary results from the Warramunga array. Bureau central séismologique Int. Ser. A, Zürich (1967).
1842. Muirhead, K. J., and Simpson, D. W. A three-quarter watt seismic station. *Bull. Seismolog. Soc. Amer.* 62, No. 4, 985–990 (1972).
1843. Wright, C. Longitudinal waves from the novaya zemlya nuclear explosion of October 27, 1966, recorded at the Warramunga Seismic Array. *J. Geophys. Res.* 74, No. 8, 2034–2047 (1969).
1844. Macleod, I. D. G. On the bandwidth of carrier-type DC amplifiers, *IEEE* CT-17, No. 3, 367–371 (1970).
1845. Hughes, J. L. Experimental conditions required to verify the existence of a photon medium. *Opt. Commun.* 3, No. 6, 374–378 (1971).
1846. Hughes, J. L. The potential of present day lasers as scientific probes for investigating the structure of matter, using and exponential amplifier. *Appl. Opt.* 6, No. 8, 1411–1415 (1967).
1847. Straubel, H. Optische Verfolgung des Kristallwassereinbaus an frei schewebenden Mikrokristallen. *Physil Blätter* 11, 498–501 (1972).
1848. Beraud, J., and Taquet, B. The generation of large currents of constant amplitude using capacitor banks. *Symp. Eng. Probl. Thermonucl. Res., Munchen* pp. 18–24 (1964).
1849. Knobloch, A., and Herppich, G. Generation of rapid rise current pulses in the mesc range for fast compression experiments. *Symp. Fusion Technol., 5th, Oxford*, paper 36 (1968).
1850. Plummer, K. M., and Skelton, D. E. The design, construction and control of a large pulsed mirror machine MTSE II. *Symp. Fusion Technol., 5th, Oxford*, paper 65 (1968).
1851.† Koch, W., and Salge, J. Über die Erzeugung von Stoßstromen mit Sprengtrennern. *ETZ-A* Band 87, pp. 697–700 (1966).
1852.† Koch, W. Inductive storage for high currents experiments. *Symp. Eng. Probl. Thermonucl. Res., Frascati-Rome* (1966).
1853.† Reed, N. E. A novel switching device for use in an HVDC circuit breaker. IEEE Summer meeting, Switchgear Session, San Francisco, California (1972).

1854.† Lafferty, J. M. Triggered vacuum gaps. *Proc. IEEE* **54**, No. 1, 23–32 (1966).
1855.* Electro Optic Development Ltd. Laser-Optronic, München, 01 Pockelzellen für die Verwendung als Q-Schalter oder Modulator. Industrial paper.
1856.† Pfeiffer, W. Versteilerung von Sprungimpulsen durch Snap-Off-Dioden. *Int. Elektron. Rundschau* Nr. 8, 191–193 (1972).
1857.* Kerry's (Ultrasonics) Ltd., Stud Welding Div., Stratford, London. Capacitor discharge portable stud welding equipment. Industrial paper.
1858.* Wiesinger, J. Blitzforschung und Blitzschutz. Deutsches Museum Abhandlg. u. Berichte 40. J. H. 1/2. Verlag Oldenbourg, München (1972).
1859.* Ewanizky, Th. F., and Wright, R. H. Jr. Coaxial Marx-Bank driver and flash lamp for optical excitation of organic dye lasers. *Appl. Opt.*, **12**, No. 1 (1973).
1860.* Laflamme, A. K. *Rev. Sci. Instrum.* **41**, 1578 (1970).
1861.* Eccosorb, H. F. Resistive-type waveguide absorber, manufactured by Emerson and Cuming, Inc., Canton, Massachusetts.
1862.* Bernotat, S., and Umhauer, H. Application of spark tracing method to flow measurements in an air classifier. *Opto-electron.* **5**, 107–118 (1973.)
1863. Bömelburg, H. J., Herzog, J. R., and Weske, J. R. The electric spark method for quantitive measurements in flowing gases. *Flugwissenschaft* **7**, H. 11, 322–329 (1959).
1864. Früngel. F. Steoroskopische Funkenblitzserien, dreidimensionale Bewegungssaufnahmen instationärer Strömungen durch hochfrequente Hochspannungsfunken. *Proc. Int. Congr. High Speed Photography, 6th, The Hague* (1962).
1865. Früngel, F., Thorwart, W., and Patzke, H. G. High speed photography of rapid air currents and shock waves by means of high-frequency high-voltage sparks. *Proc 5th Int. Congr. High Speed Photography, 5th*, pp. 498–502. SMPTE (1962).
1866. König, G. Gesichtspunkte bei der Sichtbarmachung von Strömungen in kleinen Bauelementen mit der Funkenblitzmethode. 2. Konf. Pneumatische Strahlelemente, Dresden, Vortrag 3/3.
1867. Nagao. F, Makoto Ikemami et al. Air motion and combustion in a swirl chamber type Diesel engine. *Bull. JSME* **10**, No. 41, 842–843 (1967).
1868.† Ullmann, W., and König, G. See Ref. No. 5. Pneumatische Strahlelemente, Dresden.
1869. Fister, W. Sichtbarmachung der Strömungen in Radialverdichterstufen. *Brennst. Wäerme- Kraft* **18**, 425–429 (1966).
1870. Schönbach, K. H., Michel, L., and Fischer, Heinz. *Appl. Phys. Lett.* **25**, No. 10, 547 (1974).
1871. Bostick, W. *et al.* *Symp. Thermophys. Properties, 5th, Newton, Massachusetts* (1970).
1872. Bostick, W. *et al.* *J. Plasma Phys.* **8**, 7 (1972).
1873. Braun, K., Fischer, H., and Michel, L. *Proc. Topical Conf. Pulsed High-Beta Plasmas, 2nd, Garching* (1972).
1874.* Morton, A. H. *Proc. Canberra Plasma Phys. Seminar* Publ. EP-RR 32. Canberra A.C.T. Australia (March 1972).
1875.† Bowers, D. L. Studies on the heating and containment of plasma in a magnetic trap. The behavior of a slow toroidal 0-Z pinch Canberra A.T.C. Australia.
1876.* Neugebauer, A. R. Aufbau und Erprobung einer Anordnung zur Erzeugung achtzylinderförmiger kondensierter Wasserstoffnuklearstrahlen. Dissertation. Inst. f. Kernverfahrenstechnik, Kernforschungszentrum Karlsruhe. *Deut. Forschungsber.* **2**, No. 1 (1973).
1877.‡ Bailey, D. N., and Hercules, D. M. XX. Flash photolysis - a technique for studying fast reactions. *Chem. Instrum. A83* **42**, No. 2 (1965).

1878. Willets, F. W. Evolution of flash photolysis and laser photolysis techniques. *Prog. Reaction Kinetics* **6**, No. 1, 52 (1971).
1879. Annual report 14/1972. The Austrlian Nat. Univ. Res. School Phys. Sci., Dept. Eng. Phys. (1973).
1880.† Lihl, F. Ferro carbon alloys of improved microstructure and process for their manufacture. U. S. Patent No. 3, 240, 639 and Schweizer Patent No. 381, 717 (1957).
1881.† Früngel, F. Impulstechnik. "Erzeugung and Anwendung von Kondensatorentladungen," Akademische Verlagsges. Geest Portig KG, Leipzig, pp. 235–250 (1960).
1882.† Früngel, F. "High Speed Pulse Technology," Vol. 1, pp. 362–372. Academic Press, New York (1965).
1883.† Früngel, F. Verfahren zur Hochfrequenzinduktionshärtung und-lötung. Deutsches Patent No. 917, 278.
1884.† Früngel, F. Method and device for high frequency soldering and induction hardening. U.S. Patent No. 2, 799, 760.
1885.† Thorwart, W. Mikro-Induktionshärtung mittels hochfrequenter Impulse. *VDI-Z* **95**, No. 11/12, 341–344 (1953).
1886.† Ettenreich, L. Die Oberflächenhärtung von Werkzeugen und Werkstücken aus härtbarem Stahl in extrem kurzen Zeiten. *VDI-Z* **110**, No. 8, 316–320 (1968).
1887.† Früngel, F. Verfahren zum Erzeugen extrem feinkörniger metallischer Oberflächengefüge. DAS No. 1, 957 884 (1969).
1888.* Irons, F. E. *et al.* The ion and velocity structure in a laser-produced plasma. *J. Phys. B;At. Mol. Phys.* **5**, 1975–1988 (1972).
1889.* Hughes, J. L. A proposal for overcoming the plasma mirror problem in laser-induced plasmas. *Search* **3**, No. 5, 174–175 (1972).
1890. Hughes, J. L. Laser- induced super-dense. Laser group, Research school of physical sciences. Canberra, Australian Nat. Univ. (March 1972).
1891.* Carden, P. O. Features of the high field magnet laboratory at the Australian Nat. Univ., Canberra, EP-RR 19 (Jan. 1967).
1892.* Carden, P. O. Testing the ANU 30 T high field magnet at Canberra. The Australian Nat. Univ., Canberra (1971).
1893.† Carden, P. O., and Whelan, R. E. Instrumentation of the ANU 300 KG experimental magnet. Australian Nat. Univ. (Dec. 1969).
1894.* Carden, P. O. Design principle relating to the strength and structure of the ANU 30 T electromagnet. The Australian Nat Univ., Canberra (Feb. 1972).
1895.* Carden, P. O. The design and construction of the outer solenoid of the ANU 30 T electromagnet. The Australian Nat. Univ., Canberra (Feb. 1972).
1896.* Carden, P. O. Design and construction of the inner solenoid of the ANU 30 T electromagnet. The Australian Nat. Univ., Canberra (Feb. 1972).
1897.* Boreham, B. W. Propulsion by travelling conduction waves. *J. Spacecr. Rockets* **9**, No. 2, 124–126 (1972).
1898.† Schmied, H. Probleme der Impulsmagnetisierung von Dauermagnetwerkstoffen. *Elektrotech. Z. Ausgabe* A **89**, 21, 582–586 (1968).
1899.* Barber, J. P. The acceleration of macroparticles and a hypervelocity electromagnetic accelerator. The Australian Nat. Univ., Canberra, A.C.T. EP-T12 (1972).
1900.* Goodfriend, P. L., and Woods, H. P. Apparatus for flash photolysis kinetic spectroscopy in the vacuum ultraviolet. *Rev. Sci. Instrum.* **36**, No. 1, 10–12 (1965).
1901.‡ Hayworth, B. R. Specifying a flash lamp capacitor. Laser Focus Magazine, Newtonville, Maryland (Aug. 1971).

1902. Hayworth, B. R. How to tell a nanohenry from a microfarad. *Electronic Instrumentation*, pp. 36–39 (April. 1972).
1903.† Capacitor Specialists, Inc., Escondido, California. High voltage energy storage capacitors, and high voltage, low inductance plastic cased capacitors. Industrial papers.
1904.* High Energy, Inc. Malvern, Pennsylvania. Capacitors. Industrial paper.
1905.* Solitron Dev., Inc., Riviera Beach, Florida. 250 A. peak silicon NPN power transistors. Industrial paper.
1906.† Pfeiffer, W. Der Spannungszusammenbruch an Funkenstrecken in komprimierten Gasen. *Z. Angew. Phys.* **B32**, H 4, 265–273 (1971).
1907.† Pfeiffer, W. Stufenbildung bei Funkenentladungen in Kohlendioxyd. *Z. Angew. Phys.* **B32**, H. 5-6, 329–331 (1972).
1908.† Pfeiffer, W. Das Verhalten von spannungsabhängigen Widerständen bei Steilstoßbeanspruchung. *ETZ-A* Bd. **93**, H. 9, pp. 533–536 (1972).
1909.‡ Inall, E. K. Fast circuit breaker for the discharge of a storage inductor. *Nature Phys. Sci.* **231**, No. 22, 111–112 (1971).
1910.† Reeves-Saunders, R. The evolution of high current dc arcs on rotating anodes. The Australian Nat. Univ., Canberra, Dept. Eng. Phys. (1972).
1911.† Reeves-Saunders, R. Observation of a transition into a stable mode for an arc burning on a rotating anode. The Australian Nat. Univ., Canberra. *J. Phys. D. Appl. Phys.* **4** (1971).
1912.* Boeck, W., and Troger, H. SF_6 insulated metal-clad switchgear for ultrahigh voltages (UHV). *Int. Conf. Large High Tension Elec. Syst.* 23–08 (1972).
1913.† Einsele, A. Der Siemens-F-Schalter 220 kV, 15 GVA. *Siemens-Z.* **36**, J. H. 4, 225–228 (1964).
1914.† Brueckner, P., and Flöth, H. Vollisolierte gekapselte Schaltanlagen für Reihe 110 mit sehr kleinem Raumbedarf. *ETZ-A* **86**, H. 7, 198–204 (1965).
1915.† Lager, P., Rimpp. F., and Wegener, J. Vollisolierte 110 kV-Scahltanlagen. *Siemens-Z.* **40**, J. H. 4, 263–266 (1966).
1916.† Cigre 23. Working group 23.03, paper 23.04, 1972. The compilation of the international experience on installation and operation with metal-clad substation.
1917.* Ohwada, K., and Ichihara, N. High voltage vacuum contactors. *Elect. Eng.* 32–35 (June 1972).
1918.* Pfeiffer, W. Ein Hochstromimpulsgenerator im Subnanosekundenberiech. *ETZ-A.* **92**, H. 4, 242–244 (1971).
1919.‡ Salge, J., and Braunsberger, U. Inductive energy storage systems applied for the extension of current pulse-duration of capacitor banks. *Symp. Fusion Technol.*, *7th*, Grenoble (Oct. 1972).
1920.* Inall, E. K. A proposal for the construction and operation of an inductive store for 20 MJ. Dept. Eng. Phys. Res. School of Phys. Sci., Canberra, Australia, Dec. 1971. *J. Phys. E. Sci. Instrum.* **5** (1972).
1921.† Carden, P. O. Limitations of rate of rise of pulse current imposed by skin effect in rotors. Dept. of Eng. Phys. Canberra, Australia (Sept. 1962).
1922.* Schaaffs, W. et al. Investigations concerning the applicability of x-ray flash interference and laser technology to shock-induced solidification of organic liquids. *Int. Congr. Cinematogr. Ultra-Rapide, 10th,* Nice, France (Sept. 1972).
1923.* Jamet, F., and Thomer, G. Recording of x-ray diffraction patterns for the investigation of transient changes in the crystalline structure of materials subjected to the action of shock waves. *Int. Congr. Cinematogr. Ultra-Rapide, 10th,* Nice, France (Sept. 1972).

1924.* Charbonnier, F. et al. New tubes and techniques for flash x-ray diffraction and high contrast radiography. *Int. Congr. Cinematogr. Ultra-Rapide, 10th, Nice, France* (Sept. 1972).
1925.† Stenerhag, B. et al. Exploding wires as a source of flash x-rays. *Int. Congr. Cinematogr. Ultra-Rapide, 10th, Nice, France* (Sept. 1972).
1926.† Mattsson, A. A versatile flash radiation system. *Int. Congr. Cinematogr. Ultra-Rapide, 10th, Nice, France* (Sept. 1972).
1927.† Rapp, H. K. Technical and experimental investigations of a plasma focus neutron source. *Int. Congr. Cinematogr. Ultra-Rapide, 10th, Nice, France* (Sept. 1972).
1928.* Tonon, G. et al. Laser interaction with matter as a source of UV and soft x-ray radiation: Application to x-ray cinematography. *Int. Congr. Cinematogr. Ultra-Rapide, 10th, Nice, France* (Sept. 1972).
1929. Ebeling, D. Improved x-ray flash technique broadens field of application. *Int. Congr. High Speed Photogr.*, Denver (1970).
1930. Schaaffs, W. Ergebnisse der exakten Naturwissenschaften. Bd. 28, S. 1–46 "Erzeugung und Anwendung von Röntgenblitzen" (1954).
1931. Fünfer, E., and Schall, R. Erzeugung und Untersuchung extremer kurzdauernder Zustände. *Phys. Rev.* B1, **8**, pp. 305–312 (1952).
1932. Charles, M. et al. 1 million sec radiography and its applications. *Proc. IRE Waves Electrons Sect.* June 1946).
1933.† Green, R. E. Dynamic x-ray diffraction systems. Transactions, advances in x-ray analysis, Denver (1970).
1934. Kennedy, S. W. Rapid x-ray diffraction studies using image intensification. *Nature (London)* 936 (1966).
1935. Fainberg, Y. B. *Sov. Phys. -Usp.* **10**, 750 (1968) (see Veksler's work).
1936. Veksler, V. I. *At. Energ. (USSR)* **2**, 247 (1957).
1937. Eastlund, B. J. unpublished.
1938. *IEEE Trans. Elec. Insulat.* **E1-3**, No. 2 (1968).
1939. *Mater. Res. Std.* 405 (1962).
1940.‡ Hayworth, B. R. How to tell a nanohenry from a microfarad. Electronic Instrumentation, pp. 36–39 (Apr. 1972).
1941.† Glasstone, S., and Lovberg, R. G. "Controlled Thermonuclear Reactions," pp. 164–165. van Nostrand-Reinhold, Princeton, New Jersey (1960.) (Rogowsky Belts).
1942.† Willets, W. The evolution of flash photolysis and laser photolysis techniques. Kodak Ltd. Pergamon, Oxford.
1943.‡ Bowers, D. L. et al. A slow toroidal 0-Z pinch experiment. *Plasma Phys.* **13**, 849–883 (1971).
1944.* Irons, F. E. et al. Spectroscopic study and energy balance of pulsed toroidal discharges. *Plasma Phys.* **14**, 717–728 (1972).
1945.* Morton, A. H., and Srinivasacharya, K. G. Investigation of electron-runaway in the slow toroidal 0-Z pinch, LT1. *Plasma. Phys.* **14**, 687–700 (1972).
1946.† Fahlenbrach, H. Grundlagen der modernen Dauermagnete. *ETZ-B* **19**, 345–349 (1967).
1947.† Fahlenbrach, H. Hertsellung und Eigenschaften von Dauermagnet-Werkstoffen. *ETZ-B* **19**, 477–482 (1967).
1948.† Schüler, K. Probleme des dauermagnetischen Kreises. *Z Angew. Phys.* **21**, 119–125 (1966).
1949.† Deutsche Edelstahl-Werke (DEW) Druckschrift Nr. 1141/5 (1966).

1950.† Heimke, G. Magnetkeramik.—Die Spinelle und verwandte Strukturen, *ETZ-B* **20**, H.2, 34–37 (1968).
1951.† Cooper, R. W. *et al.* Applications of magneto-optics. *IEEE Trans. Magn.* **MAG-5** (3), S.475 (1969).
1952.† Wild, P. J. A phasemeter for photoelectric measurement of magnetic fields. *Rev. Sci. Instrum.* **41**, No. 8, 1163 (1970).
1953. Furth, H. P. *et al.* *Rev. Sci. Instrum.* **27**, 195 (1965); **18**, 949.(1957).
1954.* Shneerson, G. A. *Zh Tekhn. Fiz.* **32**, 1153 (1962) [*English transl.: Sov. Phys.— Tech. Phys.* **7**, 848 (1963)].
1955.* Gordienko, V. P., and Shneerson, G. A. *Zh. Tekhn. Fiz.* **34**, 376 (1964) [*English transl.: Sov. Phys.—Tech. Phys.* **9**, 296 (1964)].
1956.† Fowler, C. M., Garn, W. B., and Caird, R. S. *J. Appl. Phys.* **31**, 588 (1960); "High Magnetic Fields," p. 269, MIT Press, Cambridge, Massachusetts (1962).
1957.† Caird, R. S., Garn, W. B., Thomson, D. B., and Fowler, C. M. *J. Appl. Phys.* **35**, 781 (1964).
1958. Herlach, F. *et al.* Laboratorio Gas Ionizzati Rep. 65/16 (June 1965).
1959.† Vollrath, K., and Schall, R. Piezoelektrische Funkengeneratoren. *Int. High Speed Congr.*, 6th, The Hague, pp. 403–408 (1962).
1960.† Assard, G. L., and Hassell, B. C. Measurements of the spatial correlation of ambient noise using a deep-submergence vehicle (Diving Saucer SP-300). USL Rep. No. 714, pp. 1–19 (1966).
1961. Barham, E. G. Siphonophores and the deep scattering layer. *Science* **140**, pp. 826–828 (1963).
1962. Barham, E. G. Deep scattering layer migration and composition: observations from a diving saucer. *Science* **151**, 1399–1403 (1966).
1963.† Fisch, N. P., and Dullea, R. K. Acoustic research studies with deep submergence vehicles. USL Tech. Memo (1966). (unpublished).
1964.* Schmied, H. Die elektrische Unterwasserentladung als Stoß-Schallgenerator hoher Leistung. *Acustica*, **19**, No. 2 (1967/68).
1965.* Arndt, G. Ultrahigh-speed machining. *Ann. CIRP* **21/1**, 3–4 (1972).
1966. Arndt, G. Ballistically induced ultrahigh speed machining. Ph.D. thesis, Monash Univ., Clayton, Victoria, Australia (Sept. 1971).
1967. Arndt, G., and Brown, R. H. Design and preliminary results from an experimental machine tool cutting metals at up to 8000 ft/sec. To be presented at the *M.T. D.R. Conf., 13th, Birmingham, U.K.*
1968.† Early, H. C., and Dow, W. G. Experimental studies and applications of explosive pressures produced by sparks in confined channels. A.I.E.E. Winter Meeting (Jan. 1953).
1969.† Yutkin, L. A. "Electroqidravlicheskiy efft." Mashgiz, Moskva, Leningrad (1955).
1970.† Kegg, R. L. A study of energy requirements for electrical discharge forming. ASME paper No. 63, Prod. -4 (1963).
1971.† Christiana, J. *et al.* Capacitor discharge metal forming. ASD INT. Rep. 7-844, Pts. I-VII (1962–1965).
1972.† Pugh, H. LL. D., Watkins, M. T., and Hodgson, G. Electrohydraulic forming. N. E. L. Report No. 151 (1964).
1973.† Kirk, J. W. Impulse forming by electrical discharge methods. *Sheet Metal Ind.* **39**, No. 424 (1963).
1974.† Duncan, J. L., and Johnson, W. The free-forming of sheet aluminum using an electric discharge method. *Proc. Mach. Tool Design Res. Conf.*, 3rd, Birmingham (1962).

1975.† Kapitza, P. L.   *Proc. Roy. Soc. Ser. A* **105**, 69 (1924).
1976.† Wiederhold, P. R.   *New Scientist* **23**, No. 406 (1964).
1977.† Früngel, F., and Keller, H.   *Z. Angew. Phys.* **9** (3) 145 (1957). "Stoß-Schallquellen, Grundlagen und Analogie zu Sprengstoffumsetzungen."
1978.† Buntzen, R. R.   The use of exploding wires in the study of small scale underwater explosions. "Exploding Wires," Vol. 3. Plenum Press, New York (1962).
1979.† Kersavage, J. A.   Pressure environments created by wires exploded in water. "Exploding wires," Vol. 2. Plenum Press, New York (1962).
1980.† Bazhenova, T. V., and Soloukhin, R. I.   Pressure field occuring in water during an electrical discharge. "Physical Gas Dynamics." Pergamon, Oxford (1961).
1981.† Soloukhin, R. I.   Shock waves forming during an electrical discharge. "Physical Gas Dynamics." Pergamon, Oxford (1961).
1982.† Vorodnikova, F. I.   The effect of heat evolvement rate in an electrical discharge in water on the distribution of explosion energy. P.M.T.F. (in Russian) No. 2, pp. 110–112 (1962).
1983.† Travis, F. W.   Thesis submitted to Univ. of Manchester for Degree of M.Sc. (Tech.) (1962).
1984.* Bodenseher, H., and Schmied, H.   Über den Ablauf einer durch Unterwasserfunken erzielten Blechverformung. *Z. Angew. Phys.* 22. Band, H **1**, 23–25 (1966).
1985.† Wood, W. W.   Experimental mechanics at velocity extremes—very high strain rates. *Exp. Mech.* 441 (1967).
1986.† Orava, R. N.   The effect of dynamic strain rates on room temperature ductility. *Proc. Conf. Center High Energy Forming, 1st, Estes Park*, June (1967).
1987. Bieniawski, Z. T.   An application of high speed photography to the determination of fracture velocity in rock. *Int. High Speed Congr., 8th, Stockholm*, p. 440 (1968).
1988. Dobbs, H. S. *et al.*   A high speed photographic investigation of the controlled fracture of conducting solids and the interruption of electric current. *Int. High Speed Congr., 8th, Stockholm*, p. 435 (1968).
1989. Fayer, A. *et al.*   Dynamic study of the form of the fracture surface in the section of a plate of tempered glass. *Int. High Speed Congr., 8th, Stockholm*, p. 433 (1968).
1990. Field, J. E.   The high speed photography of fracture in sapphire and diamond. *Int. High Speed Congr., 6th, The Hague*, p. 514 (1962).
1991. Field, J. E., and Heyes, A. D.   The fracture of materials of high elastic moduli. *Int. High Speed Congr., 7th, Zurich*, p. 391 (1965).
1992. Hänsel, H.   Results of high frequency cinematography in the investigation of brittle fracture propagation. *Int. High Speed Congr., 4th, Cologne*, p. F-5, 277 (1958).
1993. Häusler, E.   A new experimental method for investigation of shock waves in solids. *Int. High Speed Congr., 6th, The Hague* Paper XVII-A, p. 573 (1962).
1994. Henschen, H.   Interferometric investigation of the stress distribution before a running fracture in a plastic. *Int. High Speed Congr., 6th, The Hague*, Paper XV-B, p. 522 (1962).
1995. Kerkhof, F.   Ultrasonic fractography. *Int. High Speed Congr., 3rd, London*, p. 194 (1956).
1996. Kerhof, F.   Fractographic studies of mechanical impulses in plates. *Int. High Speed Congr., 7th, Zurich*, Paper G-7, p. 345 (1965).
1997. Miyata, C. *et al.*   Motion picture photography of high speed impact tensile fracture. *Int. High Speed Congr., 9th, Denver*, Paper 19, p. 531 (1970).
1998. Persson, A.   High speed photography of scale model rock-blasting. *Int. High Speed Congr., 9th, Denver*, Paper 79, p. 337 (1970).

1999. Uyemura, T., and Morishige, T.   Studies on the explosion mechanism of electric blasting caps by ultrahigh speed grid framing camera.
2000. Verbraak, C. A., and De Graaf, J. G. A.   On the mechanism of brittle fracture in steel. *Int. High Speed Congr.*, *6th, The Hague*, Paper XV-C, p. 529 (1965).
2001. Wittwer, H. J.   High frequency cinematographic analysis of the rupture process of impact loaded tensile specimens. *Int. High Speed Congr.*, *7th, Zurich*, Paper G-17, p. 409 (1965).
2002. Wittwer, H. J.   Use of the afterglow of pulsating light sources to increase the information content of high speed films. *Int. High Speed Congr.*, *9th, Denver*, Paper 66, p. 263 (1970).
2003. Zandman, F.   Photoelastic study of stress propagation during the failure of plyglass. *Int. High Speed Congr.*, *2nd, Paris*, p. 352 (1954).
2004. Korbee, W. L. *et al.*   The recording of plastic-wave propagation in high velocity and tensile tests on steel. *Int. High Speed Congr.*, *9th, Denver*, Paper 91, p. 392 (1970).
2005. Carrara, G., and Zaffanella, L.   Switching impulse clearance tests. IEEE Summer Power Meeting, UHV-Lab., No. CP 692-PWR (1968).
2006.* Nisenson, P. *et al.*   Optical processing with a $Bi_{12}SiO_{20}$ electrooptic image modulator. Paper WV 13, Spring Meeting OSA 1972. Itek Corp., Lexington (1972).

Subject Index

Accelerated ions, mean velocity and energy of, 291
Acoustic air pollution monitoring, 392–393
Acoustic data acquisition systems, 401–408
Acoustic echoes, 387
Acoustic echo sounders, 388–389
Acoustic echo techniques, 380–391
Acoustic noise, ambient, 389
Acoustic pingers, 408
Acoustic pulse energy, conversion of to electric, 398–401
Acoustic pulses
    conversion of capacitive energy to, 380–437
    echo technique and, 380–391
    electroceramics and, 394–401
    hydrophones and, 395–398
    in technology, 395
Acoustic radar, 380–382
    block diagram of, 383
Acoustic ranging pulses, 401
Acoustic sounder, 386
Acoustic sounding techniques, 391
Acoustic tone burst, 384
Acoustic transducer, 382
Acoustic waves, 401
Acoustooptic deflectors, 64
Adkin collapsing foil shutter, 362
Adkin pulse transformer, 142–143
Advanced Kinetics, Inc., 38, 142, 349
Air
    breakdown phenomena in, 15
    positive discharges in, 15
Air bag, in hydrospark forming, 434
Air classifier, spark tracing and, 209
Air-cored electromagnets, 351
Air flow, spark tracing and, 203
Airtight seal welding, 307
Alkali metal vapors, optical ionization of, 339
Alternator, high-current, 110
Aluminum crystal, diffraction pattern of, 257
Aluminum-to-copper welding, 303
An echoic shielding, 389
    in acoustic air pollution monitoring, 392–393

Anemometers, in remote sensing systems, 380
Angled incidence technique, in transmission line dimensioning, 117–121
Annular arc injector, 355
Arc lamp pulsing, solid-state switch for, 126–127
ARES megavolt Van de Graaff machine, 225–228
Argon, accelerated ion velocity, for, 291
Atmospheric sounding, with acoustic radar, 380–382
Atomic number x-ray target, 279
Australian Defense Science Service, 381
Australian National University, 89, 92, 344, 351, 367, 370
Avalanche current, 14
Avalanche growth, at corona cloud surface, 100
Avalanche pulser circuit, 142
Avalanche pulses, 15

Ballistic x-ray studies, 269–270
Barium titanate, 401
Barretter, in hydrogen thyratrons, 53
Beam plasma interactions, 288
Bennett pinch, 290
Benzonitrile, as Kerr liquid, 167
Berkeley Nucleonics Corp., 139
Beryllium-copper, in pulsed magnets, 350
Beryllium exit windows, 229, 234
Beryllium target, in neutron flash procedure, 294
Beta-radiator, kryton and, 58
Biologic phenomena, acoustic ranging pulses and, 401
Biplanar shutter, 157
Bismuth silicon oxide, in Pockels effect image modulation, 176
Bitter disk magnet, 353
Blast circuit breaker, 88–92
    current carrying capacity of, 92
Blast switch, 153
Blumlein circuits, 135–136
Blumlein excitation method, 136
Blumlein transmission line, for high-voltage pulse shaping, 184, 186, 235

Bochum University, 204
Bragg angle, in x-ray emission, 234
Bragg crystal spectrograph, 234
Breakdown, laser-produced, 347–348
Breakdown phenomena, in air, 15
Breakdown plasma, laser and, 338
Breakdown probability, in pressured air devices, 16
Breakdown process, in triggered spark gaps, 71
Breakdown strength, of pressured air, 17
Breakdown time, for lightning protectors, 83
Breakdown voltage
  for liquid spark gaps, 105
  for megavolt switch, 97, 99
Brewster interface, in transmission-line dimensioning, 118, 121
Bromobenzene, dielectric discharge in, 264
Brown Boveri Company, 360
Bunsen burner flame, spark tracing in, 208
Burnett Electronics Laboratory, 408

Cable capacitors, high-energy, 9
Cadmium image quality indicator, 296
CALCOMP plotter, 404
Camera
  electronic integral image, 274
  electronic streak, 274
  high-speed radiographic, 273
Canberra homopolar generator, 89, 149, 153, 351
Capacitance
  calculation of, 2
  defined, 1
Capacitance microphones, 394
Capacitive (stored) energy, conversion of into acoustic pulses, 380–437
Capacitor(s), 1–51
  see also Insulators
  cable, 10
  charging losses for, 24
  coaxial, 21
  coaxial disk and parallel plate series, 42–43
  coaxial tank series, 43–44
  comparison of, 3
  compressed gas, 37
  containers for, 21–22
  construction of, 3
  for corona-free operation, 42–43
  defined, 1
  delay line, 75
  dielectric strength of, 18
  dimensioning of, 1
  in discharge circuits, 1–3
  electrolytic, 49–50
  energy dissipated in charging of, 24
  extrahigh-voltage series, 43
  FC (flat), 48
  flash lamp, 19–25
  heat-transfer techniques for, 22
  high- and low-repetition rate, 29–30
  high-temperature plasma machines and, 38
  internal inductance and, 39
  laser discharge, 25–29
  laser systems and, 20
  lifetime of, 19
  lifetime vs. voltage reversal in, 20
  losses in, 22–23
  low-inductance foilwound, 36
  megajoule, 78
  Magnum, 46–47
  for Marx circuits, 45–47
  nanosecond discharge, 40–42
  off-the-shelf cases for, 21
  parallel plate, 47–49
  peak charging coltage in, 27
  plastic case, 31–32
  power loss for, 22–23
  pressured air, 16–17, 37
  pulse, 29–32
  rapid discharge, 42, 44
  resonant-frequency tests for, 34
  series inductance of, 34
  series-resonant properties of, 33
  signal-to-noise ratio for, 36
  skin-effect correction in, 35
  tantalum, 50–51
  temperature range for, 3
  temperature vs. power dissipation in, 24
  transient-response technique in, 34
  transient wave for, 35
  ultralow inductance, 32–45
  vacuum, 19
  variable-inductance technique for, 35–38
  voltage reversal percentage in, 2
  water, 75
  waveshape for, 23

INDEX

Capacitor banks
  current pulses for, 145–153
  low-loss electrolytic, 49–50
  in megajoule energy storage devices, 149–150
Capacitor discharge
  in exploding wire process, 320
  high-temperature plasma generation by, 328–337
  in impulse welding, 298–303
  nanosecond, 40–42
  rapid, 42, 44
Capacitor energy
  conversion of to current impulses, 126–153
  conversion of to heat, 298–348
  conversion of to magnetic fields, 349–379
  conversion of to voltage impulses, 154–228
  conversion of to x-ray flashes and electron beams, 229–297
  transformed, 303–312
Capacitor energy welding, 303–312
  pressure and energy requirements in, 305–307
Capacitor-ignited detonations, in high-pressure physics, 437
Capacitor spark gap units, 40
Capacitor Specialists, Inc., 27
Capacitor system, life expectancy for, 27
Carbon disulfide, as Kerr liquid, 167
Carbon steel, pulse hardening of, 313–320
Carbon tetrachloride, spark gap in, 105
Cascade processes, multiphoton ionization and, 345–346
Cathode x-ray tube, grid modulation for, 239
Cavitation bubbles, in dielectric discharge, 263, 265
Ceramic transducer, 403
Cerenkov process, 288
Charge carriers, 12
Chemical reactions, intermediates in, 186
Chloroform, as Kerr liquid, 167
Chopping gap, triggered multiple, 93
Cineradiography
  applications of, 277–278
  high-speed, 270–276
Circuit breaker, blast-operated, 88–92

Circuit pulse duration, extension of, 146–147
CO_2, as insulating gas, 15
CO_2–H_2–He lasers, 340
CO_2 laser, 230, 338
Coaxial disk capacitors, 42–43
Coaxial flash lamps, with driver circuits, 180
Coaxial gas capacitor, 223
Coaxial gas line, Van de Graaff generator and, 225
Coaxial impulse cable, 156
Coaxial line test circuits, 141
Coaxial Marx-bank driver, 181
Coaxial pulsed accelerator, 288–289
Coaxial pulse generator, 113
Coaxial spark gap, 72
  light source elements and, 170
Coaxial tank series capacitors, 43–44
Coaxial water line, high-voltage, 122–123
Collapsing foil shutters, 362–363
Compact-arc xenon lamps, 329
Compressed gas capacitor, 37
Compression factor, in megagauss fields and accelerators, 371
Concentration magnets, high field flux, 352
Concentric flash-tube sample-cell configuration, 193–194
Conduction wave accelerator, 355
Connecting lines, impedance transformation by, 113–122
Constant impedance dielectric transition, 116–117
Copper-targets (anodes), 220
Copper-to-copper welds, 305
Copper-to-silver welds, 305
Corona cloud surface, avalanche growth at, 100
Corona-free operation, 96
  capacitors for, 42–43
Corona pulse, in pressured air capacitors, 16
Creepage discharge switch, for insulators, 4
Creepage tracking, in insulators, 7
Cricket songs, 393–394
Cronex screens, 258
Crowbar switches, 74, 93–95
  in power-crowbar circuits, 145–146
Crystals, continuous deformation of, 257
Culham Laboratory, United Kingdom Atomic Energy Agency, 92

486 INDEX

Current impulses, conversion of capacitor energy to, 126–153
Current pulse generator, 137
Current pulses, metallic phase transformations by, 312–313
Cyclotron radiation, wavelengths of, 378

Debye–Scherrer patterns, 229–241
  recording of, 267–268
Deep Scattering Layers, of ocean, 402
Deep-sea fisheries, electronic equipment for, 404
Delay lines, water-insulated, 75
Deuterium
  accelerated ion velocity for, 291
  heating of by laser pulse, 341
Deuterium gas, 40
Dielectric boundary, for insulators, 7–8
Dielectric discharge
  expanding compression ring in, 268
  spray flash equipment for recording of, 263–264
Dielectric fluids
  flash point of, 18
  high-voltage, 17–18
  properties of, 18
  viscosity of, 18
Dielectric materials, gaseous and liquid, 12–17
Dielectric strength, of capacitors, 18
Dielectric transmission, constant taper and impedance in, 116
Diesel engine cylinder, rotary flow in, 205
Digital light beam deflectors, 64
Diode base welding, 307
Diode laser, as high-powered pulse generator, 133–134
Discharge circuits, capacitors in, 1–3
Discharge switches, low-inductance-capacitor, 42
Disk electrodes, spark gaps and, 71
Disk-plane gap, 71
Displacement, in holography, 173
Diver-held sonar, 408
Doppler experiment, acoustic echo sounder in, 388–389
Doppler radar, radial velocities in, 384
Doppler shifts
  in acoustic air pollution monitoring, 392
  in laser-produced plasmas, 343

Double flash generation, of x rays, 242
Double gap tube, in thyratron operation, 54
Double pulse generation, 245
Drift tube, 290
Drum cameras, 246
D–T pellets, thermonuclear reactions in, 345
Dump resistors, 125
Duratrons, 111
Duroquinone, 189

Echo sounder, 380–382
  Mark I, 383
  Mark II, 390
Edge–plane gaps, 104–105
Electroacoustic efficiency, 417
Electroceramic blocks, 399
Electroceramic spark generator, 399–400
Electroceramic transducers, 399
Electrode holders, in ultrapulse welding, 304
Electrodeless spark, 413, 416–418
Electrodes, titanium hydride, 77
Electrodynamic transducers, 401
Electrohydraulic metal forming, 423–437
  chemical explosive processes in, 430
  pulsers in, 142
Electrolytic capacitor banks, low-loss, 49–50
Electromagnet
  air-cored, 351
  superconducting, 351
Electromagnetic plasma propulsion, 355–359
Electromagnetic pulse drives, 44
Electromagnetic pulse generators, 107
  stripline system for, 121
  transmission line dimensioning in, 113
Electromagnetic pulse simulation, 222–228
  oil- and gas-filled enclosures in, 226
  pulse decay time in, 226
Electromagnetic shock tube, 356
Electromagnetic solid-state joining or welding, 301
  see also Welding
Electromagnetic waves, high-voltage pulses and, 222
Electron beam pulses
  high-energy, 279–288
  in x-ray emission, 235
Electronegative gases, 14
Electron energies, degradation of, 236
Electron guns, pulse generators for, 160
Electronic camera, in cineradiography, 273

INDEX

Electronic deflector-type camera, 270
Electronic integral image camera, 274
Electronic streak camera, 274
Electron ion separation, in laser-produced plasmas, 346
Electron pulses, generation of, 214–215
Electrons, cyclotron radiation of, 377–378
Electron transit time, 159
Electrooptic cell, 170
Electrooptic crystals
  base resonant circuits for, 65–66
  half-wave voltages of, 64
  other voltages for, 63–64
Electrooptic fringe pattern measurements, 173
Electrooptic modulation, 173
Electrooptic recording, in fast motion analysys, 261
Electrooptic shutter, 177, 222
Electrooptic switches, in plasma mirror problem, 343
Electrospark forming, 435–436
Electrostatic image tube, 258
Elliptically polarized light, 165–166
EMP, see Electromagnetic pulse (adj.)
English Electric Valve Co., Ltd., 52
Environmental Science Services Administration, 382
Epoxy resin
  predischarge channels in, 6
  in insulators, 5, 9
Essex, University of, 339
European Research Group, 96, 103
Excited singlet state, 187
Exploding aluminum foil, 74
Exploding bridge wire, 56
  for triggering of x-ray discharge, 325
Exploding wires
  applications of, 320–328
  pulsers in, 142
  as light sources, 327
  in metal forming, 424–426, 434
  striations in, 323–324
  x-ray flashes in, 230
Exploding wire shutters, 320–328
Explosive charge
  as energy source in magnetic flux compression, 371–375
  piezoelectric effect of, 399

Explosive flux compression, 371–375
Explosive forming, 424–425
Explosive processes, x-ray flash investigation of, 261
Explosive type switches, 147
Eye protection shutters, in nuclear explosions, 362

Fansteel VP capacitors, 50–51
Faraday effect, 360–365
Faraday rotational isolator, 360–361
Fast circuit breakers, 88–90
Fast tungstate, 261
FC capacitors, 48–49
Ferroelectrics, 399–401
Fexitron flash x-ray system, 253–254
Field emission cathode, 280
Field emission tube operating point, 225
Film-fluorescent screen receptor, 261
Fine wire welding, 300–301
Fire-resistant fluids, 17
Fischer–Nanolite capacitor, 41–42
Fischer–Nanolite spark lights, 134–135
Fishing industry, electronic equipment for, 404
Flash lamp capacitor, specifications for, 19–25
Flash lamps, for coaxial capacitors, 21
Flashover protection, wave launchers and, 119
Flashover voltages, for insulators, 4–5, 93
Flash photolysis, 183–187
  flash discharge times for, 199
  heavy-duty, 195
  pulsers for, 180
  spectroscope monitoring of, 188
Flash photolysis kinetic spectroscopy, 193
Flash point, for capacitors, 18
Flash tubes
  end-on configuration in, 194
  high-speed, 53
  operating characteristics of, 193
  subnanosecond flashes with, 191
  triggering of, 188–189
  types of, 192
Flash x-ray discharge, exploding bridgewire and, 325–327
Flash x-ray pulser, 255
Flash x-ray system, 286–287
  300-megawatt, 254

Flash x-ray tube
  in megavolt region, 251
  with Siemens electrode arrangement, 252
Flat capacitors, 48–49
Flextensional ceramic transducer, 403
Fluorescence, in flash photolysis, 187
Fluorescence decay, 154
Fluorescent dyes, 176
Fluoroscopy, medical, 257
Flux generator coils, 351
FM-cw radar sounder, 391
Foil capacitors, 42
Foilless diode, 236
Formative delay, 155
Fourier-plane filter, 176
Freon, in transmission line dimensioning, 114
Fuse link, 92
Fusion technology, Fifth Symposium on, 92
  hypervelocity macroparticles in, 366

Gamma-ray flashes, 279–288
  triggering of, 284
Gap-control mechanism, for spark gaps, 226
Gaps
  long, 100–101
  switching, *see* Spark gap; *see also* Megavolt switch gaps, etc.
Gap system, multiple, 74
Gas accumulators, magnetically driven, 370
Gas breakdown
  in plasma behavior, 329
  in xenon, 338
Gas discharge laws, 82–83
Gas discharge tubes
  see also Thyratrons
  controlled, 52–61
  ignitrons, 58–61
  krytrons, 56–57
  as lightning protectors, 80–83
  thyratrons, 54–56
Gaseous dielectric materials, 12–17
Gases, charge carriers for, 13
Gas lasers, pulsing, 134–142
General Electric Company, Ltd., 53
Getter, titanium hydride, 77
Glass laser, mode-locked, 230
Green light pulses, 176

Hard-tube circuits, pulse capacitors for, 29–32
Hard x-ray flashes, 279–288
Hastelloy-to-stainless steel welding, 303
"Haybale" experiment, 389
Heating and electromagnetic pressure joining, 301–303
Helium, accelerated ion velocity for, 291
Helmholtz coils, pulse generation and, 137
HFM coils, 349–350
High-current pulses
  in high-temperature plasma experiments, 145
  other applications of, 142–143
High Energy, Inc., 17, 25, 45
High-energy ion generation, 288
High-energy plasma generators, 93
High field magnets, 143, 352
High power laser interactions, 340
High pressure physics, shock waves in, 420–437
High-speed photography, in metal forming or cracking processes, 434–437
High-speed radiography, 245
High-temperature arcs, rail gun and, 370
High-temperature plasma experiments, 145
High-temperature plasma machines, 38
High-voltage equipment, 17
High-voltage impulse circuit, 218
Holography, stroboscopic, 173
Homopolar generator, 89, 149–150, 153, 351
  in electromagnetic acceleration, 367
Horn-reflector antennas, 382, 389
Humidity, vertical profile of, 385
Hydraulic actuator, 226
Hydrobeacons, 409
Hydrocarbon transformer oil, 56
Hydrogen, accelerated ion velocity for, 291
Hydrogen atom, magnetic compression of, 378–379
Hydrogen thyratron
  ceramic, 53
  data for, 52
  modulation with, 53
Hydrophones, acoustic pulses and, 395–398
Hydrospark forming, 430–432
Hypersonic wind tunnel, spark tracing in, 207
Hypervelocity flight simulation, 367

INDEX

Hypervelocity macroparticle acceleration, peak currents in, 368
Hypervelocity macroparticle research, 365–366

Ignitrons, 58–61
  capacitors for, 50
  peak reverse anode voltage for, 60
Image-converter camera, 102, 437
Image-converter measurements, 97
Image-converter photography, 219, 237
Image-converter recording, 245
Image converters
  subnanosecond exposures with, 158
  ultrafast, 157–158
Image-converter spectroscopy, 219
Image-converter tubes, in high-speed photography, 110
Image intensifier tube, 229, 257
Impedance mismatch, in transmission line dimensioning, 121–122
Impulse corona development, 101
Impulse leader development, 101
Impulse welding, 298–303
Impulsphysik GmbH (IPH), 164, 201, 211, 364
Induction heating, in electromagnetic solid-state joining, 301
Inductive store, current supply to, 151
Inductors
  in pulse hardening, 315–316
  superconductivity, 92
Institut für Plasma Physik, 149
Insulating gases, 12–17
  ac breakdown of, 14
  electronegative, 14
  sphere size and, 15
  sulfur hexafluoride breakdown in, 13–15
Insulating oils, 12, 56
Insulators
  see also Capacitors
  atmospheric pollutants and, 6–7
  behavior of, 4–12
  corona discharges and, 11
  cooling of, 10
  creepage discharge systems for, 4
  creepage tracking and, 7
  discharge current in, 7
  electron strength of, 8–9
  epoxy resin, 9

  erosion resistance and weatherability in, 6–7
  flashover voltages for, 5
  mechanical stress in, 7–8
  mica bonding in, 5–6
  organic polymers as, 6
  paper, 10
  plastic, 5
  pollution and, 4
  polyethylene, 11
  R-insulators and, 12
  thickness of, 10
  ultraviolet in predischarge of, 8
  voltage stabilizers and, 11
Integral-image cineradiography, 271
Interface mismatch, transmission line model for, 122
Interferometer, two-beam, 175
International Congress on High Speed Photography, 229
International Electrochemical Commission Standards, 216
Inverse bremsstrahlung absorption, 342
Ion energy, determination of, 290
Ion generation, high-energy, 288
Ionization energies, 12
IPC (Ion Physics Corp.), 44, 109, 113, 222, 279, 288
IPC machines, jitter in, 109
IPH, *see* Impulsphysik GmbH
Isopuls, 10, 164

Jitter
  in laser-controlled spark gaps, 86
  nanosecond, 109
Jitter spark gap, 71–72, 86
Jitter time of thyristors, 63
Johns Hopkins University, 268

K α-radiation flashes, 229
Kapton film, 20
Kapton window, 240
KDP (potassium dideuterium phosphate) crystals
  Kerr effects and, 173
  two-cut, 176
KDP electrooptic material, 173
  laser induced damage in, 339
KD^{x}P electrooptic material 64

Kerr cells, 65
  blocking factor for, 168
  data for, 168
  polarizers for, 168
  tail-cutting, 135
  theory of, 165
Kerr cell circuit, for nanosecond range, 177
Kerr cell placement, 168
Kerr cell pulse forming network, 169
Kerr cell pulsers, 165–180
Kerr cell shutters, 322, 327
Kerr constant, 166
Kerr effect, 110
  KDP and, 173
Kerr liquids, 167
Kerr system polarimeter, 172–173
Kraft paper capacitor, 25, 30
Kruskal–Shafranov limit, 334
Kryton pulse generator, 58
Krytron tubes, 56–58
  circuits for, 56
  peak current ratings for, 59–60
  technical data for, 59–60
Kyoto University, 205

Laboratoires d'Electronique et de Physique Appliquées, 157
Laser(s)
  double-discharge, 183
  giant pulse, 265
  high power, 92, 345
  interactions of with matter, 339
  mode-locked glass, 230
  organic dye, 180
  2000-J 3-nsec Nd type, 343
Laser beam energy, plasma heating by, 337–348
Laser calorimeter, 182
Laser-controlled spark gaps, 84–88
Laser-created plasma, x-ray emission and, 232
Laser-discharge capacitors, 25–29
  technical data for, 26
Laser Doppler lidars, 386
Laser excitation, 180–186
  pulsers for, 180
Laser-generated plasmas, 337
Laser-ignited spark gaps, 85
Laser photolysis techniques, 199
Laser-produced breakdown, in matter, 345

Laser-produced plasma, 85, 345–346
  electron ion separation in, 346
  highly stripped ions in, 347
  ion and velocity structure in, 342
  as positive ion source, 340
Laser Q-switch, electrooptical cell as, 172
Laser station, front-lighted, 266
Laser system performance, 182
Laser triggering, 84
Laue patterns, 229–241, 257, 268
  recording of, 266–267
Laue reflection, symmetrical, 259
Lawson critical current criteria, 290
LC generator, coaxial water line and, 123–125
Lead azide detonator, 399
Lead niobate, 401
Lecher system, 178
"Les Renardières" group, 96, 103
Lichtenberg figures, in transformer oils, 12
Light beam deflectors, 64
Light coagulators, 126
Light deflectors, 64
  optical memories and, 173
Light intensity oscillogram, 196
Lightning flashes
  overvoltages due to, 80
  simulation of, 215–222
Lightning protectors, 80–83
  circuitry for, 83
  electric data for, 83
  glow to arc transition in, 81–82
Light source, exploding wire as, 327
Light source spark gap, 170
Liley torus, 348
Linear acceleration, thyratrons in, 53
Line conductors, 113–125
Liquid dielectric materials, 12–17
Liquid spark gaps, 74, 104
Lithium fluoride thermoluminescent dosimeter, 285
Lithium target, in neutron flash technique, 293
Long gaps, breakdown of, 101
Long pulse mode, 236
Lorentz force, 356
Los Alamos Scientific Laboratory, 377
Low-loss electrolytic capacitor banks, 49–50
Low-Z plasmas, 238
Lyman source, 195

Machining, ultrahigh-speed, 423
  see also Metal forming
Mach number, in plasma propulsion, 357
Mach–Zehnder interferometer, 330
Macroparticle accelerators, hypervelocity, 365–370
Magnet(s)
  concentrator, 352
  high-field, 352
  permanent, 359–360
Magnetically driven gas accelerators, 370
Magnetic atoms, production of, 377–378
Magnetic compression, of hydrogen atom, 378–379
Magnetic fields
  conversion of capacitive energy into, 349–379
  flux compression method in, 372–376
  photoelectric measurement of, 360
  pulsed, 142, 349–354
  transient, 349
  ultrastrong, 377–378
Magnetic flux compression, explosive force in, 371–375
Magnetic focusing, 426
Magneto-Impulsa Q, 359
Magnetooptic (Faraday) effect, 360–365
Magnum capacitors, 46–47
Marx-bank driver, 180
Marx circuits, 45–47
  pulse generators and, 250
Marx generator, 47–49, 110–111, 185
  disadvantages of, 200
  LC generator and, 124
Material testing, neutron flashes in, 292–297
Matter, laser produced breakdown in, 345, 347–348
Maxwell Company, 38
Meat grinder knife, pulse hardening of, 319
Megagauss fields, 371–378
Megajoule capacitors, 78
Megavolt flash x-ray system, 286–287
Megavolt pulse techniques, 215–228
Megavolt switching spark gap, 95–103
  see also Spark gap
  breakdown voltages and, 99–100
  gap length for, 98
  point–plane gaps and, 100
Metal cracking process, 437

Metal forming
  chemical explosive processes in, 430
  electrohydraulic, 423–437
  pulsers in, 142
  short compression curves in, 420
Metallurgical processing, thermal pulser for, 128
Metaniobate, 401
Meteroid flight damage, 367
Mica bonding, in insulators, 5–6
Mica papers, in capacitors, 21
Micro-induction hardening, 313
Microstripline circuit, 141
Microphones, capacitance, 394
Mode-locked Nd:glass laser, 176, 230
Molybdenum targets, 229
Molybdenum-to-steel welding, 303
Monash University, 423
M-O Valve Company, Ltd., 54
Multichannel spark gaps, 103–112
Multichannel switches, 111
Multi-ionized atoms, 232
Multimegajoule capacitor banks, 366
Multimegavolt switches, 215
Multimegavolt switching gaps, 101–102
Multiphoton ionization processes, 345–346
Multiple chopping gap, circuit for, 94
Multiple gap system, 74
Multispark transmitter, 163
Mylar foils, in laser excitation, 181–183
Mylar MP capacitors, temperature range for, 3

Nanolite capacitors, 41–42
Nanolite pulser, 135
Nanosecond discharge capacitors, 40–42
Nanosecond pulsed streamer chambers, 160–161
Nanosecond pulse generators, krytrons for, 56
Nanosecond spark sources, 265
Nanosecond switching spark gap, 73, 103–112
Nanosecond temperature jump machines, 154–157
Naval Ordnance Laboratory, 265
Nd:glass laser systems, 237
Neptune electron accelerator, 122
Neutron detector, 238

Neutron emission, in plasma population, 358
Neutron flash cinematography, 292
Neutron flashes
  ion beam and target selection in, 292–294
  for material testing, 292–297
  pinching discharge generation and, 230
Neutron generating system, 295
Neutron-produced film contrast, 295
Neutron protectors, 81
nH capacitors, *see* Ultralow inductance capacitors
Nickel isotope, krytron and, 58
Nitrobenzene, as Kerr liquid, 167
Nitrogen, accelerated ion velocity for, 291
Nitrogen circuit breaker, 89–91
Nitrogen laser, 180, 239
  superradiant, 186
Nuclear DD reactors, 341–342
Nuclear explosions
  collapsing foil shutter for, 364–365
  eye protection shutters for, 362

Oerstit 450 magnets, 360
Oil paper capacitors, temperature range for, 3
OMARK welding, 298
Omnidirectional (OMNI) scanning sonar, 406
Optical excitation, of organic dye lasers, 180
Optical image intensifiers, 253–254
Optical memories, 173
Optical printers, 173
Optical shutter, plasma mirror as, 344
Organic dye lasers, 180
Organic molecules, excited-state reactions in, 186
Ostwald absorption coefficient, 12
Overvoltage protection devices, 77–78

Pancake-type spark gaps, 181
Paper bushings, 10
Paper insulation, 10
Parallel-plate capacitors, 47–49
Parallel-plate series capacitors, 42–43
Parallel-plate transmission line, 113, 136
Particle-particle interactions, spark tracing and, 210

Paschen's law, 13
Paschen sparking potential, 54
Peak electric fields, in EMP simulation, 222
Peak inverse voltage, of thyratrons, 61–62
Peak inverse anode voltage, for ignitrons, 61
Percentage voltage reversal, 20
Permanent magnets, charging of, 359–360
PFN-type capacitors, 29
Phosphorescence maximum, 189
Photocathode, in image converters, 159
Photocell, in ultraviolet radiation measurements, 8
Photochemical reactions, mechanisms of, 187
Photodissociation laser work, 139
Photographic flash tubes, high speed, 53
Photography, image-converter tubes in, 110
  see also Camera; Cineradiography
Photon absorption, 187
Photochemical change, 198
Photoelectric x-ray tube, 239
Photographic shuttering, Faraday effect in, 362
Photolysis flash, 195
Physiology, acoustic pulses in, 393–394
Picosecond interactions, in plasma production, 340
Picosecond light pulses, 176, 178
Piezoceramics, ultrasonic atomizer and, 401
Piezoelectric effect, shock-wave excitation and, 398–399
Pinched plasma, x-ray emission and, 233
Pingers, acoustic, 408–409
Pinhole camera, for x-ray spot imaging and x-ray emission, 230–234
Planar field emissiondiodes, 236
Plan position inductor, 407
Plasma
  high-density, 346
  laser-produced, 85, 337, 345–357
  magnetic confinement of, 348
  neutron production of, 331–333
  nonstationary, 330
  speeds of, 356
  thermonuclear, 330
  theta-pinch, 330–331
Plasma behavior, 328–331
Plasma electron temperature, 330
Plasma focus, 230, 236

Plasma generation, of high-energy capacitor discharge, 328–337
Plasma generator
  c batteries for, 39–40
  high-energy, 93
Plasma gun, pulsed coaxial, 359
Plasma heating, by laser beam energy, 337–348
Plasma machines
  blast circuit breaker for, 88–92
  high-temperature, 38
Plasma jets, erosive, 330
Plasma mirror, 343
  as optical shutter, 344
Plasma propulsion, electromagnetic, 355–359
Plasma potential, fluctuating, 357
Plasma research, capacitors for, 45
Plasma temperature, 357
Plasma-x laser system, 238
Plasmon field, 357
Plastic case capacitors, 31–32
Plastic insulators, treeing breakdown in, 5
Plexiglas insulators, 5
Pockels cells, 65
  data for, 174
  at lowered voltages, 170
Pockels cell driver, 172
Pockels cell modulators, for Q-switch applications, 171
Pockels cell pulsers, 165–180
Pockels effect, 110
Pockels effect electrooptic image simulation, 176
Point–plane gaps, 100
Polarized light, Kerr cell and, 165, 168
Polarizers
Polycarbonate dielectrics, 20
Polychlorinated byphenyl, 17
Polycrystalline screens, 258
Polyester-film capacitor, 19–20
Polypropylene capacitors, 19
Polypropylene dielectrics, 20
Polythene insulation, 10–11
Potassium dideuterium phosphate, *see* KDP
Powder patterns, recording of, 267
Power-crowbar circuits or bank, 40, 145
Power-crowbar source, in toroidal plasma experiment, 148–149
Power transistors, 63

Predischarge time, for megavolt switch gap, 97
Pressured air capacitors, 16–17
Projection welding, 308–309
Pulse burst generator, 245
Pulse capacitors
  "crowbarring" in, 30, 40, 145
  dissipation factors for, 30
  peak ripple voltage in, 30
  for soft and hard-tube circuits, 29–32
Pulsed accelerator, 288–289
Pulsed electric power, rise times in, 142
Pulsed electron systems, peak electron voltage range in, 281
Pulsed gas discharges, 135
Pulse-discharge lifetime, 20
Pulsed-Lyman source, 195
Pulsed magnet chargers, 359
Pulsed magnetic fields, 38, 349–354
  generation of, 145
Pulsed plasma dynamic processes, 329
Pulsed spiral magnet, 349–351
Pulsed streamer chambers, 160
Pulsed x-rays, generation of, 214–215
Pulse-forming network, 145
Pulse generator
  circuit for, 138
  current, 137
  feedback-controlled, 137
  free-running, 141–142
  high-speed, 140
  for large electron guns, 160
  Marx circuit for, 250
  as Marx generator, 200
  precision, 140
  in pulsed electron systems, 281–284
  Scandiflash, 251–253
  SCR, 137–138
  subnanosecond, 144–145
  for testing equipment, 164
  types of, 131–132
  Van de Graaff generator and, 223
Pulse hardening
  of carbon steel, 313–320
  conditions for, 317
  experience with, 316–317
  inductors for, 315–316
  machine for, 314–315
Pulse heating system, in metallurgical processing, 129

Pulse height analyzer, 139
Pulse lasers, giant, 265
Pulse rise time, in switching gap, 107
Pulsers, 142–143
   high-current, 139
   for laser excitation and flash photolysis, 180–200
   megavolt, 215–216
Pulse scanning systems, 405
Pulse shaping, Blumlein line for, 184
Pulse spiral generator, 269
Pulse techniques, in simulated lightning flashes, 215–222
Pulse transformers, 244
   high-voltage, 200–215
   spark tracing, 211
   specifications for, 143
Pulse voltage protectors, 82
Pulse welding, 298
   thyristors in, 63
Pulse welding machines, specifications for, 310–311
Pulsing diode lasers, 134
Pulsing gas laser, 134–142
Push–pull circuits, charge storage and, 69
Push–pull high-voltage switch, 68
Push–pull switching circuits, for rectangular pulses, 63–69
Push–pull transistor switching circuit, 67–69

Quartz flash lamps, 198
Quenched spark gaps, 88–92
   in spark tracing, 211
Quenchotron, 88, 246
Q-switch applications, Pockels cell modulations for, 171
Q-switched ruby laser, spark gap and, 87

Radar pulse drivers, 52
Radar pulse modulators, 52
Radiation detection, 130
Radiation flux density, 339
Rail gun, in hypervelocity macroparticle acceleration, 368–370
Rapid discharge capacitors, 42
Rayleigh–Taylor instabilities, 369
Rectangular pulses, push–pull switching circuits for, 63–69
Rectangular voltage pulse, 155
Reflection-horn antenna, 382

Relaxation kinetics, 154
Relaxation time constants, for insulators, 7
Remote Sensing of the Environment, Symposium on, 386
Resonant-frequency tests, for capacitors, 34
Reversal voltage, dipole stress and, 2
Rhodamine 6G solution, 182
R-insulator, 12
Ring projection welding, 307–309
Ring transmitter, single-spark, 161–164
rms current, power loss and, 23
Rod–plane gaps, 15, 71
Rogowsky belt, 34
Rogowsky-shaped electrodes, 183
Rotating prism camera, 241
Rotational Directional Transmission, in long-range sonar operation, 407
Ruby laser, in laser-controlled spark gaps, 85

Sampling oscilloscope, 159
Scandiflash pulse generators, 251, 253
Schlieren method, 201, 338
Scintillation tube, 259
Scintillator-photodiode system, 285–286
SCR pulse generator, 137–138
Scylla devices, 40, 330
Sector-scanning sonar, 405
Self-cleaning switch, 111
Semiconductor switches, controlled, 61–63
Shadowgraphy method, 201, 241
Shock heating, 40
Shock pressures, material behavior and, 420–423
Shock tube
   pressure-driven, 94
   pulsers and, 142
Shock tube crowbar switch, 95
Shock wave
   in high-pressure physics, 420–437
   in plasma propulsion, 356
   recording of, 266–267
   reflected, 93
Shock wave processes, x-ray flash radiography in, 262
Shutters
   collapsing foil, 362–363
   uncapping, 363–364
Siege-Installation system, 121

INDEX

Siemens tube configuration, 251–252, 259
Simmer resistor, 126
Single spark ring transmitter, 161–164
  radial transmission line for, 163
Sliding spark, in x-ray emission, 233–234
Slit cineradiography, 271, 274
Soft-tube circuits, pulse capacitors for, 29–32
Solid gap, 74
Solid-state switch, for arc lamp pulsing, 126–127
Sonar
  advanced techniques in, 401–409
  diver-held, 408
  omnidirectional scanning, 406
  sector-scanning, 405
  stabilization of, 408
Sonar transducer, 409
Sound stimuli, physiology of, 393
  see also Acoustic (*adj.*)
Space propulsion, plasma gun as propeller in, 359
Spark(s)
  electrodeless, 413, 416–418
  underwater shock waves from, 409–420
Spark chamber
  sealed, 72
  streamer chambers and, 160–161
Spark discharge capacitor, specifications for, 19–25
Spark electrodes, water conductivity between, 410–415
Spark gap
  breakdown of, 100–101
  breakdown voltage and jitters in, 105–106
  characteristics of, 194
  coaxial, 72
  demountable, 111
  edge–plane, 104–105
  EMP simulation and, 222
  four-electrode, 72
  high-pressure, 156
  image-converter photography and, 219
  laser-controlled, 84–88
  lightning discharge and, 82
  light source, 170
  liquid, 104–105
  megavolt, 95–103
  multichannel and nanosecond, 103–112
  multimegavolt, 101–103
  multiple, 162
  nanosecond switching, 73, 112
  pancake type, 181
  prebreakdown currents and, 101
  quenched, 88–92, 245
  rise time capability of, 107
  rod–plane sparks in, 221
  streamers and leaders in, 100–101
  triggered, 69–70, 221–222
  types available, 194
  ultraviolet light for, 217
  vacuum, 147
Spark gap capacitor unit, 40
Spark gap overvoltage, 181
Sparkover voltages, 15
Spark path, in spark traving, 202
Spark ring transmitter, 161
Spark tracing, 201–214
  in air classifier, 209
  air velocities in, 212
  in Bunsen burner flame, 208
  flow visualization in, 213
  high-voltage pulse transformers in, 200–215
  in hypersonic wind tunnel, 207
  limestone particles in, 214
  luminescent particle paths in, 213
  luminous ionized plasma in, 202
  in overpressure and underpressure conditions, 207
  particle clouds in, 210
  plasma channel in, 203
  pulse transformer and, 211
  quenched spark gap in, 211
  in turbine flow conditions, 203
Spectroscopy, image-converter, 219
Sphere size, in insulating gases, 15
Statistical delay, 155
Steel
  pressure-elongation diagram for, 430
  pulse hardening of, 313–320
  welding of, 298–307
Steered Directional Transmission, in long-range sonar, 407
Streak camera, 276, 341
Streamer chambers, nanosecond pulsed, 160–161, 185
Streamers and leaders, in long spark gaps, 100–101
Striations, in exploding wires, 323–324

Stripline system, in EMP generation, 121
Strip transmission line, 115
Strobokin control unit, 88, 202
Strobokin–Strobo x-ray flash unit, 247
Stroboscopic holography, 173
Studwelder, portable, 299
Stud welding system, applications of, 299–300
Subnanosecond pulse generators, 144–145
Subnanosecond switching gap, 110
Subpicosecond region, pulse durations in, 231
Sulfur hexafluoride (SF_6)
  breakdown of, 13–15
  direct and impulse breakdown voltages in, 220
  in transmission line dimensioning, 114
Sulfur hexafluoride gas
  in high voltage pulse transformer, 201
  in lightning flash simulation, 217
  in pulsed electron systems, 281–284
Superconducting inductor, 92
Superconducting solenoids, 354
Suprasil window, 88
Surface charges, on insulators, 8
Swim bladders, of fishes, 402
Switch cell, high-voltage transistor, 67
Switching circuits
  high-voltage, 65–67
  push-pull, 63–69
Switching gaps
  see also Spark gap
  high power, 69–74
  liquid and solid, 74
  multimegavolt, 101–102
  pulse generator and, 178
  pulse rise time in, 107
  trigger test setup for, 108
Switching means, 52–112
  see also Gas discharge tubes; Spark gap; Switching gaps
Switching systems
  EMP pulse in, 107
  pulse rise time in, 107

Tantalum capacitors, 50–51
TEA (tetraethyl ammonium) CO_2 laser, 338
Telescopes, collapsing foil shutter for, 364–365
Television picture tube scanning, 139

TEM mode, in EMP simulation, 223
Temperature, acoustic echo technique in remote sensing of, 380–391
Temperature inversions, acoustic techniques in, 386
Temperature jump machines, 154–157
Temperature sensors, in thermal pulsing, 130
Terbium aluminum garnet, in Faraday rotation isolator, 360–361
Tesla coil, in x-ray emission, 234
Thermal neutron beam, in material testing, 295–297
Thermal plane activity, 387
Thermal pulser
  for metallic phase transformation studies, 128
  operating characteristics of, 312
Thermocouple, in thermal pulsing, 130
Thermonuclear fusion, 237
Thermonuclear reactions
  self-sustained, 346
  in D-T pellets, 345
Theta-pinch discharges, 39
Theta-pinch plasma, 330–331
  experiments in, 333–337
Thin objects, radiography of, 249
Three-photon absorption process, 339
Thunderstorm flash, protection against, 95
Thyratron, 52–56
  ceramic, 52
  hydrogen, see Hydrogen thyratron
  for linear accelerators, 53–54
  triggering characteristics of, 54
  trigger power amplifier, 245
  trigger pulse voltage in, 56
Thyristor
  capacitor energy and, 126
  defined, 61
  jitter time of, 63
  leakage current for, 62
  peak inversion voltage of, 61–62
  silicon, 62
Time-delay generator, 167–168
Time jitter, in laser-controlled spark gaps, 87
Timing capacitors, 159–160
Titanium hydride getter, 77
T-jump machines, 154–157
Tobe Deutschmann Laboratories, 42
Toepler laws, 82
Tokomak process, 333, 337

Toroidal coils, 334
Toroidal magnetic field, 335–336
Toroidal plasma experiment, 148
Toroidal θ-Z pinch, 336
Toroidal vacuum vessel, 334
Toshiba, Ltd., 75
Transformed capacitor energy, welding by, 303–312
Transformer oil
  hydrocarbon, 56
  Lichtenberg figures in, 12
Transformers, high-voltage, 17
Transient response technique, for capacitors, 34
Transistors
  low-voltage, high-current, 63
  peak base current for, 63
Transistor switch cell, high-voltage, 67
Transmission line
  at ARES facility, 227–228
  in EMP simulation, 224–226
  impedance matching in, 115
  inclined incidence in, 117–121
  normal incidence in, 115–117
  parallel plate, 136
  pulse-forming, 250
Transmission line dimensioning, 113
  impedance in, 121
Triggered discharge switch, 111
Triggered multiple chopping gap, 93
Triggered spark gaps, 69–70, 221–222
Triggered vacuum gaps, 77–78
Trigger power amplifier thyratron, 245
Trigger pulse
  rise time of, 103–104
  voltage of, in thyratrons, 56
Trigger test strip, for switching gap, 108
Trigger voltage, minimum, 70
Triplet state, 187
Triplet-state lifetimes, 189
Triplet–triplet absorption, 187
Turbine flow conditions, spark tracing in, 203–204
Turbulence, acoustic pulses and, 391
Tympanal nerve, threshold curve of, 393

Ultrafast image converters, feeding of, 157–160
Ultrahigh speed machinery, metal behavior in, 423

Ultralow inductance capacitors, 32–45
  standing-wave method for, 33
Ultrapulse welding, 303–304
  characteristic features of, 309–311
Ultrasonic testing, 395
Ultraviolet light, from spark gap, 217
Uncapping shutter, 363–364
Underwater pinger beacons, 409
Underwater shock waves, from sparks, 409–420
Underwater Sound Laboratory, U.S. Navy, 401
Underwater sound sources, pulsers in, 142
Underwater spark gap, 427
Underwater sparks, 409–420
  physical behavior during, 420
  steep rise, 419–420
Uranyl oxalate actinometer solution, quanta per flash for, 197
U.S. Navy Underwater Sound Laboratory, 401

Vacuum capacitors, 19
Vacuum contactors, high-voltage, 75–80
Vacuum spark gaps, 77–78, 147
Vacuum switch, for motors, 75
Vacuum ultraviolet, flash photolysis in, 193
Van de Graaff accelerator, 296
Van de Graaff generator, 223, 279, 289
  coaxial gasline and, 225
Variable-inductance technique, for pulse capacitors, 35–38
Velonex high power pulse generator, 132–134
Verdet constant, 360
Vertical velocity field, acoustic echoes in, 387
Videcon tubes, x-ray emission and, 268
Videotape recorder, 139
Viscosity, of dielectric fluids, 18
Voltage flashover, in transmission line dimensioning, 121
Voltage pulse generators, for scientific applications, 154–165
Voltage reversal
  vs. lifetime, 20
  percentage, 2
Voltage stabilizers, 11
VP tantalum capacitors, 50–51

Water
  holdoff strength of, 124
  as Kerr liquid, 167
Water electrodes, underwater spark gap and, 412–415
Water-filled cable, coaxial, 122
Water hammer, in hydrospark forming, 432–434
Waveform photos, in capacitor studies, 19
Wave launchers, in Siege 1.3 system, 119
Welding
  airtight seal type, 307
  capacitor discharge, 298–303
  fine-wire, 300
  heating and electromagnetic pressure in, 301–303
  projection, 308–309
  pulse, see Pulse welding
  ring projection, 307–309
  solenoid gun in, 298–299
  by transformed capacitor energy, 303–313
  ultrapulse, 303–304
Wind, remote sensing of by acoustic echo techniques, 380–391
Wind speed, profile of, 385

Xenon arcs, pulsed, 136
Xenon lamp
  in arc lamp pulsing, 126
  compact arc, 329
  with thyratron switches, 137
Xenon lamp discharge, capacitors for, 25
X-radiography, submicrosecond, 241
X-ray absorption method, 342
X-ray bremsstrahlung, 237
X-ray diffraction
  dynamic, 254
  low-energy level light images in, 256
  time-resolved, 255, 260

X-ray diffraction patterns, recording of, 266
X-ray electron generator, 246
X-ray emission
  helium-like spectra in, 235
  of laser-created plasma, 232
  monochromatic x-ray pulses in, 235
X-ray filter absorption method, in plasma neutron production, 332
X-ray flashes
  in exploding wires, 230
  generation of, 229–241
  polychromatic, 230
X-ray flash penetration, 241
X-ray flash radiography, 262
X-ray fluorescent screens, 258–259
X-ray framing camera, 232
X-ray generation, double-flash, 242–245
X-ray hot spots, 333
X-ray image intensifiers, 253
X-ray image recording, 268
X-ray laser
  output of, 231
  soft, 231
  subpicosecond region of, 231
X-ray machine, flash, 224
X-ray mixing equipment, 239
X-ray neutron generation, 246
X-ray photons, in high-speed cineradiography, 270
X-ray pulsed electron systems, 285
X-ray pulsers, 229, 255
  high-voltage pulse transformers and, 200–215
  in hot vacuum, 214–215, 233
X-ray single flasher, 248
X-ray source size, 249

Zeeman effect, longitudinal, 354
Zeiss light coagulator, 126